This book belongs to Mark Burleton

Tree Maintenance

The senior author examining the trunk of a very old London planetree at the New York Botanical Garden. The tree has a diameter of 4 feet, 2 inches at the 4½-foot level, a height of 90 feet, and a branch spread of 60 feet.

Tree Maintenance

Sixth Edition

P. P. Pirone

J. R. Hartman
M. A. Sall
T. P. Pirone

New York   Oxford
OXFORD UNIVERSITY PRESS
1988

Oxford University Press

Oxford   New York   Toronto
Delhi   Bombay   Calcutta   Madras   Karachi
Petaling Jaya   Singapore   Hong Kong   Tokyo
Nairobi   Dar es Salaam   Cape Town
Melbourne   Auckland
and associated companies in
Berlin   Ibadan

Library of Congress Cataloging-in-Publication Data

Tree maintenance.

Bibliography: p.
Includes index.
1. Ornamental trees—Diseases and pests. 2. Trees,
Care of. I. Pirone, Pascal Pompey, 1907–   .
SB761.T7   1988   635.9'77   87-20418
ISBN 019-504370-7

9 8 7 6 5 4 3 2 1

Printed in the United States of America
on acid-free paper

Foreword to Original Edition

The care of trees is a never ending obligation, more insistent each year because appreciation of them is increasing. The owner of the tree carries the responsibility of his own account and he then naturally extends his concern to trees that stand on public property and need the interest of citizens.

One's responsibility is met only when one recognizes the cause of injury and disease and is informed on the remedies and proper procedures. Even if one has been taught these subjects in school he nevertheless needs to be informed on the newest difficulties and the latest approved practices. The difficulties increase. Thus it comes that new and authoritative books are always needed.

Therefore I am glad to see this book on Maintenance of Shade and Ornamental Trees by Dr. P. P. Pirone, whose work I have known for many years. He is investigator and teacher. He has an eye to the immediately practical, as evidenced by the simplicity of the writing, clearness of the directions, and the significant pictures. It is an important presentation of the subject.

Ithaca, N.Y.   L. H. BAILEY
March 1, 1941

Preface

In the preface to the fifth edition of this book, my father indicated that in all probability it would be his final edition of *Tree Maintenance*. While, at 80 years of age, he was not quite up to producing this sixth edition singlehandedly, I am pleased to report that he has been an active participant in its preparation, contributing the wisdom gained in over fifty years of working with trees.

Since I am a plant scientist involved in research, teaching, and writing, my father and I had discussed the possibility of collaborating in the preparation of this edition for some time. I felt, however, that I lacked the extensive experience with trees that would be required to evaluate the tremendous amount of information which has accumulated since the last revision. Hence I was fortunate to have as colleagues two individuals who are eminently well qualified in this respect and who agreed to collaborate on this edition. Dr. John Hartman is Extension Professor of Plant Pathology at the University of Kentucky and has been involved in diagnosing and advising on tree problems for over 15 years. He is an award-winning member of the International Society of Arboriculture. Dr. Mary Ann Sall's family operates a nursery business in Oregon, and she herself has a Ph.D. in forest pathology from Oregon State University. Her experience and training thus cover a wide range of tree-related practices and problems. Formerly a faculty member at the University of California at Davis, she brings to the book a west-coast perspective that complements the east-coast, midwestern, and southern experience of the other authors. My primary role in the preparation of this edition has been organizational and editorial, although several years spent in diagnosing and advising on tree problems while a faculty member at Louisiana State University allowed meaningful input in a number of areas.

In keeping with previous editions, we have attempted to follow a course between the technical and the popular in order to serve the needs of as broad a clientele and readership as possible. While we have retained the general structure of the fifth edition, some material has been assigned to different chapters, and a number of chapter headings have been changed to reflect more accurately their new content.

A major change in Section I occurs in the approach taken in Chapter 4. Previous editions contained a chapter that recommended specific trees suitable for various locations. Since the selection of suitable trees depends on so many factors, and since such a variety of trees is available, we felt that it would be impossible to discuss or recommend specific trees in a single chapter. Hence we have emphasized the *principles* that are important for tree selection and have included in Chapter 4's bibliography a range of references on tree selection.

Major changes in content and approach occur in Section II. Since there are general principles that apply to groups of insects or of disease-causing agents, we have elected to discuss these in broad terms in Chapters 11 and 12, respectively, while Chapter 13 deals with strategies for coping with pests and diseases. Information on specific diseases and pests of particular trees, which in previous editions was included in several chapters in Section II, is now incorporated in Section III. Section III also contains information on trees that are relatively problem-free and on trees that may be suitable for situations which would be adverse for most species.

We wish to express our gratitude to the professional arborists and scientists whose publications are listed in the various bibliographies in this book. Without their contributions, a book like this could not have been written. We wish to single out particularly the *Journal of Arboriculture* for its excellent articles on all phases of tree care. Suggestions for topics and critical reviews of portions of the manuscript by R. A. Scheibner, C. A. Kaiser, and D. E. Hershman are also much appreciated.

Finally, we wish to thank the following persons and organizations for the use of illustrations:

Barlett Tree Expert Co., Stamford, Connecticut: Figs. 12-6, 12-12, III-4, and III-48.

Dominck Basile, Lehman College, New York, New York: Frontispiece.

John Bean Div., FMC., Tipton, Indiana: Fig. 13-4.

J. C. Carter, Illinois Natural History Survey, Urbana, Illinois: Figs. III-43 and III-92.

R. A. Cool, Lansing Michigan: Fig. 5-5.

Thomas Corell, Cooperative Extension Association of Suffolk County, New York: Fig. III-100.

Robert d'Ambrosio, Ambrose Laboratories, Ltd., Eastchester, New York: Figs. 10-5, 10-11, 11-4, 11-5, 13-3, and III-35.

R. W. Doherty, Sleepy Hollow Restorations, Tarrytown, New York: Fig. 8-8.

T. H. Everett, New York Botanical Garden, New York: Figs. 1-1, 1-4, 4-2, 5-3, and III-114.

E. F. Guba, Massachusetts Agricultural Experiment Station, Amherst, Massachusetts: Fig. III-46.

J. W. Hendrix, University of Kentucky, Lexington, Kentucky: Fig. 2-4.

George Hepting, Asheville, North Carolina: Fig. 12-14.

Ulla Jarlfors, University of Kentucky, Lexington, Kentucky: Fig. 12-8.

Cheryl Kaiser, University of Kentucky, Lexington, Kentucky: Figs. 9-6, 10-13, 12-2, 10-25, 12-10B, III-6, III-45, III-47, III-88, III-89, and III-107.

Cheryl Kaiser, University of Kentucky, Lexington, Kentucky: Figs. 9-6, 10-13, 12-2, 10-25, 12-10B, III-6, III-45, III-47, III-88, III-89, and III-107.

Robert McNiel, University of Kentucky, Lexington, Kentucky: Figs. 5-4 and 12-5.

Nassau County Cooperative Extension Association, New York: Fig. III-119.

D. A. Potter, University of Kentucky, Lexington, Kentucky: Figs. 11-2, 11-3, 11-6, 11-7, 11-9, 11-10, 13-1, III-1, III-36, III-41, III-55, III-60, III-65, III-76, III-104, III-108, and III-117.

A. J. Powell, University of Kentucky, Lexington, Kentucky: Fig. 6-2.

William A. Rae, Frost and Higgins Co., Burlington, Massachusetts: Fig. 5-6.

Bob Ray, Bob Ray Co., Louisville, Kentucky: Figs. 10-1 and 10-2.

Edward Scanlon Associates, Olmstead Falls, Ohio: Figs. 1-5, 4-4, 4-7.

John C. Schread, Connecticut Agricultural Experiment Station, New Haven, Connecticut: Figs. III-9, III-26, III-27, III-68, III-115, III-129, III-130, III-132, and III-133.

Robert Southerland, University of Kentucky, Lexington, Kentucky: Fig. 1-2.

John Strang, University of Kentucky, Lexington, Kentucky: Fig. 10-19.

Richard Stuckey, University of Kentucky, Lexington, Kentucky: Fig. 3-4.

Alfred G. Wheeler, Jr., Bureau of Plant Industry, Pennsylvania Department of Agriculture, Harrisburg, Pennsylvania: Fig. III-66 and III-67.

Lexington, Kentucky   T. P. P.
December 1987

Contents

I. General Maintenance Practices

1. *The Value of Trees*   3
Value of individual trees. Determining values of large trees. How tree values are used. Arboricultural practitioners. Arboricultural periodicals.

2. *The Structure of the Tree and Function of Its Parts*   15
Parts of a tree. Structure and function of roots, stems, and leaves. Reproductive organs.

3. *The Soil and Its Relation to Trees*   33
Soil components. Elements essential for tree growth. Influence of soil on nutrient availability. Adjusting soil reaction (pH). Soil reaction adaptations of specific trees. The soil as an anchor. Urban soils. Soil improvement prior to planting. Soil improvement for established trees.

4. *Factors Important for Tree Selection*   53
Characteristics of the site. Adaptation of tree species to particular sites. Problems peculiar to streetside trees. Considerations in choosing trees for city streets.

5. *Transplanting Trees*   65
Importance of good transplanting. Season for transplanting. Preparing trees for transplanting. Digging, transporting, and planting trees. Postplanting operations.

6. *Fertilizers and Their Use*   91
Benefits of fertilizer application. Determining the need for fertilizer. Specific nutrient elements. Fertilizer formulations. Treatment area. Soil application methods. Soil-applied fertilizer rates. Foliage or dormant twig spray. Trunk implants and injections. Recommendations for newly transplanted trees. Timing of tree fertilization.

7. *Pruning*   115

Need for pruning. Response of trees to pruning. Types of pruning schemes. When to prune. Equipment. How to make the pruning cut. Safety in pruning. Wound dressings. Specific pruning situations. Pruning for disease control. Pruning for line clearance. Use of chemical growth retardants. Scheduling pruning for municipalities.

8. *Tree Preservation and Repair*   139

Preventing and treating structural damage. Types of artificial support. Treating fresh wounds. Cavity filling. Preserving trees at construction sites. Adjusting to sidewalk lifting by tree roots.

II. Diagnosis and Control of Tree Problems

9. *Diagnosing Tree Problems*   159

Responsibilities of a tree diagnostician. Importance of correct diagnosis. Diagnostic procedures. Diagnostic tools. A sample diagnostic questionnaire.

10. *Damage Due to Nonparasitic Factors*   175

Damage caused by grade changes and construction around trees. Injury by girdling roots. Gas injury. Injuries caused by low temperatures. Sunscald. Leaf scorch. Lightning injury. Hail injury. Mechanical injuries. Damage by lichens, vines, rodents, birds, termites, ants, and tree banding. Air pollution injury. Chemical injuries. Chlorosis.

11. *Insect and Mite Pests of Trees*   217

Structure of insects. Life cycle of insects. Descriptions and examples of types of insects and mites, with control recommendations: Borers, leaf miners, caterpillars, sawflies, beetles, plant bugs and lace bugs, aphids, scales, spider mites, and eriophyid mites.

12. *Parasitic Diseases of Trees*   233

Disease symptoms. Nature of parasitic agents. Dissemination and survival of parasitic agents. Factors required for disease development. Descriptions, examples, and measures for control of some common diseases: Fire blight, crown gall, bacterial wetwood and slime flux, bacterial scorch, powdery mildews, anthracnose diseases, Verticillium wilt, wood decay, and Armillaria root rot.

13. *Coping with Tree Pests and Diseases*   263

Integrated pest management. Biological control of insects and diseases. Chemical pesticides. Chemical application methods and equipment. Pesticide labels. Specific pesticides for insect, mite, and disease control.

III. Abnormalities of Specific Trees   289

Acacia. Alder. Arborvitae. Ash. Athel Tamarisk. Australian-Pine. Avocado. Bald Cypress. Banana. Banyan. Beech. Birch. Boxwood. Buckthorn. Cajeput. California Laurel. Camellia. Camphor-Tree. Carob. Catalpa. Cedar. Cedar, Incense. Chaste-Tree. Cherry. Chestnut. Chinaberry. Chinese Jujube. Chinese Pistachio. Chinese Tallow Tree. Citrus. Corl Tree. Crabapple, Flowering. Crapemyrtle. Cryptomeria. Cucumbertree. Cypress. Deseı Willow. Dogwood. Douglas-Fir. Elm. Empress-Tree. False Cypress. Fig. Fir. Firewhe

Tree. Franklin-Tree. Fringe-Tree. Ginkgo. Golden-Chain. Goldenrain-Tree. Guava. Gumbo-Limbo. Gum-Tree. Hackberry. Hardy Rubber Tree. Hawthorn. Hemlock. Hickory. Holly. Hop-Hornbeam. Hop-Tree. Hornbeam. Horsechestnut. Jacaranda. Japanese Lilac Tree. Japanese Snowbell. Juniper. Katsura-Tree. Kentucky Coffee Tree. Larch. Linden. Locust, Black. Locust, Honey. Magnolia. Mango. Maple. Mayten Tree. Mesquite. Mountain-Ash. Mulberry. Norfolk Island Pine. Oak. Olive. Orchid-Tree. Osage-Orange. Pagoda-Tree, Japanese. Palms. Paper-Mulberry. Pear, Ornamental. Pecan. Pepper-Tree. Persian Parrotia. Persimmon. Pine. Pittosporum. Planetree, London. Poinciana. Poplar. Quince, Flowering. Redbud. Redwood. Redwood, Dawn. Russian-Olive. Sassafras. Screw-Pine. Serviceberry. Silk-Oak. Silk-Tree. Silverbell. Smoke-Tree. Snow-in-Summer. Sorrel-Tree. Spruce. Stewartia. Strawberry-Tree. Sumac. Sweetgum. Sycamore. Tanoak. Tree-of-Heaven. Tuliptree. Tung. Tupelo. Turkish Filbert. Walnut. Willow. Wingnut, Caucasian. Yellowwood. Yew. Zelkova, Japanese.

Index   485

General Maintenance Practices

The Value of Trees

Almost everyone likes trees for one reason or another. They provide cool shade during hot summer days, and whether in the yard or the park, they offer a serene setting to relieve the tensions of modern life. They add to the beauty and value of property. The splendor of a springtime floral display and the pageantry of autumn coloration make trees a delight to the eyes (Fig. 1-1). Trees evoke sentiment perhaps because grandpa planted them when he and grandma moved into the neighborhood many years before or because some historic event took place beneath their boughs. No matter what the reason—aesthetic, financial or sentimental—a tree is a sound investment. A house surrounded by large trees is worth more money than a house without them. A house with no trees near it looks hot in summer, appears unbalanced, and suggests, however unjustly, a lack of interest on the part of those who live in it. Established residential areas are usually more inviting and more restful than new housing developments for one important reason: they are well supplied with large trees.

Trees have other beneficial effects:

First, they help supply the oxygen we need to breathe. Each year an acre of trees can produce enough oxygen to keep eighteen people alive. They also help refresh our air supply by using up some of the carbon dioxide that we exhale and that factories and engines emit. Prehistoric tree forms that made up the immense forests of the Carboniferous period (whose fossil remains we now mine as coal) are thought to have been important as air purifiers. The huge equisetums and arboreal ferns, now mostly extinct, fixed large amounts of carbon dioxide and released oxygen for millions of years. The resulting shift of gas concentrations and the production of cleaner air influenced the development of present-day plant and animal forms. Indeed, the current worldwide loss of trees has the potential to increase atmospheric carbon dioxide concentrations to levels that could create for the earth an undesirable "greenhouse effect."

Second, trees reduce noise pollution by acting as barriers to sound. Each 100-foot width of trees can absorb about 6 to 8 decibels of sound intensity.

Third, trees trap particulate air pollutants and absorb others, some of which are used as nutrients for growth.

Fig. 1-1. A splendid springtime display of flowering dogwood.

Fourth, trees alter the microclimate of the site where they grow. Trees may be effectively used to keep a house warmer in winter and cooler in summer. Evergreen trees planted on the north side of a house will shield it from cold winds. Deciduous trees planted to the south of a house allow much of the winter sun's rays to reach the house, thus providing some warmth. In summer, the same trees will shade the house and make the rooms on that side cooler. When air temperature is 84°F (29°C), surface temperature may be as high as 108°F (42°C), but on a street lined with trees, the surface temperature is just 88°F (31°C) because heat rays are reflected off the surface of leaves, thus making it more comfortable for pedestrians and travelers in automobiles.

Finally, the character of a city is changed by an abundance or a dearth of trees. Cities that spend liberal amounts of money to maintain their old trees and to plant new ones are generally considered nicer places in which to live. The urban forest, although not a natural ecosystem, is nevertheless a biological community of species highly dependent on the physical environment, much of it created by people (Fig. 1-2). Trees, environment, and every form of life within an ecosystem are interrelated and interdependent. Implicit in the concept of an urban forest ecosystem, then, is the idea that people have an effect on

Fig. 1-2. Trees along Independence Mall in Philadelphia.

trees and trees an effect on people. In addition, trees planted along streets and parkways of towns and cities represent a considerable financial investment. New York City alone has over 2.5 million trees.

Value of Individual Trees

Trees on private property contribute to the value of the real estate, perhaps increasing values 5 to 20 percent. On public or commercial property, trees have a value of their own apart from the real estate value.

When a tree is suddenly or unexpectedly damaged or destroyed, tree owners in some circumstances can seek reimbursement from insurance companies. To obtain reimbursement, it is necessary to appraise the value of a tree. Appraisal may also be needed to determine the damages one must pay for injuring or killing someone else's tree. Homeowners may want to know how much to deduct from their income tax for a tree damaged or destroyed by hurricane, ice storm, lightning, or fire.

To aid in this process, the Council of Tree and Landscape Appraisers produced a booklet entitled *Guide for Establishing Values of Trees and Other Plants*. Experienced and knowledgeable individuals representing five organizations in the landscape plant industry collaborated in the development of the *Guide* and its revisions, the most recent of which was published in 1988. The five organizations contributing expertise to the *Guide* are American Association of Nurserymen, American Society of Consulting Arborists, Associated Landscape Contractors of America, International Society of Arboriculture, and National Arborists Association. Use of the *Guide,* which is available from any of the sponsoring organizations, enables the tree evaluator to proceed systematically. Although some phases of tree evaluation are still subjective, the authors of the *Guide* have introduced enough objective criteria so that different evaluators might arrive at similar values.

The Council of Tree and Landscape Appraisers developed a second publication, *Manual for Plant Appraisers,* to provide further guidance in tree evaluation.

Tree Evaluation Based on the Guide. The *Guide* first makes a case for the concept that trees have value. Trees are subdivided into two groups, those that can be replaced because of their relatively small size, and those that are too large to replace. Detailed evaluation procedures for each group are then presented. The following information describes the general procedure used. One who is professionally involved in this activity will want to obtain and study the *Guide* for details.

Establishing Replacement Value. If transplantable trees are available, trees can be replaced. The tree evaluator needs to compute the value based on local costs and labor rates. Typical replacement costs for transplantable trees are presented in Table 1-1. This approach is straightforward and would not likely be disputed. Factors to consider when establishing replacement value are:

1. Costs related to transplanting:
   • Cost of delivery of transplant;
   • Cost of removal of damaged or dead tree;
   • Expenses related to preparation of transplant hole, including soil amendments and fertilizers, and to placement of the transplant; and
   • Cost of postplanting maintenance.
2. Availability of the same tree species. The species being replaced may not be available locally. If a substitute is acceptable, the cost of it can be used. If not, increased costs based on additional transportation charges need to be considered.
3. Availability of a tree of the right size. Trunk diameters of transplantable trees are measured 6 inches above ground (trunks up to 4 in. in diameter)

Table 1-1. Cost of replacing transplantable trees of various sizes

Trunk diameter (in.)	2	3	4	5	6	7	8	9	10	11	12
Basic replacement cost (dollars)*	195	325	480	660	865	1145	1410	1680	2055	2455	2855

*Approximate costs based on nursery industry estimates in 1981–1982.

Source: Adapted from ISA, Guide for establishing values of trees and other plants, 7th ed. International Society of Arboriculture, Urbana, Ill., 1988.

and 12 inches above ground (trunks 4 to 12 in. in diameter). Although trees 12-inches in diameter are considered transplantable for a loss estimate, in reality, finding a tree and transplater locally could be difficult if replacement is really needed.

Once the basic replacement cost is established, the evaluator can proceed and obtain an appraisal value as is done for larger trees. This value of the damaged or lost tree is scaled down by taking into account species, location, and condition of the tree, and it is the appraisal value that is used to determine compensation.

If a tree is actually replaced, there is no way to compensate for the loss in future growth that the transplanted tree will experience. For example, healthy 4-inch red oak tree is lost and replaced by a healthy transplant of the same species in the same location. During the next several years, while adjusting to transplanting, the new tree will not grow as fast as the old tree would have had it not been damaged. In reality, the owner of the damaged tree will have suffered a loss of future growth, and the optimum mature size reached by the replacement will have been delayed compared to the original tree. This loss is not accounted for in most tree evaluation schemes because it is impossible to quantify.

Determining Values of Large Trees

A formula is used to compute the value of large trees. Using the formula, the four factors that determine the value are:

1. Size
2. Species
3. Condition
4. Location

Collecting the information needed to make use of the formula requires knowledge, precision, keen observation, and varying amounts of subjective judgment. Training and experience are also essential and can be obtained at tree evaluation workshops that are offered to arborists by professionals in the industry.

Tree Size. Tree size is expressed as the number of square inches one would calculate for the area of the trunk cross section at 4.5 feet above ground. A trunk cross section is normally a circle, and its area *(a)* can be calculated by knowing the diameter *(d,)* or caliper, of the tree. It is known that d squared (multiplied by itself, d^2) times the number 0.7854 equals area $(a = 0.7854 \times d^2)$. Since the diameter of a tree cannot easily be measured, it can be calculated by finding the circumference (girth) of the tree by encircling the tree with a measuring tape. Diameter equals circumference divided by pi $(\pi, 3.14)$. Most tree measuring tapes, called diameter tapes, already have this calculation completed and give the tree size as diameter.

The tree value based on size is calculated by multiplying the cross-sectional

Table 1-2. Calculated size-based values of trees of selected diameters

Trunk diameter (in.)	15	20	25	30	35	40
Cross-sectional area (sq. in.)	177	314	491	707	692	1,257
Value (dollars)	4,771	8,482	13,254	19,085	25,977	33,929

area times a basic price ($27/in.^{2} in 1988). The basic $27/in.^{2} cost for large trees was probably extrapolated from actual replacement costs of smaller trees. Table 1-2 illustrates typical size-based values of large trees having selected trunk diameters.

Abnormal trunk conditions such as low branching and fluted, buttressed trunks will influence measurements, so a degree of caution and common sense as well as suggestions from the *Guide* should be employed when determining size. There is no firm line between the size of transplantable and large trees, and although the *Guide* suggests 12 inches as the division, one needs to be flexible and use the method best suited to the circumstances.

Tree Species. It is generally agreed that not all types of trees of the same size have the same value. In fact, most arborists working in a specific geographic area tend to agree on which species should be considered high- or low-value trees. Trees that are long-lived, hardy, durable, strong, adaptable, clean, beautiful, and pest-resistant are considered high-value trees. Tree species and cultivars can be assigned percentage values based on whether they have desirable traits. These values are generally given in increments of 5 or 10 percent and in some regions can be obtained from charts that often give a percentage range for each tree type.

Good examples of species differences that could account for different rating percentages are easy to imagine. For example, in most circumstances, Chinese elm could deserve a 90 percent rating whereas Siberian elm would not be worth more than a 30 or 40 percent rating because the latter is structurally weak and subject to storm damage. On the other hand, a single species, American elm, may be worth 90 percent where Dutch elm disease is not likely to be a problem but worth only 10 percent where the disease exists (Fig. 1-3). Honey locust, with its formidable trunk thorns, would deserve a 20 percent rating, but its thornless cultivars have a higher value, perhaps 60 percent. Flowering crabapple cultivars such as 'Hopa' or 'Almey', which are extremely susceptible to scab disease in humid temperate regions, might deserve a rating of only 30 percent, whereas more tolerant or resistant cultivars such as 'Mary Potter' or 'Snowdrift' may rate 70 to 90 percent. These percentages are used to modify the size-related values previously determined.

To remove some of the subjectivity inherent in making these kinds of quality ratings, published lists of trees with their ratings have been developed. The Ohio Chapter of The International Society of Arboriculture (ISA) has developed a guide for tree- and plant-rating in Ohio, the guide reflects the collective wisdom of that particular chapter and is most useful in that part of the country.

Tree Condition. Determining the condition of a tree requires an analysis

Fig. 1-3. The American elm, one of our most beautiful trees, has been virtually eliminated in many areas by the Dutch elm disease.

based on careful examination of the tree. Many aspects of this examination are utilized when diagnosing tree problems (Chapter 9).

First, an overview of the situation needs to be made to see whether the tree is typical for that species in that area. The crown should be examined for dead wood, folliage density, and structural weakness. The foliage should be scrutinized for abnormal color, size, and shape, which indicate problems, as well as for symptoms and signs of diseases and insects.

Yearly twig growth and bud size and condition need to be considered as well as twig abnormalities relating to attack by disease organisms and insects. Trunk problems including cavities, cracks, injuries, and presence of diseases or insect pests need to be noted. Condition of roots and soil-related growing conditions that might affect tree condition should also be considered. In any case, the evaluator needs to systematically determine whether a tree is in good condition and, if not, the nature and seriousness of the problem.

A percentage can be assigned (100 = perfect condition) and used in calculating the overall value. The *Guide* provides a detailed checklist of factors to consider in determining plant condition.

Tree Location. The tree location value refers to what it is that the tree does and the impression it makes in the place where it is growing. The kind, value,

Fig. 1-4. Norway maple 'Cavalier' has a compact round top and is suitable for large open areas.

and use of the property where the tree is growing should be determined. A tree growing in the woods or along the highway would have less value than a tree in a residential neighborhood. Inherent attractive features of the tree such as flowers and foliage are also important. How the tree affects the environment of the persons living near it also is important (Fig. 1-4). This can involve every-thing from providing shade where needed to providing architectural features such as screening out undesirable views and accenting buildings. Again, a per-centage is assigned, in this case subjectively, because what one might call a beautiful character or attribute of a tree in a certain location may not appear that way at all to another individual. The *Guide* provides a detailed checklist of what to look for in establishing a percentage relative to location.

Fig. 1-5. Multitrunked birch trees are a pleasant addition to the landscape.

The value of the tree is calculated by multiplying the size-based dollar amount by the percentage allowed for species, and that amount in turn by the percentage allowed for condition. This amount is then multiplied by the percentage allowed for location. For example, a 25-inch ($13,254) silver maple (35%) in fair condition (50%) and in a good location (75%) would have a value of $13,254 \times 0.35 \times 0.5 \times 0.75 = 1739.59.

Additional skill and judgment are required for tree evaluation when the following circumstances occur:

- Trees have negative value because they are entirely inappropriate or a hazard;
- Trees are located in historic or very important locations;
- Rare or unusual specimen trees are being considered;
- Multitrunked trees are being evaluated (Fig. 1-5);
- Trees are no longer at the site but are being evaluated after the damage has been done;
- It will not be known for several years whether a tree has suffered permanent injury;
- Trees are declining; and
- Potential exists for biased evaluation. Since three of the four evaluation criteria are based largely on subjective judgment, it is important that evaluators honestly arrive at the same value no matter what their professional orientation

or which side of a dispute they favor. Unfortunately, in practice, different evaluators using the same criteria often do not arrive at similar values.

How Tree Values Are Used

Tree evaluation that is conducted with the hope of collecting damages or tax relief following a casualty loss is often unrewarded, for not all circumstances dictate that there be such compensation. The professional who is involved in tree evaluation must be knowledgeable about insurance and tax laws and know what is and is not allowed. The results of court cases involving tree losses suggest that the involvement of professionals who are qualified witnesses is important.

Arboricultural Practitioners

Tree owners often are faced with the problem of finding someone competent to care for their trees. Knowing that their tree has value, and knowing that it is possible for an incompetent individual to harm their tree, the selection of the right person or company becomes an important decision. Consumers need to be wary and take the time to examine an arborist's previous work and ask for recommendations before any tree work is agreed upon. In almost any urban area, persons can be seen working with trees. Some, just new in the business or operating part-time or during periods of umemployment, may be equipped only with a pick-up truck and a chain saw. Others, having a more substantial investment, may utilize bucket trucks, climbers with ropes and saddles, and chippers and chipper trucks. Still others work for multi state corporations and are often involved in utility line clearance. They can frequently be recognized from one area to another by the similarly painted and equipped trucks and machinery. Many of these tree workers belong to firms that hold membership in organizations of arborists and tree-related professionals within state or national areas. Although size and equipment do not provide a good means of determining whether a tree-care firm is competent, sometimes membership in at least one professional organization is a clue to a firm's competence. The primary activities of members of the following organizations include many of the major tree-maintenance operations. The brief description of each organization does not describe all of its activities.

American Society of Consulting Arborists (ASCA)

This organization, located at 700 Canterbury Road, Clearwater, FL 33546, consists of professional arborists who evaluate shade trees, diagnose tree problems, and consult on all matters relating to planting, maintenance, and preservation of trees. This well-trained and experienced group has about 150 members, primarily in North America. Some members own tree-care companies, thus offering services related to their consulting recommendations. ASCA members also conduct training sessions related to shade tree evaluation.

International Society of Arboriculture (ISA)

The ISA—whose address is P. O. Box 71, Urbana, IL 61801—primarily devotes itself, through its scientific journal, annual conference, and research funding, to educational programs on all phases of tree maintenance. The various interest groups within ISA reflect the organization's diversity. These groups include the Utility Arborist Association, Society of Commercial Arboriculture, Municipal Arborists and Urban Foresters Society, Arboricultural Research and Education Academy, and Student Society of Arboriculture. In addition, ISA is represented geographically by chapters consisting of individual or groups of states, provinces, or countries. An ISA chapter or international conference is likely to be attended by individuals in all phases of tree care, from the tree climber who has participated in an arborists' jamboree to the scientist who has delivered a research paper. ISA consists of more than 4000 members, most of them from North America.

In some states, certified arborist programs have been developed to establish meaningful arboricultural standards, encourage continuing arboricultural education, and identify competent arborists. Certified Tree Workers and Certified Arborists are required to pass written examinations and attend approved educational sessions to obtain and maintain their certification. The Western Chapter of the International Society of Arboriculture has developed a study guide for this certification. A certified arborist has at least demonstrated some competence and might be considered a good choice for tree work.

National Arborists Association (NAA)

The NAA—174 Route 101, Bedford Station, Box 238, Bedford, NH 03102—consists of commercial arborists interested in improving the way professionals care for trees. The NAA series of slide, cassette, and video programs provide valuable instruction for tree workers. These educational programs and the NAA Book of Standards, which are used by the tree care industry, define proper tree care techniques for pruning; cabling, bracing, and guying; fertilizing; installing lightning protection; and applying pesticides. NAA membership is open to all commercial arborists, and the organization has been a strong advocate for them.

Other Organizations

The following groups also have an interest in preserving and maintaining trees:

- American Association of Nurserymen, 1250 I Street N.W., Suite 500, Washington, DC 20005. This is the main voice of the nursery industry, having interests in education and research on growing and marketing of trees and other plant materials.
- American Forestry Association, 1319 Eighteenth Street N.W., Washington, DC 20036. This group publishes a national register of big trees.
- Associated Landscape Contractors of America, 405 North Washington Street,

Falls Church, VA 22046. Landscape installation is an important activity of their membership.

· Council of Tree and Landscape Appraisers, 1250 I Street N.W., Suite 504, Washington, DC 20005. The council, representing several groups, published the shade tree evaluation guide.

· Metropolitan Tree Improvement Alliance, Nursery Crops Research Laboratory, Delaware, OH 43015. The organization promotes better trees in the urban landscape through education and research.

· Society of Municipal Arborists, 7747 Old Dayton Road, Dayton, OH 45427. Municipal arborists and others belong to this group, which promotes street tree management.

· National Arbor Day Foundation, Arbor Lodge 100, Nebraska City, NE 68410. This foundation sponsors the Tree City USA program.

· Society of American Foresters, 5400 Grosvenor Lane, Bethesda, MD 20814. This society for professional foresters has members with interests in urban forestry.

Arboricultural Periodicals

Arborists and others interested in trees can learn more about trees and keep up with some of the latest developments by subscribing to, or obtaining from their municipal or college library, periodicals dealing with tree care. These publications have information ranging from popular or practical articles on tree maintenance to scientific studies of trees. The following are suggested:

Arbor Age
Arboricultural Journal, The International Journal of Urban Forestry
American Forests
American Nurseryman
Grounds Maintenance
Journal of Arboriculture
Journal of Forestry
Weeds, Trees, and Turf

Selected Bibliography

Chapin, R. R., and P. C. Kozel. 1975. Shade tree evaluation studies at the Ohio Agricultural Research and Development Center. OARDC Research Bulletin 1074. Wooster, Ohio. 46 pp.

Council of Tree and Landscape Appraisers. 1986. Manual for plant appraisers. CTLA, Washington, D.C. 57 pp.

Grey, G. W., and F. J. Demke. 1978. Urban forestry. Wiley, New York. 299 pp.

International Society of Arboriculture. 1988. Guide for establishing values of trees and other plants, 7th ed. ISA, Urbana, Ill. 49 pp.

Robinette, G. O. 1972. Plants/people/and environmental quality. U.S. Department of the Interior, National Park Service, Washington, D.C. 139 pp.

2

The Structure of the Tree and Function of Its Parts

The successful establishment and maintenance of trees are greatly aided by an understanding of the anatomy and functions of a normal, healthy tree. Without such knowledge it is difficult to diagnose accurately the many abnormalities that commonly occur in trees or to appreciate various practices undertaken to maintain or rest re their vigor.

Normality in any living organism is so vague and relative that it almost defies definition. Perhaps a scientifically accurate definition might be derived by applying to the tree something of the psychologists' concept of the normal human being. By this rule, the normal tree would fall within certain limits of health and vigor that include the majority of trees in a specified range.

For our purposes, however, we cannot go far astray if we are guided in our evaluation of a healthy tree by the desirable external appearances with which we are all familiar. Since a tree is most admired when it has a symmetrical branch system and abundant, attractive foliage, we include these two characteristics among our standards of perfection. Healthy roots must also be included, for without them proper branch and foliage growth cannot continue.

To gain a working knowledge of the proper care of trees, we must know something of the structure of each of the different parts that constitute a normal shade tree, the functions these parts perform in the growth process, and the relation between the parts.

Parts of a Tree

The roots, stems, and leaves are vegetative structures that perform specific functions to maintain the life, health, and vigor of the tree. The flowers, fruit, and seeds are reproductive structures and have evolved to ensure the continuation of the tree species.

Roots

There is no organ more vital to a healthy tree than its roots, but because the root system is largely unseen, its importance seldom is appreciated by laymen.

Many believe that root distribution somehow mirrors the branching pattern of the aerial portion of the tree or that roots extend out only to the edge of the tree crown, as far as the ''dripline.'' In fact, for most trees, roots extend laterally far beyond the tree crown but do not extend to depths equivalent to the tree height.

There are two general types of genetically determined rooting patterns: tap and surface. Taproot systems are characterized by a fast-growing primary root that penetrates deeply. Hickory, walnut, tuliptree, and many oaks have taproot systems. Surface root systems are characterized by slower growing shallow primary roots with more rapidly growing lateral branches. Elm, pine, spruce, and maple are surface-rooted species.

Although the root systems of seedlings exhibit the genetically inherited pattern, roots of older trees are greatly influenced by soil and site characteristics, so the inherited pattern is often obscured. The two most important environmental factors that determine root distribution are oxygen and water. In general, tree roots will extend to greater distances both vertically and laterally in well-aerated soils. Observations on fruit trees have revealed lateral spread of roots equivalent to three times tree height in sandy soil, two times in loam, and one and one-half times in clay. Likewise in well-aerated soils, roots may extend down 30 feet; however, under most conditions 70 to 90 percent of roots will be found in the upper 18 inches of soil. Site characteristics such as hard pan, clay layer, or perched water table restrict root development. Root development also is impeded by surface covering such as concrete or asphalt.

Under favorable circumstances the root system of some trees is enormous. It may comprise one-third to one-half of the entire volume of the tree. The total length of all roots of a large spreading oak runs into hundreds of miles. Roots branch and rebranch. One researcher estimated that a mature red oak had at least 500 million root tips!

Structure of Root Systems. At the point of attachment to the trunk, the roots are relatively few and large. These woody roots taper rapidly into long ropelike strands from which lateral roots branch. The lateral roots divide and redivide, becoming progressively smaller in diameter until they become fine, absorptive, nonwoody roots at some distance from the trunk (Fig. 2-1). The woody roots are essentially perennial, whereas the nonwoody roots are short-lived and frequently replaced. Nonwoody roots may convert to permanent woody roots in response to injury to the main root.

The tip of nonwoody roots is covered by a group of cells called the *root cap*. It functions to protect the growing tip from injury, and cells near the outside of the cap have mucilagenous walls that help ease the progress of the tip through the soil. A short distance back from the tip, the rootlet is covered by numerous, very fine outgrowths of thin-walled cells, called *root hairs* (Fig. 2-2). More than 10,000 root hairs per square inch have been found on certain roots. The absorptive, nonwoody roots are composed of tissues derived by cell division within a special area called the *apical meristem*. The apical meristem is located just behind the root cap.

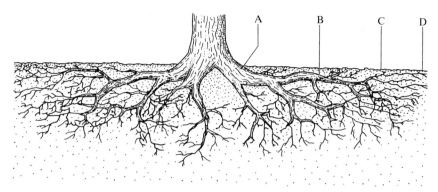

Fig. 2-1. The root system of most trees includes A, a few large woody roots; B, long ropelike lateral roots; C, fine absorptive nonwoody roots; and D, a zone of root hairs near the root tips.

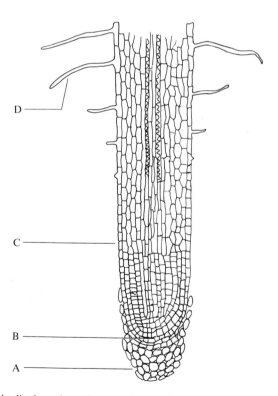

Fig. 2-2. Longitudinal section of a root tip showing A, root cap; B, apical meristem where cell division occurs; C, zone of cell elongation; and D, root hair.

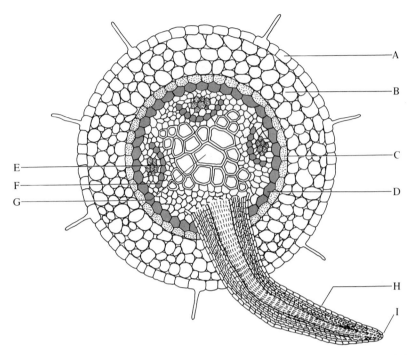

Fig. 2-3. Cross section of a young root showing A, epidermis; B, cortex; C, endodermis; D, pericycle; E, phloem; F, cambium; G, xylem; H, lateral root; and I, root cap. Note the lateral root arises from internal tissues.

The tissues of the root are organized more or less concentrically (Fig. 2-3). At the center of the cylinder are *xylem* cells, specialized for conducting water and minerals, and *phloem* cells, specialized for conducting food materials. This core is surrounded by the *pericycle, endodermis, cortex,* and *epidermis* on the outside. Lateral roots arise from the pericycle. The endodermis is a single row of cells with a bandlike thickened suberized area, known as the *casparian strip*. The strip prevents water from entering the pericycle except by passing through the protoplast of the endodermis. Thus the endodermis may exercise control over water and dissolved substances entering and leaving the root. The cortex is composed chiefly of storage cells with large intercellular spaces. Finally, the epidermis serves as covering over all. Young epidermis has root hairs and older epidermis may develop a cuticle and become lignified as well.

In cross section woody roots resemble woody stems. Annual rings are less pronounced in roots than in stems, partly because seasonal changes in environment are buffered by the soil. Long-lived woody roots develop a distinct core of heartwood just as stems do. Root tissues are adapted for storage and have less need for strong supporting tissues.

Root Growth. The root tip is pushed through the soil by the process of cell division, which takes place in the apical meristem and by elongation of the newly formed cells.

The root system of trees includes long-lived large roots and short-lived small roots. The large, main, woody roots are essentially perennial, and unless killed by unfavorable environmental conditions, insects, or disease, they will endure as long as the tree. The longevity of the small absorptive roots varies with species, environmental conditions, and tree health, but generally they live only a few weeks to a year or sometimes several years.

Some roots undergo secondary growth which causes an increase in root diameter and may start during the first or second year of root life and continue thereafter. These roots become the perennial woody component of the root system. Roots near the soil surface are the first to begin growth each spring; those at greater depths begin later. More wood is produced near the soil surface, so older roots are often not circular in cross section.

Function of the Root System. There are four primary functions of roots: anchorage, absorption, conduction, and storage.

Roots ramify through the soil, thus firmly anchoring trees. Diseased or pruned roots fail to anchor trees, just as water-soaked soil does.

Trees derive most of their water and mineral salts from soil. Water primarily enters through root hairs and epidermal cells of the absorptive roots, although some enters through older roots as well. The root hairs act to increase the surface area of the roots and come in close contact with the film of water that surrounds soil particles. Water diffuses into the root cells; however, the uptake of mineral salts is a relatively selective process dependent on metabolic activity.

Roots conduct water, mineral salts, and sometimes stored foods to stems and leaves. Conversely they also conduct foods from the leaves and stems to all parts of the system underground.

When sugars move into roots more rapidly than they can be used by growing cells, they may be converted to starch and stored for a time. During the dormant season, large quantities of starch are stored in the woody roots of trees. In the spring when active growth resumes, this food reserve is mobilized and conducted back into the stems and leaves.

Specialized Roots. The roots of many vascular plants harbor filamentous fungi, forming symbiotic associations called *mycorrhizae* (Fig. 2-4). Mycorrhizae fall into two categories: *ectotrophic* forms in which the fungus exists both inside and outside of the root and *endotrophic* forms in which the fungus lives entirely within the host cells. Ectotrophic mycorrhizae are commonly found associated with the absorptive roots of many tree species. The fungus penetrates the cell wall of the cortex and covers the outside of the root with a network of fungal strands. Ectotrophic forms are found on many important trees including alder, beech, birch, chestnut, oak, pecan, poplar, willow, cedar, Douglas-fir, fir, larch, pine, and spruce. The endotrophic form is associated with finely divided roots and the fungus lives intracellularly in cortical cells. The endotrophic form is less prevalent and is commonly found on tuliptree, maple, and sweetgum.

The mycorrhizae are important in tree physiology. They act to increase the absorption surface area for minerals, particularly phosphorus. In most sites,

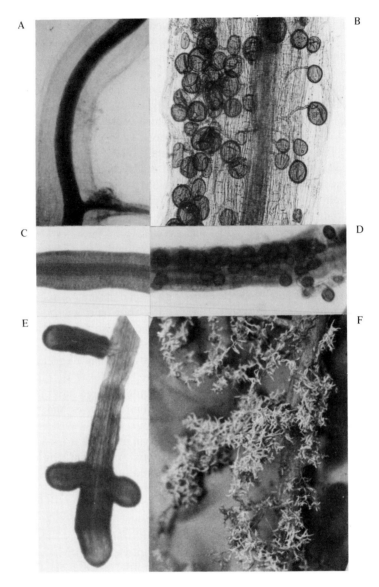

Fig. 2-4. Mycorrhizae: A, uncolonized black cherry root; B, black cherry root infected with endomycorrhizae; C, uncolonized sweetgum root; D, sweetgum root infected with endomycorrhizae; E, ectomycorrhizae formed on Norway spruce (note presence of fungus mantle around the root); F, ectomycorrhizae on pin oak (note presence of abundant short roots). (From Maronek, D. M., J. W. Hendrix, and J. Kiernan. 1981. Mycorrhizal fungi and their importance in horticultural crop production. Horticulture Reviews 3:172–213.)

mycorrhizae appear to be essential for tree survival. The fungus benefits by getting carbohydrates from the tree. The tree benefits by the increased mineral absorption and perhaps by some degree of root disease protection.

Natural grafts between roots are common. The graft involves a vascular connection which is the result of a union of cambium, phloem, and xylem of previously unconnected roots. For a graft to occur the roots must be compatible, usually of the same species, although interspecies grafts occur occasionally, and there must be growth pressure. Often a community of trees is functionally united by numerous within-species grafts. Water, minerals, metabolites, and sometimes disease organisms pass freely through the graft unions.

Root nodules form on many leguminous woody plants and many other trees, notably alder and ceanothus, as well. Associated with these nodules are microorganisms which have the ability to fix nitrogen. In the case of the legumes, the organisms are bacteria of the genus *Rhizobium*. For nonlegumes, the organisms are most commonly Actinomycetes. Usually the microorganisms invade the roots through root hairs or small wounds and infect the cortical parenchyma cells.

Stems

The stem—the woody framework of the tree—includes the tree trunk, or bole, beginning just above the root flare and the branches. It is constituted to endure the strains placed upon it by the varying conditions of its environment. For a seemingly rigid structure, the stem is amazingly flexible and will often bend to a surprising degree before breaking.

Structure of the Stem. The largest part of the stem or trunk is composed of a wood cylinder made up of tissues called *xylem* (Fig. 2-5). The xylem is composed of long tubelike cells through which water containing dissolved nutrient salts is conducted from the roots to the leaves; *parenchyma* cells, which are involved in storage of food materials and lateral transport of materials; and *wood fibers,* which lend tensile strength to the tree. The tubelike cells are of two types: thick-walled *tracheids* and thinner walled *vessel elements*. Both types of cell are found in angiosperms, whereas in gymnosperms only tracheids are present. Functional vessels and tracheids have no living contents. Indeed in the functional area of the xylem only about 10 percent of the cells are alive.

In most mature trees, a dark inner zone called *heartwood* can be seen. The heartwood is composed entirely of dead cells and as such is physiologically inactive, although it may be chemically reactive when invaded by decay fungi. It serves only as a mechanical support for the tree. Many trees are able to survive for years after the heartwood is completely rotted out.

The outer portion of the wood cylinder is called the *sapwood*. The sapwood is usually lighter in color and in density than the heartwood. The functional water-conducting and physiologically active cells are located in the sapwood. As the cells on the inner edge of the sapwood mature and die or become plugged they are converted into heartwood. The boundary between the sapwood and the heartwood is often irregular, extending several annual rings closer to the out-

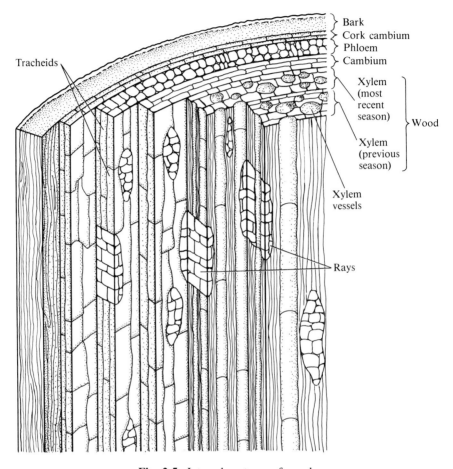

Bark
Cork cambium
Phloem
Cambium

Xylem (most recent season)

Wood

Xylem (previous season)

Tracheids

Xylem vessels

Rays

Fig. 2-5. Internal anatomy of wood.

side at some points than others. The inner part of the sapwood is continuously transformed into heartwood. Thus the heartwood constitutes a progressively expanding inner core. The sapwood, however, in mature trees tends to remain about the same thickness.

The *pith* occupies the center portion of the stem. The pith has no specific function in mature trees and is frequently so crowded by growth pressure that in older stems it is nearly lost.

Immediately outside of the xylem is a narrow band of cells known as the *cambium*. During the growing season, the cells of the cambium are stimulated to divide, giving rise to the xylem tissue on the interior and *phloem* tissue on the exterior. Thus new xylem cells are added to the outside of old xylem cells and new phloem cells are added to the inside of old phloem cells. Although the cambial cells, by repeated divisions, are differentiating into xylem and phloem, some daughter cells remain as cambium, so a cambium layer is always between the xylem and phloem. As the stem increases in circumference, the cambium keeps step with this increase by radial divisions, thus increasing the number of cells and enlarging the circumference of the cambium layer. Generally, the thicker this layer of cambium, the more vigorous the tree.

All tissues lying outside of the cambium are considered to constitute the *bark*. The bark has a more complex tissue structure than the wood and includes the phloem, cortex, and periderm.

The phloem constitutes a fairly narrow band composed of a variety of cells, most prominently *sieve-tube members,* or *sieve cells, companion cells,* fibers, and parenchyma. The sieve tubes or sieve cells serve as passageways through which food materials are conducted. Most of the normal cell contents are missing from mature sieve-tube members. With the exception of a few trees such as palms and basswood, sieve-tube members live for only one to three years. In most species, nonfunctional phloem cells are crushed by growth pressures.

The *cortex* is a complex tissue located between the phloem and outer tissues. Its function in young stems is to store food and support the stem. This tissue is not found in mature stems.

The *phellogen,* or cork cambium, produces outer bark tissues. It is located outside of the cortex and can be found in mature trees. The phellogen functions in much the same way as the cambium, except that it produces *cork* externally and *phelloderm* internally. Some species produce no phelloderm. The phellogen and the tissues it produces are collectively called the *periderm*.

The cork cells die soon after formation and form a waterproof protective tissue. Their walls consist largely of suberin, a waxy substance. The continued increase in circumference of the tree causes the cork to rupture and fall off. Eventually the loss of dead bark about equals the formation of new bark. Many species can be identified by their unique bark appearance.

Inspection of the cross section of any tree trunk (Fig. 2-6) reveals numerous concentric lines, the *annual rings,* which are plainly visible because of the alternating large and small cells. These rings represent the annual increase in the diameter of the tree. The large cells are formed by rapid growth in the spring; the smaller, by slower growth in the summer. As the season progresses, the new cells become smaller and smaller, until growth stops altogether in the fall. This compact region of small cells, known as *summer wood,* makes the annual ring. It is in sharp contrast to the large cells, or *spring wood,* formed by rapid growth during the spring.

Across the annual rings many radial lines are visible, running from the outer edge of the wood cylinder toward the center. These lines, called *medullary rays,* are groups of cells extending through the many cylinders of wood; they provide for radial transfer of water and food.

Stem Growth. The stem increases both in length (primary growth) and diameter (secondary growth) by the functioning of actively dividing cells known as *meristematic tissues* and by the enlarging of newly formed cells before secondary wall formation. The meristematic tissues make up only a small portion of the total mass of the tree. They are located in buds and stems. In the stems,

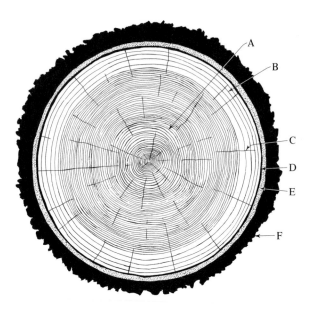

Fig. 2-6. Cross section of a tree trunk, showing A, the heartwood, composed of dead cells; B, the sapwood, which has many living xylem cells that conduct water and minerals upward; C, medullary rays, which transfer water and food radially; D, the cambium, from which are formed xylem cells on the inside and phloem tissue on the outside; E, the phloem cells, which conduct elaborated food; and F, the outer bark.

the cambium and phellogen are meristematic tissues. Meristematic tissues are the main source of the hormone auxin, which plays an essential role in regulating tree growth.

The growth of the shoot begins with division of cells of the apical meristem, which is located in the terminal bud. Subsequently the cells, which do not continue dividing, will elongate, differentiate into specialized cell types, and mature. The increase in shoot length results almost entirely from the elongation process. The mature cells cease to enlarge. Thus the apical meristem is constantly carried upward and outward by the growth of cells it produces.

Trees produce several types of buds which are categorized by position (terminal or lateral), contents (vegetative, floral, or mixed), or activity (dormant or active). Within the terminal bud, cells divide to produce leaf primordia and lateral or axillary buds. Each lateral bud is potentially the terminal bud of a new branch shoot or a flower. Under some circumstances, especially after injury, buds may be produced in locations other than at a leaf axis. They are called adventitious buds and originate from a variety of tissues including wound callus, cambium, or mature nonmeristematic tissues.

The lateral buds may continue growth without interruption, thus developing into leafy branches or flowers. More frequently they become dormant after developing to a length of about a quarter of an inch. The buds may remain dormant or resume active growth after a time. The dormancy or activity of lateral buds is governed by hormones produced by the tree.

Most woody plants grow intermittently, with one or more flushes of growth within a distinct growing season, rather than continuously. Typically rapid periods of stem elongation are followed by periods of little or no elongation even when favorable environmental conditions exist. In the tropics some tree species are found to grow continuously; in the temperate zone, however, all trees grow intermittently.

Under exceptional conditions trees normally having one growth flush per year may resume growth within the same season. Young trees, particularly oak and maple, may have a second growth flush if they are heavily fertilized. Older trees may resume growth if excessive rainfall follows a period of drought.

Water sprouts arise from dormant buds on the main trunk or branches of trees. They may be stimulated to grow by sudden exposure to light, as when an adjacent tree or shading branches are removed.

Branches are often shed by trees. This occurs by two mechanisms: abscission by a physiological process similar to leaf abscission and natural pruning by death of branches, during which no abscission zone is formed. Abscission of twigs is a response to adverse environmental conditions and to aging and subsequent loss of branch vigor. It occurs in a well-delineated abscission zone and is preceded by periderm formation. Mature white oak, poplar, willow, maple, walnut, and ash commonly undergo abscission. Most shed twigs in the autumn, but maples shed largely in the spring and summer.

Natural pruning generally begins with branches of the lower crown. Death usually is caused by low light intensity. The dead branch is sealed off from the living tree stem by gums or plugs called tyloses in angiosperms or resins in gymnosperms. The tree has no mechanism to cause the dead branch to fall off, but saprophytic fungi and insects attack the dead branch and in most cases it will fall off in time.

In the spring the resumption of stem cambial activity follows renewed activity in the buds and the development of leaves. Activity begins at the top of the tree and proceeds downward. During the early stages of cambial activity, the cells enlarge because of water uptake and their walls become thinner. The bark may be easily peeled at that time.

The onset of cambial activity appears to be closely related to the beginning of shoot growth, although shoot growth in most tree species stops before diameter growth. Even in trees that have only one flush of shoot growth in the spring, cambial activity continues until late summer or early fall. Cambial activity apparently ceases first in the upper crown, then in the stem, and finally in the roots.

Stem Functions. The functions of the stem, or trunk, of a tree are provision of structural support; conduction of water, food, and nutrients; and storage of elaborated food materials. One of the chief services rendered by the much-branched stem or trunk is support of the leaves to provide them maximum benefit from the sun's energy. It is essential to the growth of the tree that as many leaves as possible be exposed to the sun. This exposure is accomplished by means of multiple branching of the tree trunk into limbs, branches, and finally twigs. Structural support may fail if the stem is decayed or severely injured.

The trunk and its subdivisions are the connecting links between the roots, where most water and mineral nutrients are absorbed, and the leaves, where food is manufactured. It is the function of these connecting links to translocate the water and mineral elements in water solution from the roots to the leaves and to distribute elaborated food materials to every growing part of the tree, including twigs and root tips.

The large size of trees indicates that they have highly developed translocation systems. Water movement is along the path of least resistance in wood. Most trees show spiral grain to some extent; thus water usually ascends in a helical pattern. The angle of the spiral grain changes from year to year so that over a few years' time a single root may be connected to a number of different branches.

Sugars manufactured in the leaves are translocated in the phloem. The direction of movement changes during the season. In the spring, sugars are initially transported into growing leaves. At a certain stage of maturity when leaves produce more sugars than are needed for growth, the leaves begin to export sugars and the direction of flow in the supporting twig is reversed. Leaves near the stem tip may export sugars both up to the expanding terminal bud and down to the roots.

In all deciduous trees, food materials are gradually conducted from the leaves to twigs. These foods are partially assimilated for immediate growth. The surplus is stored in twigs, trunk, and roots. Late in the season, when the leaves become less active in the manufacture of food, part of this surplus is reabsorbed for assimilation by the growing parts. The initial twig and root growth and, frequently, flower production, which takes place in the spring before leaves are expanded and active, are entirely dependent upon food stored in twigs, stems, and roots.

Leaves

Leaves function as solar radiation receptors and as sites of sugar manufacture. A very large surface area is necessary to perform these functions. In forests it has been estimated that the surface area of leaves is four to six times greater than that of the ground.

Structure of Leaves. The typical leaf of a broad-leaved tree consists of a blade and a stalk called the *petiole.* In cross section (Fig. 2-7), we see the surface cells, or *epidermis,* the outer walls of which are coated with a waxy covering, or *cuticle.* The cuticle is a noncellular layer consisting of wax and cutin and is attached to the epidermis by a layer of pectin. This covering is perforated by small openings called *stomata* (singular *stoma*). Each stoma is surrounded by two kidney-shaped movable cells, the *guard cells.* Through these pores moisture evaporates from the leaf and gases are exchanged between the

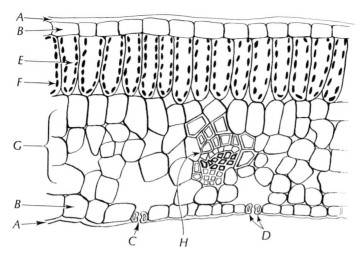

Fig. 2-7. Cross section of a leaf, showing A, the cuticle; B, epidermis; C, stoma; D, guard cells; E, palisade cells; F, chloroplasts; G, the spongy mesophyll; and H, the vascular tissues.

atmosphere and the interior of the leaf. Moisture loss from the leaf is also controlled by the leaf waxes.

Beneath the epidermis is one layer or more of long, cylindrical cells, the *palisade cells.* These contain many minute *chloroplasts,* which contain the green substance known as *chlorophyll.* Numerous large, loosely arranged cells, the *spongy mesophyll,* lie between the palisade cells and the lower epidermis. The high proportion of intercellular space facilitates the diffusion of gases between the stomata and the palisade cells.

Embedded in the mesophyll is the veinal or vascular tissue. The veins of angiosperm leaves generally form an interconnecting branching system. A vein typically consists of a strand of xylem and a strand of phloem surrounded by a layer of cells called the *bundle sheath.* The phloem is typically on the lower side and the xylem on the upper. The arrangement of veins of leaf blades varies in detail from species to species.

The leaves of most gymnosperms are either needlelike or scalelike; however, a few, for example, ginkgo, have broad, flat leaves. The leaves are generally evergreen, except in larch, bald cypress, and a few others. Generally the leaves are firm, with small, thick-walled epidermal cells and a thick cuticle (Fig.2-8). The stomata are usually sunken and are typically found in longitudinal bands.

Leaf Growth. Leaf growth is initiated by the meristematic tissue of the shoot near the apex. In many species, cell division is complete in the fall while the leaves are still enclosed in the bud. The growth of leaves in the spring may be entirely the result of cell enlargement.

The seasonal pattern of leaf growth varies widely among species. Most commonly, new leaves appear in one or more distinct flushes of growth. Some

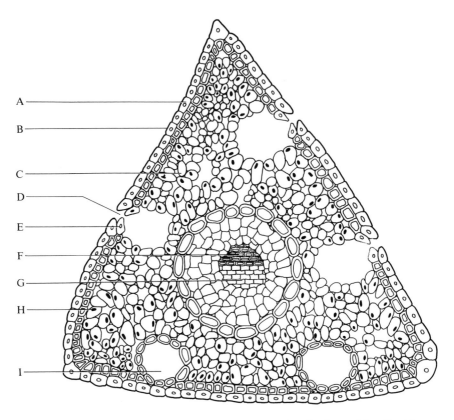

A
B
C
D
E
F
G
H
I

Fig. 2-8. Cross section of a needle, showing A, epidermis; B, cuticle; C, chloren-chyma; D, stoma; E, guard cell; F, xylem; G, phloem; H, chloroplast; and I, resin duct.

species produce new leaves more or less continually throughout the growing season. Individual angiosperm leaves tend to expand rapidly, taking from 2 to 40 days (average 14 days) to fully expand. Leaves of evergreen species grow out more slowly than those of deciduous trees. Individual leaves of gymno-sperms may continue to expand from early spring through late summer. Indeed in some species, leaves may continue to increase in length for more than a year.

Leaf life has an inherently limited duration. Leaves of deciduous species last only one growing season, whereas those of evergreen species persist at least until the leaves of the following season are produced and may last several years. The leaves of needle evergreens generally last 2 to 3 years, although some live as long as 10 to 15 years.

Leaves of most woody perennial plants are shed as a result of changes in an abscission zone at their base, often before the leaf dies. The abscission zone is the weakest part of the petiole. It contains a high proportion of parenchyma cells and few thick-walled cells. Before leaf fall, changes occur in the zone which result in the formation of a cork layer over the leaf scar and the plugging of exposed vessels. Thus the abscission zone functions to bring about leaf fall and to protect the shoot from invasion of pathogenic organisms through the leaf scar.

Leaf fall is under the control of plant hormones but is thought to be triggered by adverse environmental conditions, such as drought, temperature changes, or shortened days. These conditions stimulate leaf senescence, which always precedes natural abscission.

Function of Leaves. The primary function of the leaf is *photosynthesis* (photo = light; synthesis = manufacture), the manufacture of sugars through the use of energy provided by sunlight. Photosynthesis is the source of energy not only for trees and other plants, but indirectly for virtually all other living things through their consumption of plants. Oxygen is an important product of photosynthesis as it is essential to all living organisms, primarily for its role in *respiration.*

The manufacture of sugars is accomplished in the chloroplasts, through a highly complicated process in which carbon dioxide, which enters the leaf through the stomata, is combined with water, which is taken up through the root system. The sugars are then used by the tree as a source of energy, through the process of respiration, for further growth and development. Respiration takes place in living cells of the stem, roots, and reproductive structures as well as in the leaves.

Sugars also provide the starting material for the synthesis of other essential plant constituents such as starch, cellulose, lignin, proteins, fats, and oils. Formation of these compounds involves highly complicated processes which may also take place in living cells of organs other than leaves.

The process of *transpiration* occurs as the result of the evaporation of water from leaf surfaces, primarily through the stomata. The rate of transpiration is influenced by a number of factors including humidity, temperature, and wind, but the most important factor is the condition of the stomata. All other factors being equal, water loss is least when stomata are completely closed. Closed stomata, however, do not allow the entry of carbon dioxide essential for photosynthesis, hence for the tree to be healthy, stomata cannot remain closed. Transpiration can thus be regarded as a necessary consequence of the need for photosynthesis.

Water lost through transpiration is replaced by water taken up from the soil solution by the roots and conducted to the leaves through the xylem. The processes by which this occurs, particularly in tall trees, has been the subject of some controversy, but capillary action by which water is drawn upwards is generally thought to play a role.

Reproductive Organs

Angiosperms. Typically a flower is composed by four whorls of modified leaves: sepals, petals, stamens, and carpel(s), all attached to the receptacle, the modified stem end that supports these structures (Fig. 2-9). The stem terminates in the flower and does not continue to grow. The stamens are the male organs

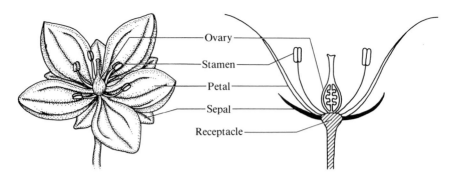

Fig. 2-9. Structure of a typical flower.

of the flower and produce the pollen. The carpels are the female organs and bear the ovaries.

Many trees have single-sex flowers which contain either stamens or carpels. In some cases these single-sex flowers are borne on the same tree, as in birch and alder. In other cases, such as holly, persimmon, poplar, and willow, they are borne on separate staminate and pistillate trees. Buckeye produces complete flowers, staminate flowers, and pistillate flowers on the same tree. Ash produces complete flowers plus either pistillate or staminate flowers on the same tree. Most fruit trees produce only complete flowers.

Trees do not flower until they are mature, typically after 5 to 20 years. The flower buds of many trees are initiated during the previous season, so abundant bloom depends upon the tree's health during the previous year.

The development of fruit usually requires fertilization of the egg, contained in the carpel, by the sperm, borne by the pollen. However, in a few species, fruit may mature without fertilization and/or seed development. These *parthenocarpic* fruit occur in some species or varieties of fig, apple, peach, cherry, plum, citrus, maple, elm, ash, birch, and tuliptree. The period from flowering until fruit ripening varies greatly among species and varieties, and in some cases because of environmental conditions and cultural practices. For many species, this interval is about 15 weeks.

Gymnosperms. The flowers of gymnosperms consist of *pollen cones* and *seed cones,* which in most species are borne on the same tree, although one exception is juniper. The pollen cones are usually from 1/4 inch to 4 inches long. They are brightly colored, often yellow, purple, or red. Seed cones are much larger, ranging from 1/2 inch to more than a foot long. Cones are usually initiated during the season before pollination occurs.

Cross-pollination in gymnosperms is common because female cones are generally concentrated in the upper branches and male cones in the lower ones. Wind-transported pollen is shed over a 2- or 3-day period for most trees. Pollination occurs in most species in early spring, soon after the seed cones emerge from the dormant buds.

The cones begin to enlarge after pollination and the scales close. In the case of Douglas-fir, development from pollination to seed maturation is completed in about four months. For most pines, the process takes 2 years.

Selected Bibliography

Cronquist, A. 1961. Introductory botany. Harper & Row, New York. 902 pp.

Cronshaw, J. 1981. Phloem structure and function. Ann. Rev. Plant Physiol. 32:465–84.

Kozlowski, T. T. 1971. Growth and development of trees. II. Academic Press, New York. 514 pp.

Kramer, P. J., and T. T. Kozlowski. 1979. Physiology of woody plants. Academic Press, Orlando, Fla. 811 pp.

Thomas, H., and J. L. Stoddart. 1980. Leaf senescence. Ann. Rev. Plant Physiol. 31:83–111.

Weier, T. E., C. R. Stocking, and M. G. Barbour. 1974. Botany: An introduction to plant biology. Wiley, New York. 693 pp.

Zimmermann, M. H., and C. L. Brown. 1971. Trees structure and function. Springer-Verlag, Berlin. 336 pp.

3

The Soil and
Its Relation to Trees

No book on the maintenance of shade and ornamental trees would be complete without some discussion of that important natural body, the soil. Usually considered by the layman to be an inert mass, soil is actually a dynamic entity in which physical, chemical, and biological changes are constantly occurring. A detailed treatment of these changes is beyond the scope of this book, but a brief presentation of the more important ones, particularly as they relate to trees, is essential.

Soil serves a a storehouse for mineral nutrients, a habitat for microorganisms, and a reservoir of water for tree growth. It also provides anchorage for the tree.

Soil Components

The Soil is composed of three distinct parts or phases: gaseous, the soil air; liquid, the soil water; and solid, the soil minerals and organic matter. These parts are interrelated very closely.

The Soil Air. The pore space in soils amounts to 30 to 50 percent of the volume. The quantity of air contained in the soil is governed primarily by the extent to which this pore space is filled with moisture. Ordinarily, air comprises about 20 percent of the volume of a well-tilled soil.

Oxygen, one of the most important components of soil air, is essential for respiration in roots and supports the activity of numerous beneficial microorganisms. Nitrogen, another component of the soil air, is used as a raw material by nitrogen-fixing bacteria to manufacture protein materials, which later decompose to yield nitrates, which nourish the tree.

All tree roots and soil microorganisms respire; that is, they continuously consume oxygen and emit carbon dioxide. In poorly aerated soil, where gaseous exchange between the soil air and the atmosphere is retarded, the oxygen supply decreases, whereas the quantity of carbon dioxide increases. This unfavorable oxygen-carbon dioxide balance may retard or, in extreme cases, completely check root growth, since roots do not thrive in such areas. This, in fact, explains why the more active roots of city trees are often confined to soil areas between the curb and the sidewalk and to nearby lawn areas.

Although oxygen is absolutely necessary for a healthy root system, an insufficiency of this element for relatively short periods will not result in serious damage. Flooding of soil in winter when roots are dormant is not as serious as it is during summer when the roots are making their most active growth. Roots of apple trees can grow slowly with as little as 3 percent oxygen, but they require at least 10 percent for good growth.

Because the concentration of carbon dioxide must be very high, over 20 percent, to cause direct injury to tree roots, lack of oxygen probably is the more common cause for restricted growth or death of roots.

Capillary and gravitational movements of water and the daily changes in temperature and barometric pressure are responsible for the aeration of soils. The oxygen supply is often renewed too rapidly in sandy soils and too slowly in heavy clay soils. In either case, the situation is improved by adding organic matter or by mixing the two classes of soil. Aeration of soils is also increased by structural changes caused by the use of lime and certain fertilizer salts. The channels remaining after the roots of plants decay and those caused by the burrowing of earthworms also aid in ventilating the soil.

Some trees can thrive under conditions of relatively poor soil aeration. Willows, for example, will grow along streams and ponds where the oxygen supply is relatively low. Some scientists believe that in willows oxygen produced by photosynthesis in the leaves is transferred internally to the root cells. The roots of one species of willow, *Salix nigra,* were found to be capable of growing and absorbing nutrients in almost complete absence of oxygen. Cypress grows naturally in swamps where the soil is flooded for long periods. The roots of this tree have pyramid-shaped protuberances, called "knees," which extend as high as 10 feet into the air. The knees are composed of light, soft, spongy wood and bark, which permit air to reach the roots during the weeks and even months when the swamps are covered with water.

On the other hand, plants like yew and cherry thrive only in soils that are well drained and well aerated. Such trees are killed or severely damaged if the soil is waterlogged for only a few days during the growing season.

Some trees can survive prolonged periods of soil flooding; others cannot. New York state nurseryman Philip M. White observed the response of trees whose roots were flooded in from 4 to 15 inches of standing water for 10 days as the aftermath of hurricane Agnes in western New York state in 1972. Listed among the survivors were:

Acer rubrum	Red maple
Cornus mas	Cornelian cherry
Fraxinus americana	White ash
Gleditsia triacanthos inermis	Thornless honey locust
Juglans nigra	Black walnut

Juniperus chinensis 'Pfitzeriana'	Pfitzer juniper
Juniperus virginiana	Red cedar
Malus dolgo	Dolgo crabapple
Morus alba	Mulberry
Platanus occidentalis	Sycamore
Populus deltoides	Cottonwood
Salix alba	White willow
Salix discolor	Pussy willow
Tilia cordata	European littleleaf linden

Trees that did not survive the prolonged flooding were:

Acer saccharum	Sugar maple
Acer platanoides	Norway maple
Betula papyrifera	White birch
Betula populifolia	Gray birch
Cercis canadensis	Redbud
Cladrastis lutea	Yellowwood
Cornus florida	White flowering dogwood
Cornus florida 'Rubra'	Red flowering dogwood
Cornus florida 'Cloud 9'	'Cherokee Chief' dogwood
Crataegus phaenopyrum	Washington hawthorn
Crataegus lavallei	Lavalle hawthorn
Magnolia soulangiana	Saucer magnolia
Malus sp. 'Hopa', 'Lodi', 'McIntosh', 'Radiant'	Apple and crabapple
Picea abies	Norway spruce
Picea pungens	Colorado spruce
Picea pungents 'Glauca'	Colorado blue spruce
Prunus persica	Flowering peach
Prunus serotina	Black cherry
Prunus subhirtella 'Pendula'	Weeping cherry
Quercus borealis	Red oak
Robinia pseudoacacia	Black locust
Sorbus aucuparia	European mountain-ash
Taxus cuspidata	Upright yew
Taxus cuspidata 'Expansa'	Spreading yew
Taxus media 'Hicksi'	Hick's yew
Thuja occidentalis	American arborvitae
Tsuga canadensis	Canadian hemlock

Inadequate aeration of the soil may affect the roots in an indirect manner. It may change the kind of organisms in the soil flora and make the roots more susceptible to attack by parasitic fungi. For example, the fungus *Phytophthora cinnamoni* may become so abundant in a poorly drained soil that it will cause severe damage to tree roots. Moreover, certain organisms thrive under conditions of little or no oxygen. They produce compounds such as hydrogen sulfide and nitrites, which may harm the roots.

Aside from direct injury to the roots, poor aeration also prevents the roots from absorbing water and minerals. A deficiency of oxygen and an excess of carbon dioxide reduce the permeability of the roots to water and prevent the roots from absorbing minerals from the soil solution.

Poor aeration of the soil results in development of roots near the soil surface. Such a shallow root system makes a tree more susceptible to toppling by winds and to droughts during the summer.

The single most critical factor in the success or failure of tree planting is soil drainage and aeration. Trees are killed more frequently by poor soil drainage than anything else. We can avoid this to some extent by selecting species more tolerant of poor soil drainage, but no tree grows best in a totally saturated soil.

Soil Water. Trees depend on the water contained in the pore space of soil in which they are growing. The roots absorb the water and pass it on to the leaves, where it functions both as a nutrient and in the absorption of carbon dioxide from the air. Water is an essential part of the protoplasm, or living matter, of every cell. It also functions in the soil as a solvent of the necessary mineral nutrients and as a medium by which they enter the plant. Further, water transports other raw materials and elaborated food from one part of the tree to another.

Water may exist in any soil in the *hygroscopic, capillary,* or *free water* state. Hygroscopic water occurs in the form of a very thin film over the soil particles where the moisture in the soil is in equilibrium with that in the air. Capillary water exists as a thicker film, which is attracted to the particle by its own surface tension. This tension is stronger than the pull of gravity. When two soil particles are in contact, and one has a thinner film of capillary water than the other, the drier surface has a greater attraction for the water. As a result, the water moves from the wetter to the drier surface, until both particles are surrounded by films of equal thickness. When the film of capillary water reaches the maximum thickness that can be maintained by surface tension, any additional water responds to the pull of gravity and flows downward into the soil to become free water. Free water occupies the spaces between the moist soil particles, just as the air does when only hygroscopic and capillary water are present. When free water occupies all the pore spaces in the soil or any horizon of the soil, its upper surface is known as the *water table.* Because of lack of air, roots cannot survive long in such a water-saturated zone (Fig. 3-1).

When soil water comes in contact with the complex organic compounds that comprise the cell walls of the root hairs, it is absorbed by *imbibition.* From this point the water now passes on, by a process known as *osmosis,* through the plasma membrane lining the cell walls and becomes part of the cell sap.

Fig. 3-1. These trees died because the high water table excluded air from their roots.

The concentration of soluble substances in the cell sap is usually greater than in the soil solution. The movement of water into the cell tends to equalize the concentrations on both sides of the cell wall. The pull exerted by the water lost from the leaves during transpiration further aids the movement of water from cell to cell and the influx of water from the soil into the root hairs.

When the soil surrounding the root hairs becomes very dry, as during a drought, the concentration of the solution in the soil may be greater than that within the cell. Then the reverse process takes place. Water moves out of the root hairs and plant cells, causing their collapse, and the continuous water column from roots to leaves cannot be maintained. Consequently, the plant wilts. Wilting also occurs if the quantity of water lost from the leaves is greater than that gained by the roots. Such reverses are often evident in shade trees during the summer.

Proper and timely application of water is critical to the survival and growth of newly transplanted trees, which have a limited root system. This topic is discussed in Chapter 5. Established trees growing in climates that typically receive summer rains rarely need irrigation except during periods of extended drought, unless they are growing at a site that limits penetration of water into the soil. Tree owners wishing to optimize tree growth should see to it that 1 inch of water per week is applied to the root zone either by rainfall or by sprinkler. Trees growing in regions with hot, dry summers are more likely to experience drought stress. Since the stresses caused by prolonged lack of water may ultimately cause a tree to decline and die, installation of a permanent irrigation system is needed in such locations.

Symptoms of lack of water include wilting of leaves and scorching, in which areas of the leaf between the veins or along the margins turn light or dark brown. See Chapter 10 for a further discussion of this topic. Prompt application of water to such trees is necessary to alleviate the drought stress and prevent long-term damage to the tree.

Trees vary with respect to their moisture requirements. Red maple, pin oak, sweetgum, and most species of willow prefer soils with a high moisture content, whereas red oak, chestnut oak, ash, and poplar grow best in soils containing smaller amounts of water. Spruces can tolerate wetter soils than can most other species of conifers.

A change in the water table often seriously affects a tree's health. Although not evident at the surface, a rise in the water table may cause drowning of the deeper roots. This happened to London planetrees planted in Battery Park at the southern tip of Manhattan Island and to Hyssop crabapple trees planted in New York City's Stuyvesant Town housing development many years ago.

Soil Minerals. The solid body of the soil consists of varying proportions of coarse grains (sand), medium-sized grains (silt), and fine particles (clay), all of which are mineral particles originating from the disintegration of rocks. The relative proportions of three components determine the soil type (Fig. 3-2). In addition, the soil contains varying amounts of humus substances formed by the microbial decomposition of plant remains.

The smaller sized particles of the clay portion of the soil are known collec-

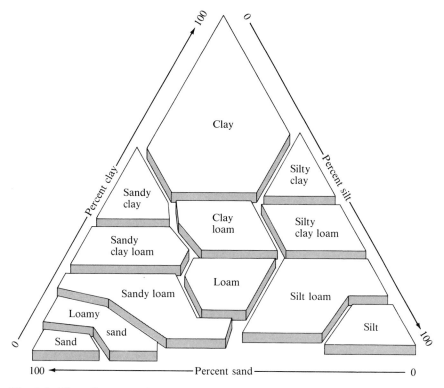

Fig. 3-2. The soil texture triangle shows the percentage of sand, silt, and clay in each of the textural classes. (From Stefferud, A. (ed.). Soil: The yearbook of agriculture 1957, p. 32. U.S.D.A. House Document No. 30. Washington, D.C.)

tively as *colloidal clay*. These particles aid greatly in the retention of water and plant nutrients in the soil. The working qualities of a soil depend largely on its content of colloidal clay and the condition in which this exists.

Chemically, colloidal clay consists mainly of iron and aluminum silicates and oxides. The silicates are the seat of retentive or "exchange" processes in soils. Surrounding each tiny fragment of the exchange mineral is a layer of absorbed water containing both *cations* and *anions*. Cations are positively charged (+) and anions are negatively charged (−) particles which are produced when soluble soil nutrients dissolve in water. The cations consist chiefly of hydrogen (H), calcium (Ca), magnesium (Mg), potassium (K), and sodium (Na); the anions are hydroxyl (OH), phosphate (PO_4), nitrate (NO_3), sulfate (SO_4), and chloride (Cl) ions. Most of these ions are plant nutrients.

As a root hair enters the zone of absorbed water, an exchange of ions takes place between this water layer and the root hairs. As a result, the mineral nutrients of the soil solution move onto the surface of the roots and the carbonic acid of the plant is released into the soil. The uptake of mineral nutrients by plant cells is relatively selective, and the process depends on metabolic activity.

Soil Organic Matter. The second solid component of soil, organic matter is usually concentrated in the upper layers. It is essential for the maintenance of soil microorganisms and it makes the soil porous and friable, which increases its water-holding capacity and improves its aeration. In its decay, organic matter yields up mineral nutrients and nitrogen for the use of the currently growing plants. In addition, organic matter has a high "exchange capacity," acting much like the exchange minerals in colloidal clay.

Soil Organisms. In addition to the familiar macroorganisms, such as earthworms and insects, the soil is inhabited by a wide variety of microorganisms, numbering in the billions per cubic inch of soil. These microorganisms, which include protozoa, bacteria, actinoycetes, fungi, and nematodes, all are involved in the decomposition of organic matter, making nutrients available for uptake by roots. Some bacteria and actinomycetes can fix nitrogen, either alone or in symbiosis with roots of certain plants. Serious diseases of trees and other plants are caused by soil-inhabiting fungi, bacteria, and nematodes. These are discussed in Chapter 12.

Certain fungi are able to form symbiotic relationships with roots; the structure and function of these mycorrhizae are discussed in Chapter 2, under Specialized Roots. The failure of certain tree species, notably beech, to grow well when planted in particular soils has been attributed to the absence of mycorrhizal fungi in these soils. Mycorrhizal fungi can be introduced into such soils by incorporating soil from natural habitats in which these tree species thrive.

Elements Essential for Tree Growth

For growth and development trees require the major elements carbon, oxygen, hydrogen, nitrogen, potassium, phosphorus, sulfur, calcium, magnesium, and the trace elements iron, boron, manganese, copper, zinc, chlorine, and molybdenum. Plants use elements in four primary ways: to form structural units, such as carbon in cellulose and nitrogen in protein; to incorporate into molecules important in metabolism, like magnesium in chlorophyll; to act as catalysts in enzymatic reactions, as several of the trace elements do; and to help maintain osmotic balance, such as potassium in stomatal guard cells. Each essential element functions uniquely and cannot be replaced by another.

As already discussed in Chapter 2, carbon and oxygen are provided by the carbon dioxide in the atmosphere, which enters the leaves through the stomata. Hydrogen is obtained from water absorbed through roots and carried to the leaves by the water-conducting system.

Nitrogen is also a component of air but higher plants cannot use the form in which it occurs in the atmosphere. In nature it is taken from the air and assimilated by certain types of soil bacteria and bluegreen algae. Trees in turn obtain nitrogen from decaying organic material in the soil.

All other elements required by trees are taken from the soil. They are absorbed by the roots only in solution, in the manner already described in Chapter 2. From the standpoint of their relative value as raw materials for plant use, the elements in the soil may be divided into three classes: nonessential; essential and abundant; and essential and critical.

Nonessential elements include silicon, aluminum, sodium, and some of the rarer elements. These elements are believed to have no role in the nutrition of plants.

The second group, essential and abundant, includes calcium, iron, magnesium, and sulfur. They are usually present in ample amounts in soil, although occasionally soil may be deficient in one or more of these elements, or these elements may be unavailable.

Elements that are essential and critical include nitrogen, phosphorus, and potassium, and in some sites magnesium, manganese, and boron. These elements are sometimes present in soil in insufficient quantities and may be the limiting factor in plant growth. This is especially true of nitrogen. Commercial fertilizers have as their main constituents the first three elements of this important group.

Influence of Soil on Nutrient Availability

Soil Type. The physical and chemical properties of soil influence the amount of nutrients held by the soil and to some extent the availability of nutrients to plants. Essential plant nutrients, present as inorganic ions, are absorbed onto the surface of soil particles and organic matter or are free in the soil solution. The soil-absorbed ions are not readily leached but are available for plants to absorb. The capacity of a soil to absorb cations (positively charged ions) is called cation exchange capacity and this is a good measure of soil fertility. Soils containing a large percentage of clay have a high cation exchange capacity because clay, consisting of large numbers of small particles, has a large surface area. Soil organic material, humus, also has a high nutrient exchange capacity (five times more than clay). Sandy loam soils therefore must be fertilized more frequently than clay loam soils or soils with a high percentage of organic matter.

Soil Reaction (pH). Soils are acid, neutral, or alkaline in reaction. The degree of acidity or alkalinity is expressed in terms of pH values, very much as temperature is expressed in degrees Centigrade or Fahrenheit. Temperature scales center on the freezing point of water, whereas the acidity–alkalinity scale is divided into fourteen major units centering on a neutral point (pH 7). Values greater than pH 7 are alkaline, and those less than pH 7 are acid. The intensity of the alkalinity increases tenfold with each unit increase in pH. For example, the alkalinity at pH 8 is ten times as great as that at pH 7. Similarly, the acidity increases ten times for each unit decrease in the pH scale. Thus the acidity at pH 6 is ten times as great as that at pH 7, and at pH 5 it is ten times as great as that at pH 6 and one hundred times as great as that at pH 7.

The pH scale refers not to the total amount of acid or alkaline constituents in the soil, but to those portions of the acid and alkali that are in a free or dissociated state. A pH reading gives an indication of the relative number of free *acid* (H+) and *alkaline* (OH−) ions in the soil, and not of the combined or undissociated molecules.

Most tree soils are acid in reaction, because a large portion of the alkaline exchange cations (those of calcium, magnesium, potassium, and sodium) have been either absorbed by the trees or lost through leaching, leaving acid exchange minerals behind. In addition, the continued use of acid-forming fertilizers and sprays (especially those containing sulfur) tends to increase acidity. In cities, certain gases, such as ammonia and sulfur dioxide, which are carried down by rainwater, further tend to increase the acidity of the soil. The pH values of most eastern tree soils lie between 5 and 6.5. They are seldom higher than 7, unless their content of limestone is naturally high or large quantities of lime have been added to them.

Although soil acidity is not the only factor governing the availability of nutrients, it nevertheless plays an important part (Fig. 3-3). For example, an increase in acidity beyond certain limits means a decrease in the supply of certain plant nutrients, the hydrogen having taken the place of the cations in the ex-

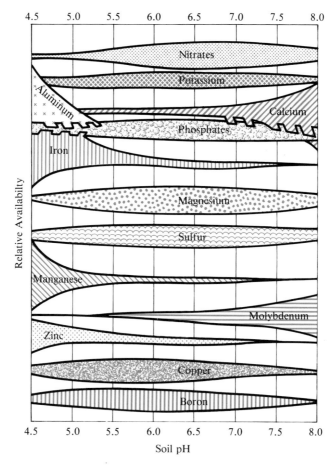

Fig. 3-3. Soil pH influences the availability of nutrients to growing plants. (From Publication AGR-19, Liming acid soils, University of Kentucky Cooperative Extension Service, University of Kentucky, Lexington, Ky. 40546.)

change complex. The exchanged cations may have been removed by the trees or may have been carried away by the drainage water. At the other extreme, as the alkalinity increases, the roots are less able to absorb iron, manganese, zinc, copper, and boron, because these minerals are insoluble under such conditions. A good case in point is that of the availability of iron salts, which are absorbed by the plant at a pH of 6.7 or below. If a tree is growing in an alkaline soil (above pH 7), there is a strong possibility that it will suffer from iron deficiency and that the leaves will become chlorotic (see Chapter 10).

In general, most plants grow best on soils having a slightly acid reaction. A pH value between 5.5 and 7 seems most favorable for nearly all trees. Growth is best when the soils in which trees are planted and grown are maintained within the pH range to which they are adapted. Relatively few trees seem to prefer a distinctly acid soil. Among those that do are silverbell *(Halesia)*, sorrel-tree *(Oxydendrum)*, yellowwood *(Cladrastis)*, and Franklin-tree *(Franklinia)*.

Adjusting Soil pH. Soil acidity or alkalinity can be changed by the use of chemicals applied to the surface or incorporated into the soil. To increase soil pH (sweeten acid soil), lime is used either as ground limestone (calcium carbonate or calcium magnesium carbonate), quicklime (calcium and magnesium oxides), or slaked lime (calcium and magnesium hydroxides). Limestone is the most convenient form to use, but the other forms are faster acting. The amount of lime needed to produce the desired change in soil pH depends on soil texture, organic matter, and the material being used. Soils with fine texture or with high levels of organic matter have a greater buffering capacity and require more lime to achieve the same adjustment as coarser soils.

Soil test results often give two values: the water pH and the buffer pH. The buffer pH takes into account the buffering capacity of soil and thus gives a better estimate of how much lime is needed. Table 3-1 gives the amount of lime needed for the indicated pH change based on water pH values. These values are for a silt-loam soil and would be somewhat different for soil with more or less clay. Table 3-2 gives information based on buffer pH values, which is applicable to any soil type.

Soil pH may be decreased (reduced in alkalinity) by the addition of sulfur. Agricultural sulfur is the most convenient form to use, but finer ground dusting

Table 3-1. Amount of agricultural limestone needed to raise the pH of a silt-loam soil based on water pH measurement

Water pH Value	Required Amount of Agricultural Limestone to Raise the Soil pH to 6.4 (lb/1000 ft²)
Above 6.4	0
6.4–5.8	0–100
5.8–5.2	100–200
Below 5.2	200

Table 3-2. Amount of agricultural limestone needed to raise soil pH, based on buffer pH measurement

Buffer pH Value	Required Amount of Agricultural Limestone to Raise the Soil pH to 6.4 (lb/1000 ft²)
6.7	70
6.6	100
6.5	115
6.4	140
6.3	160
6.2	190
6.1	210
6.0	230
5.9	245
5.8	255
5.7	280
5.6	315
5.5	325

Source: Publication ID-72, Principles of home landscape fertilization, University of Kentucky Cooperative Extension Service, University of Kentucky, Lexington, KY 40546.

or wettable sulfurs are faster acting. Aluminum sulfate is often recommended, but it may have a toxic effect on plants if used in large amounts. About 7 pounds of aluminum sulfate are required to achieve the same effect as 1 pound of sulfur. To avoid toxic effects on vegetation sulfur should not be used at temperatures above 80° F. The amount of sulfur needed to make desired pH changes is given in Table 3-3.

Lime and sulfur both need to be applied over a broad area to be effective; application should not be confined to the area near the trunk or planting hole. If applications are made before planting, the material should be worked into

Table 3-3. Suggested application of powdered sulfur to reduce the pH of an 8-inch layer of soil, as indicated, in pt/100 ft²

Original pH of Soil (Based on Water pH Value)	Pints of Sulfur for 100 ft² to Reach pH of									
	4.5		5.0		5.5		6.0		6.5	
	Sand	Loam	Sand	Loam	Sand	Loam	Sand	Loam	Sand	Loam
5.0	⅔	2	—	—	—	—	—	—	—	—
5.5	1⅓	4	⅔	2	—	—	—	—	—	—
6.0	2	5½	1⅓	4	⅔	2	—	—	—	—
6.5	2½	8	2	5½	1⅓	4	⅔	2	—	—
7.0	3	10	2½	8	2	5½	1⅓	4	⅔	2

Source: University of Kentucky Cooperative Extension Publication ID-72.

the soil down to a depth of 8 inches. After planting, the materials should be spread evenly over the soil surface and may be worked into the top 2 to 4 inches, but care must be taken to avoid damaging the tree's roots. If the soil needs pH adjustment and material is incorporated only into the backfill soil when planting a tree, then nutritional deficiency symptoms could appear as soon as the new roots spread into the untreated soil. Lime and sulfur do not effect permanent pH changes. Retreatments must be made from time to time.

County agricultural agents, agricultural experiment station chemists, and field representatives of agricultural chemical companies offer soil-testing services and supply information as to the amount of lime, sulfur, or aluminum sulfate that must be added to produce the desired change in pH on a specific soil. Alternatively, inexpensive soil-testing kits can be purchased for individual use.

Table 3-4 lists the most favorable pH range of most shade and ornamental trees.

Table 3-4. Soil reaction adaptations of specific trees

Common Name	Botanical Name	Most Favorable pH Range
Acacia	*Acacia*	6.5–7.5
Alder	*Alnus*, named sp.	6.5–7.5
Almond, flowering	*Prunus glandulosa*	6.5–7.5
Apple	*Malus* sp.	6.5–7.5
Arborvitae	*Thuja* sp.	6.5–7.5
Ash	*Fraxinus* sp.	6.0–7.5
Bald cypress, common	*Taxodium distichum*	6.5–7.5
Beech	*Fagus*, named sp. and var.	6.5–7.5
Birch, sweet	*Betula lenta*	4.0–5.0
Box, common	*Buxus sempervirens*	6.5–7.5
Buckeye, Ohio	*Aesculus glabra*	6.5–7.5
Buckeye, red	*Aesculus pavia*	6.0–6.5
Catalpa	*Catalpa*, named sp.	6.5–7.5
Cherry	*Prunus* sp.	6.5–7.5
Cherry, choke	*Prunus virginiana*	6.5–7.5
Chestnut, American	*Castanea dentata*	4.0–6.0
Chinquapin	*Castanea pumila*	4.0–6.5
Cypress	*Chamaecyparis*, named sp. and var.	4.0–7.5
Dockmackie	*Viburnum acerifolium*	4.0–5.5
Dogwood, flowering	*Cornus florida*	5.0–6.5
Douglas-fir	*Pseudotsuga menziesii*	6.0–6.5
Elm	*Ulmus* sp.	6.5–7.5
Empress-tree	*Paulownia tomentosa*	6.5–7.5
Eucalyptus	*Eucalyptus* sp.	6.5–7.5
Fir	*Abies*, named sp.	4.0–6.5
Franklin-tree	*Franklinia alatamaha*	4.0–6.0
Fringe-tree	*Chionanthus virginica*	4.0–6.0
Ginkgo	*Ginkgo biloba*	6.0–6.5
Hackberry	*Celtis* sp.	6.5–7.5
Hawthorn	*Crataegus* sp.	6.0–7.5
Hemlock, Canada	*Tsuga canadensis*	4.0–6.5
Hemlock, Carolina	*Tsuga caroliniana*	4.0–5.0

Table 3-4. Soil reaction adaptations of specific trees (continued)

Common Name	Botanical Name	Most Favorable pH Range
Hickory, shagbark	*Carya ovata*	6.0–6.5
Holly, American	*Ilex opaca*	4.0–6.0
Hop-Hornbeam, American	*Ostrya virginiana*	6.0–6.5
Hornbeam	*Carpinus*, named sp.	6.5–7.5
Horsechestnut	*Aesculus hippocastanum*	6.0–7.0
Juniper	*Juniperus*, many sp.	6.5–7.5
Juniper, common	*Juniperus communis* and var.	6.0–6.5
Juniper, creeping	*Juniperus horizontalis*	4.0–6.0
Juniper, mountain	*Juniperus communis saxatilis*	4.0–5.0
Kentucky coffee tree	*Gymnocladus dioica*	6.5–7.5
Larch	*Larix*, named sp.	6.5–7.5
Linden (basswood)	*Tilia* sp.	6.5–7.5
Loblolly bay	*Gordonia lasianthus*	4.0–6.0
Locust, black	*Robinia pseudo-acacia*	5.0–7.5
Locust, honey	*Gleditsia*, named sp.	6.5–7.5
Magnolia	*Magnolia* sp.	4.0–7.0
Maple	*Acer*, many sp.	6.5–7.5
Maple, mountain	*Acer spicatum*	4.0–5.0
Maple, red	*Acer rubrum*	4.5–7.5
Maple, striped	*Acer pennsylvanicum*	4.0–5.0
Mock-orange	*Philadelphus* sp.	6.0–7.5
Mountain-ash, American	*Sorbus americana*	5.0–7.0
Mountain-ash, European	*Sorbus aucuparia*	6.5–7.5
Mountain laurel	*Kalmia latifolia*	4.0–6.5
Mulberry	*Morus*, named sp. and var.	6.5–7.5
Oak, black	*Quercus velutina*	6.0–6.5
Oak, blackjack	*Quercus marilandica*	4.0–5.0
Oak, chestnut	*Quercus prinus*	6.0–6.5
Oak, English	*Quercus robur* and var.	6.5–7.5
Oak, European turkey	*Quercus cerris*	6.5–7.5
Oak, overcup	*Quercus macrocarpa*	4.0–5.0
Oak, pin	*Quercus palustris*	5.5–6.5
Oak, post	*Quercus stellata*	4.0–6.5
Oak, red	*Quercus borealis maxima*	4.5–6.0
Oak, sand blackjack	*Quercus catesbaei*	4.0–5.0
Oak, scarlet	*Quercus coccinea*	6.0–6.5
Oak, shingle	*Quercus imbricaria*	4.0–5.0
Oak, shrub	*Quercus ilicifolia*	5.0–6.0
Oak, southern red	*Quercus falcata*	4.0–5.0
Oak, swamp white	*Quercus bicolor*	6.5–7.5
Oak, white	*Quercus alba*	6.5–7.5
Oak, willow	*Quercus phellos*	4.0–6.5
Oak, yellow	*Quercus muhlenbergii*	4.0–5.0
Peach	*Prunus persica*	6.5–7.5
Pear	*Pyrus* sp.	6.5–7.5
Pine	*Pinus*, many sp.	4.0–6.5
Planetree	*Platanus*, named sp.	6.5–7.5
Plum, American	*Prunus americana*	6.5–7.5
Plum, beach	*Prunus maritima*	6.5–7.5
Plum, common	*Prunus domestica*	6.5–7.5

Table 3-4. Soil reaction adaptations of specific trees (continued)

Common Name	Botanical Name	Most Favorable pH Range
Plum, purple-leaf	*Prunus cerasifera pissardi*	6.5–7.5
Poplar	*Populus,* named sp.	6.5–7.5
Quince, flowering	*Chaenomeles japonica*	6.5–7.5
Redbud	*Cercis* sp.	6.5–7.5
Red cedar	*Juniperus virginiana* and var.	6.0–6.5
Savin	*Juniperus sabina* and var.	6.0–6.5
Serviceberry	*Amelanchier* sp.	6.0–6.5
Silverbell	*Halesia tetraptera*	4.0–6.0
Sorrel-tree	*Oxydendrum arboreum*	4.0–6.0
Spruce	*Picea* sp.	4.0–6.5
Spruce, Colorado	*Picea pungens* and var.	6.0–6.5
Stewartia	*Stewartia malacodendron*	4.0–6.0
Storax	*Styrax americana*	4.0–6.0
Sweetgum	*Liquidambar styraciflua*	6.0–6.5
Sweetleaf, horse sugar	*Symplocos tinctoria*	4.0–6.0
Tree-of-heaven	*Ailanthus altissima*	6.5–7.5
Tuliptree	*Liriodendron tulipifera*	6.0–6.5
Tupelo	*Nyssa sylvatica*	5.0–6.0
Walnut	*Juglans,* many sp.	6.5–7.5
White cedar	*Chamaecyparis thyoides*	4.0–5.5
Willow	*Salix,* many sp.	6.5–7.5
Willow, creeping	*Salix repens*	4.0–5.0
Yew	*Taxus* sp.	6.0–6.5

Soil Temperature. Well-established trees, as well as those recently transplanted, are adversely affected by extremes in soil temperature. Low soil temperature retards or even prevents water and mineral absorption by the roots. If the air temperature is high enough to allow loss of water from the leaves and twigs while the soil temperature is low, serious injury may result. The loss of more moisture than can be replaced through the roots often results in drying and browning of both narrow- and broad-leaved evergreens (see Winter Drying of Evergreens, Chapter 10).

Because soil temperature normally drops slowly with the approach of fall and winter, heat can be conserved for a long period by early fall mulching. This practice prevents deep freezing of the soil and reduces the amount of winter drying of small trees. The mulching materials should be removed early the following spring; otherwise they will screen out the warm rays of the sun.

The Soil as an Anchor

Embedding themselves in the soil, the roots are able to hold the tree erect. Trees tend to develop anchorage where it is most needed. The tendency of isolated trees is to develop anchorage rather equally all around, with perhaps a slightly stronger development on the side toward the prevailing strong winds. The better protected a tree is from wind, the less it depends on secure anchorage and the fewer roots it seems to provide for such anchorage.

Fig. 3-4. Poor soil anchorage contributed to the demise of this wind-thrown tree.

The factors that are important to a tree from the standpoint of anchorage are the depth and the texture of the soil, the relative abundance of such mechanical allies as stones, and the roots of other trees and shrubs. Sufficient room for the roots to spread horizontally must also be available.

Trees growing in wet soils are more easily blown over during windstorms than are trees growing in well-aerated soils (Fig 3-4). One reason is that the lubricating action of the wet soil particles lowers the trees' resistance. Another is the absence of deep anchor roots, which are unable to develop because of lack of air in the deeper layers. In wet soils most of the tree roots are confined to the upper 18 inches; in well-aerated soils the roots may penetrate much deeper.

Urban Soils

Soils in urban areas are usually very different from rural soils. Disturbances caused by human activities may result in a high degree of variability (Fig. 3-5). Thus a tree planted at one site may be in a very different soil environment than one planted only a few yards away due, for example, to the presence of construction debris at one site and not another. Modification of the soil structure can lead to compaction and restricted aeration and water drainage. Mate-

Fig. 3-5. Urban soils often include construction debris.

rials unfavorable or even toxic to trees and to soil microorganisms may be present in such soils.

These and other negative features associated with urban soils should be considered when selecting sites for tree planting. An awareness that such factors can cause poor growth or death of trees will also aid in the diagnosis of tree problems.

Soil Improvement Prior to Planting

Probably the best way to improve the physical condition of the soil before trees are planted is to incorporate large amounts of organic matter (Fig. 3-6). Such material not only improves the structure of the soil but increases its water-holding and ion-exchange capacity as well. Disappointing or unprofitable results may ensue, however, if the organic matter is added to infertile, very wet, or extremely acid soils. Such adverse conditions must be rectified before much benefit is derived from the organic matter. Wet soils, for instance, may be improved by the insertion of drain tiles to draw off the excess water, thus increasing the aeration. Shallow soils similarly will respond better to organic

Fig. 3-6. Effect of peat moss on root development of dogwood. **Left:** Tree grown in soil without the addition of peat moss. Notice the few scraggly roots. **Right:** Tree grown in soil into which peat moss has been added. Notice the heavy mass of fibrous roots.

matter incorporations if any underlying hardpan or impervious subsoil is first broken up by the use of special digging tools or by blasting with dynamite. Incorporating ground limestone into the subsoil also markedly improves the physical condition of the soil, especially where the subsoil is very acid.

Of the many sources of organic matter, those most commonly used for soil improvement are well-rotted stable manure and two distinct types of peat, peat moss and sedge peat.

Peat moss, also known as sphagnum peat or peat, is very acid in reaction (pH 3.0–4.5) and is used principally for acid-loving plants, such as rhododendrons and azaleas, hemlocks, pines, and spruces. It is also used in backfill for almost all trees. It is fibrous in nature, light brown in color, low in nitrogen content (around 1% on a dry basis), and has a high water-absorbing capacity. It should never be applied in the form in which it is purchased, but must be thoroughly soaked beforehand, sometimes up to several days. When large quantities of this type of peat are used, some nitrogen in the form of nitrate salts should be applied the following summer to supply both the trees and the bacteria that decompose the peat. This practice will compensate for the leaching due to the frequent watering necessary for transplanted trees, and for the consumption by the bacteria.

Sedge peat, also known as reed peat, muck, or humus, may in time have more extensive use than peat moss. Large deposits of this type of peat are available for exploitation. As offered for sale, sedge peat is dark brown to black and relatively high in nitrogen (2.0–3.5%). It is less acid in reaction (pH 4.5–6.8) and has a lower water-absorbing capacity than peat moss. It can be safely used around most trees.

Soil Improvement for Established Trees

Trees in heavily used landscapes such as parks, playgrounds, and even home yards frequently suffer from the effects of soil compaction. Compacted soils are poorly aerated and are difficult to water and fertilize. Trees in compacted soils gradually show symptoms of poor vigor, dieback, and decline.

Many arborists use a technique called vertical mulching to alleviate the effects of soil compaction and restore tree vigor. Vertical mulching involves drilling holes in the soil throughout the tree root zone and backfilling with porous soil amendments. Holes, 18 inches deep, are made in the soil of the tree root zone using a power drill having a 2- to 4-inch auger (Fig. 3-7). These holes are spaced 2 to 3 feet apart in a grid pattern where feeder roots grow both inside and outside the extent of the tree canopy. The holes are filled with sand, perlite, or other porous material. Organic amendments such as peat moss or wood chips may be used but they eventually break down and can become compressed. Fertilizer can also be added to the holes, if it is needed.

Vertical mulching should provide more air to the soil's lower depths and should permit greater penetration of water and mineral nutrients into the soil. Arborists using the technique report improvement in tree health, but as far as is known, no scientific research documents the benefits of this procedure. Some

Fig. 3-7. A power auger can be used for vertical mulching.

root injury is likely to occur when an auger is used throughout the root zone. However, the benefits of improved air, water, and mineral nutrient movement should also be realized.

Selected Bibliography

Anonymous. 1957. Soil: The 1957 yearbook of agriculture. U.S. Superintendent of Documents, Washington, D.C. 784 pp.

Craul, P. J. 1985. A description of urban soils and their desired characteristics. J. Arboric. 11 (11):330–39.

Hole, F. D., and J. B. Campbell. 1985. Soil landscape analysis. Rowman and Allanheld, Totowa, N.J. 196 pp.

Kozlowski, T. T. 1985. Soil aeration, flooding and tree growth. J. Arboric. 11 (3):85–88.

Thompson, L. M., and F. R. Troeh. 1978. Soils and soil fertility. McGraw-Hill, New York. 516 pp.

Tisdale, S. L, W. L. Nelson, and J. D. Beaton. 1985. Soil fertility and fertilizers. Macmillan, New York. 754 pp.

4

Factors Important
for Tree Selection

Tree species vary greatly in their ability to grow in different environments. The selection of an appropriate tree for a specific site requires consideration of the characteristics of the site, the adaptation of tree species to these characteristics, the function of the tree in the landscape, and its availability.

Characteristics such as soil type, drainage, exposure, elevation, climate, air pollution, and surrounding vegetation are important considerations in determining which tree species can be successfully grown at a specific site. Although horticultural selections may vary, some tree species are better adapted to growing, for example, under wet conditions, in very acid or alkaline soils, in the presence of air pollutants or salt spray, or in shade. Some species are extremely sensitive to frost, whereas others require chilling to break dormancy. Many species are hardy in a broad range of climatic zones. However, growers should be aware that biotypes of many species have, through the process of natural selection, arisen in specific localities.

The presence of endemic insect or disease problems may limit the success of some species in some areas. Chestnut blight, Dutch elm disease, western spruce budworm, and bronze birch borer have drastically reduced the planting of the affected species over large regions.

Plant ecologists refer to the ability of a plant to grow within a certain set of conditions as its *fitness*. In natural plant communities, the species present are those that are most fit for the circumstances. For example, in some parts of the eastern North American forest, oak and hickory predominate in shallow, relatively dry soils of ridge crests or southwest slopes; beech, linden, and maple thrive in deeper and moister soils present in valleys and northeast slopes; and in moist, alluvial soils, sweetgum, bald cypress, pin oak, pecan, and swamp chestnut oak may be present.

One of the most important considerations in tree selection is cold-hardiness. A system of designating cold-hardiness zones was developed by the U.S. Department of Agriculture (Fig. 4-1). These zones are defined by average minimum (winter) annual temperatures and trees grow in particular zones because of their inherent hardiness to winter cold. In North America, zones 1 (cold,

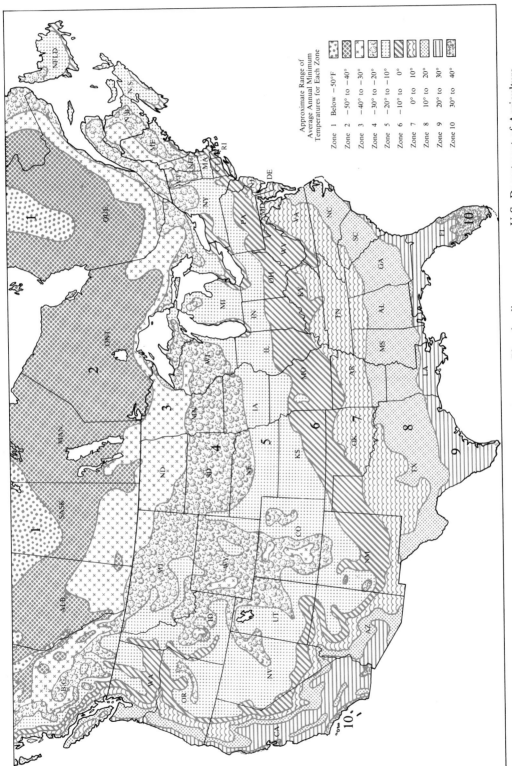

Fig. 4-1. Zones of plant hardiness. (From U.S. National Arboretum. Plant hardiness zone map. U.S. Department of Agriculture. Miscellaneous Publication No. 814.)

Approximate Range of
Average Annual Minimum
Temperatures for Each Zone

Zone 1   Below −50°F
Zone 2   −50° to −40°
Zone 3   −40° to −30°
Zone 4   −30° to −20°
Zone 5   −20° to −10°
Zone 6   −10° to 0°
Zone 7   0° to 10°
Zone 8   10° to 20°
Zone 9   20° to 30°
Zone 10   30° to 40°

below − 50° F) through 10 (mild, 30–40° F) are defined roughly from north to south (with coastal regions milder, and higher elevations colder) in 10-degree increments. In the Eastern North American deciduous forest, flowering dogwood *(Cornus florida)* is not hardy in zone 4 where minimum temperatures typically range from − 30 to − 20° F but thrives in zone 6 (− 10 to 0° F) and in warmer locations. This system is not perfect because for some species an inability to tolerate sudden changes in temperature such as a cold period following a warm period can cause more winter injury than the same temperature following a cool period. Furthermore, the cold-hardiness zone concept deals with average minimum temperatures and not with the extremes that really test a plant's hardiness. Topographical features may make an area warmer or colder than the general climate zone. Landscape designs involving aboveground planters and berms may expose plants to harsher than natural conditions. Nevertheless, these hardiness zones are helpful in designating the kinds of trees that will grow in a geographic region and where exotic trees introduced from other continents will survive. Lists of adapted trees for these zones are available from nurseries.

By observing the selection of species that are already established and thriving in a given area, one can get a rough idea of which trees are likely to be successful. Care must be taken to make observations at sites which are similar in as many aspects as possible to the intended planting location. Species native to an area may usually be relied on to grow well, require less maintenance, and have few insect and disease problems. However, many well-adapted exotics also are available (Fig. 4-2). Of course, advice from county extension agents, local arborists, or nurserymen can help prevent costly mistakes.

Special situations existing at a site may also influence tree selection. If the tree is to be located near sidewalks or underground utility conduits, root distribution must be considered. Willow, poplar, and silver maple have particularly invasive roots. Trees planted near parking areas should not produce dripping exudates or messy fruit and should not raise the soil around the tree trunk (Fig. 4-3). Plantings under utility lines should be selected to minimize interference with these lines. Low- or slow-growing species are preferred. Advanced planning is necessary to select trees that will be compatible not only with the environment of the site but also with its functions.

Consideration of the function of the tree in the landscape will further narrow the list of candidate trees. Functions include providing shade, wind break, noise barrier, visual screen, or fruit. The function may be strictly aesthetic: trees may be selected for their desirable growth habit such as crown shape, size, and density of branches or foliage or for their color of flower, foliage, or bark (Fig. 4-4).

Because the selection of suitable trees depends on so many factors and because such a variety of trees is available (some suitable only for specific regions), we do not attempt to discuss or recommend specific species. The bibliography at the end of this chapter includes a wide range of approaches to tree selection.

Fig. 4-2. The cork tree, *Phellodendron lavallei,* from eastern Asia is an excellent tree for streetside or lawn planting. Female cork trees produce fruit that litters the street, so only male trees are recommended.

Fig. 4-3. Tree roots damaging a sidewalk.

Fig. 4-4. A superior cone-shaped form of ornamental pear *(Pyrus calleryana)* is the cultivar 'Chanticleer'.

There are, however, some trees that have characteristics which make them less desirable. These are listed below, with the reasons for so classifying them:

Boxelder *(Acer negundo):* a weedy tree attractive to insects.

Silver maple *(Acer saccharinum):* soft and brittle wood highly susceptible to storm damage.

Yellow buckeye *(Aesculus octandra):* loses foliage early, becoming unsightly.

Tree-of-heaven *(Ailanthus altissima):* male flowers have offensive odor.

Southern catalpa *(Catalpa bignonioides):* flowers and seed pods produce unsightly litter, subject to caterpillar attack.

Western catalpa *(Catalpa speciosa):* seed pods produce unsightly litter, short-lived weak wood.

Persimmon *(Diospyros virginiana):* fruits produce unsightly litter.

White ash *(Fraxinus americana):* subject to diseases difficult to control.

Honey locust *(Gleditsia triacanthos):* objectionable thorns and seed pods (thornless, seedless clones good).

Black walnut *(Juglans nigra):* nut husks stain sidewalks and shoes, roots secrete substance toxic to rhododendrons, tomatoes, and other species.

Osage-orange *(Maclura pomifera):* large useless fruits that interfere with lawn mowing.

Fruiting mulberries *(Morus* sp.): fruits produce unsightly litter that attracts insects.

Empress tree *(Paulownia tomentosa):* unsightly seed pods, coarse leaves.

Most species of poplar *(Populus* sp.): weak wood, roots clog sewers.

Black cherry *(Prunus serotina):* fruits produce unsightly litter, subject to tent caterpillars.

Black locust *(Robinia pseudoacacia):* subject to borers.

Some of the less desirable trees such as tree-of-heaven, silver maple, poplars, and mulberry are well adapted to air pollutants, drought, low humidity, heat, wind, and other rigors of the inner city and industrial areas. They will thrive with little to no maintenance and might be considered for use in vacant lots, industrial areas, and various bits of "leftover" land. Professor Russell A. Beatty of the Department of Landscape Architecture at the University of California, Berkeley, says their lack of aesthetic appeal is balanced by their ability to "bring a measure of amenity to otherwise ugly and unhealthy landscapes."

Problems Peculiar to Streetside Trees

Since trees are more difficult to grow in the unnatural environment of cities, successful culture of streetside trees involves many problems not usually encountered in forest and country plantings.

It is particularly difficult for trees in large cities to do their best. City conditions are so unfavorable for most trees that it is a wonder they grow at all. Smoke and other air pollutants, mechanical injuries, a disrupted water table, highly compacted soil, lack of organic matter, limited root space, reflected heat from buildings, visitations by dogs, lack of water, use of salt on sidewalks to melt ice—these are but a few of the external forces that make the lot of a city tree so difficult. Add to these invasions by insects and infection of the roots by soil-inhabiting parasitic fungi, and one is almost ready to forgo planting trees in cities. Despite these hazards and handicaps, however, millions of trees do grow in large metropolitan areas, including Brooklyn.

In cities, the runoff of rainfall is nearly 100 percent. Hence the small open soil area between the street curb and the sidewalk, where most city trees are planted, has little chance to absorb rainwater.

Another important point to bear in mind is that soil which may have had adequate aeration at the time a tree was planted will deteriorate until the aeration is entirely inadequate. Trees growing along city streets offer excellent proof of this. When the tree is first planted in the area between the street and the sidewalk, conditions may be satisfactory. As the roots extend into the areas covered by the street and sidewalk, conditions become less favorable. In addition, trampling of the open soil area by pedestrians compacts the soil and makes

Fig. 4-5. Sidewalk trees on a city street in Lexington, Kentucky.

it as nearly impervious to air and water as concrete. Street trees are more likely to decline from lack of soil fertility than are lawn trees, because the soil mass in which their roots can develop is more restricted than it is for lawn trees (Fig. 4-5).

Three factors largely govern success in planting trees along public rights-of-way: selection of the proper species, good growing conditions, and reasonable aftercare and protection.

Considerations in Choosing Trees for City Streets

Tall-growing trees such as the London planetree, the ginkgo, and the honey locust are known to tolerate city conditions better than trees like the sugar

Fig. 4-6. Sixty-year old pin oaks *(Quercus palustris)* line a street in Lexington, Kentucky.

maple. The explanation is often that the foliage of the first three is more tolerant of air pollutants. The authors are of the opinion that some deciduous trees are more tolerant because their roots can survive under the extremely poor soil conditions prevalent in most cities; that is, they are better able to tolerate excessively dry soils, poor soil aeration, high concentrations of soluble salts, and other adverse below-ground factors.

Tall- Versus Low-Growing Trees. When the senior author wrote his first book on trees about 45 years ago, the consensus among shade-tree commissions, park department officials, arboricultural firms, and other agencies empowered to regulate the selection, planting, and care of trees on public property was that tall-growing trees such as elms, oaks, planetrees, and maples were best suited for planting along city streets (Fig. 4-6).

Times have changed and so have ideas on the proper kinds of trees to plant. Many officials now realize that it is best to select trees that will give the least trouble and require the lowest maintenance costs in future years. While many of the tall-growing or standard shade and ornamental trees have long been favorites on wide city streets, they are rapidly losing their popularity because of high maintenance costs. Such trees are more expensive to spray and prune, and during severe wind, ice, electrical and snowstorms or hurricanes they cause great damage to power and telephone wires. Moreover, tall-growing trees have a greater tendency to crumble curbstones or push them out of line, crack and raise sidewalks, and clog sewer pipes. Tall-growing trees should be planted where they have plenty of room to grow and where they will not interfere with public and private utility services and solar collection devices. The modern city planner realizes that trees should be fitted into the available space.

Tall-growing trees such as elms, pin oaks, and planetrees were popular in the past for several reasons. They are easy to reproduce either from seeds or from cuttings, they grow rapidly, transplant readily, and require relatively little aftercare. Certain low-growing trees, on the other hand, may require more skill to grow and maintain. The initial cost of such trees of streetsized planting size may be greater than that of the standard tall-growing types because of the high cost of their production, as well as supply and demand factors. Overall maintenance and, when necessary, removal costs are much lower, however, than those for tall-growing trees.

Proponents of the tall-growing trees claim that low-growing trees have several serious drawbacks: their relative scarcity as compared with tall-growing kinds; their low-hanging branches, which interfere with vehicular and pedestrian traffic; and their inability to provide as much shade as tall-growing kinds, especially for cars parked along city streets. As more nurserymen grow the low-growing forms, the first drawback is being overcome. The second problem can be overcome if low-growing trees are grown properly with a central leader; for example, with Kwanzan flowering cherry, the graft is made 6 to 7 feet above the ground level (Fig. 4-7). The drawback of low-hanging branches can also be overcome if communities would permit the planting of small-type trees inside the property line instead of between the street and sidewalk. The third

Fig. 4-7. Kwanzan flowering cherries grafted on 6½-foot stems make excellent streetside trees. These, growing in Portland, Oregon, are about 30 years old and 24 feet tall.

drawback is not solved easily, but where more shade is desired, close planting of low-growing trees is suggested.

The ideal small-type tree should grow rapidly; it should reach a useful size soon after being transplanted to its permanent site and should then abruptly slow down. It should be reasonably free from insect pests and diseases and should have pleasing growth habits.

Persons entrusted with the selection and care of streetside trees should adopt programs for proper selection and maintenance. Some of the more desirable low-growing trees need more frequent spraying than the older tall-growing kinds; if available, disease- and insect-resistant types should be selected to avoid this problem.

Size and Spacing of Trees. The best size for street planting varies from ¾ to 1½ inches in caliper at the base. Larger trees are also used but become established more slowly and are more expensive.

Most city trees are planted too closely. No trees should be planted closer than 40 feet from one another, and large spreading types should be spaced 60 to 75 feet apart. Narrow streets are best planted in ginkgo, European linden, or some of the upright varieties of maple, oak, elm, and hornbeam.

Location. Street trees are customarily placed between the walk and the street curb. Planting a tall-growing tree in this area is bound, in time, to conflict with overhead wires and with the sidewalks and curbs. This area can be properly planted with the kind of tree that will not lead to such difficulties. The decision on where to locate the tree depends on site factors.

Many subdivisions are being developed with underground electrical utilities so that overhead wire interference is not a problem. Potential sidewalk and curb damage by tree roots can be prevented by using mechanical barriers which force the roots to grow deeper into the soil before they grow in the direction of the walk. These root deflectors can be installed at planting or fit into an existing tree lawn following root pruning, if needed. The barriers have vertically oriented ribs which deflect tree roots downward. Air and water supplies for those deeper rooting trees must be ensured by installation of gravel just outside the barriers.

Where tall-growing trees are used, planting in the area between the walk and the residence is more satisfactory (Fig. 4-8). This, of course, is feasible only where the house is set back some distance from the walk. The advantages of such an arrangement are many. It removes trees from the area where they are continually subject to injury by street traffic, the trees have more favorable soil conditions, they are less likely to tangle with overhead wires, the crowns do not form over the street area, and the street looks wider. Moreover, the tree will be damaged less severely if the street is widened at a later date. A disadvantage of locating trees on private property is that the jurisdiction over such trees is taken away from the Shade-Tree Commissioner or Department of Parks. In Massachusetts, municipally managed trees can be planted as far as 20 feet within private property.

However, there are those who feel that planting trees between the sidewalk and curb is a good idea if only because a row of tree trunks provides a barrier

Fig. 4-8. Sometimes the ideal location for street trees is not along the curb but back in the front lawn. Trees here are less subject to vehicular injuries and have more favorable growing conditions.

between pedestrians and vehicular traffic. Even if the barrier is only psychological, strollers, joggers, and parents of children playing on the walk have an increased sense of security from busy street traffic when trees are planted in the "treelawn" space between sidewalk and street.

Selected Bibliography

Anonymous. 1973. Trees for polluted air. U.S.D.A. Forest Service, Misc. Publ. 1230, Washington, D.C. 12 pp.

Anonymous. 1974. Trees for New Jersey streets. New Jersey Federation of Shade Tree Commissions, New Brunswick, N.J. 42 pp.

Anonymous. 1977. Trees for New York City. New York Department of City Planning, New York Botanical Garden, New York. 27 pp.

Baker, R. L., J. E. Kissida, Jr., W. Gould, and D. G. Pitt. 1979. Trees in the landscape. University of Maryland Bull. 183:16 pp.

Daniels, R. 1975. Street trees. Pennsylvania State University. University Park. 47 pp.

Ferguson, B. (ed.). 1982. All about trees. Ortho Books, San Francisco. 112 pp.

Grey, G. W., and F. J. Deneke. 1986. Urban forestry, 2nd ed. Wiley, New York. 299 pp.

Harris, R. W. 1983. Arboriculture: Care of trees, shrubs and vines in the landscape. Prentice-Hall, Englewood Cliffs, N.J. 688 pp.

Hudak, J. 1980. Trees for every purpose. McGraw-Hill. New York. 229 pp.

Koller, G. L., and M. A. Dirr. 1979. Arnoldia 39 (3):237 pp. (entire issue).

Krussman, G. 1984. Manual of cultivated broad-leaved trees and shrubs. Timber Press, Beaverton, Ore. (3 vols.)

Littlefield, L. 1975. Woody plants for landscape planting in Maine. University of Maine Bull. 506:58 pp.

May, C. 1972. Shade trees for the home. U.S.D.A. Agriculture Handbook 425, Washington D.C. 48 pp.

Nelson, Eileen, ed. 1980. Source book for shade tree management. Cornell University Cooperative Extension Service, Ithaca, N.Y. 50 pp.

Perry, B. 1981. Trees and shrubs for dry California landscapes. Land Design Publications, San Dimas, Calif. 184 pp.

Rehder, A. 1940. Manual of cultivated trees and shrubs hardy in North America, exclusive of subtropical and warmer temperate regions. Macmillan, New York. 996 pp.

Reisch, K. W., P. C. Kozel, and G. A. Weinstein. 1975. Woody ornamentals for the midwest. Kendall/Hunt Publications, Dubuque, Iowa. 293 pp.

Whitcomb, C. E. 1976. Know it and grow it: A guide to the identification and use of landscape plants in the southern states. Whitcomb, Tulsa, Okla. 500 pp.

Wray P. H., and C. W. Mige. 1985. Species adapted for street-tree environments in Iowa. J. Arboric. 11(8):249–52.

5

Transplanting Trees

Transplanting is the moving of a tree from one location to another in such a manner that it will continue to grow. Success, however, is not determined by whether the tree merely survives but rather by its ability to resume growth and development with the least interruption. Transplanting includes moving both field-grown and container-grown trees.

Importance of Good Transplanting

Although the time it takes to transplant a tree is only a fraction of its anticipated life span, the future health, beauty, and utility of the tree can be greatly influenced by the methods used in digging, transporting, and planting. No matter how carefully transplanting is accomplished, root disturbance is inevitable. Many roots are killed in the process of digging and moving field-grown trees. Transplant shock—the retardation of growth and development occurring after transplanting—is caused largely by the physiological stress placed on the tree by the destruction of a large portion of its root system and consequent reduction in uptake of water and minerals. For normal growth and development to resume, new roots must be generated to replace those damaged during transplanting. For container-grown trees, new roots must grow out into the surrounding soil if the tree is to become successfully established. Thus careful handling will minimize physiological stress, and faulty transplanting can cause considerable retardation in growth or even death of the tree.

Transplantability

Some trees can be moved more successfully than others. The transplantability of an individual tree is governed by a number of factors including inherent characteristics of its species or clone, its size and health, and its former habitat. As a rule, trees with compact, fibrous root systems are more successfully moved than those with large taproots or scraggly root systems. Deciduous trees are more easily transplanted than evergreens and small trees are easier to move than large trees.

Table 5-1. Transplantability

High	Medium	Low
Alder	American holly	American hornbeam
American elm	Apple, Crabapple	Beech
American linden	Buckeye	Birch
Callery peach	Bur oak	Dogwood
Callery pear	Black oak	Hawthorn
Catalpa	Bald cypress	Hickory
Common pear	Cherry	Japanese pagoda-tree
Cork tree	Chestnut oak	Magnolia
Eastern red cedar	English oak	Pawpaw
Empress-tree	Ginkgo	Pecan
Fringe-tree	Golden-chain	Pin cherry
Fir	Golden larch	Sassafras
Green ash	Hemlock	Scarlet oak
Hackberry	Katsura-tree	Sourgum, Tupelo
Honey locust	Kentucky coffee tree	Swamp oak
Hop-hornbeam	Larch	Sweetgum
Jack pine	Redbud	Tuliptree
Locust	Red maple	Walnut
Mountain ash	Red oak	White oak
Osage-orange	Shingle oak	Willow oak
Pin oak	Shumard oak	
Planetree	Spruce	
Poplar	Willow	
Red pine	Yellow chestnut oak	
Silver maple	Yellowwood	
Sugar maple		
Sumac		
Tree-of-heaven		
Yew		

This table in an adaptation from published lists, reflecting the likelihood of successful transplanting of various tree species.

Experience indicates that some species require more care during and after moving to ensure survival. Table 5-1 lists a number of commonly grown species according to their success in being transplanted. Those in the high category require normal care and are most likely to survive a move in suboptimal conditions. Those in the medium category require normal care and are less likely to survive suboptimal conditions. Those in the low category require great care and should not be moved in suboptimal conditions. Individual experience may differ because of different circumstances surrounding the transplant, so the list should be taken only as a guide.

Large trees can be moved less successfully than small trees for obvious logistical reasons and some not so obvious reasons. Modern equipment enables the lifting and transporting of large trees, but such trees often suffer from prolonged periods of transplant shock. Although under standard nursery practices, the root ball of transplanted trees is proportional to the crown size, larger trees lose a much greater mass of roots. Thus trees over 4 inches in diameter often do not grow or grow at a very slow rate for several years after transplanting, while smaller trees may surpass them in size in a few years. For example, if trees 2 inches and 6 inches in diameter are transplanted at the same time using standard nursery practices and horizontal spread of roots resumes at an average 18 inches per year, then the smaller tree would recover all of the root system lost in transplanting in 2½ years, whereas it would take the larger tree 7 years. It is generally good practice to transplant the smallest acceptable tree.

Healthy trees with good reserves of stored carbohydrates are better candidates for transplanting than are spindly trees. Stout, well-formed twig growth is indicative of good levels of stored carbohydrates. The quick resumption of root growth, which is essential to overcome transplant shock, requires reserve energy. Also, healthy trees are less likely to harbor pests and diseases, which may be introduced into previously uncontaminated areas on unhealthy stock.

The site where the candidate tree has grown influences its root development and hence its transplantability. Root systems develop over a greater area and to greater depths in sandy, well-drained soils. Thus more roots will be damaged when transplanting from a sandy site. Generally, greater success is expected when plants are moved from heavy to light soils than the reverse. Nursery-grown trees usually have better root systems for transplanting than trees grown without cultivation.

Container-grown trees usually suffer less transplant shock than field-grown trees. However, trees with girdling roots or with disproportionately large tops are poor candidates for transplanting.

Season for Transplanting

By using sufficient care and properly prepared plants, most trees can be transplanted in any season. However, the time of year influences the likelihood of success. Most arborists agree that deciduous trees can be moved in the fall, in the winter before the soil freezes, or in the spring before growth begins, but fall planting has a number of advantages. In the fall the cells of most woody plants are lignified and less subject to stress due to lack of moisture. As long as leaves are still functional, but shoot growth has stopped, carbohydrates are available for root growth, which will continue while soil temperature is favorable (usually above 40° F). Tests have shown that for many species, fall-planted trees have the advantage over spring-planted trees in number of new roots, stem diameter, and height. However, a dry fall followed by a cold winter may result in the death of a high percentage of fall-transplanted trees.

Some deciduous trees survive best if they are planted in the spring. Trees with fleshy root systems, such as dogwood, magnolia, tuliptree, willow oak, and yellowwood, are more successfully moved in the spring. Thin-barked trees such as birch should be transplanted in the spring or should have protective trunk wraps if they are moved in the fall. Other deciduous trees that are best moved in the spring are American hornbeam, hickory, pecan, white fringe-tree, beech, goldenrain-tree, sweetgum, tupelo, sassafras, walnut, and white oak.

With care conifers can be moved at any season of the year. Best success, however, follows their transplanting during August and September when the soil is warm (60–70° F at 6–12 in. soil depth). Some nurserymen feel they should never be moved after October 1. Exceptions include hemlocks and Austrian pines, which are best moved in the spring after soil temperatures have begun to rise.

Broad-leaved evergreens are most successfully transplanted in the spring. They should be moved in the fall only if they can be protected through the winter.

Circumstances may dictate midsummer transplanting. When a tree is dug in midsummer, the soil in the ball of earth should be near its maximum water-holding capacity. This can be assured by trenching and watering the tree the day before the actual digging. After digging, the tree should be lifted at once and taken to a sheltered area. Make the ball 2 to 3 inches larger than normal (see Table 5-2). To reduce transpiration, trim foliage by pruning weak, injured, or broken branches or use an antitranspirant. Plant the tree as soon as possible and water well. If air temperature is high or humidity low, top sprinkle several times each day for two weeks.

Transplanting can proceed through the winter in mild climates as long as the soil does not become too muddy for operation of the equipment. Where the soil freezes, digging may proceed but the frozen ball must be protected to prevent root damage. The soil will lose moisture even when frozen and roots will be killed by desiccation. Japanese cherry, tuliptree, magnolia, and willow oak should not be transplanted in frozen balls. Most oaks, elm, linden, sugar maple, and beech requiring 6 to 14-inch balls can be successfully moved in frozen soil.

Although ball and burlapped and container-grown trees can be transplanted in any suitable season, bare-root trees are best planted in the spring. As buds swell in the spring, trees begin developing roots. If trees have already leafed out, the stimulus to root development is lost. Thus trees should be planted before the buds break.

Preparing Trees for Transplanting

Nursery-Grown Trees. Properly grown nursery trees are root-pruned every few years, not only to confine the greater part of the root system within a small area, but also to encourage the roots to branch and develop a greater abundance of small roots and rootlets. Trees that have been root-pruned can be transplanted more efficiently and with a greater chance of success.

Collected or Wilding Trees. Many trees growing naturally in fields or woods without previous cultivation are transplanted to new sites to provide shade or ornament. Small trees need no special preparation, but large trees should be root-pruned to stimulate the formation of a compact root system. Best results are obtained when such trees are root-pruned, as shown in Figure 5-1, for two successive years before they are moved. In the spring or fall a circular area around the tree at some distance from the trunk (approximately 5 in. for each

Fig. 5-1. Method of root-pruning wilding tree before it is moved. The soil is dug from sectors A, exposed roots are pruned, and the soil is replaced after it has been amended with organic material. One year later the same procedure is followed in sectors B. The tree should be moved a year after the pruning at sectors B. **Upper left:** Sketch shows the extent of the roots before pruning. **Upper right:** After root-pruning.

inch of trunk diameter) is marked off. Three ditches, covering half the circumference, are dug with a spade and all the exposed roots are severed. A year later, the remaining area is trenched. The soil excavated each time should be mixed with manure, leaf mold, or a commercial fertilizer before it is replaced in the open trenches. The tree is ready for moving one year after the second series of trenches is dug and refilled. Trenching and root-pruning in this manner encourage greater fine root development within the trench area. Also, the shock to the tree is spread over a 3-year period.

Antitranspirants. Loss of water from evergreen trees or deciduous trees in leaf can be partly overcome by spraying with an antitranspirant just before the tree is moved. Trees sprayed with these materials must still be watered in their new location, but they may become established more quickly than untreated trees. Many professional arborists and nurserymen use antitranspirants whenever they move trees in the summer or whenever they move large trees in leaf.

There are two types of antitranspirants: (1) those that form a film on the leaves, blocking stomatal pores or coating the cells inside the leaf with a waterproof film, and (2) those that chemically close stomatal pores. These materials have not been found to be effective under all circumstances.

Most of the antitranspirants can be toxic to leaves under some conditions. If used, they should be applied at low dosages and not to the entire canopy. Covering only part of the leaves will still reduce water loss but will ensure that some leaves will be unaffected if toxicity symptoms develop.

Digging Trees

Proper digging includes the conservation of as much of the root system as possible, particularly of the smaller and finer roots. Any appreciable reduction in such roots, which are the roots most active in absorbing water and nutrients, results in a considerable retardation of the tree's growth and reduces the speed of reestablishment.

Trees may be moved with bare roots or with a soil ball. The former method leads to greater transplant shock and slower reestablishment; however, handling bare-root trees is often simpler and less costly. Bare-root planting is mainly used for deciduous trees up to 2 inches in diameter planted in the fall or spring. Larger deciduous trees, evergreens, and trees moved in the winter or summer should be dug with a soil ball adhering to their roots.

Before any digging begins it is advisable to tie the tree's branches with heavy twine or quarter-inch rope. This will reduce the chances of damage or breakage of the branches.

Bare Root. Transplanting trees with bare roots is usually limited to small (less than 2-in. diameter) deciduous trees. A trench is dug around the tree to a depth just below the greater part of the root system. With most small trees this depth is 15 to 18 inches. The trench should be dug just beyond the last root-pruning zone for well-grown nursery trees or at a distance from the trunk of 12 inches for each inch of trunk diameter. The roots are then combed free of soil by the use of a spading fork.

As the roots are exposed, they should be covered with wet burlap to prevent drying and mechanical injury. The soil close to the trunk should be left intact to hold the tree erect until the lateral roots have been uncovered. Then the tree may be tipped over gradually and the soil removed from beneath the trunk. Taproots should be cut at a depth of 15 to 18 inches.

The tree should be gripped by the trunk at the root crown and lifted from the soil. After the tree has been lifted, care must be taken to prevent root death. Roots should be covered with wet burlap, damp straw, or loose soil. Research has shown that lateral roots are easily damaged under conditions of high temperature and low humidity, even if the drying period is only 30 minutes. However, damage can be reduced if dried roots are soaked for several hours before planting. Trees should be stored in a sheltered place out of direct sun and wind.

An alternative method, pulling the tree from moist soil using a pin and clevis inserted through the trunk, has been tested for use with small trees. This method works only in light soil. A high degree of top and root damage may occur if this method is used improperly.

Ball and Burlap. Whenever possible, deciduous trees over 2 inches in diameter, evergreens, and small trees in leaf should be moved with soil adhering to the roots. The diameter and depth of the soil ball depend largely on the type of soil, the root habit, the type of tree, and the size of the tree. As a rule, trees with shallow roots require a flat ball, whereas those with deeply penetrating roots require a deeper ball with a smaller diameter. The main objective governing the size of the earth ball is the protection of the greatest number of roots in the smallest soil mass.

The soil ball for deciduous trees is usually 9 to 12 inches in diameter for each inch of trunk diameter. Table 5-2 gives the general guidelines. As a rule, evergreens require a ball of smaller diameter.

Since most soils weigh about 110 pounds per cubic foot, no more soil than is absolutely necessary should be allowed to remain around the ball. A rapid

Table 5-2. Recommended minimum diameter and depth of soil balls for deciduous trees of various sizes

Trunk Diameter in Inches of Tree 1 Foot above Ground	Diameter of Ball (in.)	Depth of Ball (in.)*
1¼–1½	18	14
1½–1¾	20	15
1¾–2	22	16
2–2½	24	17
2½–3	28	18
3–3½	33	20
3½–4	38	23
4–4½	43	26
4½–5	48	30
5–5½	53	31
5½–6	58	33
6–7	65	35

*Depth is measured from the depth of the uppermost feeder roots. Scrape away excess topsoil to expose roots before digging.

Fig. 5-2. After trenching around the tree, the soil ball is shaped using the back of the spade facing the tree.

method for computing the weight of a soil ball is to square the diameter of the ball in feet, multiply this figure by the depth in feet, then take two-thirds of this total and multiply by 110. The resulting figure will give the approximate weight of the soil in pounds.

Digging should not be attempted unless the soil is moist. If the soil is dry, it should be watered thoroughly at least 2 days before digging. Unless this is done the soil ball will break and the roots will be completely separated from the soil. Such a tree will have much less chance of surviving in its new location.

The first step in digging of the soil ball is the cutting of a circle around the tree 6 to 8 inches farther away from the trunk than the intended soil ball. Trench down to approximately three-fourths of the final ball depth on the outside of the initial circle. This will allow space for shaping the ball. It is best to work with the back of the spade toward the tree to avoid loosening the soil ball (Fig. 5-2). Large roots should be cut cleanly with hand shears. Moist burlap should be used to cover exposed roots during the digging operation.

Even when mechanical trenchers are used, final shaping of the soil ball is done by hand. Beginning at the top of the soil ball, excess soil is carefully shaved off to form an oval ball.

If the tree is small (less than 2-in. diameter), a sharp spade can be used to undercut the soil ball and sever any remaining roots. Then the soil ball may be lifted from the hole and burlapped. However, for larger trees burlapping should be done before the ball is undercut and lifted.

There is a great temptation to grap the top of a tree with a small soil ball and yank it from the hole. Many soil balls are jarred loose and root systems damaged by this practice. Small balls should either be levered out of the hole using a spade or lifted out using a burlap sling.

After lifting, the ball should be placed in the middle of a burlap square and opposite corners should be tied across the top of the ball. Loose sections should be pinned snugly using sixpenny or eightpenny balling nails to make a neat package. If the soil is loose or sandy, the ball should be reinforced with rope or twine.

For larger trees, burlap strips are pinned around the sides of the soil ball before the tree is lifted. The strips should overlap the top of the ball by several inches and be sufficiently wide to cover the bottom of the ball after it is undercut. Again the burlap should be pleated and pinned to make a neat package. Rope lacing further reinforces the ball (Fig. 5-3). When the soil is loose, chicken wire, hog wire, or wire fencing should be used to cover the burlap before the rope is laced on or as a substitute for the rope.

Instead of burlap, wooden boxes may be built in the hole surrounding the soil ball. This method is used where soil is very sandy or when trees are to be held for an extended time before planting (Fig. 5-4).

All roots beneath the soil ball must be severed before the tree is lifted out of

Fig. 5-3. Balled and burlapped evergreen tree ready for transplanting.

Fig. 5-4. In sandy soils, a wooden box may be used to contain roots and soil during transplanting.

the hole. Undercutting is accomplished by running a small steel cable beneath the burlap at the base of the ball. Both ends of the cable are attached to a hook on another cable running to a winch. The winch will pull the cable to sever the roots and detach the soil ball from the ground. Keep the ends of the cable low in the hole to get a flat cut. After cutting, secure loose burlap to the underside of the ball with nail pins.

Large soil balls will require mechanical assistance for lifting. A chain sling attached to a front-end loader or crane works best under most circumstances.

If ropes or chains are secured to the base of the tree, pad it well. Some arborists lift the tree using one or two heavy steel pins inserted through the base of the trunk. This method is very risky and often results in trunk and ball damage.

After lifting, the trees should be protected against excessive drying. Move the trees to a sheltered area, place them close together, sprinkle, and cover the soil with wet mulch or plastic. If deciduous trees are dug in full leaf, they should be held 24 to 48 hours before replanting to acclimate them to root loss and thus reduce transplant shock.

Frozen Soil Ball. Where the soil freezes to depths of 1 foot or more, trees can be moved with frozen root balls. Such balls are less likely to be damaged in handling and thus require less wrapping. The tree to be transplanted and the planting site should be mulched before the soil freezes. The tree is dug in the conventional manner, but it should not be undercut until just before the move. The ball can easily crack, so care must be taken during handling. The planting hole should be dug only shortly before the move to prevent the inside of the hole from freezing. After moving, the tree should be watered only enough to settle the soil around the ball and mulch should be applied to prevent further freezing. Move trees only when the temperature exceeds 20° F.

Mechanical Tree Digging. Mechanical spades are available which are designed to dig and lift trees with a ball of soil surrounding their roots. In some cases the tree spade may also be used to transport the tree, eliminating the necessity of burlap covering. However, when several trees are to be dug before transporting, the tree spade is used to set the trees into burlap squares for conventional wrapping. Trees larger than the machine can adequately handle should not be moved this way. Workers must be specially trained in the maintenance and operation of the equipment to ensure its safe and efficient use.

The tree spade encircles the tree and four pointed blades are hydraulically forced into the soil (Fig. 5-5). The blades form the soil ball container when they meet below the tree. The soil ball is longer than on conventionally dug trees, although the diameter should be the same for both (Table 5-2). Most of the roots are cut off cleanly. Occasionally a taproot is caught between the blade tips at the bottom or a large lateral root is wedged between two blades. The soil ball can be badly damaged and roots mangled when this happens. When roots are shredded and split rather than cut cleanly, they will die back several inches from the cut end. This death intensifies transplant shock and delays the establishment of the tree. Thus it is vital that the blades be kept sharp.

When properly done, mechanical tree digging can be more successful than bare-root and comparable to ball and burlap transplanting. Tests conducted by the Forestry Division of the city of Lansing, Michigan, revealed that the mortality rate for streetside trees planted bare root was 28 percent, whereas only 1 percent of the trees planted with a tree spade died.

In-Field Container-Grown Trees. Containers set in the ground allow trees to be lifted from their growing site with much less effort than conventional digging methods. The trees are grown in wire baskets lined with filter fabric and filled with field soil. Although some roots penetrate the fabric bag, they

Fig. 5-5. Tree spade digging a maple tree in a nursery.

are restricted in size. Most of the roots are confined to the container. Trees are mechanically lifted using a device that grips both the trunk and the wire basket. Trees up to 10 feet tall have been moved by this method.

Transporting Trees

While small trees are transported very easily, large trees usually require special types of transportation equipment. When trees are to be transported relatively short distances, a sled with wooden runners can be used. For longer distances,

Fig. 5-6. Loading a large multistemmed Japanese maple, 22 feet high and 24 feet wide, with a 9½-foot root ball, at a nursery.

a number of well-designed and efficient tree-moving machines are now on the market. These are used mainly by commercial arborists, park superintendents, and other people who move large trees. The machines are mounted either on trucks or tractor trailers (Fig. 5-6).

The limit to the size of tree that can be moved is governed by the amount of road clearance over which the tree is to be transported and the weight the roads and bridges will stand.

Considerable injury may occur during the moving of large trees unless the following precautions are taken:

1. Pad unprotected areas with burlap, canvas, or some other material to avoid bruising and slipping the bark of the trunk and branches.
2. Carefully tie in all loose ends with a soft rope to avoid twig and branch breakage.
3. Keep the earth ball moist or cover bare roots with wet burlap or moist sphagnum moss.
4. Avoid excessive drying of the tops, especially of evergreens, while the trees are in transit by covering them with canvas or wet burlap or spraying them with an antitranspirant.
5. Protect from drying out, if trees are not to be planted immediately at the new site.

Planting Trees

Preparation of the Hole. The care taken in the preparation of the new site and the hole is a major factor ensuring the ultimate success in transplanting. The site was discussed in Chapter 3. Three factors should be considered in preparating the hole: size and depth, soil quality, and drainage.

For bare-root trees, the hole must be sufficiently deep and wide to accommodate the full root system without bending or cramping the roots. For ball and burlapped or container-grown trees, the hole should be at least 1 foot wider than the soil ball to allow backfull to be worked in around the ball and to improve soil aeration for the new roots. In well-drained soils the bottom of the hole should be just deep enough to place the top of the ball at the soil surface. Digging deeper would allow settling and disturbance of the new roots.

Trees planted in heavy or poorly drained soils will not thrive unless some precautions are taken prior to planting. The drainage capacity of a site can be determined by drilling a hole 2 feet deep and filling it with water. If the hole does not drain appreciably within 30 minutes, steps must be taken to improve drainage.

Improving soil drainage is usually beneficial; however, efforts to improve drainage may be expensive and do not ensure success. Agricultural drainage tiles, perforated flexible plastic tubing, or gravel may be placed in the bottom of the planting hole (Fig. 5-7) to guide away excess water. If clay fill-soil is encountered at the bottom of the planting hole, drainage holes should be drilled through the clay layer and filled with pea gravel or the clay should be dug out entirely and replaced by topsoil.

If the drainage problem is not too severe or is localized, the tree should be planted above grade within a soil mound.

Where it is not physically or economically possible to correct drainage, the best alternative is to plant trees that will tolerate saturated soils, such as bald cypress, willow, or tupelo. Avoid planting evergreens, especially yew and broad-leaved types.

There is a temptation to avoid hand labor by using the mechanical tree spade to dig the hole before transplanting a tree with the spade, but tests have shown that this may have undesirable effects. The mechanical tree spade can glaze the sides of the hole and the face of the soil ball. Soil glazing may retard root penetration and development into surrounding soil. However, the glazed surface can be roughened to avoid the problem.

Setting the Tree. The depth of planting must be close to the original so that the soil-stained ring at the base of the trunk is level with the surrounding soil surface. Experienced arborists have found that if a tree is set a bit higher in its new location than in its former site, the chances of continued good growth are increased. Trees set too deeply will not thrive. Feeder roots require oxygen to function and thus will die if placed at a soil depth where oxygen levels are too low. Flowering dogwood, American beech, and narrow-leaved evergreens are especially sensitive to planting that is too deep.

Because the roots of trees grow downward at a slight angle from the hori-

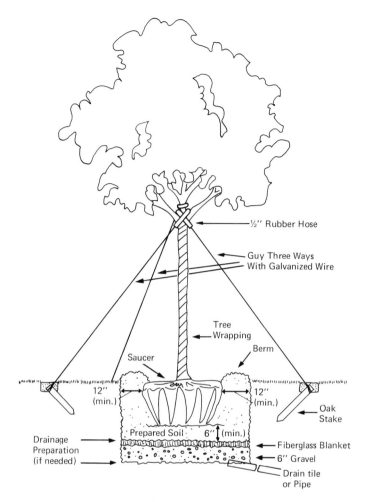

Fig. 5-7. To improve drainage in some wet sites, gravel and drain tiles are placed in the planting hole. The base of the tree trunk should be just at or above ground level.

zontal, it is wise to make a cone-shaped mound of soil at the bottom of the hole when planting bare-root trees (Fig. 5-8). Before planting, diseased, kinked, and broken roots should be pruned. The root crown can then be set on this mound and the roots spread over and down the sides, thus assuring close contact with the soil along their entire length.

Balled trees need to be set into the hole at the proper depth (Fig. 5-9). Then the rope holding the burlap should be removed and the burlap loosened. The burlap need not be removed from the bottom of the ball; indeed removal may damage the ball and the roots. It should be rolled back from the top and slit along the sides. Plastic wrapping material must be completely removed.

Conventional container-grown trees should be tapped out of their containers carefully. Root systems of trees grown in containers have been forced to con-

A

B

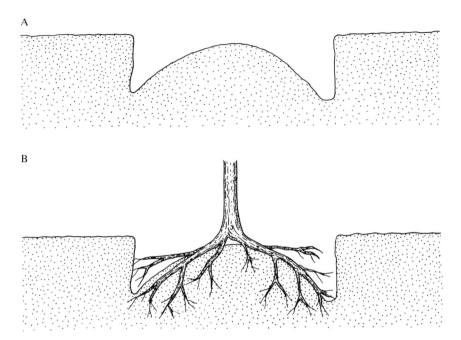

Fig. 5-8. Bare-root trees are placed on a cone-shaped mound of soil at the bottom of the planting hole. This allows closer contact of the roots with soil and thus reduces air pockets.

form to the shape of the container (Fig. 5-10A). This encircling outer layer of roots should be cut in several places (Fig. 5-10B), or removed to encourage new roots to grow out into the surrounding soil. If not removed, such roots may eventually girdle the tree.

The wire basket and filter cloth sack must be removed from in-field container-grown trees prior to planting. Prune any upper roots that show circular growth habits. No other root pruning is necessary.

Any wires that hold labels should be loosened or removed so that they do not girdle the trunk or branches as they increase in girth.

Hormone Treatments. Hormones applied as sprays, root dips, and soaks have been successfully used to promote root regeneration in tests. The hormones stimulated root regeneration from medium and large roots and resulted in significantly greater new leaf growth. Results were most favorable when fall-dug trees were treated in the spring. Attempts to stimulate roots with exogenously applied chemicals have had mixed results, but the root spray method may be more practical than others attempted.

Backfilling. There are two schools of thought about backfilling. Traditionally it was recommended that the soil removed from the hole be improved by amending with organic or coarse mineral materials such as peat, pine bark, sand, and vermiculite. However, some recent studies have shown that under

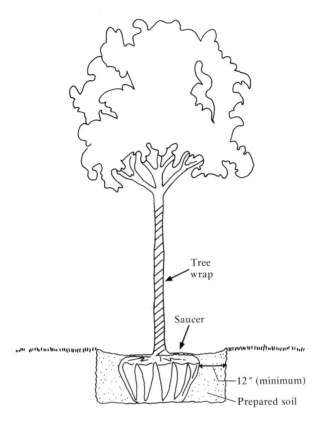

Tree
wrap

Saucer

12″ (minimum)

Prepared soil

Fig. 5-9. In well-drained soils, the planting hole for ball and burlapped trees should be just deep enough to place the top of the soil ball at the soil surface. The burlap should be rolled back from the top and slit along the sides. Plastic wrapping and twine must be completely removed.

certain circumstances backfilling with unamended soil results in better tree growth. If the surrounding soil is heavy clay, it is less favorable to root development, so roots will tend to grow only in the amended soil of the planting hole. The roots may have too much water in wet periods and too little in dry periods. It is important, however, to improve aeration of unamended soil before using it as backfill. Break up clods and compacted soil to provide small pores without large air pockets.

Once the tree has been set in the hole at the proper depth and orientation, the soil should be added gradually. For bare-root trees, work the first lot in firmly at the base of the roots. Then add more soil and work it under and around the lower roots. The tree may be gently raised and lowered during the filling process to eliminate air pockets and bring the roots in close contact with the soil. When the roots are just covered, tamp the entire area firmly with the feet. Heavy soils should not be tamped too firmly around the roots, since this

Fig. 5-10. A, encircling roots of a container-grown tree; B, encircling roots severed before planting.

will prevent oxygen and water from reaching the roots. Add water, let the soil settle, and finish filling the hole with loose soil.

For trees transplanted with a soil ball, work the soil beneath the ball to leave no air pockets. Add soil until the hole is about half filled, tamp firmly, then water to settle the soil and finish filling with loose soil.

After filling is completed, rake a 4-inch dike around the circumference at the planting hole (Fig. 5-11) to form a shallow basin for catching and preventing runoff of water. The ground should be leveled off, however, before winter sets in, since water collected in the basin may freeze and injure the trunk.

Postplanting Operations

Supporting. Newly planted trees may need artificial support to prevent excessive swaying in the wind, to promote upright growth, or to guard against mechanical damage. However, staking is not necessary for many trees and can have undesirable effects. Research has shown that staked trees develop a smaller root system and a decreased trunk taper. They are often injured by ties which may rub against or girdle the trunk or limbs. Their tops offer more wind resistance than tops which are free to bend. The decision to stake a tree should not be automatic but rather should be made after consideration of the strength of the trunk, wind conditions, and traffic patterns. Proper staking techniques can lessen or prevent problems.

Fig. 5-11. A shallow basin may be formed around newly transplanted trees to facilitate watering.

Trees with strong trunks may need stakes only to prevent mower damage and to provide anchorage until the roots grow into the surrounding soil. Two or three stakes 3 to 4 feet long are driven 1 foot into the ground (Fig. 5-12). The stakes should be placed about a foot from the trunk. Ties should be fastened to allow some movement of the trunk.

Trees with weak trunks may need stakes for support. For trees up to 20 feet in height, one or two strong stakes or poles, 6 to 8 feet in length, can be driven 2 feet or so into the ground and 6 to 12 inches from the trunk. A short length of old garden hose with a wire run through it is then placed around the trunk to protect the bark, and the wire is twisted around the stake in the form of a figure 8 to complete the support (Fig. 5-13). A wide cloth tape may be substituted for the wired hose. This is wound around the tree and the loose ends are nailed to the supporting stake.

A number of devices are commercially available to attach trees to supporting stakes. Some automatically adjust as the tree grows, eliminating the necessity of frequent changes of ties.

Guy wires or cables must be used to support larger trees. They should slope from about halfway up the trunk to the ground at an angle of about 45 degrees. The upper ends of wires are attached with lag screws or by encircling the trunk with a loop encased in a piece of garden hose or other protective cover. The lower ends of the wires are anchored to stakes driven deeply into the soil. From one to four wires or cables are usually placed around a large tree (Fig. 5-14).

Fig. 5-12. Trees with strong trunks need stakes only to protect them from mower damage and to provide anchorage. The stakes should be fastened to allow some trunk movement.

Fig. 5-13. Wire run through a piece of garden hose and the trunk and supporting stake in the form of a figure 8 supports the tree and protects the bark.

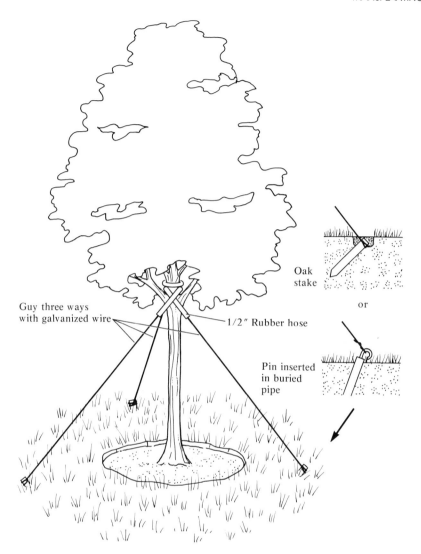

Guy three ways
with galvanized wire

Oak
stake

or

1/2" Rubber hose

Pin inserted
in buried
pipe

Fig. 5-14. Guy wires should be placed to allow some crown movement.

Guy wires or cables are rarely placed around trees set along the city streets or in public parks. Unless protected or readily visible, especially at night, they are a constant source of danger to passersby.

Pruning. The tops of transplanted trees frequently are pruned to compensate for the loss of roots and to maintain a balance between the two parts. However, recent studies indicate that this practice may not be beneficial in preventing transplant shock and may be detrimental by reducing the capacity for photosynthesis. Apparently in many cases the added moisture stress resulting from leaving the plant top intact is offset by more rapid development of supporting root system from additional carbohydrate produced. No more than 15 percent of the top at time of planting should be pruned. The amount of pruning depends not only on the condition of the tree at transplanting time but also on the care the tree is to receive while becoming established. Less pruning is needed when most of the roots have been conserved during digging and when adequate water and aftercare are assured. Of course, broken, weak, diseased, or interfering branches should be cut. Additional pruning should be delayed until after the first growing season.

Little or no pruning of balled evergreens is needed except to remove broken or severely injured branches. If the leading shoot is broken during moving, one of two practices may be followed: (1) one of the next nearest lateral branches may be bent upward and held in place by tying to a small piece of wood; or (2) all the nearest laterals but one may be removed, so that the lone lateral will tend to grow upright and will soon replace the broken leader.

Watering. The soil around the tree must be watered thoroughly after the tree is set in place. Additional water must be applied from time to time until the tree is firmly established. The frequency of watering depends on the type of soil, the size of the tree, and the amount of rainfall. However, in general the soil should be wetted to a depth of 15 inches in well-drained soils at approximately weekly intervals, or less often for heavier soils. Deciduous trees moved during late fall or winter require very little water, at least until new leaves emerge the following spring.

The water must soak into the ground slowly to allow it to moisten down to the roots. The depth of water penetration can best be determined with a soil auger or by digging with a narrow spade. The tree benefits little from frequent light waterings which moisten the soil to a depth of only a few inches. If water must be applied with a pail, it is best to pour on two or three pails at a time and allow it to soak in completely before more is added.

One common error of inexperienced gardeners is to overwater the soil around the newly transplanted tree. This drives out essential oxygen from the soil and it also favors the development of root-decaying fungi. The soil around newly planted trees should not be heavily watered in spring until root growth starts and the leaves begin to emerge.

The soil ball around a large newly transplanted tree can dry out in a surprisingly short time. Summer showers cannot be depended upon to supply sufficient moisture to keep the soil ball moist. Hence an occasional good soaking with the garden hose is necessary. To get good penetration into the ball itself, it may be necessary to punch holes into the ball with a crowbar or to bore holes of varying depths with a soil auger before applying the water.

Frequent syringing of the leaves of newly planted evergreens on cloudy days helps to cut down water loss by the leaves. and in some cases washes off spider mites and soot.

Surfactants or wetting agents may enable water to penetrate soil more quickly and more uniformly. Use of such materials may be of particular benefit in hard-to-wet soils.

Mulching. With particularly valuable trees, especially evergreens planted in

Fig. 5-15. Mulch placed around the base of a newly planted dogwood.

the fall, a good practice is to mulch the area above the roots with straw, leaf mold, or well-rotted manure (Fig. 5-15). Such mulches prevent wide fluctuations in soil temperature, keep the soil warmer, and help conserve moisture. Mulching suppresses growth of grass and weeds near the trunk, eliminating the need for mowing near the base of the tree. Mulches are usually left on over the winter and either removed or worked into the soil in spring. A newly planted deciduous tree collected from the woods should be mulched to a depth about equal to that of the mulch in its former location.

Sand is sometimes used as a mulch in the depressed area around small newly planted trees along streets. Water is thus absorbed more rapidly and is held longer. The sand also keeps the area free of weeds and may increase the chances of the tree's survival.

Wrapping. The trunk and larger branches of a transplanted tree may be wrapped with burlap, specially prepared fine mesh wire screen, paper, or some other material to prevent sunscald and drying of the bark and to reduce the possibility of borer infestation (Fig. 5-16). A protective covering is especially desirable on trees collected from shaded woods, since the bark of such trees is

Fig. 5-16. Base of a young pin oak covered with kraft paper wrapping.

extremely susceptible to sunscald. The covering should be securely fastened in place with cord and left on through one summer and then removed.

Wrapping the trunk of a newly transplanted tree has one disadvantage. In rainy seasons the trunk is kept unduly wet beneath the wrapping, which fosters the development of fungus cankers. The senior author has found that pin oaks are especially sensitive to such cankers.

To protect young dogwood, flowering crabapple, and other susceptible trees from rabbit and mouse feeding, wrap some hardware cloth or metal screening around the base of the tree in late fall.

Trees planted in lawns may need lawn mower guards to prevent basal injuries (Fig. 5-17).

Fertilizing. As a rule, fertilizer should not be applied until new roots are formed. Indeed, research has shown that high levels of added nitrogen have a negative effect on root growth, so may prolong transplant shock. One study

Fig. 5-17. Lawn mower guard.

found no benefit from nitrogen fertilization during the first three seasons of growth. Depending on soil conditions, applications may begin at the end of the first growing season. However, where plantings are inconvenient or expensive to fertilize regularly or are in sandy soils receiving heavy rainfall, slow-release fertilizers may be used. They should be placed in the hole and covered by 4 to 6 inches of soil before the tree is placed in position.

Phosphorus and to a lesser extent potassium stimulate root growth. In deficient soils, phosphorus may be added at planting by incorporating a high-phosphorus fertilizer.

Selected Bibliography

Berdel, R., C. Whitcomb, and B. L. Appleton. 1983. Planting techniques for tree spade dug trees. J. Arboric. 9(1):282–84.

Gowin, F. R. 1983. Girdling by roots and ropes. J. Environ. Hortic. 1(2):50–52.

Himelick, E. B. 1981. Tree and shrub transplanting manual. Int. Soc. Arboriculture, Urbana, Ill. 76 pp.

Shoup, S., R. Reavis, and C. E. Whitcomb. 1981. Effects of pruning and fertilizers on establishment of bareroot deciduous trees. J. Arboric. 7(6):155–57.

Watson, G. 1985. Tree size affects root regeneration and top growth after transplanting. J. Arboric. 11(2):37–40.

Watson, G. W., and E. B. Himelick. 1983. Root regeneration of shade trees following transplanting. J. Environ. Hortic. 1(2):52–54.

Whitcomb, C. E. 1979. Factors affecting the establishment of urban trees. J. Arboric. 5(10):217–19.

Whitcomb, C. E., R. Reiger, and M. Hanks. 1981. Growing trees in wire baskets. J. Arboric. 7(6):158–60.

6

Fertilizers and Their Use

Woodland trees thrive despite the absence of artificial fertilization, but many of our shade and ornamental trees do not grow in such a favorable environment. Under forest conditions, decayed leaves and dead plants replace mineral elements (referred to here as nutrient elements, or nutrients) taken up by living plants. However, in urban settings, leaves and lawn clippings are usually raked up and disposed. The soils near buildings are often disturbed by construction activities causing the loss of topsoil, compaction, and the presence of clay subsoil within the rooting zone. Turf grass competes with trees for nutrients and water, and pavement frequently covers all or part of root systems. In woodlands, natural selection ensures that native trees can survive under local fertility conditions; however, people often attempt to grow species that are not adapted to the soil composition of their landscape.

Benefits of Fertilizer Application

The effects of undernourishment may be expressed as failure of a plant to grow as rapidly as expected or as a gradual decrease in the vigor of the tree. This loss of vigor can weaken the tree so that it is less able to withstand biological and environmental pressures that can result in tree decline. In some cases, the deficiency of a specific nutrient will cause a reduction in the beauty and utility of trees and may cause symptoms that can be confused with infectious diseases. These special cases are discussed more fully in Chapter 10.

Although fertilizer application does not constitute the whole of soil improvement, it is an important part of any successful tree maintenance program. The objectives of tree fertilization are fourfold:

1. To increase the size of small trees as rapidly as possible;
2. To maintain the healthy appearance and vigor of mature trees;
3. To rescue declining trees; and
4. To cure specific nutrient deficiencies.

Species response is variable; however, nearly all trees respond to nitrogen fertilizer applications by making significant growth increases. Research trials show that young trees receiving regular nitrogen fertilizer applications are taller, have greater canopy spread, have larger and darker green leaves, and have greater root area than unfertilized trees. For example, treated 8-year-old tulip-trees produced four times as much shade as untreated ones. An increase in growth rate of 170 percent was observed in swamp oak, 61 percent for red maple, 30 percent for red oak, and 15 percent for linden. Generally, trees that have the slowest growth rates when untreated have the greatest relative response when fertilized.

Mature trees also may benefit from nutrient addition. Not only do leaf size and color improve, but the speed of wound closing increases as well. Trees responding to attack by wound-invading decay organisms are sometimes able to limit the invasion. These responses require energy, and healthy trees are able to respond more efficiently than those which are chronically undernourished.

Occasionally, declining or diseased trees can be rescued by timely fertilizer application. Although long-neglected trees can be an unpleasant sight in the landscape, careful pruning and fertilizing may be all that is needed to improve tree appearance. Before a declining large, sound, well-placed tree is removed from a landscape, rescue should be attempted. There are two opposing viewpoints regarding the fertilization of declining trees. One the one hand, it is thought that nitrogen fertilizers primarily promote shoot growth at the expense of root growth and therefore may do more harm than good. On the other hand, without additional foliage to produce food and energy reserves, how can the tree be expected to grow additional roots?

Similarly, applications of high levels of nitrogen fertilizer have been used with success to "rescue" Verticillium wilt-infected maples and goldenrain-trees. However, fertilizer application is not a cure for trees affected by fungal leaf spot, heart or root rot, or bark or cambial diseases.

There are well-documented examples of declining trees being deficient in a specific mineral element. Iron deficiency chlorosis affects many shade tree species, especially under high soil pH situations. Similarly, tip dieback and small crinkled leaves are associated with zinc deficiency, and chlorosis and decline have been associated with manganese deficiency. Applications of specific nutrient compounds may alleviate such conditions, although in some instances deficiencies may result from an improper balance of nutrients or from improper soil pH rather than lack of the specific nutrient in the soil.

Determining the Need for Fertilizer

Fertilizers are sometimes used to excess, causing unnecessary expense, occasional damage to sensitive plants, and potential water pollution. Before deciding whether to fertilize, arborists should determine what is to be accomplished by fertilization. In the landscape, a young sapling, a mature healthy tree, a declining tree, and a chlorotic tree would each have different needs.

Look for Subnormal Growth. The most direct way to determine if fertilizers are needed is to observe the growth rate and leaf color of the trees. Young shade trees with twig growth of 9 to 12 inches per year and mature trees with growth of 6 to 8 inches per year will probably not be helped by additional nutrients. Leaf color should be typical of the species; watch for unusual yellowing, especially between the veins, which may indicate a specific nutritional deficiency. Note also whether new or old foliage is affected by the problem.

Use a Soil Test. Soil tests are also helpful in determining if fertilizer application is warranted. A soil test can be obtained for a nominal fee from most state land-grant university county extension offices or from various private laboratories. A typical agricultural soil test will show soil pH values as well as levels of potassium and phosphorus. Additional tests of soil samples can determine calcium, magnesium, and other element levels. Nitrogen amounts are normally not determined because soil nitrogen levels change relatively rapidly. The pH value is probably the most useful information obtained from soil tests.

Since different laboratories use different tests which yield different nutrient values, interpretations of results may require the assistance of a knowledgeable expert. Precise soil sampling is important to accurate soil testing. Determine the extent of the root system of the tree(s) in question (usually twice the distance of the branch spread) and take a small amount of soil from each of eight to ten representative areas in the root zone of the tree; combine and mix the samples and use a pint of the mixture for testing. Soil sampling tubes or augers can be used to obtain soil for testing. Collect samples at the depth at which the roots are growing. Although the roots are confined to the top 1 to 6 inches of wet or compacted soil, most tree roots will be in the top 1 to 12 inches or deeper in some cases.

Analyze Leaf Tissues. Nutritional problems may also be diagnosed through analysis of tree foliage. Usually, recently matured leaves are taken from various parts of the tree with a suspected problem and from similar healthy trees nearby. These samples are then analyzed by an appropriate laboratory for mineral content.

Each laboratory has specific sampling procedures, but the following suggestions generally apply:

1. Use the mailers and information supplied by the laboratory.
2. Although collecting newly matured leaves of current season's growth is often suggested, some laboratories prefer mature leaves from the middle or base of the current season's growth.
3. Take samples from comparable healthy trees nearby.
4. Wash dust and debris from freshly collected leaves.
5. Do not collect diseased, injured, or dead leaves. Try to collect leaves as soon as possible after symptoms become visible.
6. Supply the laboratory with all the information requested and send partially air-dried specimens to the laboratory.
7. Collect soil samples for each tree at the same time leaves are collected for analysis.
8. Get professional help in interpreting the results of tissue and soil tests.

Table 6-1. Amounts of mineral elements typically found in leaves of normal and deficient woody plants*

| | Proportion of Dry Leaf Tissue | | | |
| | Normal | | Deficient | |
Mineral	Percentage	Parts per Million	Percentage	Parts per Million
Nitrogen	2.0–4.0		1.5	
Potassium	0.75–2.5		0.3–0.6	
Calcium	0.7–2.5		0.2–0.5	
Magnesium	0.2–0.6		0.05–0.2	
Phosphorus	0.12–0.5		0.08–0.1	
Sulfur	0.2–0.5		0.12–0.14	
Chlorine	0.01	100		
Iron		50–400		33
Manganese		20–800		7–18
Zinc		15–100		<15
Boron		15–100		8
Copper		5–20		<5
Molybdenum		0.1–1.5		<0.1

*Values were derived from a variety of reference sources and are composite figures. Mineral levels for specific healthy and deficient landscape trees may or may not fall within these ranges.

Typical levels of mineral nutrients found in healthy and deficient leaves of a range of woody plants are listed in Table 6-1. The range for healthy leaves can be considered a sufficiency range and values below and above the range may be deficient or in excess, depending on the species being examined. For trees growing in good soils at the correct pH, most foliar nutrients will be sufficient if the soil test shows them to be sufficient. Again, comparison of analyses of affected and healthy trees can be of much more value than comparison of an analysis to published average values. In addition, normal mineral element values for most tree species are not known.

It is important to remember that fertilizers will not solve problems caused by inadequate sunlight or water, air pollution, plant diseases, or insect attack.

Specific Nutrient Elements

Nitrogen is the element most frequently in short supply for optimum tree growth, although the other nutrients may be deficient under special circumstances. These are summarized in Table 6-2. The deficiency or sufficiency of nutrients in the soil is affected by the mobility of the nutrients and is strongly influenced by soil factors such as texture and pH. More nutrients become mobile as the soil becomes coarser and the pH decreases. Mobility also determines how nutrients should be applied for maximum effectiveness. Nutrients that move readily can be applied to the soil surface and watered in. Those that bind tightly to soil particles need to be incorporated into the soil or sprayed onto foliage. The

Table 6-2. Soil conditions under which nutrient element deficiencies are likely to occur

Soil	Unavailable Nutrient
High pH (alkaline)	Boron, calcium, copper, iron, manganese, phosphorus, zinc
Low pH (acid)	Boron, calcium, molybdenum, phosphorus, potassium
Coarse sandy	Boron, calcium, copper, magnesium, manganese, nitrogen, potassium, zinc
Leached (high rainfall)	Calcium, molybdenum, nitrogen
Wet (poorly drained)	Iron, manganese, nitrogen
Parent mineral material lacking	Phosphorus, calcium, boron
Serpentine (high in magnesium silicate)	Calcium, molybdenum
Organic	Copper, zinc, manganese
Low organic matter	Potassium
Sodic	Calcium
Calcareous	Copper, iron, manganese, zinc, magnesium, molybdenum

following brief discussion describes the availability of the essential elements and how a deficiency of the element affects trees.

Nitrogen. This element is nearly always in short supply for maximum tree growth. Lack of nitrogen results in small yellow foliage and generally reduced growth. Nitrogen is readily lost by leaching, volatilization, or denitrification. Other plants, notably turf, compete with trees for available soil nitrogen.

Phosphorus. Adequate levels (30 or more pounds per acre [lb/A] soil test) of phosphorus for tree growth are present in nearly all soils. However, when the soil pH is out of the range 5 to 7, phosphorus becomes increasingly unavailable to plants. Phosphorus, which is relatively immobile in soil, is involved in many essential plant processes. Root growth and development are particularly affected by phosphorus availability.

Potassium. Most soils contain enough potassium (165 + lb/A soil test) for tree growth. Deficiencies of potassium occur where soils are acid, sandy, low in organic matter, and low in cation exchange capacity.

Potassium is involved in the regulation of a number of essential plant processes. Its deficiency usually produces no distinctive symptoms other than a reduction in growth. Potassium is relatively immobile in soil.

Calcium. Levels of calcium adequate to support tree growth are found in most soils. Deficiencies occur in very acid soils, sandy soils in areas with high rainfall, serpentine soils, and alkaline or sodic soils. Calcium is important both in the cell structure of the tree and in the soil, where it affects pH and availability of other nutrients.

Magnesium. Most soils have adequate levels of magnesium to support tree growth. Magnesium may be leached out of sandy, acid soils, and in calcareous soils it is tied up in insoluble forms that are not available for root uptake. Magnesium is a constituent of chlorophyll and its deficiency results in yellowing of leaves. In most cases the most efficient method to increase the availability and retention of magnesium is to adjust soil pH and to increase the cation exchange capacity by adding organic matter.

Sulfur. Sulfur dioxide from air pollution provides most plants with adequate levels of sulfur. Sulfur is an essential constituent of enzymes and its deficiency results in a general retardation of physiological processes, often manifested as chlorosis.

Micronutrients. Most soils contain sufficient amounts of available trace elements to promote tree growth. Deficiencies, however, do occur. They are most commonly observed in trees growing in unusual soil circumstances, especially on soils outside of a pH range 6.0 to 7.0 (see Table 6-2). Under these conditions many elements are fixed into insoluble compounds and are thus unavailable for root uptake. Iron deficiency is probably the most common and conspicuous micronutrient deficiency in trees, and it occurs primarily in soils of high pH.

Less commonly, deficiencies occur in sandy soils where heavy rainfall depletes mobile ions by leaching. Boron and molybdenum are particularly prone to loss in this way. In some soils the presence of other elements may interfere in the uptake of a micronutrient. Finally, some soils may intrinsically lack particular trace elements. It is important to differentiate between soils actually deficient in an element and those where the element is present but in a form unusable by higher plants. Generally, where soil pH is causing a nutrient to be unavailable, a longer lasting solution to the problem is to adjust the soil pH by liming or adding sulfur. However, some soils are difficult to treat and pH adjustment is not always successful. Direct application of the nutrient to the soil or to tree foliage or trunk may be needed to avoid problems associated with inappropriate soil pH. It is extremely important to accurately identify the specific micronutrient deficiency before applying corrective elements, because applying the wrong material may make the problem worse.

Fertilizer Formulations

Organic Concentrates. Activated sewage sludge, tankage, blood, fish and seed meals, and bat guano are used as specialty fertilizers. In general these materials are richer in nitrogen than in phosphorous or potassium and contain a higher percentage of nitrogen than most manures, crop residues, or composts (see Table 6-3). With the exception of urea, organic nitrogen must be decomposed into forms usable by higher plants. The activity of soil microorganisms, which are decomposition agents, is dependent upon favorable soil temperature and moisture. Thus nutrients become available in flushes whenever soil microbe growing conditions are near optimum. Often these conditions also favor leaching and loss of nutrients. Research has shown conclusively that organic nutrients are not superior to inorganic nutrients for encouraging plant growth. The concentrations of nutrients, soluble salts, and trace elements in organic

Table 6-3. Average analysis of organic materials

Organic Concentrate	Nitrogen (%N)	Phosphorus (%P$_2$O$_5$)	Potassium (%K$_2$O)
Dried blood	13.0	1.5	
Fish meal	10.4	5.9	
Activated sewage sludge	6.5	3.4	0.3
Tankage	7.0	8.6	1.5
Cottonseed meal	6.5	3.0	1.5
Linseed meal	6.0	1.0	1.0
Bat guano	13.0	5.0	2.0

Source: Western fertilizer association. 1980. Western fertilizer handbook, 6th ed. Interstate Printers and Publishers, Inc., Danville, Ill. p. 132.

fertilizers vary tremendously, making optimum application rates difficult to calculate. Some organic concentrates such as sewage sludge contain heavy metals, which may accumulate in the soil to levels toxic to trees.

Bulky Organic Materials. Organic amendments most commonly used by home gardeners and horticulturists are animal manures, digested sawdust, leaf mold, peat, and compost. These materials increase the quality of mineral soils by adding organic matter, which increases aeration, water penetration, and cation exchange potential. They are, however, low-grade fertilizers that contain less than 1 to 2 percent nitrogen, phosphorus, or potassium and hence are of little benefit to soils with adequate organic material. Wood residues are low in plant nutrients and if incorporated into the soil, they can increase the need for nitrogen and phosphorus, which are needed for their decomposition.

Commercial Dry Fertilizers. The form of fertilizer most familiar to homeowners and arborists is dry inorganic fertilizer. These commercial products are either made by chemical reactions or taken from mineral deposits. They may be complete fertilizers, which provide all three primary plant nutrients—nitrogen, phosphorus, and potassium—with or without a selection of trace elements. The percentage by weight of each is indicated by the formula on the label such as 10-10-10 or 10-8-6. The first number denotes the percentage of nitrogen; the second, phosphorus (P$_2$O$_5$ or equivalent); and the third, potassium (K$_2$O or equivalent). Alternatively, they may be simple fertilizers providing only one essential element (Table 6-4). There is a broad range of essential element ratios available. Since there is also a range of percentage by weight of these elements, cost comparisons between formulations is sometimes difficult.

Commercial fertilizers may be formulated as homogeneous products or bulk blends. In homogeneous products each granule or pellet has the same analysis. Bulk blends are mixtures of two or more fertilizer materials.

Commercial inorganic fertilizers are formulated to provide plant nutrients in an immediately available form. Water to put the nutrients into solution and soil temperatures favorable for root metabolism are required for root uptake. Thus inorganic fertilizers applied when soil temperatures are too low for root metabolism may be lost by leaching before the plant is able to use them.

Table 6-4. Composition of common fertilizer sources

Compound	Formula	Nitrogen (%N)	Phosphorus ($\%P_2O_5$)	Potassium ($\%K_2O$)
Ammonium nitrate	NH_4NO_3	35.0		
Ammonium sulfate	$(NH_4)_2SO_4$	21.2		
Sodium nitrate	$NaNO_3$	16.5		
Potassium nitrate	KNO_3	13.8		46.6
Urea	H_2NCONH_2	46.6		
Natural organic matter	—	4.0		
Monoammonium phosphate	$NH_4H_2PO_4$	12.2	61.7	
Diammonium phosphate	$(NH_4)_2 + CaSO_4$	21.2	53.8	
Superphosphate	$Ca(H_2PO_4)_2 + CaSO_4$		20.2	
Triple superphosphate	$Ca(H_2PO_4)_2$		48.0	
Monopotassium phosphate	KH_2PO_4		52.2	34.6
Potassium chloride	KCl			60.0
Potassium sulfate	K_2SO_4			54.0

Source: Adapted from L. F. Rader, Jr., L. M. White, and C. W. Whittaker. 1943. Soil Science 55:201–218.

Liquid Formulations. Liquid formulations are designed to be applied to the soil in much the same way as commercial dry formulations, to be sprayed on the foliage, or to be injected in the trunk. When applied to the soil they behave in a manner similar to the dry materials.

Liquid formulations applied to the foliage or injected or implanted in the wood effectively provide nutrients to trees growing in soils where these nutrients would be unavailable because of soil factors and are not as subject to leaching as soil-applied materials. They are, however, too costly and inconvenient for general use.

Foliar sprays allow nutrient elements to be absorbed by the leaves. Response to nutrients is evident within a few days. Usually the response does not last more than one year and reapplications are needed. The effectiveness of foliar feeding is governed by environmental factors, plant age and species, and the formulation of the material. Results vary depending on circumstances.

Response to water-soluble implants or injection of liquid formulations into the trunk may be evident within a few weeks. For large trees which have inaccessible root systems and which are difficult to spray with liquid fertilizers, trunk treatments provide a good alternative. Response to trunk feeding usually continues for several years.

Controlled-Release Formulations. To combat the problem of leaching losses, fertilizers have been formulated which are able to release nutrients slowly over a long period of time. Since only a small percentage of the nutrients will be lost at any one time, more nutrients are available for plant uptake and application frequency can thus be reduced. The controlled-release formulations include three main types: (1) low-solubility coating over soluble fertilizer; (2) fertilizer of low solubility; and (3) fertilizers of low solubility which require microbial action to release nutrients.

All forms of controlled-release fertilizers provide plant nutrients at a fairly uniform level over a longer period of time from a single application than do most other formulations. Timing of application is less critical because such fertilizers are less vulnerable to leaching. The controlled-release products are expensive and their use may not be justified for ordinary landscape tree care. However, in areas of low fertility and high rainfall or where economics or inaccessibility dictates an infrequent fertilizer schedule, these materials have the advantage over other forms of fertilizer.

Treatment Area

The area to be fertilized is governed by the location of the absorptive roots. It was once thought that most tree roots occurred within the dripline, the line defined by the outer edge of the foliage. However, it is now known that tree roots commonly exist within a circle with a radius two times the crown radius, which will encompass an area well beyond the dripline. Species vary greatly in root extension. Roots of narrow or columnar trees extend well beyond twice the crown radius. Measurements of sugar maple indicate that the root area is 1 ¾ times larger than the crown area and for tuliptree it is 2½ times larger, whereas for pin oak root area is about the same as crown area. With most trees, absorptive roots lie mainly in the outer two-thirds of the circle. Trees growing in sandy well-drained soils have more extensive root systems than those in finer textured soils. Roots of large urban trees commonly occupy an entire lawn or backyard of urban lots and beyond if physical barriers are absent.

Fertilizers should be applied throughout the estimated root zone. For example, if a tree has a trunk-to-dripline branch spread of 15 feet, then the roots probably extend outward from the trunk 30 feet. The circular area around the tree to be fertilized would be approximately 2800 square feet (Table 6-5). For most homeowners having large trees the entire lawn will be treated (Fig. 6-1).

Table 6-5. Estimated fertilizer application area around trees of selected branch spread

Total branch spread (ft)	10	20	30	40	50	60
Distance (ft) from the trunk to the extended branch trips	5	10	15	20	25	30
Probable extent of root zone (distance [ft] from the trunk)	10	20	30	40	50	60
Landscape area to be fertilized around the tree ($\times 1000$ ft^2)*	0.3	1.2	2.8	5.0	7.9	11.3

*Fertilizer rates are normally given as pounds of nitrogen or other mineral element per 1000 ft^2.

Fig. 6-1. To fertilize the trees in this typical front yard, the entire lawn area should be treated.

However, do not apply fertilizers within a foot of tree trunks. The high concentration of mineral salts may damage the root collar and trunk base.

Be aware that soil surface fertilizer applications will affect turf grass. Amounts desired for trees in some circumstances can be harmful to the grass. Similarly, fertilizer applied to turf grass (Fig. 6-2) may affect tree performance. Turf growing in dense shade either should not be treated or should be treated only sparingly. If high levels of nitrogen are applied in dense shade, the grass is forced to metabolize that nitrogen, thus requiring more light energy than is available, so that grass may die. It is also best to treat the lawn beyond the root zone area to avoid a green island appearance.

Soil Application Methods

The method of application that is most efficient under a given set of conditions depends largely upon the objectives of fertilization, the soil conditions, and economics. Although many methods may work equally well, it is foolish to waste money on complicated techniques which give no better results than simple ones.

Soil Surface—Broadcast. Application of dry fertilizer to soil surface is referred to as broadcast. This method is appropriate for nutrients that move readily in the soil, notably nitrogen and most chelated micronutrients. Fertilizers may be applied using a cyclone seeder or lawn spreader. To ensure uniform

Fig. 6-2. Lawn care company applying fertilizer.

coverage, it is best to divide the amount to be applied in half and apply the first half in even swaths across the area. Then apply the second half at right angles to the first. After application, the area should be watered thoroughly to move the nutrients into the root zone. A second watering on the next day will further aid distribution of the material, especially in heavy turf.

Soil Surface—Drench. Liquid fertilizers may be incorporated in irrigation water and applied to the soil surface. This method is appropriate for nutrients that are mobile in the soil. A proportioner permits soluble fertilizer to flow accurately into irrigation lines. Fertilizer is applied over the same area as described for broadcast applications.

Liquid formulations have no special advantages over dry formulations when applied to the soil surface. Both require water to carry them to the rooting zone. Liquid formulations do not penetrate any deeper in the rooting zone than do properly watered-in dry fertilizers. They are generally more expensive than the dry formulations and when used with irrigation water there is the risk that the nutrients will be flushed to depths below the root zone.

Soil Incorporation—Drill Holes. Some nutrients, notably phosphorus and potassium, lack mobility in soil and thus are best incorporated into the soil to be available for root uptake. Also, where trees are growing in turf, it may be desirable to fertilize the tree but not the grass, so material would be placed in the soil beneath turf roots. One method to incorporate either dry or liquid formulations into soil is to place them in drill holes. Holes spaced about 2 feet apart and about 18 inches deep are made in concentric rings around the tree

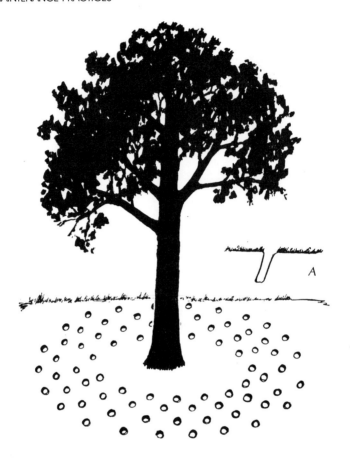

Fig. 6-3. Drill holes for soil incorporation of fertilizer should be placed concentrically at 2-foot intervals. They should not be placed within 1 foot of the tree trunk and should extend out to a distance twice the radius of the tree crown. The holes should be 18 inches deep and slanted toward the tree as shown in inset A.

throughout the rooting area (approximately 250 holes per 1000 square feet) (Fig. 6-3). They may be made with a punch bar, manually operated soil auger, or power-operated auger. The augers are preferred over the punch bar because the punch bar compacts the soil at the bottom and sides of the hole, which impedes nutrient dispersal. The holes should have 2-inch diameters. Avoid making holes within 2 feet of the trunk for trees 12 inches in diameter and within 3 feet for trees above 18 inches in diameter. This avoids damaging the root collar with high concentrations of fertilizer and severing main transport roots. The material to be applied is divided between the holes and poured into the holes through a funnel. Either liquid or dry formulations may be used. Automatic devices for drilling holes and inserting fertilizer have been developed. After application, water thoroughly. Soil plugs may be replaced before watering but if they can be left out, better aeration will result.

Studies have shown that in most cases this method of applying nitrogen provides no more tree growth stimulation than broadcast treatments. However, where the tree is growing in thick sod or compacted soil, the added aeration may be as beneficial to tree growth as the fertilizer.

A similar fertilization method involves the use of tree spikes. These are made from concentrated solid fertilizer and are driven into the ground. These spikes are more expensive than other forms of dry fertilizer. At recommended rates, they produce an uneven distribution of the immobile nutrients, e.g., phosphorus and potassium. They do have the same advantage as drill holes of enabling one to provide nitrogen fertilizer to tree roots with only a minimum to turf, and they may require less labor to apply.

Soil Incorporation—Injection. Air or water pressure may be used to inject fertilizer into the soil (Fig. 6-4). The air injection method forces dry fertilizer along with sand into the soil to a depth of 15 to 18 inches. After treatment the

Fig. 6-4. Tree feeding needle attached to a pressure sprayer may be used to apply fertilizer.

area should be well watered. The distribution of the injection sites should be the same as with the drill holes.

Liquid formulations or any soluble dry fertilizer may be injected into the soil with water. Injection sites should be 2 to 3 feet apart, about 160 sites and 200 gallons per 1000 square feet. The material should be injected to a depth of 15 to 18 inches. Application rates should not exceed 1½ gallons per hole. Liquid should be forced into the soil under moderately high pressure of 150 to 200 pounds per square inch, but not higher because this will cause air pockets to form around the root system.

Injection into soil is an efficient method of incorporating fertilizer into the soil. The equipment may be used for other purposes and the method is less tedious than drilling holes and funnelling dry fertilizer. However, water-soluble forms of phosphorus and potassium are more expensive than the less soluble compounds. Also, fertilizer solutions are notoriously corrosive to equipment. On balance, the benefit of ease of application probably outweighs the disadvantages.

Again, for optimum tree growth, the injection methods have generally not been shown to be superior to surface broadcast of nitrogen fertilizer. Soil injection does provide improvement of aeration and water penetration in compacted soils.

Soil Incorporation—Trenching. The trench method was developed in an effort to place nutrients nearer to the root system. A trench about 2 feet wide is dug in a circle beneath the outer spread of the branches. This area is then filled with rich composted soil to which fertilizers have been added. This method has three obvious disadvantages: (1) the soil is enriched in only a small part of the total area covered by the roots; (2) many roots are injured when the trench is dug; and (3) large amounts of sod must be lifted and replaced where trees are growing in lawns.

Soil-Applied Fertilizer Rates

Fertilizer rates are based on the number of pounds or ounces needed per 1000 square feet of root area. Whether wet or dry, watered in from the surface or incorporated, the amount of fertilizer to use can be calculated from the estimated root area (Table 6-5). For large trees, assume the area to be the entire part of the yard where the tree is located.

The amount of any fertilizer to apply largely depends on the objective of fertilization, the soil type, formulation of the fertilizer, and tree species. The following section describes appropriate rates under most circumstances and conditions under which greater or lesser amounts should be given. However, these rates are general guidelines only; advice from local arborists or cooperative extension personnel should be sought to determine appropriate rates for local conditions.

Nitrogen. There are numerous forms of nitrogen fertilizer, but to be usable by higher plants, all forms must be converted to nitrate (NO_3), ammonium (NH_4), or urea.

Urea is an organic form of nitrogen that may be absorbed directly by roots and is soluble in water. Two different synthetic organic nitrogen sources containing urea are urea formaldehyde (UF) and isobutylidene diurea (IBDU). These are mixtures of bound and free urea, making some urea available immediately while the rest is made available over a period of time. After urea is watered into the soil, it is converted by microorganisms first into ammonium and then into nitrates. Although urea can easily be leached, ammonium cannot, so some of the nitrogen applied as urea will remain in the soil even if heavy rain follows application.

Various formulas for computing the ratio of nitrogen fertilizer needed in relation to trunk diameter have been devised. However, in landscapes where trees of different sizes are to be treated at the same time, it is simpler to apply fertilizer per unit area over the entire range of tree root extension. Thus trees with larger root systems will receive larger amounts of nitrogen.

For most soils, the typical moderate rate of shade tree fertilization is 2 to 4 pounds of nitrogen per 1000 square feet (87–174 lb/A) per year. Experimentally, trees have been shown to respond to rates from 1 to 12 pounds, but response is not proportionately greater when more than 6 pounds of nitrogen is applied. In some circumstances, rates over 6 pounds can be injurious. Soils with high levels of organic matter need less nitrogen, whereas sandy soils, especially in areas with high rainfall, require greater amounts. Where turf is present, the total yearly amount of fertilizer should be calculated to include the amount that is applied for the turf grass. If the lawn is already being fertilized, it is unlikely that the tree will need much more fertilizer. If trees are being fertilized at a high rate to correct a specific problem or to promote rapid growth, the amount of fertilizer used should be divided into two or more applications to avoid grass burn. Evergreens require lower levels of nitrogen than deciduous trees. Overfertilization leads to open growth with widely spaced branches, and thus may ruin the typical crown shape of conifers. Overfertilization of deciduous trees may increase foliage density to the point where interior foliage may be weakened by heavy shade.

The number of pounds of fertilizer required to get a specific rate of nitrogen can be calculated easily. For example, urea is 45 percent nitrogen; to get 2 pounds of nitrogen one would need to apply 4.4 pounds of urea (2 divided by 0.45 = 4.4). Ammonium sulfate is 33.5 percent nitrogen and sodium nitrate is 16 percent nitrogen. If one is using a complete N-P-K fertilizer, the percentage of nitrogen is given by the first number of the analysis.

Table 6-6 points out the consequences of different levels of annual fertilization on turf grass and woody plants in a typical landscape. It is important to determine the fertilization objectives before deciding how much nitrogen to use.

Phosphorus and Potassium. These elements are rarely in short supply and overuse of fertilizers containing phosphorus or potassium may lead to toxicity symptoms on plants and to water pollution. Therefore, it is important to use these fertilizers only when needed and in the right amounts. Soil tests showing phosphorus levels below 30 pounds per acre or potassium levels below 165

Table 6-6. Consequences of applying different annual levels of nitrogen to a typical urban landscape

Annual Nitrogen Application ($lb/1,000\ ft^2$)	Effect on Turf Grass	Effect on Woody Plants
0	Quality and growth minimum. Weeds may become dominant. Tall fescue more tolerant of low fertility.	Mature healthy plantings will continue moderate growth. Immature woody plants will not make rapid growth.
2	Good quality and growth. Optimum for most turf grass sites that cannot be irrigated.	Mature healthy plantings will make good growth. Little effect on attempts to "rescue" declining plants. May stimulate some growth of young plants.
4	Lush, high-maintenance turf grass. May be detrimental if timing and culture are mismanaged or heavy shade exists.	May push unneeded growth for mature plants. When nitrogen fertilizer is needed to reverse a decline this rate may help. Young plants can more rapidly attain size at this rate.
6	Problematic for bluegrass and fescue lawns. Thatch accumulates in bluegrass. Excessive growth is succulent and susceptible to disease. Root system tends to be shallow, thereby increasing drought susceptibility. Excessive top growth is at the expense of good root development. Without irrigation during dry periods, bluegrass lawn will die. Unless applied in three equal doses at three separate intervals (minimum of 6 weeks apart), or unless applied during cold weather, lawn will be burned with excess fertilizer.	Will cause shoots to lengthen considerably and the succulent growth may be more susceptible to disease. Difficult to know how the woody plant actually uses the available nitrogen in root vs. shoot growth. When extension growth is important (e.g., small landscape tree needing maximum extension growth to provide shade for house), this amount of fertilizer may be warranted. Regular irrigation is a must when following this program.

Source: Publication ID-72, Principles of home landscape fertilization, University of Kentucky Cooperative Extension Service, University of Kentucky, Lexington, Ky. 40546.

pounds per acre indicate these elements are in short supply. Application of phosphorus is also warranted when foliar analysis indicates that the current season's foliage is below 0.1 percent phosphorus. Foliar analysis will not give reliable indications for potassium because it is very mobile in the plant. Phosphorus and potassium are largely immobile in soils; hence these fertilizers should be incorporated into the soil rather than applied to the surface.

There are various forms of potassium fertilizer. Potassium chloride (KCl) may be used where salinity is not a problem, but chloride may be toxic if used

Table 6-7. Rates of phosphorus and potassium fertilizer needed to correct deficient soils

Soil Test Results (lb/A)	Fertilizer Needed (lb/1,000 ft²)		
Phosphorus	15%P	20%P	46%P
0–9	52	34	15
10–19	34	23	10
20–29	17	11	5
30+	0	0	0
Potassium	15%K	60%K	
0–99	52	14	
100–149	34	11	
150–199	17	10	
200+	0	0	

Source: After E. M. Smith and C. H. Gilliam. 1980. How and when to fertilize field-grown nursery stock. Am. Nurseryman 151(2):8, 68–69, 72–73.

in arid regions. Potassium nitrate (KNO_3) may cause nitrogen excess problems if it is used to supply large amounts of potassium. Potassium sulfate (K_2SO_4) is the preferred form for most uses. Wood ashes are a traditional source of potassium, but they can raise the pH of alkaline soils to undesirable levels.

Table 6-7 gives the amount of fertilizer of various percentages of phosphorus and potassium needed to correct deficiencies. These rates are for middle-range soil textures. Sandy loam soils require about 50 percent less and clay loam soils need about 50 percent more.

Other Essential Elements. Materials containing the other essential elements should be applied only if foliar analysis and/or symptoms reveal a deficiency. If needed, soil-applied nutrients should be applied as follows:

Calcium: Apply according to soil test. If deficiency exists, up to 150 pounds of gypsum ($CaSO_4$) or 30 pounds of sulfur may be needed per 1000 square feet on alkaline soils. Apply lime (Tables 3-1 and 3-2) if soil is below pH 6.0. Calcium is relatively immobile in soil and should be incorporated into the soil to be effective. Treatment should be made prior to planting and the material should be incorporated to a depth of 18 inches. If treatment is made after planting, then materials should be worked into the top 4 to 6 inches of soil.

Magnesium: Apply according to soil test. If deficiency exists, up to 50 pounds of Epsom salts ($MgSO_4 \cdot 7H_2O$) per 1000 square feet may be needed. Apply dolomitic limestone (Tables 3-1 and 3-2) to acid soils.

Sulfur: Present in many fertilizers, additional sulfur is probably not needed, but it can be applied as gypsum at 15 pounds per 1000 square feet.

Iron, Manganese, Zinc: Chelated forms may be applied to soils at 0.5 to 4 pounds per 1000 square feet. Consult product labels for exact rates. Sulfates of iron or manganese can also be used at rates ranging from 4 to 30 pounds per 1000 square feet, depending on need.

Boron: Borax may be used at the rate of 1 pound per 1000 square feet.

Copper: Copper sulfate, applied at 3 pounds per 1000 square feet, is normally sufficient.

Molybdenum: Sodium or ammonium molybdate may be applied at 4 to 40 pounds per 1000 square feet.

Arborists should be sure of the identity of the specific problem before applying nutrients. In addition, whenever possible, adjustment of soil pH is preferable to and usually more effective over time than application of specific mineral elements.

Foliage or Dormant Twig Spray

Application to aerial parts of the tree is an efficient way to provide some micronutrients. Soil treatments will not supply materials to roots where soil conditions render the nutrients immobile.

Dilute solutions are sprayed on the foliage until they just begin to drip off. Coverage needs to be uniform to correct deficiencies of elements that are not readily transported between plant parts. Nitrogen, potassium, phosphorus, chlorine, and sulfur are very mobile within the plant. Zinc, copper, manganese, iron, and molybdenum are only partially mobile. Calcium and magnesium are largely immobile. The presence of a surfactant in the mixture will increase the effectiveness of most sprays. The amount of surfactant used is important because too much will reduce absorption and may cause burning of leaves. Consult the label for recommended amounts.

Materials penetrate the leaf cuticle and epidermal walls by a diffusive process which is influenced by temperature and the concentration of the spray. They are absorbed onto the surface of the plasma membranes and pass through the membranes to enter the cytoplasm via active pathways. Thus environmental factors which influence the rate of plant metabolism will also influence the efficiency of absorption. In general, high temperatures (above 85° F) and low humidity reduce absorption, and young leaves absorb more readily than old leaves.

The chemical form of nutrients may also influence absorption. Chelated forms of micronutrients are generally absorbed well. However, different chemical forms of phosphorus give dramatically different test results. In descending order of efficiency of absorption the forms of phosphorus are H_3PO_4, K_2HPO_4, NaH_2PO_4, KH_2PO_4, $Ca(H_2PO_4)_2$. Sprays of 2600 parts of nitrogen per million parts of water in the form of urea, calcium nitrate, and ammonium sulfate all give the same results, which implies that the nitrogen source does not influence absorption. The addition of urea to spray solutions increases the effectiveness of foliar sprays containing phosphorus, manganese, sulfur, magnesium, and iron.

Chelated forms of zinc and zinc sulfate may be applied to dormant twigs to correct zinc deficiency. Because zinc is only partially mobile in plants, coverage must be very thorough. Dilute formulations are sprayed until the material just begins to drip off.

Spray treatments are effective for temporarily correcting nutrient deficiencies. Improvement in foliar color is usually evident within a few days. Thus sprays can be used to assist in diagnosing mineral deficiency problems. Spray treatments must be repeated annually. They are best used for correcting micronutrient deficiencies where soil conditions preclude efficient soil treatments.

For nutrients such as nitrogen, which are required in larger amounts foliar applications are inconvenient, messy, and expensive. Uniform coverage of mature trees is nearly impossible. Fertilizer applications to foliage can cause leaf or fruit burn under high termperature; if the solution is too concentrated, they leave a whitish residue on foliage. Although many chelated forms can be combined with routine pesticide sprays, urea is incompatible with some pesticides. For major nutrients this method of application should be considered as a temporary measure that supplements soil treatment. Rates for solutions of selected mineral nutrients applied as sprays to tree foliage follow:

Magnesium: Use 3.3 ounces (19 teaspoons) of Epsom salts ($MgSO_4 \cdot 7H_2O$) per gallon of water. Magnesium sulfate is incompatible with some pesticides.
Boron: Use 1 to 2 teaspoons of boric acid per gallon of water.
Copper: Use 3 to 6 teaspoons ($\frac{1}{2}$ to 1 oz) of copper sulfate ($CuSO_4 \cdot 5H_2O$) per gallon of water.
Iron: Use 4 teaspoons of ferric sulfate ($FeSO_4 \cdot H_2O$) per gallon of water.
Manganese: Use 2 to 8 teaspoons of manganous sulfate ($MnSO_4 \cdot 2H_2O$) per gallon of water.
Zinc: Use 4 to 6 teaspoons of zinc sulfate ($ZnSO_4 \cdot 7H_2O$) per gallon of water.
Molybdenum: Use $\frac{1}{10}$ to $\frac{1}{2}$ teaspoon of sodium molybdate ($NaMoO_4 \cdot 2H_2O$) or ammonium molybdate [$(NH_4)_2MoO_4 \cdot 2H_2O$] per gallon of water.

Trunk Implants and Injections

Injection or implants in the stem or trunk can provide a rapid response and are especially useful where other methods fail or where roots are inaccessible. The process creates tree wounds and should not be considered as a first alternative. These methods are appropriate only for application of micronutrients.

Implants are made by placing gelatin capsules containing small amounts of soluble nutrients into holes drilled into trunks or branches (Fig. 6-5). The holes are made in a spiral pattern around the trunk. Their number depends on the size of the tree (see Table 6-8), and their diameter depends on the size of capsule used, in most cases $\frac{3}{8}$ to $\frac{1}{2}$ inch. The holes should be drilled deep enough to seat the entire capsule plus a plug entirely in the wood. Do not leave an air pocket around the capsule. Some capsules are manufactured to seal the hole, so do not require a plug. However, in most cases a good-fitting plug

Fig. 6-5. Trunk implant used for correcting mineral deficiency.

placed with its surface flush with the cambium layer will promote wound covering and exclude most rot organisms. The holes are usually covered by tree callus within 6 months.

Material may also be implanted in trees without gelatin capsules. Ferric citrate (an iron chelate) has been used to correct iron chlorosis in this manner. A $\frac{7}{16}$-inch spiral wood bit is used to drill holes into the wood to a depth of $1\frac{1}{2}$ inches. A plastic syringe with the tip removed injects $1\frac{1}{2}$ grams of powdered ferric citrate into the hole. The hole is filled with a tight-fitting cork. The placement and number of holes is the same as with gelatin implants.

Soluble forms of micronutrients may be injected in solution into the trunk either by gravity feed or under pressure. Holes are drilled in the trunk root flare region using a standard high-speed drill bit. Sawdust must be cleaned from the holes before they are fitted with lag screw injection heads. It usually takes two or three turns of the screw in the wood to seal the hole and hold the screw in place.

Table 6-8. Trunk implants

Trunk Diameter (in.)	Capsule Size	Distance between Holes (in.)	Number of Capsules per Tree
1–4	No. 2	2	2–6
4–12	No. 000	3	4–12
>12	$\frac{1}{8}$ oz	4	>9

Source: After D. Neely. 1976. Iron deficiency chlorosis of shade trees. J. Arboric. 2(7):128–130.

Table 6-9. Low-pressure injections of iron sulfate

Tree Diameter (in.)	Number of Holes per Tree	Amount of FeSO$_4$ (oz)	Number of Refills
1–4	4	0.5	1
5–10	4–5	2.0	1
10–15	5–6	4.5	1–2
15–20	6–7	5.5	2–3
20–25	7–8	6.5	3–4
25–30	8–9	6.5	4–5
30–35	9–10	7.5	5–6

Source: J. W. Fischbach and B. Webster. 1982. New method of injecting iron into pin oaks. J. Arboric. 8(9):240.

With the gravity feed system the holes should be 2 inches deep. The soluble material is placed in an elevated reservoir filled with about ½ gallon of water. It is allowed to trickle into the hole. After the reservoir is nearly empty, plain water should be added to flush the residue into the tree. It takes about 24 hours to treat a tree in this manner. The number of holes needed depends on the size of the tree; holes should be 3 inches apart for trees less than 4 inches in diameter and 8 to 10 inches apart for larger trees (see Table 6-9).

It takes much less time to inject the solution into the tree when the solution is under pressure. Using pressures of 100 to 200 pounds per square inch, 1 quart can be injected in from ½ to 5 minutes, depending on the tree species and time of year. Holes are drilled ⅗ to 1½ inches deep; if deeper, the injection time is increased. Injection holes should be placed direcly under scaffold branches at or just above the root flare. For trees over 16 inches in diameter, holes should be made every 6 inches around the trunk. For most trees, 1 to 2 quarts of solution can be injected. After material is injected, leave the hole exposed; apply no sealants. Many species have been successfully treated in this manner; however, butternut, shagbark hickory, black cherry, white ash, maples, firs, and pines may not accept injection solutions readily.

Even smaller volumes of liquid having greater nutrient concentrations can be injected into trees under pressure using a plastic capsule and feeder tube system (Fig. 6-6). In this case, small-diameter holes are drilled into the root flare area, a thin plastic tube is inserted into each hole, and a pressurized plastic capsule (Mauget system) containing the nutrient solution is impaled on the tube. The punctured plastic capsule delivers a small volume of solution into the tree.

Injection rates are higher in trees after new leaves expand than during dormancy or early in the growth period. Presumably negative xylem pressure potential associated with transpiration is responsible for this advantage. Likewise, injection rates are faster from midmorning to early evening and with mild wind. Root injection produces more uniform response in the top of the tree but is often inconvenient.

The injection and implant methods require holes to be drilled in the trunk. These holes, aside from being unsightly, may be ports of entry for rot organ-

Fig. 6-6. Mauget injection.

isms. Even when rot organisms do not invade, the wood is often discolored. The injured tissue associated with all wounds is limited by the tree to the wood present at the time of injection (see Chapter 7). Thus dieback, staining, or infection is usually confined to a small column. However, retreatment may be required every 3 to 5 years, so considerable damage could be caused by injection over the life of a tree. When injecting or implanting, (1) keep the holes as small as possible; (2) drill as low on the tree as possible; (3) avoid vertical alignment of holes from previous treatments; (4) remove external fixtures as soon after treatment as possible; and (5) implant or inject only those substances that have been proven to be effective and noninjurious to the tree. Damage to the tree will be minimized if these precautions are followed.

Implants and injections have been used successfully to treat deficiencies in micronutrients, most notably iron and zinc. Response to treatment is usually evident within 3 weeks and retreatment is required only every 3 to 5 years. Some researchers have found that injections produce more consistent response than implants. Injection of dilute zinc sulfate, for example, was judged superior in treating zinc deficiency over implants of chelated zinc cartridges. The cartridges, when used according to manufacturers' directions, delivered too little zinc to overcome symptoms. However, implants require less equipment and can be successfully made by less highly skilled workers. Labels and instructions of implant and injection products should be studied and understood before applications are made.

Recommendations for Newly Transplanted Trees

Historically, if trees were fertilized at all, they were treated with a complete fertilizer added to the backfill when transplanted. Researchers have recently found, however, that for most normal soils newly transplanted trees do not respond to such fertilizer until several years after moving. High levels of nitrogen may have a negative effect on root growth, thus prolonging transplant shock. Phosphorus has a slightly stimulatory effect on root growth.

Once established, trees respond significantly to nitrogen except in very fertile soils. In one study, fertilization had little effect on tree growth until the third year after transplanting. Then the trees responded with vigorous growth to additions of nitrogen but not to phosphorus or potassium. In most soils, nitrogen is the only element that is chronically in short supply.

Trees newly planted in sandy soil where rainfall is high may require additions of nitrogen before establishment. Slow-release formulations added to the planting hole and then covered with several inches of soil will provide a continuous supply of nutrients for up to 18 months. Slow-release forms should also be used where inaccessibility or expense necessitates an infrequent application schedule.

Timing of Tree Fertilization

Applications of fertilizer should be timed so that nutrients are available for periods of rapid growth. Some trees, for instance, spruce and fir, have a single flush of growth in the spring. Many more—yew, holly, and most shade trees, for example—have two or more periods of growth, the most important in the spring and another in midsummer. A few, such as juniper, grow continuously. For spring growth, fertilizer applications made to the soil should be made in the late winter or early spring 4 to 6 weeks before budbreak. Foliar sprays should be made after leaves begin to expand, whereas trunk implants and injections are most successful when done after leaves are fully expanded. For later growth flushes, fertilizers may be applied after the spring flush but no later than the first of July. Applications made in late summer or early fall may stimulate new growth in some tree species so late in the season that hardening off will not occur before the autumn freeze. Such late growth would be very susceptible to damage by cold weather. Note that fertilizing lawns even at some distance from trees could cause some trees to produce late season growth. Fertilizers may be applied to the soil after the top of the tree is fully dormant in the fall. Roots will continue to absorb nutrients until soil temperatures fall below 40° F.

Although fertilizers may be applied in the early spring, early summer, and late fall, researchers have shown that for nitrogen trees benefited most from the early spring application. Many horticulturists recommend fertilizing in the later fall. However, nutrients such as nitrogen, if not absorbed by the roots at that time, can easily be lost by leaching and thus may not be available for spring growth. In sandy soils where summer rainfall is high, a second midsummer application is beneficial.

The frequency of application depends on the kind of tree, the objective of fertilization, and other factors. It may vary from an annual treatment to one made every 3 or 4 years. Most trees will respond to annual applications of nitrogen. Trees being treated for micronutrient deficiency using foliar sprays need annual applications. Where trunk injections or implants are used, retreatment should be made when symptoms of deficiency reappear. For phosphorus and potassium, soil tests are needed to time the frequency with accuracy. It is possible to kill trees with too much fertilizer, so it is better to wait for symptoms of deficiency for nutrients other than nitrogen to appear than to apply fertilizers indiscriminately.

Selected Bibliography

Hamilton, D. F., M. E. C. Graca, and S. D. Verkade. 1981. Critical effects of fertility on root and shoot growth of selected landscape plants. J. Arboric. 7(11):281–90.

Himelick, E. B., and K. J. Himelick. 1980. Systemic treatment for chlorotic trees. J. Arboric. 6(7):192–96.

Maynard, D. N., and O. A. Lorenz. 1979. Controlled-release fertilizers for horticultural crops. Hortic. Rev. 1:79–140.

Neely, D. 1976. Iron deficiency chlorosis of shade trees. J. Arboric. 2(7):128–30.

Neely, D. 1980. Trunk and soil chlorosis treatments of pin oak. J. Arboric. 6(11):298–99.

Neely, D., E. B. Himelick, and W. R. Crowley, Jr. 1970. Fertilization of established trees: A report of field studies. Illinois Nat. Hist. Surv. Bull. 30(4):235–66.

Reil, W. O. 1979. Pressure-injecting chemicals into trees. Calif. Agric. 33(6):16–19.

Swietlik, D., and M. Faust. 1984. Foliar nutrition of fruit crops. Hortic. Rev. 6:287–355.

Van de Werken, H. 1984. Fertilization practices as they influence growth rate of young shade trees. J. Environ. Hortic. 2(2):64–69.

7

Pruning

The pruning of trees is probably the most noticeable and important of all tree maintenance practices. Thoughtful pruning produces a structurally sound tree which can better withstand adverse environmental conditions. In addition, properly pruned trees require less cabling, bracing, and sometimes less managing of pests to maintain good health. Appropriate pruning will add to the aesthetics and prolong the useful life of trees.

Need for Pruning

Trees are pruned principally to preserve their health and appearance and to prevent damage to human life and property.

Pruning for Health. Broken, dead, or diseased branches are pruned to prevent pathogenic organisms from penetrating into adjacent parts of the tree and to reduce inoculum for spread to other trees. Indeed pruning is an important control measure for fire blight, black knot, several twig blights, some cankers, and dwarf mistletoe. It can be of some benefit in controlling Dutch elm disease and crown gall.

Live branches are removed to permit penetration of sunlight and circulation of air through the canopy. An open canopy makes a less favorable site for powdery mildew, favors formation of flower buds, and allows pesticide sprays to cover leaves more evenly.

Proper pruning of the tree crown can reduce wind resistance and help prevent breakage. Branches that form an acute angle of attachment are removed because they are especially prone to breakage. When branches having narrow crotch angles enlarge, stress on the crotch increases. Also, acutely angled branches often have bark embedded in the branch attachment, which causes a weak joint. These weak attachments are prone to splitting and breaking out during ice storms or high winds. Thus pruning can increase the structural stability of trees.

Pruning for Appearance. For centuries trees in formal gardens have been carefully pruned to produce desired—often fantastic—shapes. In most landscapes, however, pruning is used to maintain or restore the characteristic crown form. Occasionally it is desirable to keep normally large-growing shade or ornamental trees within restricted bounds of small properties. Judicious pruning of new growth during summer will produce the desired dwarfing effect with no harm to the tree. The Japanese art of bonsai is the extreme of this practice. Pruning influences the balance between vegetative growth and flower bud production. In some species, this influence can be used to invigorate a declining tree or to produce fewer but larger flowers and fruit.

Pruning for Safety. Dead, split, and broken branches are a constant hazard to human life and property. Danger from falling limbs is always greatest in trees along city streets and in public parks. Low-hanging live branches must be removed to a height of 10 to 12 feet when they interfere with pedestrian and vehicular traffic. Branches that obscure clear vision of warning signs or of traffic must also be removed.

Trees are pruned to prevent them from interfering with energized lines. Branches touching lines can interrupt service and wind-thrown limbs can knock down electrical and telephone lines.

Response of Trees to Pruning

Effect on Tree Shape. The growth habit of a woody plant is largely genetically determined. The characteristic shape is influenced by the location of leaf and flower buds, the pattern of budbreak along the branches, and differential elongation of branches in various positions on the stem. These factors produce three basic tree forms: columnar, excurrent, and decurrent (Fig. 7-1).

In the technical sense the term columnar shape is used to describe trees whose terminal growth is concentrated in a single growing point, for example, single-trunk palm trees. Columnar is also used to describe growth shown by deciduous trees whose branches are parallel and erect, with tree shape tapering to a point, for example, Lombardy poplar.

Most conifers and some deciduous trees have an excurrent form. In these trees the terminal bud exerts weak apical dominance over the lateral buds. Thus various numbers of lateral buds will grow in the same season as the shoots on which they are formed. The terminal shoot, however, is usually able to elongate more than the lateral shoots, so a conical shape develops.

The decurrent form is produced by most shade trees. Terminal buds exert strong apical dominance so few or no lateral buds develop in the year that the shoot on which they grow is produced. However, in the following year the lateral buds are released and may outgrow the terminal shoot. Thus the trees become round-headed with many shoots elongating to a similar degree.

Pruning interferes with the hormonal dominance system of trees and thus has a significant influence on tree shape even beyond the simple removal of wood. Although the one-time removal of the terminal bud cluster on an excurrent tree will not change it to a decurrent form, repeated removals will alter its growth habit. The shoots from buds nearest to the point of removal will elongate to a greater degree than shoots farther down, giving the tree a bushy appearance. Similarly, the removal of the terminal bud cluster on a shoot with strong apical

Fig. 7-1. Crown shapes: A, columnar (palm); B, excurrent (fir); C, decurrent (zelkova).

dominance will allow one or more buds just below the cut to grow. However, buds farther down the shoot will still be repressed. Consequently, the new branches will be concentrated right below the cut.

The removal of a lateral branch with an acute angle of attachment from a young leader branch will often force an accessory bud to grow from the same node. The resulting branch is usually attached at a wider angle to the leader than was the original branch.

When a lateral is removed at its point of origin or the length of a branch is reduced by cutting just above a lateral which is large enough to assume the terminal role, new growth will be distributed over the entire remaining branch. The bushy regrowth that follows removal of terminal buds will be avoided and a more natural crown shape will be retained.

Effect on Growth. Lush regrowth on young trees following pruning is a familiar sight. Since the same root system is providing water and minerals to fewer branches, the remaining shoots are able to grow vigorously. This gives rise to the notion that pruning is invigorating. However, when a substantial portion of the top is removed from young trees, less food is produced by the trees for root and future top growth. If the tree's carbohydrate reserves are already depleted, heavy pruning could jeopardize tree health. Severe pruning, especially of young trees, while invigorating to individual branches may be dwarfing to the entire tree.

Moderate pruning of mature trees is generally invigorating. Trees that produce fruit on one-year-old wood show marked shoot growth increase when pruned during dormancy. The pruning removes potential fruit, allowing more of the tree's energy to be channeled into foliage and shoot production. Crown thinning allows more light to penetrate to interior and lower leaves. Experiments have shown that the photosynthesis rate within the crown of thinned apple trees is four times higher than that of unthinned trees. The increased photosynthesis rate of these leaves makes up for removed foliage and the remaining branches are able to respond vigorously following pruning.

Effect on Flowers and Fruit. Pruning influences the number, size, and quality of flowers and fruit. Dormant pruning of trees such as apples, which produce fruit on one-year-old wood, reduces the number of flowers and fruit, but those remaining grow larger. Thinning of these trees designed to increase light penetration to the inner crown will increase flower bud formation and hence the crop in the following year.

Dormant pruning of trees such as goldenrain-tree, which produce flowers on current season's growth, increases the number and size of flowers and fruit on the remaining branches.

Wound Reactions. Pruning produces wounds; thus the tree becomes vulnerable to attack by wood-rotting fungi. However, trees are not passive and are able to marshal both chemical and physical defenses. The work of Dr. Alex Shigo and his colleagues at the U.S. Forest Service has led to a better understanding of these processes. According to Dr. Shigo chemical boundaries are formed in the tissues present at the time of injury. Food reserves in sapwood are converted to resinous materials and phenols, which discolor the wood and cause it to become more decay resistant. This localized resistance keeps the decay from spreading internally. Vessels are also blocked, which slows vertical spread. The cambium responds by laying down a thin layer of unique cells called a barrier zone. This zone separates the wood formed before wounding from that formed after wounding. The zone acts as a barrier to protect the new wood and restricts the development of decay organisms to the tissue formed before wounding. Callus tissue forms at the margin of the injury and may eventually close the wound, but trees do not heal in the conventional sense of replacement of damaged tissue. The pruning wound will remain embedded in the tree as new tissue grows over it.

The success of the response in preventing decay depends on the vigor of the tree and the virulence of the invading organisms. The genetic constitution of individual trees may enable them to marshal their defenses more efficiently. In many cases the defenses are breached by the invaders and new barrier zones are formed.

Types of Pruning Schemes

The two basic pruning schemes are defined in terms of the tree's reaction: thinning out and heading back.

Thinning Out. There are situations where reducing the tree crown is desirable and necessary. The best way to accomplish this is "thinning out," in which entire lateral branches are removed to their point of origin. Alternatively, the length of a branch is reduced by cutting just above a lateral that is large enough to assume the terminal role, a practice also known as drop crotching. Thinning out reduces the tree's height and spread while retaining its natural shape (Fig. 7-2). Only selected portions of the tree's canopy are removed,

Fig. 7-2. Before **(left)** and after **(right)** thinning out.

Fig. 7-3. Tree after topping.

reducing the possibility of sunscald damage. Pruning cuts are made close to the trunk, leaving only the collar of the removed branch instead of stubs. The pruning cuts are less conspicuous than those left from heading back and they "close" more rapidly and completely. Regrowth tends to be scattered naturally along the length of the remaining branches. Thinning out requires greater skill and time than heading back, but in most situations it is well worth it.

Heading Back. This pruning scheme may involve pinching, tip pruning, shearing, pollarding, dehorning, stubbing, and topping. Pinching and tip pruning are used for training young trees. Cuts are made so that buds or branches below the cut are encouraged to grow in a desired direction. Shearing, or heading back, of young shoot tips of needle evergreens such as hemlocks creates bushy regrowth and a denser canopy. These are examples of desirable heading back cuts.

Topping, stubbing, and dehorning are examples of drastic removal or cutting back of large branches in mature trees (Fig. 7-3). Major branches are cut so that stubs several feet long are left without any important side branches attached to them. Topping is not desirable and usually harms the tree.

Many homeowners have their trees topped often by so-called professionals when their trees have reached heights which they consider unsafe. They fear a strong wind might blow over these large trees. This fear is largely unjustified. The extensive root system of a healthy tree, if left relatively undisturbed, provides adequate support for the tree. An old healthy tree with a good root system is actually less likely to blow over than a smaller tree with its smaller, less developed root system.

Some homeowners believe that the stimulation of new growth associated with topping is actually beneficial to the tree. Although the tree appears rejuvenated with new foliage and branches, this only masks the real damage topping does to the tree.

Trees may also be topped to remove potentially hazardous dead and diseased branches which may break off during ice storms or windstorms. Unfortunately, topping removes healthy as well as unhealthy limbs. The hazardous limbs are best removed by selective pruning instead of topping.

Large, mature trees are often topped to prevent interference with overhead utility wires. They are also topped when they block views, interfere with buildings or other trees, or shade solar collectors or other areas (e.g., lawns and gardens) where sunlight is wanted. In some of these situations, removing large limbs may be necessary; however, alternatives such as proper early training, selective thinning out of branches and limbs, or whole tree removal should be considered and adopted where feasible.

Removing much of the tree canopy upsets the crown-to-root ratio and seriously affects the tree's food supply. A 20-year-old tree has developed 20 years' worth of leaf surface area. This leaf surface is needed to manufacture sufficient food to feed and support 20 years' worth of branches, trunks, and roots. Topping not only cuts off a major portion of the tree's food-making potential; it also severely depletes the tree's stored reserves. It is an open invitation for the tree's slow starvation.

Removing the tree's normal canopy suddenly exposes the bark to the sun's direct rays, often scalding the newly exposed outer bark.

Topping removes all the existing buds that would ordinarily produce normal sturdy branches.

Large branch stubs left from topping seldom close or callus. Nutrients are no longer transported to the large stubs and that part of the tree becomes unable to seal off the injury. This leaves the stubs vulnerable to insect invasion and fungal decay. Once decay has begun in a branch stub, it may spread into the main trunk, ultimately killing the tree. Fruiting bodies of decay fungi are often visible on the bark of decaying trees.

Topping stimulates the regrowth of dense upright branches just below the pruning cut (Fig. 7-4). These new shoots, referred to as "suckers" or "water sprouts," are not as structurally sound as are the naturally occurring branches. The water sprouts often consist of succulent growth which is more susceptible to diseases (such as fire blight of rosaceous trees) and herbivorous insects (such as aphids and caterpillars).

Since water sprout regrowth is generally rapid and vigorous, a topped tree often will grow back to its original height faster and denser than a tree that has been properly pruned or thinned. This makes topping, at best, only a temporary solution to oversized trees. Some tree species (e.g., sugar maple, oak, and beech) do not readily produce water sprouts. Without the resulting foliage, a

Fig. 7-4. Prolific sprouting follows topping.

bare trunk results and the tree quickly dies. Deteriorating branch stubs, along with weak sucker growth, make topped trees highly vulnerable to wind and ice damage.

From an aesthetic aspect, topping disfigures the tree. Unsightly branch stubs, conspicuous pruning cuts, and a broomlike branch growth replace its natural beauty and form. The practice of topping probably started in areas of Europe where people did not want shade trees, but trees that would allow the sun to shine on their gardens. Species used there such as London planetree and linden can survive many such toppings. Since a good selection of low-growing landscape trees is available now, there is no need for constant pruning to maintain low growth. Topped trees are unnatural substitutes for shade trees meant to offer several lifetimes of beauty and enjoyment.

When to Prune

Although circumstances beyond the control of arborists often dictate the timing of tree maintenance procedures, there are optimum times to prune to achieve desired effects. Jobs such as removal of safety hazards, dead branches, or diseased branches, however, can be done anytime.

Large cuts should be made in the late winter or early spring before bud break when hydrostatic pressure is greater than atmospheric pressure, i.e., sap pressure is positive. A wound made at that time will cause the tree to "bleed." Although some believe the bleeding to be unsightly, it has a positive benefit in reducing the probability of colonization of the pruning cut by decay organisms. In the spring the rate of callus production is more rapid, further reducing the chance of invasion. Other acceptable pruning times include winter and midsummer.

Spring pruning after budbreak is undesirable because during this time the bark is at its tenderest and damage to the bark is most likely. In addition, the food reserves of the tree are being directed toward new growth, leaving less energy available for wound repair. Pruning in the late summer is also undesirable because it interferes with food storage necessary for growth in the following spring and slows root production. In addition, a number of important decay fungi have been observed to produce their largest numbers of spores during the fall; this is another factor favoring the early spring for major structural reshaping.

Pruning to produce a desired shape or to correct a misshapen crown is most readily accomplished during the growing season. Also, dead or diseased branches are more conspicuous then.

Equipment

Adequate equipment is essential for good pruning (Fig. 7-5). Equipment must be maintained in first-class condition both for the safety of the operator and for the good of the tree.

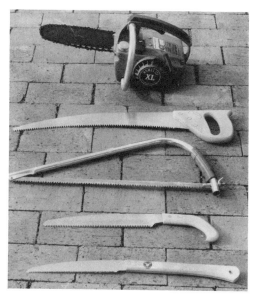

Fig. 7-5. Different types of pruning saws: gasoline-powered chain saw, hand saws.

It is important to select the right sized equipment to fit the job. One-hand shears, either scissors (curved blade) or anvil (straight blade) types, may be used to cut branches up to ½ inch across. The scissors-type is generally preferable because it can cut closer to the trunk and does not crush the wood tissue. Long-handled loppers (usually curved blade) provide added leverage and enable one to cut limbs up to 1¼ inches across. Most pole pruners can also be used to cut 1¼-inch limbs but allow the operator to work at some distance from the cut. Larger branches require a saw. One should not force shears to cut limbs larger than they were designed to cut. The twisting of the blades that accompanies the outsized cut will damage the nearby bark tissues, produce an imprecise cut, and may damage the shears.

Both hand and power saws are available in a variety of lengths and widths. For efficiency of use, precision of cut, and safety of the operator it is important that the saw be neither too large nor too small for the task.

Three types of hand saws are commonly used in tree work: pruning, pole, and crosscut. For small cuts, pruning saws with teeth that cut with the pulling motion are used. This type of saw is particularly suitable for overhead cutting.

A pole saw may be used for small cuts made at some distance from the operator. This saw also cuts with the pulling motion. Although pole saws allow cuts to be made from the ground, such cuts require more time and energy to make. Thus there is an efficiency tradeoff to be made and this should be considered in planning equipment use.

Crosscut saws, either bow or hand, are appropriate for larger cuts. Because they are most commonly used to cut green wood, a fleam tooth pattern, which cuts equally well in both directions, is preferred. To prevent the saw from binding, the set should be wider than the thickness of the blade. A finer set is used for soft woods, a wider set for hard woods.

A wide variety of powered pruning tools is available. The most common type is the chain saw powered by a direct-drive two-stroke gasoline engine. Hydraulic and pneumatic power systems, which allow for pole handle mounting of the cutting bar, are also used. Chain saws range in size from lightweight, short bar types suitable for use by climbers in the tree to powerful saws having long chain bars suitable for felling large trees. In addition, hydraulic boom-mounted circular saws extending from a large vehicle are used for trimming trees along rural rights of way.

Considerable improvements have been made in power tools recently. The best information concerning type and size of equipment is found in manufacturers' instructions.

Arborists must be equipped to ascend the tree where pruning is to be done. Sturdy stepladders and extension ladders are good for pruning low branches. Bucket trucks or ropes and saddles are needed to attain greater heights. Climbing spurs, useful for scaling utility poles, should not be used in trees. The injuries produced by spurs provide openings for tree decay organisms.

Tree climbers should know how to use their climbing equipment, including how to properly tie knots and how to lower heavy limbs, equipment, and themselves from the tree. Climbing equipment must be in good condition and well designed. One of the authors found himself dangling 30 feet off the ground in a climbing saddle which had slipped up around his chest following a fall from a limb. This frightening experience could have been prevented if the saddle in use had been designed with leg straps attached to the butt strap. Needless to say, poor or damaged equipment is not acceptable for tree work.

How to Make the Pruning Cut

Laterals. Research concerning the wound responses of trees has shown that most of the damage caused by pruning can be avoided by cutting along those lines that the tree forms for natural shedding. When a branch dies, the trunk tissue responds by walling off the dead branch. Thus proper pruning should only remove branch tissue, leaving the trunk intact.

How does one tell where the branch ends and the trunk begins? Fortunately the tree provides the necessary clues. Every branch has internal tissues that separate it from the trunk. As this tissue forms, the bark is forced upward to form a raised ridge on the trunk. This ridge is referred to as the *branch–bark ridge*. The raised or swollen area surrounding the base of the branch is made up of trunk tissue and is referred to as the *branch collar* or *shoulder ring*. When a branch dies, the collar continues to grow.

Pruning by making use of these clues is sometimes referred to as "natural target pruning" (Fig. 7-6). The pruning cut should be made at an angle from

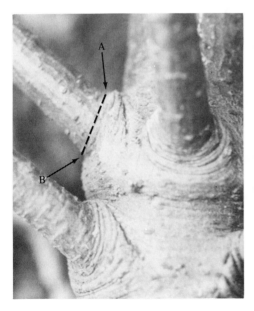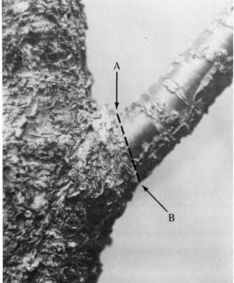

Fig. 7-6. Proper pruning cut is made just outside the branch collar (A–B). **Left:** Conifer. **Right:** Hardwood.

a point just beyond the branch–bark ridge to the junction of the lower part of the branch and the branch collar. Occasionally the branch collar is difficult to identify. Then the cut is made at an angle opposite to that of the branch–bark ridge. The resulting wound should be approximately circular in cross section. Injury to the branch–bark ridge or the branch collar should be avoided. Do not make flush cuts. Do not leave stubs. Either extreme increases the risk of internal decay for the tree and may delay callus closing of the wound.

When cutting small branches with hand-held shears, place the cutting blade beneath or on the side of the branch (Fig. 7-7). The cut is made either upward or diagonally in one stroke without twisting. The cut should not be made downward because the trunk is more likely to be damaged in this manner.

To avoid damage to the tree, large branches are removed using a three-cut system (Fig. 7-8). This method removes most of the weight of the limb before the critical final cut is made and does so in a manner to prevent the bark from tearing. The first cut is made on the branch underside out several inches from the branch collar. Cut about one-fourth of the way through the branch. The second cut is made from above and is made slightly farther out on the limb. The weight of the branch will cause the branch to fall and break off at the point where the bottom cut was made. If the limb is large and likely to cause injury

Fig. 7-7. When pruning small branches, place the blade beneath or on the side of the branch and make the cut with one stroke.

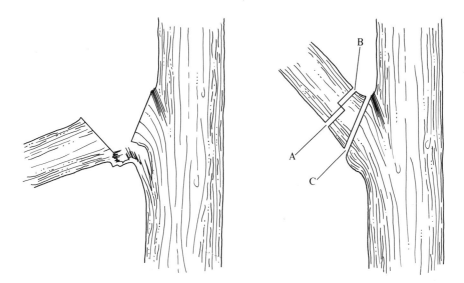

Fig. 7-8. Left: The wrong way to prune a large branch. A single cut close to the main stem may result in tearing of the bark. **Right:** The correct three-cut system for pruning a large branch: a preliminary undercut is made at A; second cut is made at B, to sever the main part of the branch. The remaining stub is removed by cutting at C.

or damage, it should be roped off before cutting and then lowered to the ground. The final cut is made as described previously.

Main Stems. It is occasionally necessary to remove a terminal branch (Fig. 7-9). Damage to the tree can be lessened by cutting the main stem off just above a lateral and at an angle parallel to the branch–bark ridge of the lateral. When cutting a young tree, leave a short stub to prevent the new terminal from splitting out. For larger trees, make an initial topping cut several inches above the final critical cut to remove the weight of the terminal branch. Be sure to select a lateral branch that is not less than a quarter to a third of the size (diameter) of the main stem that has been removed. Small branches and twigs are not sufficient to assume the terminal role for a larger tree.

Where two main stems occur, it is sometimes necessary to remove one of them, especially if they form a narrow-angled crotch. In this situation, a branch collar may not exist, and finding the branch–bark ridge can be difficult. The bark from each stem may be concealed within the crotch, so it is hard to know just where the fork begins. It is best to prune as would be done for removing a main stem as described previously. Even though a proper pruning cut is made in this situation, decay is almost inevitable, as it is for topping cuts. It is best

YOUNG TREE

Proper cut

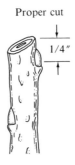

Improper cuts

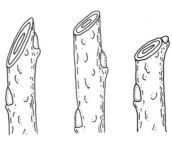

MATURE TREE

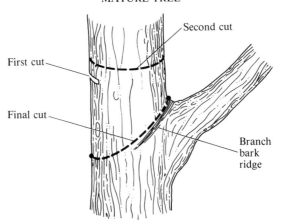

Fig. 7-9. Removal of terminal branch: for a young tree, leave a short stub; for a mature tree, use three-cut system and make the final cut at an angle parallel to the branch bark ridge of the highest major lateral branch, leaving no stem stub.

to recognize narrow crotch angles when the tree is young and remove one of the branches then.

The young terminal growth (candle) of a pine tree may be removed by snapping off the tender tissue with the fingers.

Safety in Pruning

People engaged in tree pruning have an obligation to prevent injury or damage to themselves, their crew, passersby, other trees, and surrounding property. In 1968 the American National Standards Committee on Safety in Tree Trimming Operations, Z-133 was organized in response to the efforts of a mother whose son had died while trimming trees. The committee produced a publication which details safety standards to be maintained while engaged in pruning, trimming, repairing, maintaining, and removing trees, and cutting brush. It is available from American National Standards Institute, 1430 Broadway, New York, NY 10018. Commercial operators should follow these standards.

Homeowners pruning their own trees should also be aware of safe procedures. Eye protection apparatus is the most important basic safety equipment needed during all pruning operations. When working in the aerial part of the tree, secure footing and a safety harness or line are essential to prevent falls. One should know the type of wood in the tree, since greater precautions are necessary in pruning weak-wooded trees, such as poplars, silver maple, willow, and tuliptree, than in pruning elm, oak, hickory, and planetree, which have strong, flexible wood. Bark peeling and fungus growths are signs of dead and dying branches; limbs with these symptoms should never be depended upon for support. Climbing danger is greatest when branches are wet and when temperatures are low. Large limbs should be roped and lowered to the ground and not left to fall, which could cause injury and damage. Unskilled amateurs should never climb; inexperienced professionals should not climb when carrying power tools unless they are properly trained in climbing and use of power tools (Fig. 7-10). Any power tool heavier than 15 pounds should have a separate support line.

All overhead lines and cables should be considered to be energized and should never be touched directly or indirectly. Amateurs should never work closer than 10 feet from any overhead line. Workers who prune trees for line clearance must be specially trained. Other professionals should notify the utility company before any work is done within 10 feet of a power line. After storms, damaged lines may energize metal conductors, such as wire fences, at some distance. Special care is needed under these circumstances and workers must be alerted.

Wound Dressings

Arborists and gardeners have been advised over the years to use various wound dressings ranging from commercial tree paints, orange shellac, asphalt paints, creosote paints, and house paint, to grafting wax. Current research indicates

Fig. 7-10. Professional tree worker properly equipped to prune landscape trees.

that these materials are not effective in protecting trees from wood-rotting organisms. Indeed they may prolong the period of susceptibility. When these coatings crack, as they frequently do, moisture may enter the area between the wood and the coating. Keeping the wounded area wet fosters the growth of decay organisms. Thus wound dressings should be considered only as a cosmetic treatment and should be used sparingly in a thin coat, if at all. An important exception is the use of tree wound dressings on wounds of oak wilt-susceptible trees which have been pruned in late spring. Apparently, treated trees are less likely to get oak wilt because they are less attractive to the insect vectors. This may be the only beneficial use of tree wound dressings.

Plant hormones have been found to be variable in their effect on callus production and to be ineffective in preventing decay.

Specific Pruning Situations

Newly Transplanted Trees. For years the idea prevailed that trees should be severely pruned at time of transplant to compensate for root loss and to reestablish a favorable shoot-to-root ratio. Recent studies have shown that this practice may prolong transplant shock. In many cases the additional moisture stress from leaving the top intact is offset by the more rapid development of the supporting root system from the additional carbohydrate produced. At time of transplant, pruning should be limited to removal of dead, broken, diseased, or interfering branches. It is best to leave small shoots along the trunk for later removal because they protect the trunk from sunburn and aid in the development of proper trunk taper.

Young Trees. After a tree becomes established, usually 2 to 3 years after transplanting, it may be pruned to achieve a desired shape. First, dead, diseased, misshapen, or crisscrossing interfering branches are removed. Temporary branches left along the trunk should be headed back but not removed until a sturdy trunk is established. Trees that have a strong central trunk and a conical shape usually require no additional pruning, whereas those with a round top benefit from early training to establish well-spaced scaffold branches.

The scaffold branches are the main branches that make up the framework of the tree. The establishment of a well-developed framework requires several years of selective pruning. Initially more potential scaffold branches are left than will be ultimately desired. Some will dominate, making final selection easier. During the first few years of training, remove branches growing from the same node as potential scaffolds, branches with acute angles of attachment, and those below the base of the permanent crown. Eventually select five to seven well-formed scaffold branches with even radial and vertical distribution.

Mature Trees. If the early training is properly done, trees will not need to be pruned for many years. However, they should be inspected annually to identify and remove hazards. Trees may require periodical cleaning out, that is, the removal of dead, diseased, poorly placed, or weak limbs (Fig. 7-11). The crown may require thinning to increase light penetration for better flower and fruit production, allow better coverage of spray material, or reduce water needs. Changes in land use may require that the height of the crown be raised or lowered. To maintain a natural crown appearance and to avoid bushy regrowth, mature trees should be thinned out rather than headed back.

Fig. 7-11. Left: Undesirable branches interfere with the normal development of major branches. **Right:** The same tree after pruning to remove the undesirable branches.

Conifers. Needled evergreens usually do not require pruning to maintain their characteristic crown shape. Top breakage, often caused by ice storms, may result in an undesirable forked leader. It is best to prune competing laterals and maintain a single central stem.

Some evergreens are pruned to produce a more bushy or compact plant. Pruning the growth at the ends of young shoots forces the plant to make new growth along the branches.

Not all evergreens can be pruned in the same manner at the same time. The method depends on the species and the type of growth.

Pines and spruces make only one flush of growth during the year. They can be trimmed at any time, but best results are obtained when the new growth is soft. In pruning spruces, especially the blue spruce, the long ends should be removed at the point where side shoots are formed. This will check the terminal growth for a year or two and will enable the branches to fill out. Most pines are best pruned before the new needles start to unfold, that is, when the new growth still gives the appearance of candles. No damage is done to the needles, and the tree will retain its original healthy appearance.

Most of the other evergreens, including arborvitae, junipers, yew, cypress, and false cypress, continue to develop during the entire growing season. These may be trimmed until midsummer, although early summer pruning is preferable, since new growth will then cover the cuts before the end of the growing season.

Winter pruning is sometimes preferable when an evergreen has become straggly as a result of long neglect. The long branches are cut to encourage new bud development early in the next growing season. When pruning is postponed until summer, new buds will not develop until the following spring, and in the interim the plant will appear unattractive.

Evergreens with a distinctive shape should be pruned to retain the desirable effect. Only the long shoots should be removed. Formal evergreen hedges are best pruned with hedge shears. Hemlocks may well be pruned to increase their branching and fullness. In the New York City area late June appears to be the best month, since new growth soon starts and continues throughout the summer. By early fall, the effects of pruning are well hidden by the new growth.

It is sometimes suggested that foliage be cut wet to prevent the cut ends from turning pink; however, this practice increases the risk of spreading foliar, twig, and branch diseases and should be avoided.

Special Shapes. Pruning may be used to create trees with unusual and artistic shapes. Two such methods are pollarding and topiary. Pollarding is a pruning system in which, starting at a young age, the tree is headed back every year to approximately the same place (Fig. 7-12). The ends of the framework branches become knob-shaped and most of the regrowth branches arise from these knobs. The size of the tree is kept uniform year after year. Pollarded trees make attractive formal plantings. London planetree, linden, black locust, and fruitless mulberry are good candidates for pollarding.

The fantastic shapes of trees formed into geometric, animal, or other forms

Fig. 7-12. Pollarded Ginkgo trees.

are created by a pruning and training system referred to as topiary (Fig. 7-13). The best trees for this use are small-leaved evergreens with readily growing latent buds. The trees must be trimmed several times each season to be kept in shape.

Fig. 7-13. Topiary pruning.

Pruning for Disease Control

Very often it is necessary to prune trees that are partially affected by diseases such as fire blight of apple, pear, and mountain-ash; sycamore anthracnose, Cytospora canker of blue spruce; Diplodia tip blight of conifers; wilt of maples and elms; and twig blight of chestnut oak. In such cases, the infected twigs or branches should be removed well below the point of visible infection.

Extreme caution must be exercised to avoid transmission of the causal organism from a diseased to a healthy tree by means of the pruning tools. Indeed, Dutch elm disease can be spread by contaminated sawdust on pruning saws. After use on a tree suspected of being diseased, the cutting edges of all tools should be thoroughly disinfected by application of 70 percent denatured ethyl alcohol. Furthermore, pruning when the leaves are wet should be avoided, since parasitic organisms are easily spread under such circumstances.

For further discussion of the role of pruning in disease control, see Chapter 13.

Pruning for Line Clearance

Electric utilities must maintain adequate clearance of overhead circuits to prevent interruptions to their customers' service. Tree branches can cause interruptions by contacting power lines. Sometimes inexpert pruners drastically prune trees during right of way clearance causing public outrage and increased pruning costs for the utilities. Trees do interfere with utility lines, and maintenance of these lines to prevent power outages is costly to the consumer. There are ways to minimize tree interference with utility lines.

Trees are either top, side, overhang (under), or through trimmed to allow them to grow 2 to 4 years before they again become hazardous to the line (Fig. 7-14). Careful use of the method referred to as thinning or drop crotching will direct growth away from wires while retaining a more natural crown appearance. Thinning is much preferred to topping because it does not promote unsightly bushy regrowth. The regrowth after heading back usually interfers with the electrical lines sooner and will require more cutting on revisits. One utility company discovered that switching from topping to thinning reduced their pruning expenses by 50 percent. Others have reduced pruning costs by offering to re-

Fig. 7-14. Trees can be pruned to avoid interference with power lines.

move customers' trees and replace them with desirable low-growing trees that would not become a line clearance problem.

Use of Chemical Growth Retardants

Because of the high cost of pruning trees manually, less expensive ways are being investigated. One is the use of growth regulators. They are primarily being used by utility companies to increase the length of time between trimming visits. Trees are treated just after pruning to retard sprouting. The growth regulators are particularly useful where topping is implemented.

Growth-retarding chemicals are being applied by overhead foliar sprays, trunk injection, basal spray-bark banding, and painting of cut surfaces. Some of the methods have resulted in reduced costs for utility companies.

Scheduling Pruning for Municipalities

Many cities have assumed an exclusive duty to maintain street trees. Pruning operations in these localities are more than aesthetically desirable; they are critical to ensure public safety and avoid municipal liability. Courts have ruled that trees must be maintained in safe condition and any damages arising from street trees are attributable to the city.

Cities can schedule pruning operations by the following methods:

1. Request of residents.
2. Crisis—only when an emergency exists.
3. Task pruning—performing certain functions such as clearance of right of way obstructions.
4. Species pruning—scheduling common service needs by tree type.
5. Programed maintenance—scheduling service needs by localities.

Most cities use a combination of these schedules. It has been found that using only request and crisis scheduling is inefficient and may increase the municipality's liability for damages occurring due to unidentified hazards. Modesto, California found that the programmed maintenance schedule allows the entire city to be covered in a 7-year cycle. This schedule utilizes personnel in the most efficient manner and reduces the number of calls for crisis situations. An economic evaluation of pruning cycles using computer optimization techniques found that a 4 to 5 year cycle maximizes the return from pruning while minimizing the cost for most cities.

Selected Bibliography

Anonymous. 1972. American national standard safety requirements for tree pruning, trimming, repairing, or removal. American National Standard Institute. Z 133.1-1972. New York. 16 pp.

Baker, F. A., and D. W. French. 1985. Economic effectiveness of operational therapeutic pruning for control of Dutch elm disease. J. Arboric. 11(8):247–49.

Bowles, H. 1985. Growth retardant use by utility companies. J. Arboric. 11(2):59–60.

Bridgeman, P. H. 1976. Tree surgery. David and Charles, North Pomfret, Vt. 144 pp.

Domir, S. C., and B. R. Roberts. 1983. Tree growth retardation by injection of chemicals. J. Arboric. 9(8):217–24.

Giles, F. A., and W. B. Siefert. 1971. Pruning evergreens and deciduous trees and shrubs. University of Illinois. Coop. Ext. Serv. Circular 1033. 57 pp.

Harris, R. W. 1975. Pruning fundamentals. J. Arboric. 1(12):221–26.

Hield, H. 1979. Trunk bark banding with chlorflurenol for growth control. J. Arboric. 5(3):59–61.

Leben, C. 1985. Sap pressure may affect decay in wounded trees. Am. Nurseryman 161(7):59–63.

Shigo, A. L. 1983. Targets for proper tree care. J. Arboric. 9(11):285–94.

Shigo, A. L. 1984. Compartmentalization: A conceptual framework for understanding how trees grow and defend themselves. Ann. Rev. Phytopathol. 22:189–214.

Shoup, S., R. Reavis, and C. E. Whitcomb. 1981. Effects of pruning and fertilizers on establishment of bareroot deciduous trees. J. Arboric. 7(6):155–57.

Smithyman, S. J. 1985. The tree worker's manual. Ohio Agric. Educ. Curric. Materials Serv., Columbus, Oh. 145 pp.

Steffek, E. F. 1982. The pruning manual, 2nd ed. Van Nostrand Reinhold, New York. 152 pp.

8

Tree Preservation and Repair

Mature trees, in addition to providing shade, beauty, and a relaxing atmosphere, enhance the value of property. Therefore, it is important to preserve trees when their roots, trunk, or limbs are threatened by the actions of nature or man. Physical damage to tree structure is usually obvious: broken limbs damaged bark, exposed roots. However, lives of trees are more often threatened by damage which cannot be seen on casual inspection: internal decay, boring or mining insects, root suffocation. Some forms of damage are preventable; some are not. The preservation of mature trees depends on preventing damage where possible and ameliorating the impact of damage once it has occurred.

Is the Tree Worth Preserving?

When considering a course of action, whether to prevent or treat injury, the first question should be: Is the tree worth saving? The answer depends on the value of the tree (see Chapter 1) and the cost of preserving it.

Preventing and Treating Structural Damage

Damage to the structure of the tree may result from external forces such as windstorms or ice storms or vehicle collision or from the internal action of boring insects and decay fungi. Trunk, limbs, or branches may crack, split, or break.

In many cases structural damage caused by windstorms or ice storms can be prevented by careful pruning, such as the early removal of limbs forming narrow crotches. However, in some cases artificial support such as that provided by braces and cables may be required to preserve structurally weak or injured trees. Some trees, such as oak, beech, and locust, rarely need artificial support because they consist of tough wood fibers and usually have a single major stem from which branches arise at approximately right angles. Despite the absence of tough wood fibers, even pine, spruce, and fir seldom need support because of their single stem and type of branching. Other trees, such as elm, maple, and poplar, which have inherent structural weakness resulting from the manner in which their branches arise, often require considerable mechanical support.

When Is Artificial Support Justified?

Artificial supports prevent such injuries as crotch splitting and branch breakage, both of which eventually lead to wood decay. They may also prolong the life of many trees after decay or extensive cavities have already developed. Such supports may strengthen an otherwise weakened trunk.

The proper installation of cables or braces is often helpful in prolonging the life of trees having tight crotches, split crotches, deep trunk cavities, or inherent susceptibility to breakage. Changes in the tree's surroundings may also create a need for artificial supports due to new exposure to winds not previously affecting it.

Narrow Crotches. Narrow crotches commonly occur in certain trees in the course of normal development (Fig. 8-1). In the narrow angle between two sharply forking major branches, the bark and cambium are hindered from developing normally or may be killed by the pinching action of the branches as they increase in size. This results in a weak union of the type that readily splits when subjected to stress. Bracing can prevent splitting.

Split Crotches. After some splitting, artificial supports can prevent further damage and loss of one or both branches.

Fig. 8-1. Left: Narrow branch crotch with embedded bark. The union is inherently weak. **Right:** Normal branch attachment.

Deep Trunk Cavities. Cavities are formed by extensive decay of internal wood of trunk and limbs. Because the decayed wood no longer provides mechanical support for the tree, artificial bracing may be needed.

Susceptibility to Breakage. Because of inherent properties such as heavy foliage production and brittleness of the wood, some species of trees are more susceptible than others to breakage by wind and ice storms.

Trees that are highly susceptible to breakage by winds are 'Bradford' pear, chestnut oak, honey locust, horsechestnut, mimosa (silk-tree), poplar, maple (red, silver, and sugar), sassafras, Siberian elm, tuliptree, willow, and yellow-wood. Although the following are more resistant to wind damage, they too can lose branches and be toppled by winds of very high velocity: American beech, American elm, ginkgo, hackberry, littleleaf linden, London planetree, Norway maple, red oak, sweetgum, and white oak. Cabling large limbs may prevent disfigurement of susceptible species.

Land Use Changes. Wind patterns may be changed by construction or removal of buildings or the removal of nearby trees. These changes may subject a mature tree to greater wind stresses than previously present. Without judicious pruning and artificial supports trees may suffer breakage. Also, artificial supports may be required to preserve trees after the loss of some large roots during wall or curb construction or grade changes.

Types of Artificial Support

Tree supports are of two types: rigid and flexible bracing. Rigid bracing involves the use of bolts and threaded rods for supporting weak or split crotches, long cracks in the trunk or branches, and cavities. Flexible bracing involves the installation of wire cables high in the tree to reduce the load on weak crotches and long arching limbs without hindering the normal branch sway.

Rigid Bracing. Prepared lengths of steel or duralumin rods with lag threads at both ends (screw rods) are used for rigid bracing in sound wood. The holding power of such rods is similar to that of ordinary wood screws used in fastening two boards together. The lag rod is set in a hole drilled $\frac{1}{16}$ inch smaller than the rod. The wood surrounding the hardware will discolor, but if it is sound to begin with, it will seldom decay to any great extent.

Rods or bolts with machine thread and nuts and washers on each end are used when decay is present or where considerable "squeeze" must be applied to close a gap. The bolts are inserted into holes drilled the same size as the bolt. The washers should be seated just on the wood but not sunken into it (Fig. 8-2). A wood chisel should be used to carve out a smooth-edged hole in the bark to seat the washer. Washers should be round and not diamond-shaped. Where decay is present, the bolts may break through the boundary zone (see Chapter 7) and cause the decay to spread. The spreading decay will loosen the bolt so its holding power is reduced to that of the washers seated on both sides of the limb. In most cases, removal of badly decayed limbs should be considered before bracing. However, braces may extend the time that even badly decayed trees remain safe and attractive.

Fig. 8-2. Left: Bark removed with a drill bit exposing the sapwood surface. **Right:** Washer seated on the sapwood surface. Round or oval washers are preferable.

Braces should be installed during the dormant season. In the spring, the bark is easily injured and may split. Holes made during the fall are subjected to large levels of fungal spores and thus a stand a good chance of becoming infected. Holes made directly above major roots often result in less chance of injury to the tree than holes made between major roots. Holes made closer to the base of the tree result in less injury than those made higher up on the trunk. Trees that are growing vigorously sustain less injury from bracing than slower growing trees. Wound dressings are not necessary and may promote infection.

CROTCH BRACING. The installation of artificial support may be justified on sound trees with weak crotches, although in many cases this results in crotch splitting elsewhere in the tree. To provide additional support, wire cables should be installed higher in the tree whenever the branches extend more than 20 feet above the crotch. In the past it was recommended that one or more parallel rods be inserted above the crotch. However, authorities now believe that rods should be placed at or just below the crotch (Fig. 8-3). If trees are bolted above the crotch, gusty winds may cause further splitting of the crotch.

SPLIT LIMBS OR TRUNKS. Screw rods may be used to hold split limbs together or to support the sides of long splits in the trunk. They are usually placed about a foot apart and are staggered so as not to be in the same direct line of sap flow. Again, a fastened split may simply split elsewhere.

RUBBING LIMBS. Limbs that rub against each other can be braced together or kept separated by proper bracing. A single rod screwed through the center of each limb at the point of contact will hold the limbs rigidly together and thus

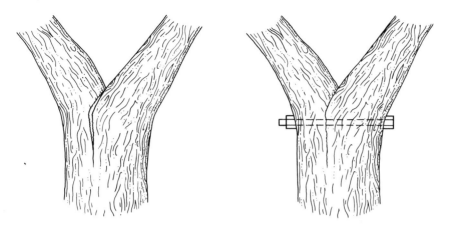

Fig. 8-3. Rods should be placed at or just below the split crotch. If two rods are used, they should be placed parallel to each other at the same level.

eliminate rubbing (Fig. 8-4). If it is desirable to keep such limbs separated, a length of pipe equal to the desired space between the two limbs can be placed over the rod before the branches are fastened together (Fig. 8-5). If the limbs are to be separated and free movement between them is desired, the installation of a U-bolt in the upper limb, which rests on wood block attached to the lower limb, will prove satisfactory (Fig. 8-6). In time the wood block will need replacement because of wear. If aesthetics and function permit, removal of one of two rubbing limbs is advised.

Flexible Bracing. Wire cables are used to brace trees together and to strengthen weak crotches. These generally are placed high in the tree to provide maximum

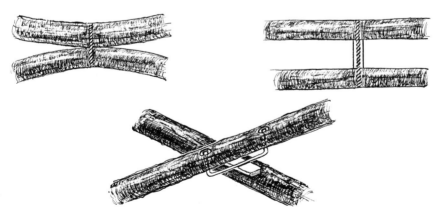

Fig. 8-4. Upper left: Method of inserting a screw rod to hold two limbs together and thus prevent rubbing. **Fig. 8-5. Upper right:** Separating two branches by means of a screw rod and a short piece of pipe. **Fig. 8-6. Below:** Method of installing a U-bolt to prevent rubbing of two limbs and at the same time allow free movement of the branches.

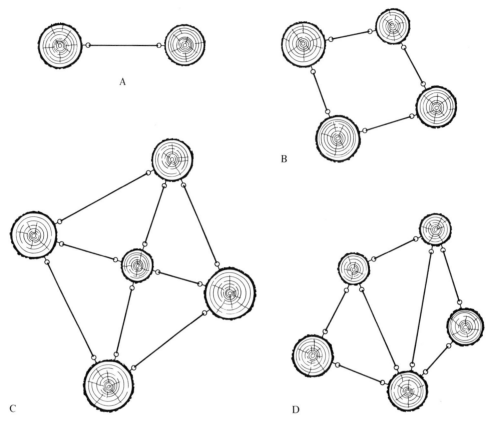

Fig. 8-7. Flexible bracing systems. A, Simple direct: a single cable supports two limbs arising from a single crotch. B, Box: cables attached to all large limbs in a rotary manner. C, Hub and spoke: cables radiate from a limb or metal ring in the center of the main branches. D, Triangular: cables placed in triangular form.

support from the use of materials of minimum size and cost. Copper-covered steel wire, although more expensive than ordinary glavanized wire, is stronger, size for size, and does not rust.

Flexible bracing systems commonly employed are of several types. In the simple direct system, a single cable is installed to support two limbs arising from a single crotch (Fig. 8-7A). In the box system, cables are attached to all large limbs of a tree in a rotary manner. This arrangement provides lateral support and maximum crown movement, but little support to weak crotches (Fig. 8-7B). The hub and spoke system, in which cables radiate from a limb or metal ring in the center of the main branches (Fig. 8-7C), provides little direct or lateral support. In the triangular system cables are placed as shown in Figure 8-7D to provide direct support to weak or split crotches and lateral support to the branches. Triangular bracing is probably the most efficient of the various types described.

The wire cables used for flexible bracing are placed at a point approximately two-thirds the distance from the crotch to the top of the tree. They should not rub against each other.

Lag hooks, hook bolts, bent eyebolts, and other materials are most often used to anchor the cables. Such anchors should be well separated (4–6 in.) when more than one is used on a single branch and should be placed in a straight line with the cable and as nearly at right angles to the branches as possible.

LAG-HOOK INSTALLATION. Cadmium-plated hooks (½ to ⅝ in. diameter and 6-in. thread) are recommended as anchors for most branches. Rods should be inserted in branches at least 4 inches in diameter and to a depth of two-thirds of the diameter of the limb. The exact size selected is governed by the amount of stress and the size of the branches. Holes about $\frac{1}{16}$ inch smaller than the hooks are bored at the proper anchorage points, and the hooks are screwed in just far enough to allow a cable to be slipped over the open ends. After the lag hooks are in place, the limbs are pulled together sufficiently so there will be no slack in the cable after it is hooked on, but they are not pulled tight. A thimble is inserted over the hook and cable and the cable is then passed through it and spliced to form a permanent union between the two anchors. All spliced areas should be coated with a metal preservative paint.

EYEBOLT INSTALLATION. Where wood decay is present or considerable stress occurs, eyebolts or hook bolts are generally used as anchor units. The hole for these is drilled completely through the wood, the bark is countersunk to a depth just below the cambium on the wood surface, and the bolt, washer, and nut are put in place. The cable ends are then inserted through the eyes and properly spliced.

Cabling systems should be checked every 10 years and either moved or removed.

GUYING. Guy cables are used when other trees are not available for cable attachment. Cabling to the ground may be necessary when other trees are removed and the retained tree is subjected to high levels of wind stress. The cables are attached to three separate points on the tree usually on the main stem at the base of the crown. Compression springs added to the cable system will allow the tree to sway and reduce the danger of breakage at the cable attachment site. The cables are then attached to a soil anchor. The type of anchor required depends on the soil type and the level of anticipated strain. Simple screw-type anchors are appropriate for loam or clay soils. Cast iron cone anchors are useful in lighter soils. The common impact anchor driven into the ground by means of a sledge hammer may be used for low-strain situations, but this is ineffective where the guy cable is greater than 45 degrees to the plane of the ground.

Treating Fresh Wounds

After a windstorm which has left broken limbs, a lightning strike which has split open the trunk, or a vehicle collision which has ripped off a large section of bark, it is natural to want to do something for the injured tree. Good intentions very frequently lead to increased damage.

Broken Limbs. Broken limbs should be removed using the same guidelines given in Chapter 7 for pruning. Topical applications of wound dressings have been shown to be ineffective and may prolong the period of susceptibility to rot organisms. After limb removal is complete, the tree crown should be examined for split crotches or off-balanced limbs that may need artificial supports.

Bark Torn from Trunk. Large wounds such as those caused by vehicle collision should be treated as soon as possible after the damage occurs. Any bark that is still moist and can be refitted on cambium should be tacked into place and covered with burlap to prevent drying out during reattachment. No other action should be taken except water and fertilizer application as appropriate to maintain the vigor of the tree. Again, wound dressings are ineffective. The dead bark and wood splinters may be removed after a year. In the past wounds were scribed to make smooth edges. In some cases, this may cause removal of cambium and bark tissues that could have aided in wound closure. Leaving even small pieces of bark and cambium on recently exposed sapwood will speed up the wound closing process.

Lightning Damage. The injured tree should be inspected carefully before any attempt is made to repair the damage. Many trees are severely injured internally or below ground, despite the absence of external symptoms, and will soon die regardless of treatment. Consequently, expensive treatments should not be undertaken until the tree appears to have a good chance of recovery.

Where damage is not great, lightning-struck trees may benefit from immediate treatment. Before they have been allowed to dry out, the long, thin pieces of bark that have been split or lifted from the sapwood should be tacked back on with small nails. Then the area should be covered with burlap for a few days to allow the bark to reattach. Torn bark should *not* be trimmed immediately but should be left for a year, when it may be removed up to the living callus.

Lightning Protection. Trees growing in locations favorable for strikes as well as large, rare, or otherwise valuable trees can be adequately protected at comparatively small cost by the installation of proper equipment. The theories and details of installing lightning-protection equipment are too involved to permit discussion in this book. This information is available in Tree Preservation Bulletin No. 5, which may be obtained from the Superintendent of Documents, U.S. Government Printing Office, Washington, DC 20402, for a small fee. Installation must be entrusted to a specialist in tree care and should never be attempted by the layman.

The principles governing the protection of buildings and other structures from lightning bolts are employed also in safeguarding trees. The major difference is in the materials used and methods of installation. Tree protection systems must have flexible cables to allow for the swaying of the trunk and branches and adjustable units to allow for the tree's growth.

A vertical conductor is fastened along the trunk with copper nails, from the highest point in the tree to the ground. The ground end is fastened to several radial conductors, which are buried underground and extend beyond the root area. These, in turn, are attached to ground rods 8 feet in length, which are driven vertically into the ground.

The system should be inspected every few years, the air terminals extended to the new growth, and other adjustments necessitated by expansion of the tree should be made. Ground terminals should also be checked periodically.

Cavity Filling

It is apparent that cavity fillings do not add strength to the tree and in most instances are not needed. However, filling can improve the appearance of some trees, so the safest method of filling is described here (Fig. 8-8).

First, decayed wood is removed from the cavity. Care must be taken to remove only the softest rotted wood without damaging the surrounding healthy

Fig. 8-8. Cavity filling.

tissues. If the boundary zone separating invaded from uninvaded tissues is broken, decay will spread rapidly and the tree will suffer. Mechanical devices have been developed to remove the punky wood, but they should not be used or only used with great care.

The cavity may then be filled with any material that will not rub against living tissue. Painting of the interior of the cavity with orange shellac, creosote, tar, or asphalt paint is of no benefit.

It is tempting to bore drainholes to remove water from tree cavities. This practice will not help the tree because the boundary zone will be breached and decay may spread.

Although often used, concrete is not a good material for filling cavities. Concrete will not bend with the swaying of the tree. As a result, the concrete may crack or crumble or the wood may wear away at the point where friction occurs. Inability of the filler to flex with tree movement may result in further spread of decay and possibly breakage of the trunk. Thus concrete should be used only where there is little possibility of movement, for instance, small cavities near the base of the tree.

Polyurethane foam available for filling cavities is more suitable for larger cavities than is concrete. This material is tough, somewhat resilient, and adheres well to both sapwood and heartwood. Moreover, it is light in weight, can be poured into the cavity, expands and solidifies rapidly, is somewhat flexible after setting, and permits callus initiation along the edges of the filled cavity.

Preserving Trees at Construction Sites

Excavation for buildings, roadbeds, or pipelines is always potentially threatening to nearby trees. Even when care is taken to avoid hitting trees with heavy equipment, many trees die from root damage. Roots may be severed during excavation, suffocated by compaction or grade raising, or suffer from drought or standing water when drainage patterns change. Trees suddenly exposed to direct sun or reflected heat from paved surfaces may suffer from heat stress and sunscald. However, many trees can be saved if proper precautions are observed.

Excavation. The installation of water or sewer pipes or underground utility lines, digging out footings for a building foundation, or cutting out space for a sidewalk or roadway near trees may cause severe root damage. The degree of damage depends largely on how close the excavation comes to the base of a tree. The most critical zone encompasses the main root plate, the area adjacent to the base of the tree containing the largest roots. A trench crossing this area will threaten the life of the tree. It will not only deprive the tree of needed water, stored food, and minerals but will also interfere with anchorage and subject the tree to the threat of wind-throw. Extending out from the root plate to the edge of the tree crown is an area that contains a high density of important conducting roots. Excavations in this area will cause a serious threat to tree health. Since tree roots may extend out as far as twice the tree crown, even trenches that sever roots beyond the edge of the tree crown will cause damage.

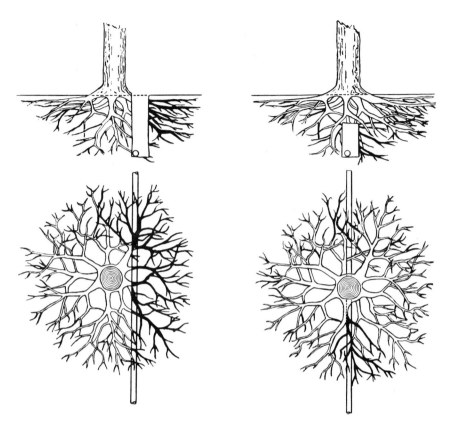

Fig. 8-9. Tunnel beneath root systems. Drawings at left show trenching that would probably kill the tree. Drawings at right show low tunneling under the tree, which will preserve many of the important feeder roots.

The degree of damage may be reduced by careful placement of trenches to avoid large roots and by tunneling under rather than cutting roots greater than 1½ inches in diameter (Fig. 8-9). Open trenches should not be routed beneath the crowns of trees that are to be preserved. If larger roots are severed, they should be cut off squarely. If the root system is appreciably reduced, water and mineral nutrient status should be monitored, especially during dry periods. Trenches should not be dug near trees during rainy periods because trees with inadequate anchorage may topple in wet soil.

Trees can be saved by tunneling under their root systems. In one case, only two out of thirty-four silver maples died after an underground utility line was placed alongside their crowns. The trees were 64-year-old street trees. The utility company tunneled under all roots greater than 2 inches in diameter. The city of Lexington, Kentucky recently installed a sewer line beneath the root system of a 200-year-old bur oak using a tunneling machine.

Location of building, roadway, and sidewalk sites needs to be carefully thought out if trees are to be preserved.

Fig. 8-10. Damage to roots may be minimized by building bridges for walkways.

There are alternatives to excavations for building foundations. Homes or walkways elevated on pilings will do less physical damage to tree root systems and allow preservation of the existing terrain (Fig. 8-10). Partial foundations to support a structure, much like a bridge over a river, physically destroys roots on only two sides of the building.

Lowering Grade. Leveling building sites often requires removal of soil around the base of trees. Little root severance is likely if the level is lowered no more than 2 inches. Injury to the tree as a result of even a slight lowering of the grade can cause injury due to soil compaction. Exposed large roots, if cut or broken, should be cut off squarely and covered with peat moss to prevent drying. If many large roots are cut or damaged, a few branches should be pruned to maintain proper balance between these parts. After grading, mulch should be applied to prevent desiccation of exposed roots. The remaining undisturbed roots should receive water and mineral nutrients as needed.

Under some conditions injury can be reduced by terracing the new slope around the tree, rather than cutting the soil away in a uniform gradient (Fig. 8-11). The terrace should extend at least to the edge of the tree crown and be supported by a retaining wall. In this manner, the original soil depth and many of the conducting roots will be preserved. Tree water status should be carefully monitored. About a year later, after the tree has become adapted to the change, fertilizer may be applied.

Trees left along the edge of road cuts often continue to die for many years

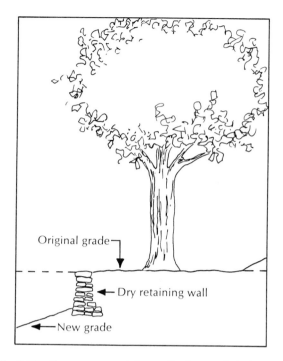

Original grade

Dry retaining wall

New grade

Fig. 8-11. Construct a retaining wall when grade is lowered.

after the initial work was done. Trees should be removed some distance back from the top of the cut in proportion to the depth of cut.

Raising Grade. Extra soil may be placed beneath trees either to deliberately level the construction site or as a convenient way of disposing of excavated soil (Fig. 8-12). In either case, serious damage to trees may ensue. Because most tree species are sensitive to the addition of soil over their roots, excavation soil should never be piled, even temporarily, over the roots of trees which are to be preserved.

The injury to trees as a result of deliberate raising of grade more than a few inches can be reduced by following the procedure illustrated in Figure 8-13. The ground around the base and underneath the branch spread is cleared of all plants and sod, and the soil is then broken up without disturbing the roots. Commercial fertilizers are applied. Four- or six-inch agricultural tile or split sewer pipes are laid in a wheel-and-spoke design with the tree as the hub. The radial lines of tile near the tree should be at least 1 foot higher than the ends joining the circle of tile. A few radial tiles should extend beyond the circle and should slope sharply downward to ensure good drainage. An open-jointed stone or brick well is then constructed around the trunk up to the level of the new fill. The inner circumference of the stone well should be about 2 feet from the circumference of the trunk. Six-inch bell tiles are placed above the junction of the two tile systems, the bell end reaching the planned grade level, and stones

Fig. 8-12. Piling soil over the root zone of trees can be very harmful.

are placed around the bell tiles to hold them erect. All ground tiles are then covered with small rocks and cobblestones to a depth of 18 inches. Next, the large rocks are covered with a layer of crushed stone, and these in turn with gravel, to a level of about 12 inches from the final grade. Cinders should never be used as a part of the fill. A thin layer of straw, hay, or synthetic filter material should be placed over the gravel to prevent soil from sealing the air spaces. Then good topsoil, which is the same or a coarser texture than the original soil, should be spread over the entire area except in the tree well and the bell tile. To prevent clogging, crushed stone should be placed inside the dry well over the openings of the radial tile.

The procedure outlined can be followed where a grade is to be raised around several trees in a group. In such cases the tile can run from one tree to another, increasing the air circulation.

The area between the trunk and the stone well should be either covered by an iron grate or filled with a mixture of 50 percent crushed charcoal and 50 percent sand, to make it less hazardous. This will prevent not only filling with leaves and consequent impairment of the efficiency of air and water drainage, but also rodent infestation and mosquito breeding. Bell tiles should be filled with crushed rock and covered with a screen.

The method outlined constitutes an ideal one to ensure the least disturbance by fills, but there are alternative methods. Where water drainage is not a serious problem, coarse gravel in the fill can be substituted for the tile. This material has sufficient porosity to ensure air drainage. Instead of bell tile in the system, stones, crushed rock, and gravel can be added so that the upper level

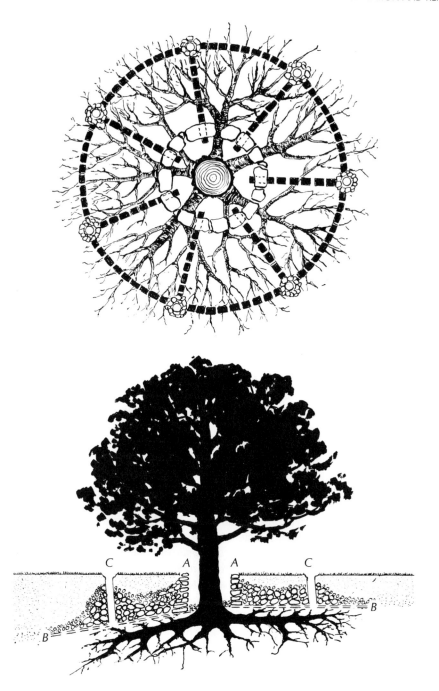

Fig. 8-13. Preventing injury by soil fills. **Upper:** Bird's-eye view. **Lower:** Side view showing A, the dry well; B, ground tile; and C, upright bell tile.

of these porous materials slants toward the surface in the vicinity below the outer branch spread.

Compaction. Soil compaction can result from continual foot or vehicular traffic. Sometimes, before a new home is landscaped, a few inches of topsoil, or even subsoil, is placed over compacted soil. Plant roots become established in the upper soil layer. However, the compacted soil beneath can form an artificial hardpan and either restrict needed deep rooting or produce a high perched water table. Eventually the newly planted trees will suffer from insufficient or excess soil moisture and decline.

Compaction damage to existing desirable trees can be prevented. Only one roadway should be used for movement of trucks and equipment to the building site. Piling thick layers of wood chip mulch over the tree root zone in the area likely to be compacted may reduce the impact of equipment traffic during construction. Grading should not be done while soil is very moist. Workers should be cautioned against trampling soil near trees. Barriers surrounding valuable trees should be erected at the dripline to keep machinery away.

In areas such as unpaved parking lots or playgrounds where compaction is unavoidable, the impact on trees can be reduced. It has been reported that expanded slate, rotary tilled into the surface soil, provided the best protection against compaction. The slate accounted for 33 percent by volume of the surface soil mixture. Expanded slate is a commercially available material which is made by a heat expansion process so that the slate becomes porous (about 50% pore space), inert, and rigid. A layer of wood chips placed to a depth of 3 to 4 inches over the soil surface was also found to reduce compaction in heavily used areas. Vertical mulching may also relieve compaction effects (see Chapter 3).

Paving. Grading and paving with either asphalt or concrete to build roads or walkways can damage adjacent trees. If larger roots extend entirely across the intended road or walkway, they may be bridged over. The roots may be wrapped in 2 to 3 inches of Styrofoam. As the root grows the Styrofoam becomes compressed. This system will allow transport of water and minerals from the functional feeder roots on the opposite side of the paved area.

Near valuable trees, paving materials that allow greater air and water penetration should be used in preference to asphalt or concrete. Alternative materials that are less damaging to tree roots include sand-set bricks, stones, or concrete pavers; gravel; decomposed granite; wood chips, blocks, or planks; and steppingstones. When using these materials, deep excavation should be avoided to minimize the number of roots served. Specialized paving bricks with aeration holes are used in the parking area of the Missouri Botanical Gardens in St. Louis. They make an attractive and useful surface while protecting the trees that shade the parked cars.

Changing a Tree's Environment. Following construction, a tree once growing in partial shade may suddenly be exposed to full sun, perhaps exacerbated by reflected heat from buildings and pavements. Such a tree, often with a reduced root system, may react to this new circumstance with leaf scorch, sunscald, and even death. On a less frequent basis, a tree growing in full sun may be shaded by a new building; trees that cannot tolerate shade will suffer. A change in soil environment to increased or decreased air, water, and mineral availability can be detrimental to tree health.

Tree Response to Construction Injury. One or more of the problems just discussed can result in gradual or sudden decline and death of the tree. The effects may not show up for several years, or they may occur within months. Frequently tree decline involves leaf growth reduction, twig dieback, epicormic foliar growth, branch and limb dieback, and death. See Chapter 9 for a discussion of diagnosing these decline problems.

Adjusting to Sidewalk Lifting by Tree Roots

As tree roots growing under concrete walks increase in size, the walks may be lifted, creating a hazard to pedestrians. There are several remedies which, depending on circumstances, may help to solve the problem. Walks can be reconstructed to curve around the tree's basal roots, or a thin concrete slab, possibly with a foam cushion beneath it, can be utilized. Paver bricks that adjust easily to changes in terrain can be used. Root pruning and root control barriers can also be used, though drastic root pruning will harm the tree.

Selected Bibliography

Felix, R. E., and A. L. Shigo. 1977. Rots and rods. J. Arboric. 3(10):187–90.

Jeffers, W. A., and R. E. Abbott. 1979. A new system developed for guying trees. J. Arboric. 5(6):121–23.

Kozlowski, T. T. 1985. Soil aeration, flooding and tree growth. J. Arboric. 11(3):85–88.

Mayne, L. S. 1982. Specifications for construction around trees. J. Arboric. 8(11):289–91.

Schoenweiss, D. F. 1982. Prevention and treatment of construction damage to shade trees. J. Arboric. 8(7):169–75.

Shigo, A. L. 1986. A new tree biology. Shigo and Trees, Associates. Durham, N.H. 595 pp.

Yingling, E. L., C. A. Keeley, S. Little, and J. Burtis, Jr. 1979. Reducing drainage to shade and woodland trees from construction activities. J. Arboric. 5(5):97–105.

II

Diagnosis and Control of Tree Problems

9

Diagnosing Tree Problems

The authors are often asked by commercial arborists, county extension agents, city park departments, nurserymen, landscape architects, public utility companies, and homeowners to diagnose the cause of decline or death of shade and ornamental trees. At the request of a public utility company, during the summer of 1956, for example, the senior author examined and then diagnosed the cause of many abnormalities among more than 300 trees in New York City. Since then he has examined several thousand more trees in New York, New Jersey, and Long Island, New York. In addition, the other authors have examined thousands of diseased and declining landscape trees in Kentucky, California, Louisiana, Wisconsin, and Indiana. These activities have provided the experience needed to address the complex subject of tree problem diagnosis.

Successful treatment of any abnormality of trees depends primarily on correct diagnosis. Treatment based on incorrect diagnosis may only make the problem worse. Some troubles occur so frequently on certain trees, and have such specific symptoms and signs, that they are readily diagnosed by the competent tree worker. Other troubles are so complex, unresearched, or deficient in reliable background information that even the most expert arborist cannot fathom them.

A reliable arborist will not hesitate to admit an inability to diagnose some abnormalities and will not hesitate to recruit the aid of a specialist to help in diagnosis, much as medical doctors often consult specialists in difficult cases. Corrective treatments of unknown value will be tried only as a last resort—after all attempts at an accurate diagnosis have failed.

Responsibilities of a Tree Diagnostician

Make Careful Observations. The primary responsibility of a good diagnostician is to make careful observations. The observer must not only scrutinize the obvious symptoms, but also look for hidden ones.

Many years ago, the senior author was asked to diagnose the cause of death of trees growing along the streets of a New Jersey community. One, a Norway maple 16 inches in diameter, was purported to have been killed by natural gas escaping from leaking mains. The only basis for this diagnosis was that a gas leak had been repaired in the general vicinity. The author removed the soil around the base of the tree and found that someone had completely girdled the tree below ground at least a year or so earlier. The uneven cuts made by an instrument were still plainly visible! Here was a case, then, where a little digging by the first ''expert'' might have resulted in an entirely different diagnosis.

Use Good Judgment. The most important qualification of a good diagnostician is plain ordinary common sense; some refer to it as knack, intuition, or good judgment. Sometimes poor judgment is exercised when one unwisely becomes so absorbed in detailed symptoms that important gross symptoms are overlooked. On the other hand, failure to notice details can result in indefinite diagnoses such as ''stress-related decline'' when the real problem was something specific such as a scale infestation.

Essential elements of good judgment in diagnosis are a desire to learn more about the problem, to persist until the problem is solved, and to avoid quickly jumping to the wrong conclusion. This may sound like a detective solving a mystery, and in most cases that is exactly what is being done. One should be able to put together seemingly unrelated pieces of information to arrive at a solution. One of the authors had to examine a pesticide bill of sale from the previous property owner in order to solve a case of suspected chemically induced shade tree phytotoxicity. In another instance, observation of aphid honeydew on the rooftop of an automobile parked some distance away from the problem tree led to questioning about the car's usual parking place and the conclusion that the problem was caused by a previous heavy aphid infestation.

Understand the Tree. The diagnostician must have a thorough understanding of a normal tree. It is necessary to know the name of the particular tree in question and its characteristics such as winter-hardiness, tolerance to dry and wet soils, and reactions to other environmental factors.

Frequently, certain symptom patterns can provide clues as to where the problem lies if the tree's structure and the function of its parts are known. For example, a swelling on a branch coupled with abnormal red leaf color suggests that the phloem just below the swelling is impaired. By knowing that the phloem transports carbohydrates manufactured by the leaves downward to other parts of the tree one would expect that phloem dysfunction would result in carbohydrate accumulation, hence swelling just above that point. In addition, excess carbohydrates are sometimes converted to red pigment-forming sugars, hence red leaf color. Furthermore, when food is not moved to the roots, they starve and the tree may begin to show symptoms of nonfunctioning roots such as mineral nutrient and water deficiency. Therefore, look for a girdling wire, canker, or some other cause of phloem obstruction.

Ask Questions. The good diagnostician will not be afraid to ask questions. The diagnostician must know the history of the tree, such as when and how it was planted. The past climatic history including information regarding drought periods, severity of previous winters, and prevalence of hurricanes or other unusual weather needs to be known. This is acquired either from the diagnostician's records or by asking the owner or person in charge of the tree about it. Critical events such as timing of construction activities and application of deicing salts may not seem important to the tree owner, so this information will not be volunteered and must be obtained through questions.

The timing of symptom occurrence is also important. Unfortunately, many tree owners do not recognize early symptoms and think the tree is healthy until the ''sudden'' appearance of severe symptoms. In addition, tree owners, for some mysterious reason, often withhold information essential for proper diagnosis. Thus questions couched in diplomatic language sometimes yield information not otherwise available. After all, if a tree's trouble is the result of someone's mistake, questions asked in an accusing way will not yield much information.

Know the Tree's Environment. The tree diagnostician must also have a thorough understanding of the relationship between the soil and the tree. Is the soil properly drained? Is it well aerated? Does it hold enough moisture? Is it fertile, based on a soil test? The expert must also have a working knowledge of tree physiology, entomology, and plant pathology. Many of these subjects are treated in some detail in various parts of this book. Additional reading and certainly extensive field experience are necessary before one can really qualify as an expert diagnostician. The primary benefit of field experience is that, when confronted with an unknown problem, at least those maladies with which one has had experience can be ruled out, thus narrowing the remaining possibilities.

Importance of Correct Diagnosis

An incorrect diagnosis at the start will naturally lead to improper treatment. For example, a number of years ago the senior author was called to explain the branch dieback in several large oaks growing on an estate in Westchester County, New York. The problem was thought to result from a lack of mineral nutrients. The trees were fertilized for several years, but branches continued to die back until the top of one of the trees was nearly bare.

A little digging at the base of the trees and probing into the bark just below the soil line showed that the Armillaria root rot fungus had invaded and killed virtually all of the important tissues beneath the bark. In all probability, this fungus had been present in the roots for years and was the prime cause of the dying back of the top. All the fertilizer in the world applied to trees whose roots were diseased would never have done any good.

In another case, the senior author was asked to determine the cause of death of a 150-year-old white oak (*Quercus alba*) on a private property in Glencoe, Illinois. The tree's owner had planned to sue the local utility on the assumption that the tree had died as a result of a natural gas leak about 45 feet away. The leaves on the tree were wilted and brown when examined in late October 1960. This indicated that the tree had been at least partially alive earlier that growing season. Many trees, including another white oak of approximately the same age located closer to the leak, appeared to be perfectly healthy.

A study of the terrain and of the past history of the tree and the property

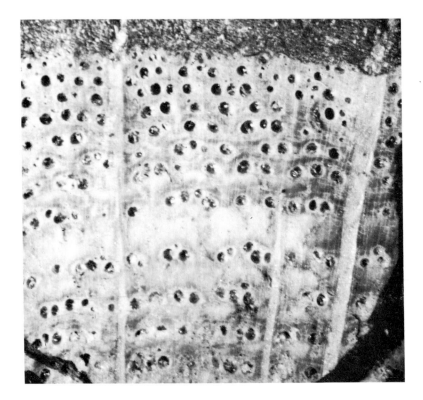

Fig. 9-1A. Annual rings of white oak. Note rings made in the last 6 years are closer to each other than those made 7 to 12 years earlier.

elicited the following information. A very expensive and beautiful house had been built about 7 years earlier, the front of which was within 25 feet of the tree. The grade had been changed. In the area nearest the house at least 2 feet of soil had been placed over the original soil level. In addition to raising the soil level, a circular driveway of asphalt had been constructed around the tree. This paved area covered 67 percent of the absorptive root area of the tree.

One of the lower branches, about 4 inches in diameter, was removed from the tree and cut into 1-foot lengths for more detailed study. Examination of the annual rings in the author's laboratory revealed that the tree had begun to do poorly about 6 years earlier, or about a year after the house and the driveway had been constructed. This is clearly indicated in the enlarged photograph of a cross section of the branch (Fig. 9-1A and B). The rate of growth of a tree can be measured by the spacing between the large-holed cells which form in the spring and are known as spring wood. The wider the space between each row of spring wood cells, the more vigorous the tree's growth.

The tree's demise was actually caused by several factors, including a change in grade, construction of the driveway around the tree, and the use of poor soil

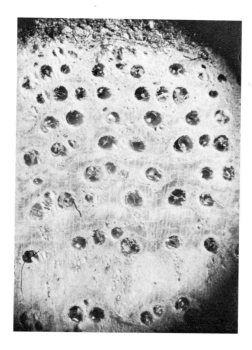

Fig. 9-1B. A larger magnification of Fig. 9-1A showing annual rings of the 8 years prior to the tree's demise.

fill. When faced with the facts and observations gathered by the author, lawyers representing the owner of the tree refrained from damage claims against the utility supplying the natural gas.

One of the more difficult diagnoses involves troubles caused by nutrient deficiencies. Research on symptoms of and cures for mineral element deficiencies has been done for just a few landscape tree species. Generally speaking, foliage deficiencies of the various elements become apparent either at the base or at the tip of the current season's growth, depending on whether the element can move from one part of the plant where it has been used to another to be reused. A deficiency of nitrogen, phosphorus, potassium, magnesium, or zinc is often apparent first on the older leaves, because the plant moves these needed elements to the newest growing tissues. On the other hand, a deficiency of iron, manganese, sulfur, calcium, boron, or copper is apparent first on the youngest leaves, because once incorporated into the older leaves these elements remain there and are not mobilized to satisfy the new growth (see Chapter 6).

Several state colleges of agriculture and private laboratories provide leaf nutrient analysis services for professional arborists. Increased use of these tests may aid arborists and tree owners in diagnosing and controlling ailments caused by nutrient deficiencies and toxicities.

Diagnostic Procedures

Standard procedures used by specialists in diagnosing tree troubles vary slightly with the individual. Some diagnosticians take more stock in symptoms above than below ground. Others, like the authors, feel that the most serious tree troubles and those most difficult to diagnose are more frequently associated with below-ground symptoms and factors.

Before examining any part of an ailing tree, one should study the general surroundings. Are nearby trees and other plants healthy? Have any special treatments been given prior to the appearance of the abnormal condition? Is the tree under diagnosis so situated that a leaf bonfire beneath it may have played a part in its decline? After these and related questions have been answered one should then proceed with the direct examination of the tree.

Examination of Leaves. The leaves constitute the best starting point because they are most accessible and are first to show outwardly the effects of any abnormal condition. Here, also, a complete understanding of a normal leaf is essential, because the size and the color of normal leaves vary greatly among the different tree species and even among trees of the same species.

Insect injuries to leaves are rather easily diagnosed, either by the presence of the pest or by the effect of its feeding. The leaves may be partly or completely chewed, or they may be yellowed as a result of sucking of the leaf sap, blotched from feeding between the leaf surfaces, or deformed from feeding and irritation.

Leaf injuries produced by parasitic fungi are not diagnosed as readily, however, because the causal organisms are usually not visible to the unaided eye. In some instances, tiny black pinpoint fungus bodies in the dead areas, visible without a hand lens, are indicative of the causal nature. Lesions resulting from fungus attack are more or less regular in outline with varying shades of color along the outer edges. They may range in size from tiny dots to spots more than half an inch in diameter. When several spots coalesce, the leaves may become blighted and die.

Atmospheric conditions preceding the appearance of spots on leaves can often be used to advantage in determining the cause of injury. For example, when leaf spots appear after a week or 10 days of continuous rains and cloudy weather, there is a greater likelihood that some parasitic organism is responsible, because such conditions are favorable for its spread. When leaves are spotted or scorched following a week or more of extremely dry, hot weather, lack of water (see Leaf Scorch, Chapter 10) may be responsible. Low temperatures in late spring may also result in much injury to the tender leaves.

Changes in leaf structure, appearance, or function may result from such widely different causes as air pollution, deficient or excessive moisture, lack of available food, poor soil aeration, root injuries, or diseases. Some of these can be disregarded if nearby trees of the same species as those under study are perfectly healthy.

Examination of Trunk and Branches. A careful inspection of the branches and trunk should follow examination of the leaves. Sunken areas in the bark indicate injury to tissues beneath. They may have been produced by fungus or bacterial infection or by nonparasitic agents, such as low and high temperatures, or even by improper pruning. The presence of fungus bodies in such areas does not necessarily indicate that the fungus is the primary cause. Only a person with considerable mycological training can distinguish the pathogenic from the nonpathogenic species. The diseased wood beneath the bark, if it is the direct result of fungus attack, shows a gradual change in color from diseased to healthy tissue. It is usually dark brown in the earlier, more severely infected portions, then changes to olive green or light brown, finally reaching a lighter shade of either in the more recently affected parts. On the other hand, injuries resulting from low or high temperatures are usually well defined, an abrupt line of demarcation appearing between affected and unaffected tissues.

The bark of the trunk and the branches should also be examined for small holes, sawdust, frass, and scars or ridges, which indicate borer infestations in the inner bark, sapwood, or heartwood. As a rule, most borers become established in trees of poor vigor; thus it is necessary to investigate the cause of the weakened condition rather than to assume that the borers are primarily responsible. Branches and smaller twigs should always be examined for infestations of scale insect. Although most scales are readily visible, a few so nearly resemble the color of the bark that they are sometimes overlooked.

Branches or twigs with no leaves or with wilted leaves should be examined for discoloration of the sapwood, the usual symptom of wilt-producing fungi. The service of a mycologist is needed for determining the species of fungus involved, because positive identification can be made only by laboratory isolations from the discolored tissue.

The appearance of suckers or water sprouts along the trunk and main branches may indicate a sudden change in environmental conditions, structural injuries, disease, or excessive, incorrect, and ill-timed pruning.

Look for galls, or swellings on the twigs and branches. Many are insect-caused galls and several such as rust swellings and crown galls are caused by fungi and bacteria. Although galls or overgrowths occasionally present on the main trunk may be caused by parasites, many are produced by factors not clearly understood by scientists.

The general vigor of a tree usually can be ascertained from the color in the bark fissures and the rapidity of callus formation over wounds. In vigorously growing trees, the fissures are much lighter in color than the bark surface. A rapidly developing callus roll over the wound indicates good vigor; however, it does not necessarily mean that the tree is good at limiting internal decay.

Examination of Roots. Because of the inaccessibility of roots, many arborists rarely inspect them. In diagnosing a general disorder in a tree, however, the possibility of root injury or disease must be carefully considered. More than half of the abnormalities in the hundreds of street and shade trees examined by the authors were found to be caused by injuries to or diseases of the root systems.

The sudden death of a tree usually results from the destruction of nearly all the roots or from the death of the tissues at the trunk base near the soil line.

The factors most commonly involved in such cases are infection by fungi such as the Armillaria fungus, the Verticillium wilt fungus, and the Ganoderma fungus, winter injury, rodent damage, lightning strikes, and toxic chemicals like gasoline, oil, salt, and certain weedkillers. Trees that become progressively weaker over a period of years may be affected by girdling roots, decay following injury caused by sidewalk and curb installations or road improvement, poor soil type, poor drainage, lack of food, changes in grade, and excessively deep planting from the start.

Nemas, more commonly referred to as nematodes, are responsible for the decline of many shade trees. The root-knot nema *Meloidogyne incognita* produces small swellings or knots on the roots of susceptible trees. But such swellings or knots are not always caused by nemas; those found on the roots of alder, Australian-pine, and Russian-olive may be natural outgrowths.

Diagnostic Tools

Certain implements are essential in diagnosing tree troubles. Most important of these is a shovel or small spade for removal of soil around the base of the trunk to facilitate detection of girdling roots, rodent damage, winter injury, and infection by the Armillaria fungus or other parasitic fungi. In addition, a spade is used for digging soil in the vicinity of the smaller roots to be examined for the presence of decay. A soil auger or soil sampling tube is essential for collecting representative samples of soil for analytical tests and for determining the nature of the subsoil and the drainage (Fig. 9-2). Pruning shears and a pocket knife are necessary for cutting twigs and small branches to determine the presence of discoloration in the sapwood caused by wilt fungi and other injuries in the inner bark, cambium, and wood. A small saw is helpful where branches larger than can be handled with pruning shears must be cut (Fig.

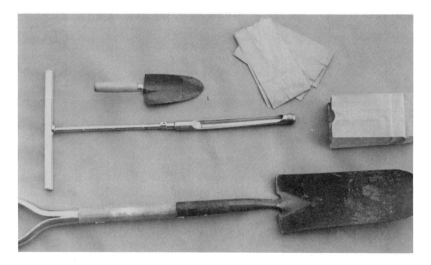

Fig. 9-2. Soil sampling tools: spade, soil auger, trowel, collection bags.

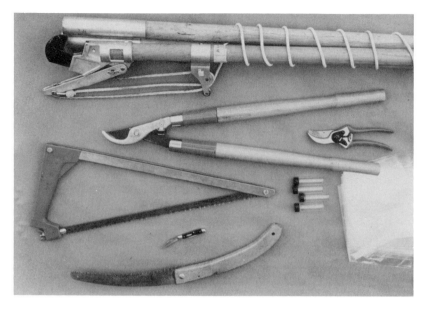

Fig. 9-3. Foliage and limb sample collection tools: folding saw, pocket knife, saw, loppers, pole pruner, collection vials, hand shears, collection bags.

9-3). The authors find that a pole pruner, either telescoping or assembled from short lengths, is excellent for cutting samples. This pruner is easily assembled and then disassembled and stored in an automobile luggage compartment.

A most important tool is a chisel or a curved gouge with a heavy handle for tapping the bark to locate dead areas and for making incisions into the bark and sapwood (Fig. 9-4). Bark and sapwood specimens needed for further study are also collected with this instrument with the aid of strokes from a 2- or 3-pound composition mallet. A sharp instrument such as an ice pick is also useful to help determine the extent of soft, decayed wood.

An increment borer facilitates sampling for study of the annual growth rate of the tree (dendrochronology) and for determining the extent of heartwood decay. One must be reasonably careful with this instrument on living trees, for it is known that discoloration and decay frequently develop around holes made by it. An increment hammer is also available for collecting shallow samples of the bark and wood.

Some diagnosticians are making use of the Shigometer, an electronic instrument which detects internal discoloration and decay in the tree and can also provide a relative measure of its vitality (Fig. 9-5). A slender probe is inserted into a small hole bored into the tree and provides data on the relative resistance or conductivity of the tree tissues being measured. Decaying wood, the cells of which are disrupted, contains higher levels of moisture and salts and therefore conducts electricity more readily than healthy wood. Completely hollowed out parts of the tree are not electrically conductive and yield an instrument reading

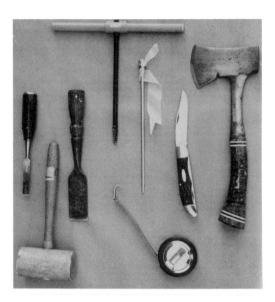

Fig. 9-4. Bark and wood sample collection tools: small wood chisel, mallet, large wood chisel, increment borer, pocket knife, hatchet, diameter tape.

Fig. 9-5. A Shigometer in use.

reflecting less electrical conductivity than healthy wood. Thus without cutting the tree, one can gain insight into its internal condition. The hole bored for the probe is so small that the technique is essentially nondestructive. Using another kind of probe utilizing two sharpened pins, electrical resistance/conductivity of

Fig. 9-6. A, pH meter; B, conductivity meter, a device for measuring soluble salts.

the cambium and outer xylem can be measured. Electrical conductivity of dead tissues examined in this way would be expected to be less than that of healthy tissues. If large numbers of trees are examined at the same time in exactly the same way, the relative vitality of individuals can be determined.

A hand lens is valuable for detecting the presence of red spider mites and tiny fungus bodies in leaf lesions. Binoculars are useful for making observations of symptoms where sampling is not possible.

Screw-top vials should be available for collecting insects to be submitted to a professional entomologist for identification. Plastic bags for holding leaf specimens and paper bags or other containers for holding bark and sapwood specimens, or soil samples, should also constitute part of the diagnostician's working kit.

If the diagnostician has a laboratory, additional tools and instruments are helpful for tree problem diagnosis. A soil pH meter and a device for measuring soil-soluble salts are needed to provide preliminary information for diagnosing some soil nutrient and fertility problems (Fig. 9-6A and B). A dissecting microscope ranging in power from 10× to 70× provides enough magnification

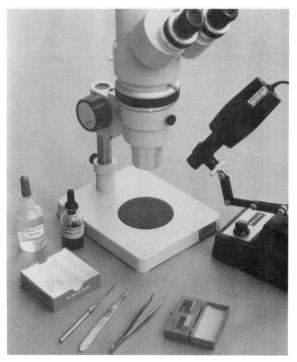

Fig. 9-7. Diagnostic laboratory tools: dissecting microscope, light source, cover slips, forceps, scalpel, pick, microscope slides, water, stain.

to observe tree rings, fungal fruiting structures, nematodes, and insects (Fig. 9-7). Dissecting tools such as scalpels, picks, and blades are also needed. A compound light microscope is helpful to people familiar with its use. This microscope, generally ranging in power from 50× to 1,000×, can be used to detect fungal spores, bacteria, virus inclusion bodies, and plant abnormalities. Transparent tape can be used to make quick microscope mounts of fungi growing on the surface of leaves and stems. Glass slides, cover slips, and dropper bottles containing water and some stains also are needed.

Portable "laboratories" containing some of the diagnostic equipment listed are commercially available.

A Sample Diagnostic Questionnaire

Following is a sample questionnaire that may be sent to tree owners who desire a tree diagnosis by mail:

Name and Address of Sender

THE PROBLEM

1. What are you observing to be wrong with the tree? Describe any abnormalities you see, with regard to size, color, shape, or death of leaves, twigs, branches, limbs, trunk, and roots.

GENERAL QUESTIONS

1. Kind, age, and size of tree?
2. Where is the tree situated? Along street, on lawn, in a park? Near body of water, salt or fresh? On level ground or on a slope?
3. How long has the tree exhibited the trouble? If the trouble appeared suddenly, describe the weather conditions occurring just previously. Describe any other unusual conditions.
4. Is the trouble visible all over the tree, on only one side, in the lower or upper branches?
5. Do any other trees of the same species in the near vicinity show the same injury? Do other species show it?
6. How much annual growth have the twigs made during the past 3 years?
7. Has the grade around the tree been raised or lowered during the past 7 years? If so, explain amount of change and describe the procedure and type of fill used.
8. Has any construction work been done nearby within the past 3 years—house, road, driveway, curbstone, garage, or ditches for laying water or sewer pipes?
9. What work has been done on this tree recently? Has it been pruned within the past 2 years? If so, how much? Fertilizer or pesticide treatments—when, what, and how much?
10. If a young tree, how long since it was planted in its present location? How deeply was it planted? (Use a spade to determine depth of roots.) What treatments were given during the first year after transplanting?

QUESTIONS ABOUT THE SOIL

Answers to these questions are extremely helpful, particularly if the tree has been dying back slowly over a period of years.

1. What kind of soil surrounds the tree? Sandy, loamy, or clayey? What soil cover—asphalt, cement, crushed rock, cinders, sand, grass, mulch, weeds, or no cover? How much open area is there?
2. What is the depth to subsoil, to rock or shale, to hardpan?
3. What is the pH reaction of the soil?
4. Does water stand on the soil after a heavy rain?

QUESTIONS ABOUT THE ROOTS

These are the most difficult questions to answer, since considerable digging may be required. The answers are most important, however, especially when a general disorder is involved.

1. Is a girdling root present? Sometimes such roots are well below ground and you may have to dig a foot or so before you can be sure.
2. Are the larger roots normal in color? Do they have rotted bark or discolored wood? If so, submit specimens.
3. What is the appearance of the finer roots? Are root hairs abundant and white? (Dig down a foot or so beneath the outer spread of the branches.)

QUESTIONS ABOUT THE TRUNK

1. Are there any long, narrow, open cracks present? If so, which direction do they face?
2. Are cavities present? If so, describe size of opening, condition of interior if unfilled. If filled, give details.
3. Is there any bark bleeding? If so, how extensive? Submit specimens including bark and sapwood.
4. Are there any swollen areas? Describe.
5. Is there a swollen area completely around the trunk? If so, cut into the swelling with a chisel to determine the possible presence of some foreign object such as a wire.
6. Are there any fungi (mushrooms or bracket-type) growing out of the bark? If so, include a few specimens.
7. Are there cankers (dead sunken areas in the bark)? Submit specimens.
8. Are there any borer holes or other evidences of insect work?
9. Is the bark at or just below the soil line healthy? (Use a curved chisel to determine this.)
10. If wood beneath this bark is discolored, describe color and extent. Submit several pieces of bark and wood.

QUESTIONS ABOUT THE BRANCHES

1. Is the bark cracked for some distance? On what side of the branches?
2. Are there any cankers in the branch?
3. Is there any discoloration in the branches or twigs which have wilted leaves or which are leafless?

SPECIFIC POSSIBILITIES

1. Did the trouble appear
   a. Immediately after a thunderstorm?
   b. After chemicals were injected into the trunk?
   c. After sprays were applied? (Name the ingredients used and indicate when applied).
   d. After weed killers were applied in the vicinity?
   e. After treating a nearby cellar for termite control?
   f. After any other chemical treatment?

SHIPPING THE SPECIMENS

By the time you have answered the questions that pertain specifically to your particular case, you are ready to collect and ship specimens.

Woody specimens without leaves should be wrapped in newspaper and packed in a pasteboard box. Twigs and small branches should be cut to approximately 1 foot in length.

Include with diseased branches or twigs a portion of the adjacent healthy parts. (When submitting branches without leaves be sure to give the name of the tree.)

Leaves should be wrapped dry, preferably individually, in ordinary wax paper, then in wrapping paper or in a pasteboard box. Never wrap leaves in wet paper toweling, wet cotton, or plastic sheets because many of the specimens will be decayed or covered with all kinds of organisms by the time they reach the diagnostician.

Trunk specimens should include a portion of the dead bark and sapwood. If possible, adjacent healthy tissue also should be included on the same specimens. Bark and sapwood specimens should be about 2 inches long and 1 inch wide and separately packed in waxed paper.

Label all specimens and have this label correspond with the description in your letter or on your questionnaire.

Certain answers to the questionnaire will result in immediate recognition of the problem by most people, especially readers of this book. Many state university colleges of agriculture have plant disease diagnostic laboratories staffed with professional diagnosticians and diagnostic forms with questions appropriate for their lab. Most county extension offices have access to such laboratories, and specimens taken there are often mailed to the university lab. Some private diagnosticians also provide questionnaires and shipping instructions.

Selected Bibliography

Anonymous. 1960. Index of plant diseases in the United States. U.S.D.A. Handbook 165, Washington, D.C. 531 pp.

Anonymous. 1973. A tree hurts too. U.S.D.A. Forest Service. NE-INF. 16–73, Washington, D.C. 28 pp.

Drooz, A. T. 1985. Insects of eastern forests. U.S.D.A. Forest Service Misc. Publ. 1426, Washington, D.C. 608 pp.

Grogan, R. 1981. The art and science of diagnosis. Ann. Rev. Phytopathol. 19:333–51.

Hacskaylo, J., R. F. Finn, and J. P. Vimmerstedt. 1969. Deficiency symptoms of some forest trees. Ohio Agricultural Research and Development Center Research Bull. 1015:68 pp.

Hepting, G. H. 1971. Diseases of forest and shade trees of the United States. U.S.D.A. Forest Service Agriculture Handbook 386, Washington, D.C. 658 pp.

Horst, R. K. 1982. Westcott's plant disease handbook, 4th ed. Van Nostrand Reinhold, New York. 803 pp.

Johnson, W. T., and H. H. Lyon. 1976. Insects that feed on trees and shrubs. Cornell University Press, Ithaca, N.Y. 464 pp.

Manion, P. D. 1981. Tree disease concepts. Prentice-Hall, Englewood Cliffs, N.J. 399 pp.

Peace, T. R. 1962. Pathology of trees and shrubs. Clarendon Press, Oxford, England. 722 pp.

Pirone, P. P. 1978. Diseases and pests of ornamental plants, 5th ed. Wiley, New York. 566 pp.

Rose, A. H., and O. H. Lindquist. 1982. Insects of eastern hardwood trees. Can. For. Serv. For. Tech. Rep. 29:304 pp.

Sinclair, W. A., H. H. Lyon, and W. T. Johnson. 1987. Diseases of trees and shrubs. Cornell University Press, Ithaca and London. 574 pp.

Streets, R. B., Sr. 1969. The diagnosis of plant diseases. University of Arizona Agric. Exp. Sta. Extension Service, Tucson. 230 pp.

Tattar, T. A. 1978. Diseases of shade trees. Academic Press, New York. 361 pp.

Wallace, H. R. 1978. The diagnosis of plant diseases of complex etiology. Ann. Rev. Phytopathol. 16:379–402.

10

Damage Due to Nonparasitic Factors

Poor growth and even death of trees are frequently caused by nonparasitic agents. Nonparasitic injuries generally affect a wide range of tree species, whereas most parasites such as disease organisms and insects attack specific tree species. Following are some of the most common nonparasitic problems.

Damage Caused by Grade Changes and Construction Around Trees

The addition or removal of soil may seriously disturb the delicate relationship between roots and soil and thus result in considerable damage or death of the tree. Yet such procedures are done when new homes and highways are built, lawns or terraces graded, or street improvements made.

The authors are often asked to diagnose the cause of the gradual weakening and even the death of beautiful oaks, maples, or beeches on properties in newly developed residential sections (Fig. 10-1). When homes are built on wooded lots, the possibility of soil fill injury is a primary consideration. First the trunk base is examined to see whether it flares out. Where no fill has been applied, the trunk is wider at the soil line than it is a foot or so above. If the trunk has no flare at its base and it enters the ground in a straight line, then a fill has probably been applied. Some digging with a good spade will then reveal the depth of fill (Fig. 10-2).

Other visible symptoms are small yellow leaves, premature fall color, the presence of numerous suckers (epicormic growth) along the main trunk and branches, many dead twigs, and in some instances large dead branches. When the owner is questioned as to what precautions were taken to minimize the shock of placing the roots in a new environment, the usual answer is that no precautions were thought necessary. The building contractor dug the subsoil from the cellar location and spread it around the trees. A foot of good topsoil was placed over this and the job was done! When symptoms become visible a few months to several years later, the results are angry and frustrated homeowners who had built the house of their dreams on a costly wooded lot.

Why Trees Suffer from Soil Fills. The moment a blanket of soil is placed

Fig. 10-1. Soil fill placed around tree roots during construction caused the death of these trees in a new housing development.

Fig. 10-2. Soil removed to show the depth of fill placed over the tree root system.

over the existing soil a marked disturbance of soil water and the amount and kind of air existing in the soil occurs. A decrease in oxygen content of soil yields an anaerobic soil environment. Air (primarily oxygen) and water are essential for normal functioning of roots. Soil microorganisms necessary to break down soil organic matter that serves as food for the roots also need air. In addition, when air is lacking, certain gases and chemicals may increase and become toxic to roots. Toxins produced by anaerobic bacteria may be as harmful as the damage resulting from asphyxiation due to lack of oxygen. Fills may raise the water table, increasing the prospect of an anaerobic root environment, or fills may impede water penetration, subjecting the tree to drought.

The roots do not readily regenerate in the new soil layer, and as a result the roots die; symptoms may become visible in the aboveground parts within a month, or they may not appear for several years. The stockpiling of soil around trees for later use may also have a detrimental effect, especially if it occurs during the growing season.

Factors Governing the Extent of Injury from Fills. The extent of injury from fills varies with the species, age, and condition of the tree; the depth and type of fill; drainage; and subsequent exposure to parasites and nonparasitic agents. Sugar maple, beech, dogwood, oak, tuliptree, pines, and spruce are most severely injured; birch, hickory, and hemlock suffer less; elm, poplar, willow, planetree, pin oak, and locust are least affected. Older trees may be more sensitive to root smothering than younger trees, although both are affected. Trees in weak vegetative condition at the time the fill is made are more severely injured than trees in good vigor.

Obviously, the deeper the fill, the more marked is the disturbance to the roots, and consequently the more serious are the effects. Clay soil fills cause most injury, because the fineness of the soil shuts out air and water most nearly completely. The application of only an inch or two of clay soil may cause severe injury. Gravelly fills cause least trouble because air and water permeate them more readily. As a rule, the application of a layer several inches deep of gravelly soil, or even of the same kind of soil in which the tree has been growing, will do little harm. Such trees eventually become accustomed to the new situation by producing additional roots near the surface. Some trees are able to survive despite the addition of a considerable amount of soil over the original level. They apparently produce a layer of roots some distance above the original ones.

Treating Soil Fill Injury. Injury caused by soil fills can often be prevented (see Chapter 8). It is much more difficult to rescue trees from the effects of existing fills. All too often, the problem is not diagnosed until severe symptoms of decline are apparent.

Lowering Grades. Depending on how much soil is removed, lowering the grade can be as disastrous as raising the soil level. Since the majority of the tree's roots are in the top 18 inches of soil, and most of the feeder roots are in the top 6 inches, removal of only a few inches can cause serious harm. Lowering the grade removes valuable topsoil, exposes feeder roots to drying out and cold temperatures, severs roots needed for anchorage and water transport,

Fig. 10-3. Many tree roots were severed to install a curb.

and lowers the water table. Symptoms of grade lowering injury are similar to those for trees having soil fills: leaf stunting and chlorosis, epicormic growth along branches and trunk, twig and branch dieback, and death. The extent of damage done to the tree by grade lowering will be greatest when the grade cut is made deep and near the trunk (Fig. 10-3). Some trees will not suffer appreciably from a removal of only a few inches of topsoil in an area well away from the trunk.

Excavations. Excavation for building foundations and utility lines damages trees by severing the roots (Fig. 10-4). Depending on how near the tree the excavation occurs, symptoms of damage like that for grade changes may range from slight for a trench made outside the tree branch spread to severe decline and death following trenching close to the trunk. There are virtually no remedial measures for the latter case, although providing adequate fertility and water to the root system remaining may help. Insofar as possible, trees should be protected when excavations are made for water and sewage lines. The trenches should be located away from the trees. If they cannot be routed around the trees, the best alternative is to tunnel under them. Power-driven soil augers are available for this job (see Chapter 8).

Soil Compaction. Roots suffer from soil compaction at many construction sites, despite physical protection of roots from cuts, fills, and excavation. Soil compaction involves physical compression of the soil from the surface so that

Fig. 10-4. Excavation for building foundation severed many tree roots.

as soil particles are pressed together, the spaces between them become smaller. Compacted soil is not easily penetrated by water and air, two essential needs for roots. Thus tree growth is affected adversely, and typical decline symptoms set in. Soil compaction results from construction equipment traffic, piles of building supplies, soil piles, and even foot traffic (Fig. 10-5). Compaction occurs more readily in clay than sandy soils and in wet more than drier soils.

Paving. Applying a layer of concrete or asphalt over the root zone of existing trees has much the same effect on the trees as adding soil fill or compacting the soil. Often paving is done following a shallow excavation or grade lowering to establish a foundation for the paving (Fig. 10-6). The extent of tree decline which results will be in proportion to the amount of damage done to the tree's

Fig. 10-5. Soil compaction during construction results from heavy equipment traffic.

Fig. 10-6. This tree, remaining after a parking lot was constructed, lost nearly all of its effective roots.

roots. Root systems prevented from getting water and air by the impervious paving layer are not easily rescued from the impending damage.

Injury by Girdling Roots

Many trees are weakened and some are killed by roots that grow closely appressed to the main trunk or large laterals in such a way as to choke the members they surround, much as does a wire left around a branch for a number of years. The most likely effect of this strangulation is to restrict movement of carbohydrates to the roots through the phloem, thus leading to gradual root starvation. Starved roots provide less water and nutrients to the top. This choking action may also restrict the movement of nutrients in the trunk or in the strangled area of the large roots. When a large lateral root is severely girdled, the branches that depend on it for nutrients commonly show weak vegetative growth and may eventually die of starvation. If the trunk of a tree is severely girdled, the main branch leader may die back. As a rule, trees affected by girdling roots do not die suddenly, but become progressively weaker over a 5- or 10-year period despite good pruning and fertilization practices.

Girdling roots that develop below ground level can often be detected by examining the trunk base. If the trunk ascends straight up from the ground, as it does when a soil fill has been made, or is slightly concave on one side, instead of showing a normal flare (swelling) or buttress at the soil line, then one can suspect a girdling root. Of course digging the soil alongside the trunk should reveal whether there is a strangling member.

Trees growing along paved streets suffer more from girdling roots than do those in open areas, and middle-aged or old trees more than younger ones. Norway and sugar maples, oaks, elms, and pines are affected most frequently. Girdling roots occur on transplanted trees, whereas trees growing in the wild do not seem to have the problem.

Development of Girdling Roots. Roots are deflected from their normal spreading course by unfavorable environmental conditions. They grow most rapidly in areas where air, moisture, and food supply are best. A good case in point is that of a tree growing along a city street. The roots of such a tree, when newly planted, are usually spread out in all directions by the planter. For a number of years the roots spread in the general direction in which they are originally placed. Because of unfavorable conditions under the pavement, the larger roots and especially the younger ones will eventually grow away from the street toward the open areas between the curb and sidewalk and toward lawns. In the course of bending, the younger roots cross over older roots and become closely appressed to the base of the main trunk. Such roots grow and expand rapidly because of the more favorable conditions and eventually press against or strangle the roots they cross or the trunk base they touch (Fig. 10-7).

Root girdling frequently occurs where a street tree is planted in a hole filled with good soil and surrounded by hard, impervious subsoil. Inability to pene-

Fig. 10-7. Treatment of girdling roots. **Top:** A girdling root. **Center:** Removing the root with a chisel and mallet. **Bottom:** The girdling root has been removed.

trate the more unfavorable soil forces the roots to develop within the filled area, thus favoring girdling.

Careless placement or overcrowding of the roots at transplanting time may also lead to root girdling and the rotting of the crowns. This has been observed not only with shade trees along city streets, but also with red, white, and Scotch pines planted in reforestation projects. In the latter case, a direct correlation exists between careless placement of roots and the amount of subsequent infection in the root crown by fungi and bacteria. As twisted roots develop, air and water pockets are formed. These make ideal breeding places for organisms that produce root and crown decay.

In a few instances, fertilization of the soil near the trunk of young street trees has stimulated new root production, produced overcrowding, and eventually resulted in the formation of girdling roots.

Girdling roots are more numerous in compacted than in loose soils.

Container-grown trees are more likely to be subject to girdling roots than are open-grown trees. Trees left too long in containers develop roots that spiral along the walls. Unless such roots are straightened or cut out at the time of final transplant, they may cause girdling of the stem later.

An increased problem with girdling roots has been associated with dense shade over the soil near the tree trunk.

Diagnosis of Girdling Roots. The best time to determine the presence of girdling roots is in late fall just before frost. At that time, the side of the tree above the girdled area will show lighter green leaves, which tend to abscise prematurely. One must always bear in mind, however, that other factors may produce similar symptoms and that the most positive sign is the discovery of girdling roots. Such roots can be uncovered by carefully removing the soil around the trunk on the side showing weak top growth, poor bark development, or a pronounced swelling at the base of the trunk. Usually the girdling root is found at or a few feet below the soil surface. Occasionally it may be deeper if the taproot is being strangled.

Treatment for Girdling Roots. If the girdling root has not yet seriously impaired the tree's chance of recovery, it should be severed with a chisel and mallet (Fig. 10-7, center) at its point of attachment to the trunk or to the large lateral root. A few inches should be cut from the severed end to prevent its reuniting with the member from which it was severed. It may not be necessary to actually remove the offending root segment, especially if removal of a deeply embedded root would cause more trunk or root wounding. Finally, the soil is replaced in its original position. If the tree has been considerably weakened, judicious watering should be practiced and, if needed, fertilizer applied.

Gas Injury

In urban areas, natural gas is commonly transported through underground pipelines, often near the root systems of landscape trees. Although natural gas (composed mostly of methane) is not directly toxic to trees, natural gas seeping from a leaking gas main can have adverse effects on the environment of the roots, thus indirectly injuring nearby trees. When a leak occurs, gas gradually fills up the air spaces in the surrounding soil, displacing oxygen as it does so. Microbial changes occur in this oxygen-depleted environment which lead to more anaerobic conditions and eventually to the production of hydrogen sulfide by soil bacteria. Hydrogen sulfide inhibits root respiration and nutrient uptake, causing death of the tree. Other toxins produced anaerobically by soil micro-organisms may also be toxic to tree roots.

Diagnosis of Gas Injury. Since the symptoms of gas injury are similar to symptoms caused by other disorders, be certain that in fact there is a gas line in the vicinity. Gas detection meters are available for detecting leaks. Gas company officials want to repair gas leaks, so they are likely to cooperate in the effort to find one. Look for a circular pattern of declining or dying plants of all types in the vicinity of the suspected leak. The plants will show the most injury near the center of the area and die more rapidly from a large leak than a small one. Dig up soil from the affected area. Soil from a serious gas leak will have a dark or black color and a characteristic sour odor. Roots present in such soil will be dead.

Treatment of Gas Leaks. First the leaky gas main must be repaired. The soil needs to be aerated by drilling auger holes or forcing compressed air through a water needle inserted into the affected soil. If affected plants have enough live tissue to warrant saving, dead branches should be removed and fertilizer and water applied as needed. Where gas such as methane is continually being produced in the soil, as in the case of some reclaimed landfills, trees and shrubs may not be feasibly planted until gas levels decrease to safe levels.

Injuries Caused by Low Temperatures

Low temperatures may injure trees in either vegetative or dormant condition. Injury is most common in regions where seasonal variations are greatest. For purposes of discussion, low-temperature injuries may be divided according to the season of the year in which they occur, that is, spring, fall, and winter.

Spring Frosts. Sudden drops to freezing temperature or below result in wilting, blackening, and death of tender twigs, blossoms, and leaves of deciduous trees, or reddening of the needles and defoliation of the newest growth of needle evergreen trees (Fig. 10-8). Swollen buds of broad-leaved trees exposed to freezing temperatures may give rise to expanding leaves having symmetrically placed holes. Because of cold air settling in local areas, trees growing in valleys and low areas are damaged more often than those on higher ground. Even in individual trees, frost injury may occur on lower branches while the treetop escapes damage.

Warm days in early spring stimulate premature growth, which is readily injured by the cold days and nights that follow. Injury is most severe when the low temperatures occur later in spring when the new growth is well advanced. The freeze of April 22 and 23, 1986, which followed a lengthy mild period, injured flowers, fruits, leaves, and twigs of many trees in the Ohio River valley region of the United States.

Fig. 10-8. Spring frost injury to yew.

Autumn Frosts. A cool summer followed by a warm autumn prolongs the growing season. Under such conditions twigs, buds, and branches fail to mature properly and therefore are more subject to injury by early autumn frosts. Unseasonable cold waves in the fall may also result in much damage. The disastrous frost that hit the Pacific Northwest on November 11, 1955, severely injured and killed thousands of trees and shrubs.

Winter Injury. Despite their dormant condition, trees frequently suffer severely during the winter. Their winter-hardiness is influenced by drainage, location, natural protection, species of tree, and character of the root system, as well as by a combination of unfavorable weather conditions.

Although more sensitive to cold temperatures than other plant parts, tree roots are not normally injured by cold temperatures because of the insulating effects of the soil. The roots of trees are more likely to freeze in poorly drained than in well-drained soils, maple, ash, elm, and pine roots being most susceptible. The effects of frozen roots are seldom noted until the following summer, when the aboveground parts wilt and die. Roots are injured most frequently during winters of little snowfall or in soils bare of small plants and other vegetation.

With the advent of container-grown trees in nurseries and the use of aboveground tree planters in urban landscapes, the possibility of tree roots becoming damaged by cold temperatures increases.

Cold winter temperatures are capable of injuring tree tissues either because the species is not locally hardy or because usual hardening off processes have

Fig. 10-9. Winter injury to trunk cambium and wood on *Prunus*.

not occurred. Properly hardened off trunk and limb tissues usually withstand lower temperatures than twig, leaf bud, or flower bud tissues. It is not unusual for some marginally adapted flowering trees to fail to bloom following extremely cold January temperatures, which kill the flower buds but not the trees. However, early winter cold coming before the tree has completely hardened can be more injurious to the trunk and branch tissues than to other parts of the tree. This kind of winter injury often follows a sudden drop in temperature and is potentially more damaging than midwinter cold. If the cold has injured the cambium, then the ability to make new xylem and phloem is lost and the tree could die. Very cold late winter weather following partial dehardening could also injure trunk tissues more than bud tissues.

We have cut into the bark and wood of numerous peach trees following such winters and observed dark, discolored trunk phloem, cambium, and xylem tissues while limb and twig tissues were healthy and green, as were root collar tissues protected from the cold (Fig. 10-9). Such trees were struggling to survive despite producing flowers and leaves in the spring. Some injured trees died by midsummer despite starting the season with healthy roots and foliage because their transport systems had been injured. Valuable trees have been rescued from this kind of damage using a practice known as bridge grafting.

The ability of the tree to withstand cold is also governed by factors influencing the degree of hardiness acquired during the weeks preceding cold weather. Trees still actively growing in late fall because of excessive nitrogen fertilization or because of mild, moist weather may be injured when the first cold weather strikes. However, there are trees not affected by late fertilization and mild weather, perhaps hardening because of gradually decreasing hours of daylight. The subject of winter-hardiness and cold temperature injuries needs further research. Aside from damage to flower buds, trunk and branch tissues, and roots, cold temperatures during the winter initiate or aggravate more or less localized injuries on the trunks and branches. These are known as frost cracks, cup shakes, and frost cankers.

FROST CRACKS. Longitudinal separations of the bark and wood are known as frost cracks. These openings may be large enough to permit the insertion of a hand and may extend in a radial direction to the center of the tree or beyond.

Frost cracks are most likely to form in periods of wide temperature fluctuations. The water in the wood cells near the outer surface of the trunk moves out of the cells and freezes during sudden drops in temperature. The loss of water from these cells results in the drying of the wood in much the same way as green lumber dries and cracks when exposed to the sun. At the same time, the temperature of the cells in the center of the trunk remains much higher and little drying of these cells or shrinkage of the wood takes place. The unequal shrinkage between the outer and the inner layers of wood sets up great tension, which is released only by the separation of the layers. The break occurs suddenly along the grain of the wood and is usually accompanied by a loud report (Fig. 10-10).

Cracks formed in this manner appear principally on the south and west sides of the trunk, since these are heated by the sun's rays and a greater gradient prevails when the temperatures drop suddenly during the night. If all exposures of the trunk were heated evenly, tension and frost cracks would not develop, as all tissues would shrink at the same rate.

Although the cracks are created and observed during cold weather, it is apparent that injuries to the trunk and roots, often made years before, are the reason the crack appears where it does (Fig. 10-11).

Deciduous trees appear to be more subject to cracking than do evergreens. Among the former apple and crabapple, ashes, beech, cherry, goldenrain-tree, horsechestnut, linden, London planetree, certain maples, pin oak, tuliptree, walnut, and willow suffer most. Isolated trees are more susceptible than those growing in woodland areas, and trees at their most vigorous age (6–18 in. in diameter) are more subject than very young or old trees. Probably because of root injury or decay, trees growing in poorly drained sites are more subject to cracking than those growing in drier, better drained soils.

When warmer temperatures arrive, the frozen tissues thaw and absorb more water, and the crack closes. The cracked zone in the wood never closes completely, even though the surface may be sealed by callus formation. Because of insufficient support, however, the same area again splits over the following winter. The repeated splitting and callusing eventually results in the formation of a considerable mass of tissue over the seam.

The most serious aspect of frost cracks is that they are a potentially hazardous defect and they provide conditions favorable for the entrance of wood-decay fungi. Avoiding unnecessary injuries and pruning properly should reduce

Fig. 10-10. Frost crack on trunk of a young horsechestnut tree.

frost cracking. Young trees susceptible to sunscald injury leading to frost cracks may be protected by attaching sisal-kraft paper or strips of burlap to the trunk in late fall or by applying a coat of whitewash. Tying a wide board upright on the south side of the trunk will also help.

CUP SHAKES. Wide temperature fluctuations within a relatively short period

Fig. 10-11. Internal extent of frost crack injury. The crack was probably initiated by an injury when the tree was very young.

produce a type of injury known as cup shakes. This results from conditions that are the reverse of those responsible for frost cracks. A sudden heating by sunshine of the outer tissues of the trunk, following low temperatures, will cause these tissues to expand more rapidly than the inner tissues, resulting in a cleavage or separation along an annual ring.

FROST CANKERS. Low-temperature injuries may be confined to small localized areas on the trunk, branches, or in crotches. The lesions or cankers resulting from this limited injury are common on maple and London planetree and are confined principally to the southern and western exposures of the trunk. Scalding by the sun's rays on these exposures may penetrate as deeply as the cambium, resulting in a well-defined lesion.

Winter Drying of Evergreens. Winter injury to such plants as rhododendron, laurel, holly, pines, spruces, and firs, expressed as winter drying, is rarely caused by excessive cold during the winter. The damage is caused, rather, by desiccation.

Evergreen plants transpire continuously. Although water loss is low during winter months, it may increase considerably when the plants are subjected to drying winds or are growing in warm sunny spots. When roots are embedded in frozen or dry soil, water lost through the leaves cannot be replaced by root uptake, and drying out occurs.

Fig. 10-12. Winter injury to rhododendrons and other broad-leaved evergreens appears as browning along the edges of the leaves.

In the earlier stages, winter injury on broad-leaved evergreens is evident as a scorching of the leaves at the tips and along the outer margins (Fig. 10-12). The color of the affected parts tends to be brown rather than yellow. On narrow-leaved evergreens, the needles are browned entirely or from the tips downward along part of their length. The terminal buds and twigs are brittle and snap readily when bent.

Prevention of Winter Injury. From the foregoing discussion it is obvious that the ability of a plant to withstand low temperatures depends on many factors beyond human control. In some instances, however, precautions can be taken to reduce the possibility of winter damage.

The selection of well-drained soils as sites for trees cannot be overemphasized. Trees considered susceptible always withstand low temperatures better when planted in soil with good aeration and water drainage. Soil aeration to improve soil drainage around existing trees may also help.

To protect broad-leaved evergreens, such as rhododendron and laurel, wind-breaks of coniferous evergreens, which as a rule suffer less severely, are to be encouraged. Heavy mulches of oak-leaf mold or acid peat moss to prevent deep freezing and thus to facilitate water absorption by the roots are of great assistance in avoiding winter injury. To ensure an ample water supply during the winter months, soils around evergreens should be thoroughly soaked before freezing weather sets in.

Winter browning can be prevented, or at least greatly reduced, by spraying the leaves with antitranspirants. These materials should be applied with a pressure sprayer on a mild day in late fall or early winter. The senior author has seen many kinds of evergreens, including rhododendrons, boxwoods, and mountain laurel, which came through the winter undamaged following an application of such materials. The argument against using antitranspirants, which could inhibit foliar gas exchange needed for photosynthesis following transplanting, would not be made for winter protection, because gas exchange is less critical in winter.

Care of Winter-Injured Plants. In any attempt to aid a winter-injured plant to regain its vigor, several precautions should be taken.

Drastic pruning of such plants is not advisable, since additional harm is likely to occur. Moreover, pruning should be deferred until the buds open in the spring, when dead wood can easily be distinguished from live. A moderately pruned tree will recover more quickly than one severely pruned or not pruned. It is impractical to prune winter-injured needles or leaves of evergreens, but all dead wood should, of course, be removed.

Some disagreement exists among arborists as to the specific fertilizer practices to be followed on severely winter-injured trees. Those who favor omission of fertilizer for the entire season following injury claim that heavy spring applications may either further injure already damaged roots or result in the increased production of foliage, whose demands for water might overtax injured root and trunk conducting tissues. These advocates say that a moderate application of fertilizer might be justified about the first of July, since some new water-conducting tissue will have been formed by that time.

Advocates of medium to heavy applications of fertilizers in the spring following winter injury claim that such treatments encourage the formation of new tissue, thus making it possible for the tree better to withstand summer conditions. After all, it is the foliage which manufactures the raw materials needed for repair of injured tissues.

Sunscald

The bark of trees in some circumstances is subject to injury through heating by the sun. The sunscalded bark will appear discolored and dried out, often in a long strip on one side (usually the southwest, the sun's direction during the hottest part of the day) of the trunk. Discolored, dead tissue is also present under the bark in the cambial tissues so the trunk is no longer expanding in that region. After a time, the injured area takes on a sunken appearance and the tree begins to produce callus tissue outside the killed area. Frequently the wood under the sunscalded tissue will decay, the decay eventually occupying the entire woody cylinder present at the time of the initial injury. In extreme cases, such weakened trees can break.

Sunscald is more likely to occur on shade-grown thin-barked trees following sudden exposure to direct sun. This occurs after transplanting a young tree from a closely planted site in a nursery to an open site in the landscape. Similarly, young, thin-barked trees on a wooded lot can be injured by sudden exposure to the direct sun following clearing for construction. Again, injuries to roots and branches during these critical operations enhance the likelihood of sunscald damage. Sunscald can occur at any time of year, even in winter, although differential heating and cooling of bark tissues from day to night may be causing the injury in winter.

Light colored or reflective tree wraps are normally used to protect newly planted trees from direct sun. White latex paint protects otherwise dark, heat-absorbing tree trunks by reflecting sunlight; painted trunks, however, are rarely accepted in the landscape. Any reasonable means of shading tree trunks on the southwest for the first years of exposure should suffice. Providing adequate water for newly planted and newly exposed trees is also essential. Certainly, avoiding unnecessary injury to trees should help, but because transplanting purposely inflicts tree root injury, these trees are especially vulnerable during their first two years.

Leaf Scorch

Although the trouble known as abiotic leaf scorch is most prevalent on Japanese red, sugar, silver, Norway, and sycamore maples, it also occurs on dogwood, horsechestnut, ash, elm, and beeches and to a lesser extent on oaks and other deciduous trees.

Symptoms. Scattered areas in the leaf, appearing first between the veins or along the margin, turn light or dark brown (Fig. 10-13). The edges of the discolored areas are very irregular. All the leaves on a given branch appear to be affected more or less uniformly. When a considerable area of the leaf surface is discolored, the canopy of the tree assumes a dry, scorched appearance. In severe cases, the leaves dry up completely then fall prematurely. When such defoliation occurs before midsummer, new leaves are formed before fall. In most cases, the leaves remain on the tree, and little damage results.

The position of the most severely affected leaves can be used, in many instances, as a means of distinguishing leaf scorch from leaf spot and blight damage caused by fungi. Leaves affected by scorch are usually most abundant on the side of the tree exposed to the prevailing winds or to the most intense rays of the sun, whereas spotted and blight leaves resulting from fungus attack are usually scattered throughout the treetrop, the greatest numbers being on the lower, more densely shaded parts.

Biotic scorch caused by xylem-limited bacteria more nearly resembles abiotic scorch, but patterns of leaf browning may differ. Xylem-limited bacteria frequently cause browning along the leaf margin and damage appears in any part of the tree. Furthermore, the problem seems limited to the southern and mid-Atlantic regions of the United States.

Cause. Leaves are scorched when the roots fail to supply sufficient water to compensate for that lost at critical periods. Newly transplanted trees are frequently victims of abiotic scorch. They are most severely affected during periods of high temperatures and drying winds. The inability to supply the

Fig. 10-13. Leaf scorch as a result of drought on flowering dogwood.

necessary water is influenced by the moisture content of the soil and by the location and condition of the root system. Even the most extensive root system cannot supply enough water to compensate for the tremendous amounts lost through the leaves if the moisture content of the soil is low because of a prolonged drought.

Though of less common occurrence, scorch may appear in trees growing in excessively wet soils. Under such conditions lack of air greatly impedes the water-absorbing capabilities of the roots.

Trees with diseased roots, those whose root systems have been reduced as a result of transplanting, curb installation, or building operations, and those whose roots are restricted by or covered with impervious materials are most subject to leaf scorch. For example, sugar maples adjacent to paved concrete driveways and near street intersections are commonly observed to have more scorch than those more favorably situated along the same street or those growing on lawns or along country roads.

Where injury to a part of the root system is severe and permanent, or where soil conditions are continuously unfavorable, leaf scorch may occur regularly. In fact, some trees show leaf scorch every year regardless of whether the season is cool and moist or hot and dry.

Heavy infestations of aphids and other sucking insects usually contribute to the severity of leaf scorch. Some investigators feel that the trouble is associated with a deficiency of potassium in the leaves. They believe that potassium-deficient leaves tend to lose water more rapidly and consequently become scorched more readily.

Control. Any practice that increases root development and improves the tree generally will help reduce leaf scorch. The application of chemical fertilizers, especially those containing potassium, is suggested. In addition, insects and fungus diseases should be kept under control.

Where soils are likely to become dry, water should be applied, especially during seasons of low rainfall. Where the soil is heavy and compacted, some improvement will result from breaking up the surface layers, vertical augering, or from inserting upright tile to facilitate penetration of water. In the rarer cases where scorch results from the inability of the roots to absorb water because of excessively wet soils, drain tiles may need to be installed.

Leaf scorching in particularly valuable trees can be prevented, or at least reduced, by spraying the leaves with antitranspirant before the regular ''scorching'' period.

Lightning Injury

Every year thousands of shade and ornamental trees throughout the country are struck by lightning. The amount and the type of injury are extremely variable and appear to be governed by the voltage of the charge, the moisture content of the part struck, and the species of tree involved.

The woody parts of the tree may be completely shattered (Fig. 10-14) and may then burn. A thin strip of bark parallel to the wood fibers down the entire length of the trunk may be burned or stripped off, the internal tissues may be severely burned without external evidence, or part or all of the roots may be killed. The upper trunk and branches of evergreens, especially spruces, may be killed outright, while the lower portions remain unaffected. In crowded groves, trees close to one directly hit will also die. In many cases, grass and other vegetation growing near a stricken trunk will be killed.

So-called hot bolts with temperatures over 25,000° F will make an entire tree burst into flames, while cold lightning—striking at 20,000 miles per second—can make a tree literally explode. On occasion, both may fail to cause apparent damage, but months later an affected tree dies from burned roots and internal tissues.

Tall trees, trees growing alone in open areas, and trees with roots in moist soils or those growing along bodies of water are most likely to be struck. Though no species of tree is totally immune, some are definitely more resistant to lightning bolts than others. Beech, birch, and horsechestnut, for example, are rarely struck, whereas elm, maple, oak, pine, poplar, spruce, and tuliptree are commonly hit. The reason for the wide variation in susceptibility is not clear. Some authorities attribute it to the difference in composition of the trees; for example, trees high in oils (beech and birch) are poor conductors of electricity, whereas trees high in starch content (ash, maple, oak, etc.) are good conductors. In addition, deep-rooted or decaying trees appear to be more subject to attack than are shallow-rooted or healthy trees.

It is commonly believed that lightning never strikes twice in the same place.

Fig. 10-14. Shattering of the trunk on a lightning-struck tuliptree.

This is not true, for some trees have been struck by lightning as many as seven times, judging from the scars on their trunks.

Wires carrying electric current may also cause some injury to trees. Suitable nonconductors should be placed over wires passing in the vicinity of the trunk and branches. Much of the injury by wires can be avoided by judicious pruning, as discussed in Chapter 7. Outdoor evergreens bedecked with lights at Christmas can be damaged if there are too many poorly placed bulbs or if worn equipment is used. The bulbs should be placed so they do not come into direct contact with the needles or twig tips.

Prevention and repair of lightning injury are discussed in Chapter 8.

Hail Injury

Hailstorms can cause considerable damage. They may completely defoliate trees or shred the leaves sufficiently to check growth sharply. They are most injurious to young trees or those with undeveloped foliage in early spring. Areas in the bark and cambium may be severely bruised or even killed from the impact of the hailstones. When bark injuries do not callus over rapidly, they often serve as entrance points for wood decay fungi. Where hail has caused damage, all bark injuries will be on the same side of the affected twigs and branches.

Mechanical Injuries

Damage to trees, especially the trunks, by motor vehicles, by children, and even by thoughtless adults occurs commonly and the cause is usually obvious.

Most mechanical injuries are confined to the outer portions of the trunk (Fig. 10-15). The effect of mechanical injuries on tree health may be negligible if the injury is small and if the tree is vigorous and good at limiting decay. Or an injury may be serious enough to cause rapid death due the trunk girdling or decline and death following invasion of disease organisms such as decay fungi.

Equally obvious to even the casual observer is mechanical injury caused by ice storms and windstorms. Tree damage ranges from broken twigs and branches to uprooting of entire trees.

Accumulations of snow, especially on evergreens such a yews, hemlocks, and junipers, can cause severe branch breakage. Such accumulations should be gently shaken off as soon as possible with a broom or bamboo rake. Ice-covered trees should be handled even more carefully. If the temperature is on the rise, rinsing the ice-coated trees with water from a hose will thaw the ice.

Some deciduous trees are prone to cracking and splitting of branches in ice storms and windstorms. Among those most subject are the ashes, catalpas, hickories, horsechestnuts, red maple, empress-tree, Siberian elm, silk-tree, tuliptree, and yellowwood.

Less obvious to many tree owners is mechanical injury caused by girdling wires and by lawn mowers and string trimmers. Wires and plastic twine are sometimes improperly used for lacing up the burlap of a balled and burlapped

Fig. 10-15. Mechanical injury to a maple trunk.

tree. Even when the girdling wire was not originally placed tightly against the root collar, later trunk expansion spells trouble for the tree. Often, as in the case of a group of white pine trees observed to be in a state of decline 10 years after planting, the wire is deeply embedded into the bark tissues of the root collar a few inches below the soil surface and is not noticed until too late.

Wires used for tree support at transplanting or even for attaching the plant label in the nursery have been known to girdle tree trunks and branches some years later (Fig. 10-16). These aboveground girdles are usually easy to diagnose. In the early stages, the tree begins to swell just above the constriction. Eventually, the entire tree above the girdle will begin to decline and die.

The lawn mower is a more destructive instrument for debarking trees than are automobiles and malicious children. Careless use of this instrument results in severe damage to the inner bark and cambium near the soil line. Parasitic

Fig. 10-16. Wire used to help support this tree caused complete girdling of the trunk.

fungi, for example, *Ceratocystis,* which infects planetrees, and wood-destroying fungi can establish themselves in wounds made by lawn mowers.

Insects such as the dogwood borer are attracted to these injuries. Removal of phloem tissues by this debarking will lead to gradual root starvation and tree decline. If mulch replaces grass planted close to the base of the tree, there is less chance of causing damage. Installation of a tree guard eliminates the need for edging and trimming grass around lawn trees.

Lichens

Lichens often appear as a green coating on the trunks (Fig. 10-17) and branches of trees. They are actually two plants in one, being composed of a fungus body and an algal body, which live together in complete harmony. The alga supplies

Fig. 10-17. Lichens growing over a tree trunk.

elaborated food to the fungus and, in turn, receives protection and some food from the fungus.

Lichens do not parasitize trees, but merely use the bark as a medium on which to grow. They are unsightly rather than injurious, although when extensive they may interfere with the gaseous exchange of the parts they cover. Because of their extreme sensitivity to sulfur dioxide released from factory chimneys, lichens seldom appear on trees in industrial cities. They rarely develop on rapidly growing trees, because the bark is shed before the lichens have an opportunity to grow over much of the surface. Because of this, lichens on certain species may indicate poor tree growth. We have noticed that in some pin oak plantings, those most vigorous have fewer lichens than those of the same age nearby in a state of decline. However, no research has been done to verify any correlation between lichen growth and tree vigor.

As a rule, lichens can be eradicated by spraying the infested parts with Bordeaux mixture or any ready-made copper spray.

Vines

Many kinds of leafy vines grow up the trunks and out of the branches of trees (Fig. 10-18). Several are woody, twining types that have the potential to stran-

Fig. 10-18. English ivy growing on pin oak.

Fig. 10-19. Mouse damage to the base of a crabapple trunk.

gle or girdle the tree. Examples include trumpet vine, Japanese honeysuckle, bittersweet, and domestic or wild grapes. Other vines climb by means of aerial roots or tendrils, some having adhesive disks. Such climbers are a threat to the tree because of shading tree foliage, bending and possibly breaking branches from their weight, and competing for soil moisture. Examples include winter creeper, English ivy, Virginia creeper, cross vine, Kudzu, and poison ivy. Vines should be kept off young trees. Some of the nontwining types can be tolerated on large tree trunks, but they should not be allowed to grow out on the branches.

Rodent Damage

Young trees with thin, succulent bark are subject to damage by rodents such as field mice, pocket gophers, and rabbits during winters of heavy snowfall, or where the trees have been mulched with some strawy material (Fig. 10-19). Placing a collar of a quarter-inch wire mesh or even ordinary window screening around the base of such trees will prevent much damage. Use of strawy mulches in which mice are likely to nest should be avoided. The trunk bases of susceptible trees can also be painted with a repellent such as thiram.

Sapsucker Damage

Yellow-bellied sapsuckers have been observed feeding on twenty-six kinds of trees in the northeastern United States. The birds under observation fed partic-

Fig. 10-20. Yellow-bellied sapsucker and feeding holes on cherry.

ularly on Atlas cedar, hemlock, red maple, yellow, paper, and gray birches, and mountain-ash. Damage due to bird pecking has also been observed on apple, cherry, maple, and Scotch pine. Sapsuckers peck holes around the tree to obtain sap (Fig. 10-20), although insects also make up a part of their diet. The bird bores even rows of holes spaced closely together. As these holes fill with sap the sapsucker uses its brushlike tongue to draw it out. Extensive and repeated pecking may cause cambium and bark injury and branch or trunk swelling. Girdling may kill portions of the tree above the boring injury.

To discourage sapsuckers from feeding on landscape trees, wrap hardware cloth or burlap around the area being injured or apply a sticky repellent material, such as Tree Tanglefoot, on the bark. Decayed aspen trees make good nesting sites for sapsuckers, so their removal may discourage the birds from using the area.

Termites and Ants

Termites are usually associated with damage to wood in buildings, but they are occasionally found feeding on sound heartwood and sapwood of living trees.

Fig. 10-21. Larval and adult stages of carpenter ants infesting wood of black birch.

Several different kinds of termites, including dry wood and the more destructive subterranean types, can cause damage. Subterranean termites nest underground and enter the tree through injured roots, or they may build protective earthen tubes along the outer bark connecting the underground shelter to their feeding area of the tree trunk. One of the authors observed serious dieback of individual branches of an upright yew caused by termite activity. Parting of the branches to expose the earthen tubes along the trunk revealed the cause of the problem.

Ants of several different kinds also infest live trees. Perhaps the most common are carpenter ants (Fig. 10-21). Ants generally form galleries in the already decayed heartwood and sapwood of trees. Only rarely do the galleries extend into healthy wood, and even then the ants do not feed on the tree.

Ants are morphologically different from termites. The principal difference is that ants have a noticeable constriction of the body between the thorax (where legs and wings are attached) and the abdomen, whereas termites have a broad connection between these two sections. Ants can be controlled by blowing Diazinon dust into the colony with a hand duster, or spraying the colony with Diazinon or Baygon. Termite extermination requires destruction of underground nests, generally by a professional exterminator. Certain termite-proofing materials are toxic to tree and shrub roots and to humans, so precautions must be taken.

Tree Band Damage

Banding trees with a sticky substance to trap insects such as cankerworm moths can be damaging unless means are taken to prevent the inner bark and cambium from absorbing the material. If the banding material is applied directly to the bark, or if the thicker outer bark is shaved or cut away and the material then applied, severe damage may follow.

Sugar maples appear to be more susceptible than other trees to banding substances applied directly to the bark.

Air Pollution Injury

Fossil fuel combustion and a variety of industrial processes produce or induce production of phytotoxic gases, aerosols, and particulates. The major groups of air pollutants are oxidants, sulfur dioxide, and industrial byproducts. Their effects on trees and other plants depend on pollutant concentration, sources and distribution, plant sensitivity, and timing of occurrence.

At one time, air pollution was thought to be a local problem with readily identifiable sources, and for some geographical locations and many industrial by-products this is still the case. However, cars and trucks operate in such large numbers, and smokestacks are built to such heights, that air currents carry pollutants to locations far removed from their sources. Some of these pollutants can become trapped by weather conditions favoring inversion layers in the atmosphere, which prevent upward movement and dilution of the pollutant. Symptoms of air pollution injury sometimes show up after several days of such air stagnation. The effects of air pollution levels below those that cause symptoms on plants are not known, but increased plant stress could occur.

Oxidant Air Pollutants. The burning of fossil fuels should ideally yield nontoxic carbon dioxide and water; however, impurities, additives, and incomplete combustion yield other products as well. When these products are exposed to sunlight, ozone and peroxyacetyl nitrate (PAN), both toxic to plants, are produced. These oxidant air pollutants are important components of photochemical smog, which affects air quality in cities throughout the world. Ozone and PAN are toxic in very small concentrations.

Needle evergreen trees exposed to ozone show chlorotic flecking, banding, or needle tip burn symptoms. Broad-leaved plants show brown, purple, or chlorotic upper leaf surface flecking (Fig. 10-22). Ozone injury does not occur commonly, but when it does happen, symptoms can be striking. Newly expanded leaves appear to be most susceptible, so when symptoms are observed, the timing of the most recent episode can be found. If the timing coincides with air stagnation periods and symptoms are also occurring on ozone-sensitive indicator plants such as grapes, petunias, and cucumbers, then the symptoms of some air pollutant can be suspected. A May 1984 air stagnation over the mid-South region of the United States caused many newly emerged pine needles to show pronounced chlorosis and banding while the previous year's needles remained unaffected.

Symptoms of some diseases, nutritional problems, and feeding of insect pests such as lace bugs and spider mites may resemble ozone injury. Ozone can also be produced by electrical discharges from sparking electric motors and lightning. Levels of the pollutant in these situations are very low, and unless somehow confined, as in a greenhouse, damage does not occur.

Fig. 10-22. Flecking caused by oxidant air pollutants.

The main symptom of exposure of broad-leaved plants to PAN is a patchy silvering or light tan glazing of the lower leaf surface resulting from collapse of the epidermal cells. The affected leaf may exhibit spots or patches of papery-thin, almost transparent tissues. Damage from PAN is less common than ozone damage. Nitrogen dioxide (NO_2), also a product of fossil fuel combustion, is important for ozone and PAN production; NO_2 is less toxic to plants but causes yellowing of leaf margins and interveinal tissue when present at high levels.

Sulfur Dioxide and Acid Rain. Sulfur dioxide (SO_2) is the most common component of smoke from fossil fuel power plants which is toxic to trees. Coal and oil commonly contain sulfur as an impurity, and it oxidizes to sulfur dioxide during combustion. When emissions are not adequately dispersed, concentrations of the pollutant can become high enough to cause plant symptoms. Conifer trees show a reddish needle tip burn, and reddish brown dead areas appear between the veins on leaves of broad-leaved trees. Mild exposure may only cause interveinal leaf yellowing. Symptoms of exposure to sulfur dioxide resemble scorch resulting from drought or xylem-limited bacteria, and because damage can occur some distance from the pollutant source, diagnosis can be difficult. Finding similar symptoms on a wide range of nearby vegetation is helpful in diagnosis of damage due to SO_2 and other toxicants, but this assumes the diagnostician is familiar with the possible causes of symptoms on plants other than trees.

Power plants are not the only source of sulfur dioxide. One of the authors once diagnosed a very local sulfur dioxide exposure which resulted from the demolition of an old-fashioned electric refrigerator which used the chemical as a refrigerant. Heavier than air, the sulfur dioxide gas migrated to lower elevations in the landscape, killing plant leaf tissue the entire distance.

Acid rain results from an increased sulfuric acid content of rainwater due to high atmospheric sulfur dioxide levels; acid rain can have a pH as low as 4.0. Acid rain has not been shown to directly damage trees, and in fact may be beneficial if trees need sulfur or soil pH needs lowering.

Aerosols of sulfur dioxide, deposited and condensed on leaf surfaces in a process described as acid deposition, may be converted to more concentrated sulfuric acids when dissolved in small amounts of water such as dew. It is thought that concentrated acid is actually causing damage, rather than the more dilute acid rain. The actual effect of acid deposition or acid rain on tree health or productivity is not known; however, these substances may be having effects on freshwater lakes and man-made corrosible structures. Because there is concern about acidic precipitation resulting from sulfur dioxide pollutants, it has been suggested that use of high-sulfur fuels be reduced and improved ways to remove sulfur from fuel and smoke be developed.

Air Pollutants from Industrial Processes. Hydrogen fluoride is a by-product of manufacturing processes that use fluoride-containing ores as raw materials. Copper smelting, cement, glass, fertilizer, and aluminum manufacturing may produce fluorides toxic to trees. Fluoride injury is similar to sulfur dioxide injury. Flourides cause water soaking, collapse, and drying out of affected leaf tissue, leaving scorched areas. Reddish brown areas on the leaf margins, between the veins, and on needle tips are characteristic symptoms. Severe episodes can cause defoliation. Since the symptoms cannot easily be differentiated from those of other pollutants, knowing the source of fluoride and its proximity to the damage will provide circumstantial evidence of its occurrence.

Injury and death of tree foliage can result from exposure to ammonia, chlorine, hydrogen chloride, and other industrial chemicals. Injury from these substances is likely to follow inadvertent exposures such as leaks and spills.

Particulate Air Pollutants. Some industrial processes, farming practices, and even gravel roads produce dust or smoke particles that coat nearby vegetation. Where tree leaves are coated with visible dust or soot, stomata may be blocked and leaves may receive little sunlight; therefore, the leaves photosynthesize less actively. Because the leaves of evergreen trees remain attached throughout the year (in some species for several years), a considerable deposit may accumulate, thus making such trees' survival difficult where particulate air pollutants occur. Some dusts or ash particles may contain substances directly toxic to the foliage; after prolonged soil accumulation, changed soil characteristics may cause toxicity. In one instance oak decline has been attributed to the accumulation of cement dust.

Arborists and others interested in the care of trees should not be too hasty in concluding that leaf abnormalities are necessarily caused by air pollutants. They should bear in mind that unfavorable weather, insects, fungi, bacteria, spray materials, growth regulators, and viruses also produce symptoms that might be confused with symptoms of air pollution.

Over the years, the authors have examined hundreds of cases of alleged air pollution injury to vegetation in highly industrialized areas and have found that only a small percentage of the damage was caused by smoke or gases. Before blaming poor growth of vegetation on air pollutants, therefore, one would do well to call in a competent diagnostician.

Chemical Injuries

Weedkillers. While the chemicals used for weed control are referred to as weedkillers or herbicides, the chemicals themselves cannot distinguish between weeds and desirable plants such as trees. When carelessly used, the various chemical weedkillers can damage or even kill trees. The authors have noted many cases of toxicity to trees caused by such materials. Weedkillers function in a variety of ways. The most commonly used materials are growth regulators, applied to existing weeds in the landscape, and seed germination and seedling inhibitors, applied to soil to prevent weeds from appearing in the landscape. Other materials include contact and systemic chemicals applied to existing weeds and soil sterilants applied to soil to kill existing and future weeds.

Herbicides containing plant growth regulators such as 2,4-D (2,4-dichloro-phenoxyacetic acid) or dicamba are commonly used for eliminating weeds from lawns. Since their chemical structure mimics plant hormones, when they are taken up by a tree, they have an effect on the size and shape of developing parts of the tree. The most general symptom is a distortion or malformation of leaves and twigs (Figs. 10-23 and 24). Leaves also roll upward or downward at the midrib or along the edges, thus becoming cupped. Leaf petioles may curl downward, in some cases forming loops. Although these herbicides are almost always used without incident, the authors have noted growth regulator chemical injury on many trees, including elm, dogwood, boxelder, hawthorn, Norway maple, tuliptree, various oaks, sassafras, and yew. Unless the trees received an unusually heavy dose of chemical, they have usually recovered.

Applying high concentrations of volatile formulations of 2,4-D near trees during warm spring weather invites injury. Even trees not yet in leaf can absorb enough hormone into their dormant buds to produce distorted leaves. To avoid possible damage, use the less volatile forms of herbicide, avoid windy days, and use a coarse droplet spray at low pressure to reduce drift. Follow herbicide label directions and do not use the herbicide sprayer for other spraying jobs;

Fig. 10-23. 2,4-D injury to mulberry.

even minute traces of 2,4-D in such sprayers can cause tree injury. Where a sprayer with a metal tank has been used, the 2,4-D can be removed as follows: Fill the tank with warm water and household ammonia at the rate of 1 gallon of ammonia or 5 pounds of sal soda for each 100 gallons of water. Then pump out a few gallons to wash pump parts, hose, and nozzles, and allow the remainder to stay in the tank for 2 hours or so. Finally, drain the tank and rinse several times with clear water. Activated charcoal can be used to rid spray tanks of 2,4-D residues.

Some growth-regulator herbicides are applied as granular formulations, often mixed with a fertilizer. Obviously, these products would not make good tree fertilizers. Furthermore, some formulations contain dicamba. In the soil, dicamba is more persistent than 2,4-D. It can move downward in the soil, be absorbed by tree roots, and move systemically to the foliage where leaf cupping and distortion can occur.

Preemergence herbicides such as crabgrass preventers are applied to the soil and generally do not injure trees.

One of the most common systemic herbicides applied to existing weeds in the landscape is glyphosate (Roundup, Kleenup). Applied to growing herbaceous and woody weeds, it is translocated to the roots, and soon after, the plants begin to die. If it is mistakenly applied to landscape trees, they too can be killed. Glyphosate applied to a nearby weed can drift to a woody plant and symptoms might not appear until the next season. In such cases, the new growth may be bunchy and distorted, resembling growth-regulator herbicide injury. The authors have observed several cases of young flowering crabapple trees suffering injury in this way. Trees may or may not be killed, depending on the

Fig. 10-24. Growth-regulator herbicide injury to yew.

dose and species involved. Diagnosing glyphosate injury on trees may require a knowledge of weed control operations the year before. In general, small, thin-barked trees are more vulnerable than large, thick-barked trees. The effect of glyphosate application on exposed tree roots has not been studied, but such roots represent a potential mode of entry.

Contact herbicides such as paraquat kill green tissues on contact. Where such materials are used near trees, avoidance of foliage and of thin green bark is essential. The chemical is not systemic, so a small amount of kill resulting from paraquat drift is seldom fatal to the tree. Small circular brown spots have been observed on leaves of trees exposed to fine drift of this chemical.

The herbicide tebuthiuron kills woody plants. This brush-killer, applied to the soil, kills the tree roots, resulting in gradual decline and death of the plant. Although excellent for control of unwanted brush, this material should not be used where desirable species of trees or shrubs are present.

Soil sterilant herbicides are often used to destroy vegetation along walks,

Fig. 10-25. Pear leaves injured by soil sterilant.

fences, and driveways (Fig. 10-25). Many users are not aware that some of these materials are long lasting, can move in the soil, and are very toxic to trees whose roots take up the chemical. Ignorance of just where tree roots grow only compounds the problem.

Herbicides such as prometon and bromacil are especially damaging when misused. These chemicals are taken up by the roots and injure foliage on the same side of the tree. This one-sided effect can sometimes be mistaken for vascular wilt diseases. Some conifers such as spruce, having a spiral xylem arrangement, will show damage in a spiral pattern up the tree. Affected branches turn yellow, then reddish brown, and eventually die while the rest of the tree remains green. Some of these materials are so long-lasting that where applied, soil has to be removed and replaced in order to get any plants to grow.

Calcium Chloride. Trees growing along country roadsides may be damaged by the calcium chloride used to control dust on dirt roads. Heavy rains wash the chemical off the roads and carry it down to the roots, which absorb it and then transport it to the leaves, where it causes injury. Correct grading of the roadside to facilitate removal of flood waters and prevent their being soaked into the soil around roots will greatly reduce the chances of damage.

Tree species vary in their tolerance to calcium chloride. Red oak, white oak, and American elm are most tolerant, whereas balsam fir, white spruce, beech, sugar maple, cottonwood, and aspen are least tolerant.

Damage to the leaves appears as a leaf scorch, which is difficult to distinguish from physiological leaf scorch caused by lack of water in the soil. Leaf cupping symptoms have been observed on emerging linden leaves in spring following a winter of sidewalk calcium chloride deicing.

Salt. Salt is frequently scattered over roads, parking lots, and sidewalks in winter to melt ice or prevent water from freezing. Each year more than 9 million tons of salt (sodium chloride) may be applied along streets and highways in the United States for deicing. In parking lots, snow mixed with previously applied salt often is repeatedly pushed into piles on the landscape strips or ends where trees are planted, thus concentrating all the salt from an entire area into a few locations. When the salty solution finds its way to open soil near roots it can be very toxic. In fact, sodium chloride (ordinary table salt) is five to ten times more toxic to some trees than is calcium chloride. High soil salt concentrations tightly bind soil water so that water becomes unavailable to tree roots. Lacking water, trees essentially undergo drought and respond similarly. Gradual decline with all the typical symptoms may occur, or in extreme cases, leaf scorch and death occur. The presence of sodium and chloride ions in the soil makes it more difficult for the tree to take up potassium and phosphorus, so deicing salts can cause nutrient problems as well. Sand or sawdust should be used to prevent slipping in areas near a tree and other plant roots.

Sand treated with salt and piled along roadsides for use in icy roads may also damage trees. The salt is leached down into the roots and causes burning. Sand piles thus treated should be placed at some distance from the root areas of the trees.

Researchers investigating the association of salt applications and tree mortality have observed that trees within 30 feet of highways are affected, whereas the trees beyond that distance were nearly always healthy. Salt injury appears as a marginal leaf scorch, early fall coloration and defoliation, dieback of twigs and branches, and in severe cases death of the entire tree.

Some trees are more tolerant to salt than others and may be classified as follows:

Very tolerant: Black cherry, red cedar, red oak, and white oak.
Tolerant: Black birch, black locust, gray birch, largetooth aspen, paper birch, white ash, and yellow birch.
Moderately tolerant: American elm, linden, hop-hornbeam, Norway maple, red maple, shagbark hickory.
Intolerant: Beech, birch, hemlock, red pine, speckled alder, sugar maple, white pine.

Cinders. Hard and soft coal cinders, especially those that have not been exposed to weathering, should not be placed over the soil beneath trees. Toxic materials in such cinders will be washed down in rainwater and harm the roots.

Salt Spray. Trees growing along the seashore are often injured by salt spray

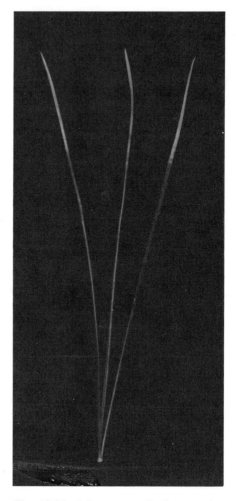

Fig. 10-26. Salt spray needle tip necrosis.

blown from the ocean. During hurricanes the spray has actually damaged leaves 50 miles from salt water. Most of the damage, however, usually is confined to within a few miles of the coast.

Among the broad-leaved trees, elm, magnolia, Norway maple, sugar maple, tuliptree, and tupelo appear most susceptible to salt spray, whereas red oak and horsechestnut are very resistant. Among the evergreens, white, jack, Balkan, Japanese red, and Swiss stone pines, hemlock, and junipers are most harmed, whereas spruce, Austrian, ponderosa, and Japanese black pines, holly, and yew suffer least.

Salt spray is also damaging to the foliage of needle evergreens planted downwind from salted high-speed roadways. Needle tips on the exposed side of the trees turn brown, principally from dehydration and salt toxicity (Fig. 10-26).

Removal of badly damaged trees and pruning of mildy affected trees are all that can be suggested.

Fungicides and Insecticides. The various sprays recommended in Part III of this book for disease and insect control may occasionally injure trees. In general, damage from fungicides and insecticides is more severe on undernourished trees and on those growing in poorly drained soils than on vigorous specimens. Insect injury and frost damage also predispose leaves to spray injuries. Cool, damp weather favors the chances of injury from bordeaux moisture and other copper fungicides. High temperatures and humidities increase the chances of injury from lime sulfur and other sulfur sprays. High temperatures also enhance the probability of injury resulting from malathion and oil sprays.

Copper Sulfate. This chemical is frequently mentioned as a control for algae (green scum) in ponds and lakes. The usual concentration is 1 part of the copper sulfate in a million parts of water, by weight. When the concentration is in excess of this figure, there is danger of harming trees growing around the edge of the pond or lake.

Chlorosis

The uniform yellowing of leaves resulting from a reduction in the normal amount of chlorophyll is termed chlorosis. This loss reduces the efficiency of the leaf in manufacturing food.

Fig. 10-27. Maple leaf chlorosis.

Chlorotic leaves may result from fungus, virus, or insect attack; low temperatures; toxic materials in the air or soil; excessive soil moisture; surpluses of soil minerals; lack of or nonavailability of nutrients.

Chlorosis of many shade trees, especially of pin oak, appears to be associated with nonavailability of iron rather than with the lack of iron in the soil. This is especially true in soils containing limestone, ashes, or other alkaline materials, where the pH of the soil ranges from 6.7 to 8.5. Iron may be present in such soils, but in a form that cannot be absorbed by the plant.

Symptoms. The leaves of affected trees first turn uniformly yellowish green, or they may remain green along the veins but turn yellow in the intraveinal areas (Fig. 10-27). The terminal growth of twigs is reduced, and the tree is generally stunted. The tissue between the leaf veins or along the leaf edge may die on trees affected for several years with chlorosis. Eventually whole branches or the entire tree may die prematurely unless the condition responsible is corrected. New growth may be more severely affected than growth produced earlier in the season.

Control. Chlorosis caused by deficiency or nonavailability of iron can often be corrected by special treatments. These treatments may involve soil acidification using sulfur, injection or implantation of iron into the tree trunk, or application of iron to the foliage. These methods are discussed in Chapter 6.

Selected Bibliography

Alden, J., and R. K. Hermann. 1971. Aspects of cold hardiness mechanism in plants. Bot. Rev. 37:37–142.

Anonymous. 1964. Protecting trees against damage from construction work. U.S.D.A. Inform. Bull. 285, Washington, D.C. 26 pp.

Anonymous. 1965. Protecting shade trees during home construction. 1965. U.S.D.A. Home and Garden Bull. 104, Washington, D.C. 8 pp.

Anonymous. 1973. Air pollution damages trees. U.S.D.A. Forest Service, Northeastern Area State and Private Forestry, Upper Darby, Pa. 32 pp.

Evans, L. S. 1984. Acidic precipitation effects on terrestrial vegetation. Ann. Rev. Phytopathol. 22:397–420.

Harris, R. W. 1983. Arboriculture. Prentice-Hall, Englewood Cliffs, N.J. 688 pp.

Hepting, G. H. 1963. Climate and forest diseases. Ann. Rev. Phytopathol. 1:31–50.

Hoeks, J. 1972. Effect of leaking natural gas on soil and vegetation in urban areas. Agric. Research Report 778. Centre for Agricultural Publishing and Documentation, Wageningen, Holland. 120 pp.

Jacobson, J. S., and A. C. Hill (eds.). 1970. Recognition of air pollution injury to vegetation: A pictorial atlas. Air Pollution Control Assoc., Pittsburgh, Pa. 100 pp.

Loomis, R. C., and W. H. Padgett. 1973. Air pollution and trees in the east. U.S.D.A. Forest Service, State and Private Forestry, Northeast Area and Southeast Area, Upper Darby, Pa. 28 pp.

Mazur, P. 1969. Freezing injury in plants. Ann. Rev. Plant Physiol. 20:419–48.

Reinert, R. A. 1984. Plant response to air pollutant mixtures. Ann. Rev. Phytopathol. 22:421–42.

Scheffer, T. C., and G. G. Hedgcock. 1955. Injury to northwestern forest trees by sulfur dioxide from smelters. U.S.D.A. Tech. Bull. 1117, Washington, D.C. 49 pp.

Sinclair, W. A., H. H. Lyon, and W. T. Johnson. 1987. Diseases of trees and shrubs. Cornell University Press, Ithaca and London. 574 pp.

Treshow, M. 1970. Environment and plant response. McGraw-Hill, New York. 422 pp.

Wood, F. A., and J. B. Coppolino. 1972. The influence of ozone on deciduous forest tree species, in Effects of air pollutants on forest trees. VII International Symposium of Forest Fume Damage Experts. Vienna, Sept. 1970, pp. 233–53.

11

Insect and Mite Pests of Trees

A working knowledge of insect pests, of how they injure plants, and of methods of combating them is essential to any tree preservationist. A detailed discussion on insect identification methods is best left to texts written specifically for that purpose; however, important characteristics of the major insect pests are presented here.

Accurate identification of any insect can be obtained by submitting the specimen to the state entomologist, usually located in the state capital, or an insect identification laboratory located in the state agricultural experiment station. The county extension office is a good place to start.

Structure of Insects

Insects belong to a larger group of animals called Arthropoda. The Arthropoda are characterized by having segmented legs, segmented bodies, and a more or less tough skin that also serves as the skeleton. Besides insects, this group includes spiders, mites, sowbugs, millipedes, centipedes, crayfish, and many other similar animals. The insects differ in structure from other arthropods in that the body segments are organized into three distinct sections: head, thorax, and abdomen.

The head segments are closely fused and appear to be a single segment that bears the feelers, or antennae; the eyes; and the mouthparts. Mouthparts of adult plant-feeding insects are adapted for chewing, sucking, or some combination of both.

The thorax, the section just behind the head, has three segments with a pair of segmented legs on each segment, for a total of six legs. Other arthropods usually have more legs. If wings are present, as they usually are in insects, they are also attached to the thorax. Other arthropods never have wings.

The abdomen is composed of a variable number of segments depending on the species. In the female of some species, a sharp-pointed organ for depositing eggs, the ovipositor, is found at the posterior end of the abdomen.

The major groups of insects, or orders, are separated from each other primarily on the basis of the number and structure of the wings, the type of mouth parts, and the type of metamorphosis they go through in their life cycles.

Life Cycle of Insects

One of the most interesting and unfathomable mysteries of nature is found in the study of the life cycle of the insect as it passes through various stages from egg to adult. The process that embodies all these changes in form is known as *metamorphosis*.

The life cycle of butterflies, beetles, sawflies, true flies, and moths comprises four separate stages: egg, larva, pupa, and adult. Insects with this life cycle are said to have complete metamorphosis.

After mating, the female deposits her eggs in a safe place that will provide a favorable environment for the usually wormlike larvae when they emerge, or she produces the eggs in such large numbers that survival of only a small percentage will assure abundant offspring. The names of the larval forms vary according to the groups to which they belong. Caterpillars are the larvae of butterflies and moths; grubs are the larvae of beetles (Fig. 11-1 [left]); false caterpillars, those of sawflies; and maggots are the legless larvae of flies.

Most insect larvae have chewing mouthparts, though their parents may not. They do not have their parents' compound eyes and wings, and many do not have six true legs, having in some cases none and in some cases many more false legs on the abdomen. Not only do the larvae look different from their parents, they often live in different places and eat different foods.

The common names of insects often reflect the first or most common plant they were associated with. The host range of an insect is often much wider than that implied by name. The two-lined chestnut borer, for example, attacks oak, beech, and hornbeam, in addition to chestnut.

Fig. 11-1. Left: Flat-headed borers. **Right:** Adult beetle of one of the flat-headed borers. Wing covers are hard.

As a larva grows it sheds its skin, or molts, several times, each successive skin being larger and better able to accommodate the animal's ever-increasing body. Finally the larva becomes full grown and is ready to enter the third stage of development, the pupa. Larvae of many insects prepare a protective covering for this stage. The cocoon of the moth and sawfly, the pupal case of the true flies, and the pupal cell of the beetle represent such protective coverings.

Although the pupal stage is considered a resting period because the pupa does not feed or move about, it is marked by considerable metabolic activity and many important changes, which culminate in the emergence of the adult insect.

A different form of development, called incomplete metamorphosis, characterizes such insects as grasshoppers, plant lice, plant bugs, lace bugs, and leafhoppers. The egg develops into a small wingless creature called a nymph. Often resembling the adult except for its ungainly head and small body, the nymph passes through several molts. Wings become more developed and prominent and the entire body assumes the appearance of the adult insect with each successive molt. Thus the last stage is attained directly, no pupa being formed in the life cycle.

Types of Insect Pest

The insect pests of plants may be grouped according to their placement in the natural classification system, or according to the habit by which they are pests. For instance, beetles are a natural grouping, but beetles are also included in such habit groups as borers and leaf miners. Because pest control methods are usually more contingent on the pest's habits than on its scientific classification, we use the habit classification as much as possible.

Borers. Any insect that feeds inside the roots, trunks, branches, or twigs of a tree is known as a borer. The borer is usually a larva or worm stage of a beetle, moth, or wasplike insect, although some beetles may bore into the tree as adults.

Eggs are deposited in bark crevices and scar tissue or, in a few instances, below the bark surface by the adult during the spring, summer, or fall, depending on the species. Tiny larvae hatch from the eggs and penetrate the bark. Once they are below the surface, they cannot be controlled by the ordinary contact insecticides.

Commonly occuring landscape tree borers include banded ash borer, *Podosesia aureocincta;* bronze birch borer, *Agrilus anxius;* dogwood borer, *Synanthedon scitula;* flatheaded borer, *Chrysobothris femorata;* lesser peachtree borer, *Synanthedon pictipes;* lilac borer, *Podosesia syringae syringae;* locust borer, *Maegacyllene robiniae;* peach tree borer, *Sanninoidea exitiosa;* and two-lined chestnut borer, *Agrilus bilineatus.* The adults of many of these insects are clearwinged moths, while others are beetles.

FACTORS FAVORING BORER INFESTATIONS. Although a few borers can attack vigorously growing trees, most species become established in trees low in vigor. Consequently, any factor that tends to lower the vitality of a tree predisposes it to borer attack. Among the more important of these factors are prolonged dry spells, changes in the environment unfavorable to the growth of the tree (see Chapter 10 for examples), loss of roots in transplanting, repeated defoliation by insect or fungus parasites, and bark injuries caused by frost, heat, or mechanical agents. In some instances, trees of low vitality and injured trees discharge chemical signals that are attractive to borer insects.

PREVENTION OF BORER ATTACK. Many borer infestations can be prevented by improving, when feasible, any unfavorable situation around the trees. Proper fertilization, adequate watering during dry spells, control of defoliating insects and diseases, and pruning of infested or weakened branches are recommended. All bark wounds should receive immediate attention to facilitate rapid healing and thus reduce the amount of exposed tissue.

Borer infestations, common in recently transplanted trees, can be materially lessened by placing a barrier over the trunk and larger limbs immediately after the trees are set in their permanent location. Newspaper, wrapping paper, burlap, or specially prepared kraft paper is used. The last is now employed extensively because it provides greater protection, is applied more easily, and is neater in appearance than the other materials. The kraft paper consists of two layers cemented together by asphalt and is sold in 4- and 6-inch widths, wound in bolts of 25 or more yards. It is wrapped around the tree in much the same way that a surgical bandage is applied, at an angle that permits sufficient overlap to make a double thickness. Binder twine is wound in the opposite direction to hold the paper in place. Both the twine and the paper remain in place for one or two years—long enough to provide protection during the most susceptible period—and then they are removed.

Asphalt-impregnated kraft paper should not be treated with an oil-based insecticide because the oil will dissolve the asphalt.

Polyethylene wrap has not proved satisfactory because it holds too much moisture and could favor mold development. Aluminum foil wrap has been tried with considerable success. It is easily applied, and it does not need to be tied down with twine.

CONTROL. The method used to control already established borers depends on the part of the tree infested and the species of borer involved. Twig and root borers are controlled by pruning and discarding infested parts. Borers that make galleries which lead to the bark surface (sawdustlike frass is a telltale sign) can be controlled by injecting a toxic paste into the holes and then sealing them with a small wad of chewing gum, grafting wax, or putty. The pastes are sold under such trade names as Bor-Tox, Borer-Kil, and Borer-Sol. Some borers inside the tree can be killed by crushing with a flexible wire or sharp knife inserted into the opening.

Application of insecticide to the bark during the adult's egg-laying period or before the eggs hatch and the borers enter the bark is standard procedure for controlling trunk and branch borers. The insecticides Dursban, Lindane, and Thiodan are most frequently used. The bark on the trunk and larger branches should be sprayed thoroughly. An understanding of the borer life cycle, espe-

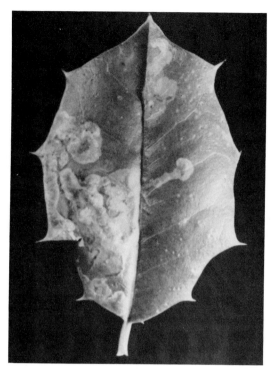

Fig. 11-2. Holly leaf miner damage.

cially the timing of egg hatching, is needed for best results. Consult the state or county extension office nearby for information on timing of borer sprays.

Leaf Miners. Leaf miners are insect larvae which feed between the upper and lower leaf epidermis of several kinds of landscape trees. This feeding activity results in unsightly light green to brown blotches or serpentine mines in the leaf (Fig. 11-2). Heavy feeding can cause the entire leaf to die, and a heavy infestation can damage a tree. Interestingly, though several different leaf miner larvae cause similar symptoms, the adults may be completely different kinds of insects.

Birch leaf miner, *Fenusa pusilla,* is a sawfly larva causing blotch mines, which may run together and blight the entire leaf. Boxwood leaf miner, *Monarthropalpus buxi,* is a midge fly larva causing blister mines of boxwood leaves. There are five holly leaf miner species; the two most common, *Phytomyza ilicis* and *P. ilicicola* (Fig. 11-3), feed on English and American hollies. These two species of fly larvae cause blotch or serpentine mines on holly leaves. Pin pricks in leaves are egg-laying and feeding punctures made by the small, black, robust flies. Locust leaf miner, *Odontota dorsalis,* is a beetle larva causing blotch mines which skeletonize leaves. The adult beetle is wedge shaped, flat, and orange with a black line down the back. Two species of moths, the solitary

Fig. 11-3. Holly leaf miner adult.

and gregarious oak leaf miners, *Lithocolletis hamadryadella* and *Cameraria cincinnatiella,* respectively, attack red and white oaks, causing blotch mines. The former has one and the latter several larvae per mine.

CONTROL. Systemic insecticides can be used to kill leaf miner larvae in the mines. Products with systemic action include Cygon and Orthene. Protectant sprays are timed to kill adults before they lay eggs in the leaf. Such materials include Sevin, malathion, Diazinon, Dylox, and Lindane.

Foliage-Feeding Caterpillars. Caterpillars can be conspicuous consumers of tree leaves. These larvae are relatively large and move freely on the tree. When they are present in sufficient numbers, they destroy a noticeable portion of the foliage and in some cases weaken the tree. Caterpillars have chewing mouthparts, and they have six true legs on the thorax and from four to ten unjointed legs on the abdomen; they may be smooth, hairy, or spiny, and may or may not be brightly colored. The adult phase is a moth or butterfly, a member of the order Lepidoptera, a large and well-known insect group. These insects undergo a complete metamorphosis during their life cycle and may have more than one generation per year. Moth larvae are more numerous than butterfly larvae as tree pests.

The gypsy moth (Fig. 11-4), *Lymantria (=Porthetria) dispar,* is a highly destructive pest of trees in the northeastern United States, where it has caused enormous losses and has been the subject of much publicity. The extent of infestation is still increasing, but whether or not it will become a pest nationwide remains to be seen.

The dark and hairy larvae may grow to a length of 2½ inches after hatching in early spring from egg masses laid the previous season on branches, buildings, or other outdoor objects such as automobiles and camping equipment. Larvae crawl up the tree and begin feeding and while still small, spin a silken thread from which they are suspended. This allows the wind to carry them

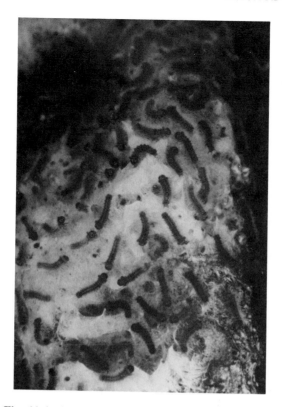

Fig. 11-4. Gypsy moth larvae emerging from egg mass.

several hundred yards, where they can feed on another tree. As the larvae mature, they may feed only at night, crawling down the tree during the day to hide in a protected place. It is then they often encounter people.

Larvae of heavily infested trees may continue to feed throughout the day. They feed for about 7 weeks, then pupate. In midsummer the adults emerge and mate, the females lay eggs, and they die without feeding. Oaks suffer from the defoliation more than many other trees, and they may not survive two or more consecutive years of leaf loss. Other trees damaged by gypsy moths include alder, apple, arborvitae, black gum, beech, elm, hawthorn, hemlock, hickory, linden, maple, poplar, pine, sassafras, spruce, and willow.

Other caterpillars damaging to trees include fall and spring cankerworms, *Alsophila pometaria* and *Paleacrita vernata,* which have wingless female adults and inch-long yellowish green, green, brown or black smooth larvae. Orange-striped oakworm, *Anisota senatoria,* and yellow-necked caterpillar, *Datana ministra,* are widespread in the East and feed on several different kinds of trees. Red-humped caterpillar, *Schizura concinna,* and white-marked tussock moth, *Orgyria leucostigma,* also have wide host feeding preferences. The larva of the io moth, *Automeris io;* the linden looper, *Erannis tiliaria;* and elm span-worm, *Ennomos subsignarius,* are all important tree leaf eaters. A few butterfly

Fig. 11-5. Pine sawfly larvae.

larvae such as mourning-cloak, *Nymphalis antiopa,* and silverspotted skipper, *Epargyreus clarus,* feed on trees. Several caterpillars construct webbed nests in the tree for protection. Fall webworm, *Hyphantria cunea;* eastern tent caterpillar, *Malacosoma americana;* forest tent caterpillar, *M. disstria;* and California tent caterpillar, *M. californicum,* are common nest makers.

CONTROL. Caterpillars have many natural enemies including birds and parasitic wasps. An insecticide specifically toxic to caterpillars is made from the spores of a bacterium, *Bacillus thuringiensis* (Thuricide, Dipel). Other insecticides used against these pests are Sevin, Dursban, Orthene, methoxychlor, Imidan, and Dylox.

Sawflies. Sawflies are nonstinging wasps that resemble flies. The larval stage (Fig. 11-5) of these insects damages trees. Some sawfly larva resemble butterfly and moth caterpillars, whereas others look like slimy, legless slugs.

Many sawfly larvae such as red-headed pine sawfly, *Neodiprion lecontei;* mountain-ash sawfly, *Pristiphora geniculata;* butternut woollyworm, *Erio-*

campa juglandis; and pear slug, *Caliroa cerasi,* are defoliators capable of devouring needles or skeletonizing and consuming leaves. Other sawfly larvae are leaf miners, for example, birch leaf miner, *Fenusa pusilla;* elm leaf miner, *F. ulmi;* and European alder leaf miner, *F. dohrnii.* The maple petiole borer *Caulocampus acericaulis,* is also a sawfly larva.

Adult female sawflies have a sawlike ovipositor at the tip of the abdomen, which is used to slit or cut plant tissue. Eggs are laid in these slits.

CONTROL. Most sawfly larvae are controlled with sprays of Diazinon, Orthene, Sevin, or methoxychlor. Insecticides should be applied to pine-and leaf-eating sawfly larvae when they appear in early spring and summer. Leaf miner control requires more precise timing because adults need to be killed at egg-laying time or larvae killed before they are protected by the mine. A systemic insecticide such as Cygon can be used to destroy these larvae.

Foliage-Feeding Beetles. Japanese beetle, *Popillia japonica,* and elm leaf beetle, *Pyrrhalta luteola* (Fig. 11-6), are important tree pests. Adult Japanese beetles are about ⅜ inch long and a shiny greenish bronze color. They often feed in groups on sunny parts of the tree. Elm leaf beetles are less than ¼ inch long, but they have long antennae. They are yellowish green with black stripes. The two represent contrasting life histories: Japanese beetle has a wide host range, whereas elm leaf beetle seeks out various elm species or zelkova, a close relative. Japanese beetle adults feed on and defoliate trees (larvae [grubs] feed on roots of grasses), whereas elm leaf beetle adults and larvae both feed on elm leaves (and the larvae are most damaging). Japanese beetle overwinters as a larva, whereas elm leaf beetle overwinters as an adult. In the United States Japanese beetle is found only in the northeastern states, whereas elm leaf beetle is found wherever elms are grown. Both pests skeletonize leaves and defoliate their hosts. Trees which lose their leaves in consecutive growing seasons may have their energy reserves depleted; trees thus stressed have less ability to withstand other potentially harmful pests or adverse environment.

Fig. 11-6. Elm leaf beetle larva.

CONTROL. Insecticides such as Sevin, Diazinon, methoxychlor, and Orthene can be used to control foliage-feeding beetles. Sprays for Japanese beetle are applied in June and July, whereas sprays for elm leaf beetle are applied in May and June.

Plant Bugs and Lace Bugs. These insects have piercing-sucking mouthparts, which in many instances case a stippling symptom on infested leaves. This stippling may involve patches of chlorotic or necrotic cells, different from the finer chlorotic flecks produced by spider mites, for example.

Tarnished plant bug, *Lygus lineolaris,* and four-lined plant bug, *Poecilocapsis lineatus,* produce small, brown, uniform, circular spots on leaves; these symptoms resemble those caused by parasitic fungi or bacteria. Sycamore plant bug, *Plagiognathus albatus,* injures leaves by producing holes, slits, and tears in the leaves, leaving tattered, chlorotic leaves. Honey locust leaves are discolored, deformed, and stunted by honey locust plant bug, *Diaphnocoris chlorionis.* Plant bugs appear as flattened, shield-shaped insects ⅛ to ¼ inch long.

In addition to chlorotic stippling, lace bugs produce dark, shiny excrement spots on leaf undersides (Fig. 11-7). Lace bugs are about ⅛ to ¼ inch long and have a broad thorax with fine, lacelike wings. Important lace bugs attacking trees include sycamore, walnut, and alder lace bugs, *Corythucha ciliata, C. juglandis,* and *C. pergandei,* respectively.

CONTROL. Sprays of Sevin, malathion, or Orthene applied in late spring and summer can control lacebugs.

Aphids. Aphids have mouthparts that pierce and suck the cell sap from fo-

Fig. 11-7. Lace bug: adults, nymphs, and tarlike excrement.

Fig. 11-8. Top: Winged aphid. **Bottom:** A group of nonwinged aphids.

liage, stems, and roots. They may be found singly or in groups on the affected tissue, often forming very large colonies. Aphids vary in size, ranging from 1/16 to 1/4 inch, and their color ranges from purple, red, or black to yellow, pink, or green. Most aphids overwinter as eggs, hatching in the spring as adults, some winged (Fig. 11-8), some not. Many generations of aphids may occur during the growing season, sometimes all on one host, and sometimes on several different hosts.

Aphid damage may include gradual weakening due to depletion of plant food, cupping, stunting or distortion of leaves or needles, defoliation, and death of branches. Aphids also cause damage by spreading viruses.

Many aphids secrete large quantities of honeydew, partly digested sweet and sticky plant juices. Honeydew is attractive to bees, wasps, ants, and especially sooty mold fungi. These fungi are black and unsightly, giving a dark, dusty appearance to affected foliage and in severe cases, interfering with photosynthesis by shading out the sun. In addition, honeydew may drip on people and property and cause dust and dirt to stick to the leaves.

Some aphids produce waxy substances which, when combined with cast molted skins, can produce a white or gray protective covering for the aphid colony. Aphids existing in such colonies are often referred to as woolly aphids.

Some aphids that are tree pests include balsam twig aphid, *Mindarus abietinus;* California laurel aphid, *Euthoracaphis umbellulariae;* cowpea aphid, *Aphis craccivora;* giant bark aphid, *Longistigma caryae;* green peach aphid, *Myzus persicae;* green spruce aphid, *Elatobium abietinum;* potato aphid, *Macrosiphum euphorbiae;* and woolly apple aphid, *Eriosoma lanigerum.* Other pests resembling aphids include adelgids (phylloxera) such as pine bark adelgid, *Pineus strobi;* and balsam woolly adelgid, *Adelges piceae.* Pysllids such as the pear psyllid, *Psylla pyricola,* also have piercing-sucking mouthparts and can cause distortion of leaves.

CONTROL. Aphids have many natural enemies which are predators, but by the time natural populations of the predators have grown to be an effective force, the damage caused by aphids already may have occurred. Insecticides commonly used to control aphids include superior oil or oil plus ethion during the early spring while trees are still dormant, and, when the aphids are observed to be a problem, summer applications of malathion, Diazinon, Thiodan, Orthene, Guthion, or Dursban.

Scales. Another important group of insects, scales have sucking mouthparts and are wingless (except for the males). There are three kinds of scale or scale-like insects that commonly attack trees: mealybugs, soft scales, and armored scales (Fig. 11-9).

Mealybug injury to infested trees appears as foliar discoloration, deformation, and wilt, and eventual death of affected plant parts caused by loss of sap. These insects may attack roots, stems, twigs, leaves, flowers, and fruits. They are white and motile, deriving their name from the white, waxy secretions that cover their bodies. Important examples include Comstock mealybug, *Pseudococcus comstocki;* taxus mealybug, *Dismicoccus wistariae;* long-tailed mealybug, *P. longispinus;* and beech scale, *Cryptococcus fagisuga.*

Soft scales vary widely in appearance, ranging from the cottony maple leaf scale to the shell-like kermes and lecanium scales. The shell or covering of these insects is part of the soft scale exoskeleton. Soft scales such as the lecaniums have two periods of motility: the crawler stage, occurring just after egg hatch, which moves to the leaves; and the second instar stage, which migrates from the leaves back down to the twigs in the fall to overwinter or to lay eggs. These scales debilitate the host tree by sucking phloem sap. They often produce

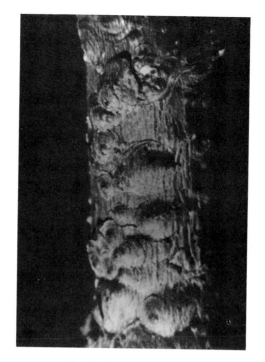

Fig. 11-9. Obscure scales.

large quantities of honeydew, which promotes growth of sooty mold fungi; these can blacken the affected leaf or shoot. Soft scales include calico scale, *Lecanium cerasorum;* cottony-cushion scale, *Icerya purchasi;* cottony maple scale, *Pulvinaria innumerabilis;* black scale, *Saissetia oleae;* tuliptree scale, *Toumeyella liriodendri;* and hickory lecanium scale, *L. caryae.*

Armored scales have a hard, shell-like covering composed of a waxy secretion and old shed skins. The covering is not part of the insect; the insect itself can be exposed by carefully lifting the shell. At this stage, the scale insect has no eyes, legs, or antennae, but it has fine sucking mouthparts, which siphon nutrients from the plant. The removal of phloem sap by large numbers of scales debilitates the tree and can cause gradual death of twigs and branches. Armored scales move only in the first instar, which is also called the crawler. Once the crawler settles and begins feeding, it remains in that place until it dies.

The following are examples of armored scales damaging to many landscape tree species: oystershell, *Lepidosaphes ulmi;* greedy and latania, *Hemiberlesia rapax* and *H. latania;* California red, *Aonidiella aurantii;* Florida red and dictyospermum, *Chrysomphalus ficus* and *C. dictyospermi;* San Jose and walnut, *Quadraspidiotus perniciosus* and *Q. juglansregiae;* white peach, *Pseudaulacaspis pentagona;* pine needle, *Phenacaspis pinifoliae;* black pineleaf, *Nuculaspis californica;* juniper, *Carulaspis juniperi;* fiorinia hemlock, *Fiorinia externa;* hemlock, *Abgrallaspis ithacae;* obscure and gloomy, *Melanaspis obscura* and *M. tenebricosa;* ivy or oleander, *Aspidiotus hederae.*

CONTROL. Most scale-infested trees can be treated with a dormant superior oil spray applied in early spring before bud break. The crawler stages are vulnerable to contact insecticides such as Diazinon, malathion, Orthene, and Sevin, but the insecticide must be applied in spring or summer when crawlers are active. It is important to obtain an accurate diagnosis of the scale to know exactly which chemical to apply and when.

Spider Mites. Spider mites (family Tetranychidae), which are not really insects—they have eight legs rather than six—feed on trees and cause damage. There are many species of spider mites and they damage both conifer and broad-leaved trees. Spider mite mouthparts are equipped with a pair of sharp stylets, which rupture cells of the leaf tissue. The spider mite mouth takes up the sap from the injured cells while the stylets dig deeper. A flecking or stippling symptom develops on the foliage as the pests destroy the chlorophyll-containing cells of the epidermis. The heavily infested leaf or needle may take on a yellowish or bronze color, which can sometimes be mistaken for ozone or oxidant air pollutant injury. Some mites produce a weft of fine webbing over the surface of the infested foilage, and eggs, egg shells, and cast skins are found on most infested leaves.

Mites usually overwinter as eggs and develop successively into six-legged larvae, protonymph, deutonymph, and eight-legged female and male adults. Adult mites are tiny, only $\frac{1}{100}$ to $\frac{1}{25}$ inch long! A diagnostic test that works well for infested conifers involves sharply shaking or jarring the foliage over a piece of white paper. Mites which land on the paper can be seen as very fine, dark flecks moving about on the paper. With the aid of a hand lens, eggs, cast skins, webbing, and crawling mites can be observed on the leaves. Reasonable field identifications can be made if their color, host plant, and geographic location are known.

Mite populations can build up rapidly, especially during dry weather. They often become a serious problem when predator mite populations are reduced by the use of insecticides such as Sevin or Lindane. Mites destructive to landscape trees include spruce spider mite, *Oligonychus ununguis;* European red mite, *Panonychus ulmi;* two-spotted mite, *Tetranychus urticae;* and southern red mite, *O. ilicis.*

CONTROL. Spider mites of evergreens are usually controlled with sprays of Vendex, superior oil, or oil plus ethion in early spring. Late spring and midsummer applications of Acaraben, Kelthane, Tedion, Vendex, or malathion are often used to control deciduous tree spider mites.

Eriophyid Mites. These mites are different from spider mites in that they have only two pairs of legs. Their bodies are soft, spindle shaped, and almost wormlike. Eriophyid mites are extremely small, requiring a hand lens for observation. Their life stages consist of egg, nymph, and adult. The latter may take on different shapes, making identification difficult.

Eriophyids, called blister, bud, gall, or rust mites, feed only on plants. They may attack leaves, shoots, twigs, stems, buds, flowers, and fruits. Symptoms include abnormal shape or distortion, blisters, discoloration, erineum (a form

Fig. 11-10. Maple bladder gall mite damage.

of gall or abnormal hair growth from leaf epidermis) galls, stunting, russeting, bronzing, withering, and witches' broom. Most of these mites are host-specific, that is, they feed on only one or a small group of closely related types of plant.

Maple bladder (Fig. 11-10) and spindle galls are familiar pouch galls formed as a result of attack by eriophyid mites *Vasates quadripedes* and *V. aceriscrumena*. The galls are a localized growth reaction of the host following mite feeding. Eggs hatch and young adults live inside the galls.

Conspicuous witches' brooms of hackberry result from *Eriophyes celtis* infestation of the tree buds. The infestation interferes with normal growth; numerous buds become clustered together on the stem. When the buds break, thin, stunted, tightly bunched twigs develop, resulting in a witches' broom.

Ash flower galls, caused by *Eriophyes fraxinivorus,* are distinctive because affected flowers are swollen, fused, and distorted.

The erineum of beech, caused by *Acalitus fagerinea,* consists of patches of pale green to yellow felty hairs located in the angles between the lateral veins and the midrib of the lower leaf surface. *Eriophyes elongatus* and *E. calaceris* form bright red erineum patches on the upper surface of maple leaves.

The eriophyid mite which attacks hemlock, *Nalepella tsugifoliae,* causes foliage to have a yellowish or brownish cast without galls being produced. Its feeding is debilitating to infested trees.

CONTROL. Eriophyid mites are not normally damaging enough to warrant control measures. If there is a need, they are usually controlled by an application of superior oil or superior oil plus ethion in early spring just before budbreak. Sevin, Kelthane, or Dursban can be applied when leaves or flowers first begin to form.

Selected Bibliography

Drooz, A. T. 1985. Insects of eastern forests. U.S.D.A. Forest Service Misc. Pub. 1426, Washington, D.C. 608 pp.

Johnson, W. T., and H. H. Lyon. 1976. Insects that feed on trees and shrubs. Cornell University Press, Ithaca, New York. 464 pp.

Keifer, H. H., E. W. Baker, T. Kono, M. Delfinado, and W. E. Styer. 1982. An illustrated guide to plant abnormalities caused by eriophyid mites in North America. U.S.D.A. Agriculture Handbook 573, Washington, D.C. 178 pp.

Rose, A. H., and O. H. Lindquist. 1982. Insects of eastern hardwood trees. Can. For. Serv. For. Tech. Rep. 29. 304 pp.

Westcott, C. 1973. The gardener's bug book. Doubleday, Garden City, N.Y. 689 pp.

12

Parasitic Diseases of Trees

Although disease control unquestionably constitutes one of the major phases of successful tree maintenance, it is often the least clearly understood by the arborist. This lack of understanding is due partly to the difficulty of recognizing disease symptoms and partly to the complex nature of the parasitic agents. Insufficient knowledge of the various chemicals used to combat diseases and of their proper applications is also a contributing factor.

Broadly speaking, a tree is considered diseased when its structure or functions deviate from the normal ones described in Chapter 2. Such a definition includes abnormalities caused by drought, excessive moisture, lack of nutrients, and chemical and physical injuries, as described in Chapter 10. In this chapter discussion is limited to diseases caused by infective agents such as fungi, bacteria, mycoplasmas, nematodes, and parasitic higher plants. The symptoms they produce, their nature, and the chemicals and arboricultural practices used to combat them are also included in the discussion.

Disease Symptoms

Tree disease symptoms occur when a part of the tree fails to function as it should following an attack by a pathogen. Often the symptoms are a direct expression of tissue death caused by the pathogen. Leaf spots, twig blights, cankers, and root rots are good examples (Fig. 12-1). However, even in these cases, the symptoms may go farther than that. Root rot not only causes root death or necrosis, but the entire top of the tree may show wilt or nutrient deficiency symptoms because the roots are not functioning as they should. Table 12-1 presents tree disease symptoms associated with disruptions of tree functions when certain tissues are attacked by pathogens.

In addition to symptoms produced on the plant attacked, certain signs, such as fruiting and vegetative structures of the causal organisms, help in diagnosing the disease. For example, many fungi produce fruiting bodies that are easily recognized and facilitate diagnosis. Such bodies vary in size from pinpoint dots, which are barely visible to the unaided eye, to bodies a foot or more in

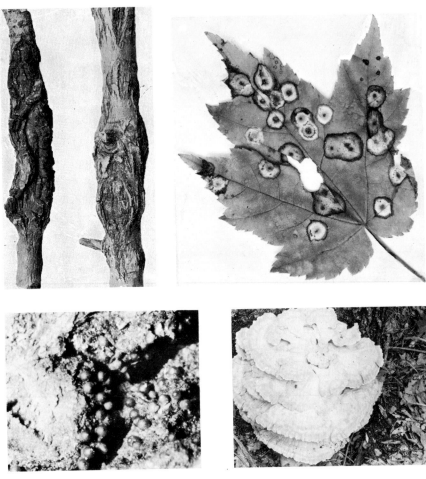

Fig. 12-1. Symptoms and signs of fungus diseases. **Upper left:** Cankers on ash stems. **Upper right:** Circular spots on maple leaf caused by the fungus *Phyllosticta*. **Lower left:** Pin-point fruiting bodies of the *Nectria* canker fungus. **Lower right:** Fruiting bodies of the fungus *Polyporus sulphureus*.

diameter (Fig. 12-1), which commonly occur in wood-decay diseases and appear as shelves or brackets adhering to the dead bark.

Additional signs that aid in determining the cause of a disease include droplets of colorless or milky ooze that appear on the bark of trees affected by fire blight disease and bright orange telia that appear in spring on rust-infected cedar twigs and branches.

Nature of Parasitic Agents

Plant parasitic fungi and bacteria, the most common tree pathogens, are microorganisms which do not contain chlorophyll, which is necessary for the manu-

Table 12-1. Typical symptoms observed when specific tissues are attacked

Tissue Attacked	Function Disrupted	Typical Symptom
Roots	Water and nutrient uptake	Root death, tree wilt, dieback, nutrient deficiency
Sapwood, xylem	Upward water movement	Sapwood discoloration, leaf wilt and death
Heartwood	Strong, sound structure	Trunk, branch breakage, cavity formation
Stem phloem	Downward food movement	Discolored, dead inner bark, root starvation, top decline
Stem cambium	Xylem and phloem formation	Discolored, dead cambium, sunken area on stem
Meristematic	Normal cell division	Stem or leaf galls
Succulent shoots	Normal terminal growth	Dead shoots, thinned or bushy growth
Leaves	Photosynthesis	Dead spots, chlorosis, defoliation, tree weakening

facture of food. As a result, they are dependent on the higher plants for their sustenance. Bacteria and fungi that depend entirely on decaying organic matter for their food source are known as *saprophytes*. These organisms are frequently beneficial, in that they decompose organic materials and liberate minerals for use in future plant growth. Forms that are able to obtain their nourishment by direct attack on living plant cells are known as *parasites* or *pathogens*. It is this group that is involved in disease production.

Bacteria. Bacteria are microscopic, single-celled organisms which multiply by simple fission, a single individual dividing to form two. Some types of bacteria possess flagellae, whiplike appendages that effect a limited amount of locomotion in liquids. Bacteria have several forms, but those that parasitize plants are usually rod-shaped.

Plant pathogenic bacteria are capable of living on organic food sources in the absence of the host plant. Many, such as leaf-spotting and shoot-blighting bacteria, are found on the host plant surface, living epiphytically but not causing disease until conditions are right. Some bacteria, implicated as causal agents of scorch diseases, are limited to the host xylem during pathogenesis.

Fungi. Fungi, as a rule, are much more complex than bacteria. They may possess two distinct stages—vegetative and reproductive. The vegetative stage usually consists of a weft or mass of microscopic, threadlike strands called *hyphae*. These hyphae may be composed of innumerable cells joined together. Certain tissues in this complex are capable of performing highly specific functions.

The reproductive stage may be simple or very complex, either microscopic or visible to the naked eye, often reaching 12 or more inches in diameter. In all cases, however, the reproductive bodies produce microscopic spores of various shapes and colors, which function much as do the seeds of higher plants.

The spores may be ejected from, drop out of, or ooze from the reproductive bodies and are carried by some means to susceptible plant parts. In a moist environment, these spores germinate by pushing forth a tiny strand known as a germ tube. Once inside the host plant, the tube branches into numerous strands, which constitute a new fungus body. Eventually the fungus produces a new crop of spores directly, or it produces fruiting bodies capable of producing spores.

Viruses. Plant viruses are infectious particles composed of protein and nucleic acid. They are introduced into plant cells by specific vectors, most commonly insects and nematodes in the case of trees, and harness the manufacturing capacity of the cell to make more virus particles. Typical virus symptoms on infected leaves include mosaic or mottling patterns, ring spots, flecks, necrotic lesions, chlorosis, line patterns, curling, and dwarfing. Stunting, decline, and excessive branching may also be symptoms. Virus diseases of landscape trees have been studied very little, although viruses of fruit trees are known and presumably occur in their landscape relatives. Mosaic viruses of ash, elm, birch, poplar, and oak have been described. Ash, birch, and poplars are subject to infections by other viruses as well.

Mycoplasma. These organisms, intermediate between bacteria and viruses in size and other properties, are important tree pathogens. Witches' brooming and yellowing are two types of symptoms commonly caused by mycoplasma. Lethal yellowing of palms and phloem necrosis of elms, previously thought to be caused by viruses, are now known to be caused by these organisms.

Nematodes. Plant parasitic nematodes are tiny unsegmented roundworms, which have well developed reproductive and digestive systems. Their life stages consist of eggs, larvae, and adults, the latter two stages spent feeding mainly on host plant roots. These nematodes have a needlelike stylet or feeding tube which allows them to feed on individual plant cells to obtain nutrients for growth. Some feed by grazing here and there on root cells, whereas others settle in one spot, continuing to feed in that location, often causing root galls. Root feeding nematode problems are generally more serious in warm temperate and tropical regions.

Root-knot nematodes feed on roots of woody plants, causing root swellings or galls, which disrupt the uptake of water, producing symptoms of wilt, nutrient deficiency, and poor growth of the top. Boxwood, catalpa, dogwood, maple, peach, pine and willow are susceptible to root-knot nematode. Root lesion nematode kills portions of the root on which it feeds. The top of the plant becomes stunted or declines in proportion to the amount of parasitism. Landscape trees are not normally heavily damaged by lesion nematodes; however, the nematodes will feed on ginkgo, maple, pine, spruce, sweetgum, and tuliptree roots and seriously damage shrubs such as American boxwood and blue rug junipers. Stubby root nematodes cause reduced top growth because of root tip feeding. They have been associated with pines having littleleaf symptoms. Dagger nematodes feed on ash, elm, maple, and spruce roots and also vector certain virus diseases such as ash ringspot virus. Pine wilt nematode

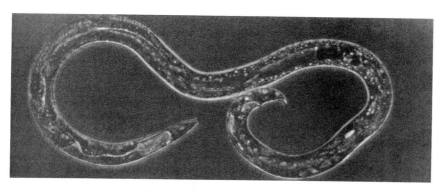

Fig. 12-2. Pine wilt nematode.

(Fig. 12-2) feeds on cells lining the resin canals of pine branches and stems, causing wilting and death.

Parasitic Higher Plants. Mistletoes and dwarf mistletoes are common parasites of some trees in some areas. These plants grow on twigs and branches of their host, extracting water, mineral elements, and food from the tree by way of a parasite nutrient-uptake organ. The tree branch is often swollen at the point of mistletoe attachment.

The true or leafy mistletoes *(Phoradendron* and *Viscum)* are most frequently associated with hardwoods growing in the southern two-thirds of the United States. These small leafy plants (Fig. 12-3)—the same mistletoe used during holidays—are capable of making much of their own food and depend on the host tree for water and minerals. Mistletoe seeds, spread by birds, germinate on the tree branch and penetrate the host. The top foliage may be harvested or may die back in extremely cold weather, but the plant grows back in a few years. True mistletoes may be harmful to the tree during times of stress, but otherwise they are not damaging. They can be controlled by pruning, if necessary.

Dwarf mistletoes *(Arceuthobium)* are more damaging to their tree hosts because they lack true leaves for food manufacture. They are parasites of pine, fir, Douglas-fir, and other conifers in western United States and of black spruce in the East. The invaded twigs, branches, or trunk tissues may be swollen or produce witches' brooms. The overall effect is a reduction in tree vigor, poor growth, and even death of older trees. The parasitic plant produces sticky seeds which are shot out at some distance, allowing spread of the plant to nearby trees. Dwarf mistletoes should be pruned and destroyed, if possible, and badly infected trees should be taken down.

Dissemination and Survival of Parasitic Agents

Bacterial plant pathogens and many fungal pathogens gain entrance into the plant tissue through the natural openings, such as the stomata and the hyda-

Fig. 12-3. Top: Clumps of mistletoe on a black walnut limb. **Bottom:** Close-up of a mistletoe plant.

thodes in the leaf tissues or the lenticels in the stem. Wounds caused by insects or by mechanical means also afford convenient openings. Many fungi are also able to penetrate directly the unbroken epidermis of the leaves.

By growth or multiplication within the host tissue, fungi and bacteria absorb the cell contents and cause death, or their presence can cause overgrowth or dwarfing of the invaded tissues.

Parasitic fungi are disseminated primarily in the spore stage. They may be splashed by rain or carried by air currents, insects, animals, on the hands of the tree worker, or on tools, from tree to tree or from place to place.

Wind-blown rain is probably the most important agent in the local dissemination of bacteria and of fungus spores, particularly of those fungi that cause leaf spots, leaf blights, and twig cankers. Spores produced on the surface of infected leaves or in cankers are splashed or washed to nearby healthy leaves and twigs, on which new infections occur.

Air currents are also important agents of dissemination. For example, the spores of some rust fungi are blown by the wind many miles from the tissues on which they are formed.

Insects are well-known agents of dissemination. Bees and flies are known to spread the bacteria that cause fire blight disease. Several species of bark beetle are chiefly responsible for the spread of spores of the fungus causing the Dutch elm disease. Insects are the chief means of spread of viruses and mycoplasma.

The chestnut blight fungus is known to be spread by birds and squirrels lighting on or traveling over extensive bark cankers, on the surface of which spores are produced in large numbers. These spores adhere to the talons or feet and may be deposited again on some distant tree.

Spores may be spread on tree workers' hands and clothing, as is the case with the Cytospora canker disease of spruce, especially when diseased branches are pruned during wet weather. The fungus responsible for planetree cankerstain may be disseminated on infested pruning saws, and even in many types of wound dressings. Spores of *Ophiostoma ulmi*, the Dutch elm disease fungus, may be spread on pruning tools.

Pruning saws, shears, and other metal tools can be disinfested by being dipped in or wiped with 70 percent denatured alcohol. Tree-climbing ropes contaminated with fungi can be disinfested by being confined for 3 hours in a 10-gallon closed receptacle containing 4 ounces of liquid formaldehyde; this chemical should be used with caution.

Soil-inhabiting fungi, bacteria, and nematodes, capable of producing diseases such as wilts and root and collar rots are carried largely by the transfer of infested soil on tools, machinery, or shoes, or by washing during heavy rains. Winds, by lifting soil particles and carrying them from place to place, may also be a means of dissemination of this group of plant parasites.

Many soil-inhabiting organisms live as saprophytes in the soil until they come in contact with susceptible tree roots. They enter such roots, even unwounded ones, and become parasites. Wilt fungi grow directly into the conducting tissues, where they cause a partial clogging or secrete toxins that are carried in the water stream to the leaves. There they accumulate and cause a collapse of the leaf tissues.

Overwintering of Bacteria and Fungi

Leaf-spotting and leaf-blighting bacteria and fungi overwinter in fallen infected leaves, on dormant buds, and in small twig cankers. Canker-producing bacteria and fungi overwinter in the margins of the cankers. Wilt fungi pass the winter in the vascular tissues of infected trees and in the soil.

Factors Required for Disease Development

Three conditions must be met for a tree disease to occur. There must be (1) a susceptible host, that is, a tree in which it is possible for a specific disease to occur; (2) a virulent pathogen, one that is capable of causing the disease in this host; and (3) an environment suitable for disease to develop (Fig. 12-4). If any one of these three factors is missing, no disease will occur.

Although most trees are resistant to most diseases, it is well established that virtually all trees are susceptible to at least one disease. Furthermore, virulent tree pathogens occur in most areas. However, merely bringing together virulent pathogens and susceptible trees is not enough. Many diseases can occur only under specific environmental conditions, such as the presence of a tree wound, a specific stage of tree growth, a long period of leaf wetness, saturation of the soil with moisture, or stressful growing conditions. Many fungi that normally are weak parasites, which do little damage to trees growing under proper conditions, can readily destroy trees growing under adverse conditions.

There are some diseases which develop more readily on lushly growing trees. The dense foliage of such trees may provide the most favorable microclimate for disease development; succulent tissues are often more susceptible to infection.

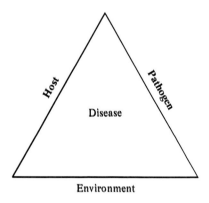

Fig. 12-4. The disease triangle.

Fig. 12-5. Fire blight canker and blight on pear.

Fire Blight

Fire blight is known primarily as a disease of apple and pear trees, and consequently it is of most concern to orchardists. It also occurs on many ornamental trees and shrubs in the rose family. Commercial arborists, estate superintendents, and even the owner of a single susceptible tree must often cope with this very destructive disease. Besides attacking apple and pear trees, including flowering crabapple and callery pear in the landscape, fire blight appears commonly on several species of cotoneaster, cockspur hawthorn, English hawthorn, and mountain-ash. It occurs less frequently on firethorn, serviceberry, flowering quince, cultivated quince, Christmas berry, flowering plum, spirea, rose, and Stransvaesia.

Symptoms. The leaves near the growing tips and the flowers suddenly wilt, turn brown and black, and look as though they were scorched by fire (Fig. 12-5). Twigs also are blighted on most of the ornamental hosts. Affected shoots with dead leaves hanging from them often curve near the tip, forming a shepherd's crook. In some, the mountain-ashes, for example, the infection spreads down to involve large branches. Extensive cankers on the trunk and main branches develop on the larger ornamental trees, as well as on apple and pear. The flowering quince is susceptible principally to blossom blight, the disease being rare on woody parts. The presence of bloom infection and the absence of typical twig blight in such cases give the infected tree the appearance of having been injured by frost.

The disease is widely distributed, but its yearly occurrence is sporadic. We observed very heavy damage to 'Aristocrat' flowering pears in 1986 in the Ohio River valley area on trees 6 to 12 inches in diameter that had shown few or no disease symptoms the previous 15 years.

Cause. Fire blight is caused by the bacterium *Erwinia amylovora*. Fire blight disease is thought to occur as follows. The bacteria overwinter in cankers from the previous year's infections. Natural, epiphytic populations of *E. amylovora* can also occur on apparently healthy shoots, leaves, and buds of the host plant tissues without producing blight symptoms. During the spring, epiphytic and canker-residing bacterial populations increase and droplets of bacterial ooze are produced in the margins of cankers. Bacteria, spread by insects attracted to the bacterial ooze and by rain and wind, invade flower parts, leaves, and shoots through nectaries, stomata, hydathodes, or wounds. They then penetrate and spread within the host tissues, causing death of blooms, fruit, twigs, and branch tissues. Secondary spread to other parts of the tree and to nearby trees by insects, wind, and rain continues until new shoot growth ceases, at which time cankers become dormant.

Fire blight disease is favored by long frost-free periods before bloom, followed by humid weather with 65 to 70° temperatures during and after bloom, with occasional rains during that period.

Control. Blighted twigs should be pruned in winter about 4 inches below the infected areas and destroyed. Large-limb cankers can also be surgically removed by cutting 4 inches beyond the margin of the canker. This practice eliminates an important potential source of inoculum for subsequent epidemics, although it is difficult to thoroughly prune large landscape trees.

The pruning must be done carefully, so that all infected branches are removed. It is not necessary to sterilize the pruning tools for dormant pruning. Trees that are badly infected should be removed, as should old, neglected pear trees, which can be sources of inoculum.

While trees are dormant, apply copper sulfate (4–5 lb/100 gal) to the twigs and branches to help reduce overwintering bacterial inoculum. This chemical may color nearby objects blue, so care must be taken if this is done in urban landscapes. Thoroughly wash the spray tank following use, since the chemical is corrosive.

In a nursery or similar planting, spray the antibiotic streptomycin (50–100 ppm solution, or 4 oz of 21% streptomycin sulfate/100 gal) at 4- to 5-day intervals during bloom, and again at weekly intervals during the period following bloom when the new shoots are elongating rapidly. Streptomycin is best applied late in the day, when the air is still and when intense light will not rapidly break the chemical down. Repeated applications help the streptomycin penetrate the plant tissues. During a severe outbreak of fire blight, just one or two spray applications will not be effective.

We prefer not to see streptomycin used in urban landscapes. *E. amylovora* resistance to streptomycin has been reported, and the possibility of resistance transfer to human pathogens, though extremely remote, is real. Fixed copper such as Basic Copper Sulfate (not copper sulfate) can be used instead of streptomycin, although it is not nearly as effective.

Sucking insects such as aphids, leafhoppers, plant bugs, and pear pyslla create wounds that bacteria can enter. Control these insects before bloom and throughout the summer, especially if the planting has had a history of an insect problem. In the landscape, do not apply insecticides for fire blight control unless it is being done for other reasons.

Excess nitrogen fertilization, which encourages rapid tree growth and increased susceptibility to fire blight, should be avoided. Removal of suckers and water sprouts may reduce disease.

Susceptible trees should be inspected frequently during the growing season and infected spurs and terminals removed by breaking them out. If pruning tools are used, sterilize between cuts using 70 percent alcohol or a solution of 1 part sodium hypochlorite bleach to 10 parts water. Prune or break back at least 12 inches from active fire blight. This job will be very difficult to do for large landscape trees. If a thorough job of blight removal cannot be accomplished, it is better to leave this task until winter.

Disease-resistant flowering crabapple cultivars are available and should be planted in areas where fire blight is likely to be a problem.

Crown Gall

Crown gall is a common disease of trees in the rose family, such as apple, cherry, pear, plum, and flowering almond. The disease also occurs on such widely differing species as chestnut, Arizona cypress, European juniper, incense cedar, sycamore maple, oleander, poplar, English walnut, hickory, willow, English yew, and, in Hawaii, on the macadamia tree.

Symptoms. Roughened swellings or tumorlike galls of varying sizes appear at the base of the tree or on the roots (Fig. 12-6). On poplar, hickory, and willow, these galls may also appear on limbs and branches. Growth of the tree may be retarded and the leaves may turn yellow, or the branches and roots may die as a result of water and nutrient transport interference caused by the presence of these galls. Although established trees are rarely damaged appreciably, young trees may be killed. Other agents such as *Phomopsis* fungi can cause symptoms resembling crown gall. We have observed mature hickory trees showing extensive phomopsis galling on twigs and branchs.

Cause. Crown gall is associated with the soilborne bacterium *Agrobacterium tumefaciens.* The bacterium does not kill the parts attacked but carries a separate genetic factor, tumor-inducing principle, which stimulates abnormal cell growth. It enters plants through wounds and is primarily a parasite of young nursery stock.

Control. Sanitation is important for crown gall control. Nurserymen should discard all young fruit trees and ornamental trees showing galls on the stem or main roots. They should follow practices which minimize wounding the stems and roots of young trees in the nursery. They should also be aware that infected plants, infested soil, insects, contaminated tools, and irrigation water are capable of carrying the crown gall bacteria.

A biological control agent, Galltrol A or Norbac 84, consisting of a certain

Fig. 12-6. Crown gall on walnut stem.

strain of *Agrobacterium radiobacter,* shows promise for reducing crown gall in nursery plantings. This bacterium competes with the crown gall bacterium for attachment to wound sites and is effective when used as a root dip or spray for cuttings and seedlings.

Gallex is a product used as a crown gall treatment after infection has already occurred. The active ingredients are 2,4-xylenol and *meta*-cresol. The material is applied directly to the galls and causes them to shrink. Gallex may need to be reapplied and is limited to those galls that can be exposed and reached with treatment. Follow manufacturer's directions for use and necessary precautions.

Tree species and cultivars not susceptible to crown gall include bald cypress, beech, boxwood, catalpa, cedars *(Cedrus),* gingko, goldenrain-tree, hemlock, holly, hornbeam, larch, linden, magnolia, pine, spruce, tuliptree, and tupelo.

Bacterial Wetwood and Slime Flux

The foul-smelling and unsightly seepage from wounds in the bark or wood of various shade trees is known as slime flux. It occurs most commonly on bacterial wetwood-infected trees, such as elm, oak, birch, and maple. Although slime flux development is seasonal, evidence of wetwood and slime flux-stained bark is visible anytime (Fig. 12-7). This slime is different from the liquid associated with alcoholic flux. Alcoholic flux emanates from shallow wounds and persists only for a short time.

Symptoms and Cause. Wetwood seepage originates from infections of the

Fig. 12-7. Slime flux oozing from a pruned branch stub of an American elm.

heartwood and inner sapwood by common soil-inhabiting bacteria such as *Enterobacter cloacae (Erwinia nimmipressuralis)*. Wetwood bacteria are capable of growing anaerobically (without oxygen) in the internal wood tissues. Methane and osmotic or metabolic liquids, two by-products of the bacterial activity, accumulate under pressure and are forced out of the tree through the nearest available opening, usually a trunk wound or branch stub. Pruning a branch or taking a core with an increment borer can sometimes release the materials under pressure, squirting the worker with foul-smelling liquid and gas.

Normally flowing to the wounded bark surface, the wetwood fluid is a clear watery liquid containing several nutrients. On the surface it soon changes to a brown, slimy ooze, as a result of the feeding by fungi, bacteria, and insects. This surface slime flux may kill bark, cambium, and bark surface organisms as well as grass growing near the base of the tree.

Wetwood-infected trees have an internal core of wood that is wet but not decayed. These infected branch, trunk, and root tissues also have a high pH. Wetwood-infected wood is resistant to decay by fungi. The extent of wetwood spread in the tree may be limited by tree defenses; however, wetwood can spread into new tissues as new injuries occur. Thus deep injection holes and pruning can expand wetwood infection. Take care to avoid pruning live branches on infected trees.

Control. Thus far no effective preventive or curative measure is known. If this problem is killing the bark, it may be helpful to drain the slime flux away from the branch or trunk so that it drips on the ground. Drilling a hole into the tree and inserting a semirigid plastic tube has helped in some cases; however, this results in additional wounding and the threat of expanded wetwood or decay should be considered. Loose dead bark should be carefully cut away so that the area can dry.

Bacterial Scorch

Xylem-limited bacteria have been associated with a leaf scorch disease of shade trees in the southeast and Atlantic Coast states. Trees shown to be affected by scorch disease include elm, mulberry, oak, and sycamore. Citrus in Florida is also affected. The bacterium *Xylemella fastidiosum* is consistently associated with leaf scorch disease, however plant pathologists have not been able to cause infections and scorch symptoms by inoculating with the bacteria. Remission of symptoms through the use of antibiotics has been done experimentally, providing circumstantial evidence for a bacterial involvement in scorch disease.

Symptoms. Oak leaf scorch is characterized by a marginal leaf necrosis bordered by a chlorotic halo. As the disease progresses, the dead leaf margin increases, extending toward the midrib, and leaves curl. Premature shedding of affected leaves may leave tufts of unaffected leaves on uninfected branches. Diseased trees may show branch dieback and decline. This disease may be more widespread than currently reported.

Cause. The rod-shaped bacteria associated with the xylem of diseased trees

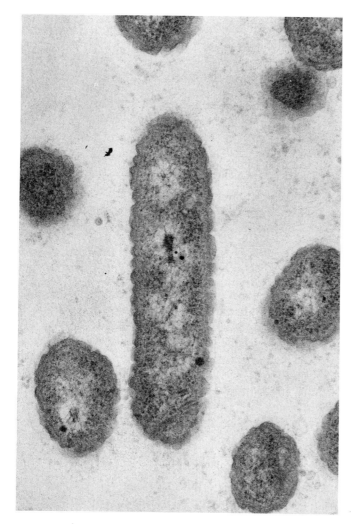

Fig. 12-8. Electron micrograph of *Xylemella fastidiosum,* cause of bacterial scorch of pin oak.

have been observed microscopically (Fig. 12-8). Their direct involvement is still circumstantial.

Control. Onset and remission of symptoms can be delayed experimentally using oxytetracycline antibiotic. Use of this technique requires further study to see whether long-term control can be accomplished and phytotoxicity avoided.

Powdery Mildews

The leaves and twigs of many deciduous trees and shrubs are occasionally covered with a grayish white dusty material known as powdery mildew. The

Fig. 12-9. Powdery mildew on oak.

mildew is made up of delicate, cobweblike strands of fungus tissue covered with microscopic colorless asexual spores. Many minute, spherical, black bodies—fruiting structures of the sexual stage of the fungus—may be visible to the unaided eye in the grayish white areas. The spores of the fungus may overwinter in these bodies. Some powery mildews overwinter vegetatively in dormant buds of the infected tree.

Symptoms. Although mildew fungi grow mainly over the surface (Fig. 12-9), they also penetrate the leaf surface with numerous fine filaments, which extract nutrients from the host plant. Heavily mildewed leaves may turn yellow, dry up, and fall prematurely. Mildews appear most commonly toward the end of summer and in late fall and are most prevalent on trees or branches growing in shaded and damp locations.

Cause. There are a number of distinct species of powdery mildew fungi, which can be distinguished only by microscopic examination. The following list includes most of the species, together with the trees they parasitize:

Brasiliomyces trina on oak.
Erysiphe aggregata on alder.
E. cichoracearum on aspen, eucalyptus, and smoke-tree.
E. lagerstromiae on crapemyrtle.
E. liriodendri on tuliptree.
E. polygoni on acacia, black locust, and serviceberry.
Microsphaera spp. on alder, ash, beech, birch, buckthorn, catalpa, chestnut, dogwood, elm, hackberry, hickory, holly, honey locust, hornbeam, hophornbean, linden, magnolia, maple, oak, pecan, planetree, sycamore, and walnut.
M. diffusa on black locust.
Phyllactinia guttata on alder, ash, beech, birch, black locust, boxelder, boxwood, catalpa, chestnut, chinaberry, crabapple, dogwood, elm, hawthorn,

hickory, holly, hornbeam, horsechestnut, linden, magnolia, maple, mountain-ash, mulberry, oak, pear, planetree, sassafras, serviceberry, sycamore, tulip-tree, walnut, willow, and yellowwood.

Pleochaeta polychaeta on sugarberry.

Podosphaera leucotricha on apple.

P. clandestina on almond, apricot, cherry, hawthorn, mountain-ash, peach, pear, persimmon, plum, and serviceberry.

Sphaerotheca lanestris on oaks.

S. pannosa on flowering *Prunus*.

S. phytophila on hackberry.

Uncinula spp. on buckeye, elm, hackberry, linden, maple, poplar, and willow.

Control. Resistant cultivars of flowering crabapples are available and offer an excellent means of control. Beneficial management practices such as pruning out infected twigs, raking up infected leaves, improving air movement around the trees, and increasing the amount of sunlight reaching them can often reduce severity of the disease. Dusting or wettable sulfurs, previously used to control powdery mildew, have largely been supplanted by Bayleton, Benlate, and Pipron. As a rule, however, only particularly valuable large specimens and small nursery trees are sprayed to control mildew. In most cases, the disease appears so late in the growing season that it does little real damage.

Anthracnose Diseases

Anthracnose diseases are common on the foliage and sometimes the twigs of ash, sycamore, maple, and white oak. The anthracnose fungi can also attack walnut, hickory, elm, birch, catalpa, linden, planetree, tuliptree, and horsechestnut.

Symptoms. Early in spring, anthracnose fungi may kill twigs and newly expanding leaves, causing symptoms that resemble frost injury (Fig. 12-10). Small, sunken dead areas, cankers which can girdle the branch, form on infected twigs and branches. Later infections of the leaves or leaflets cause dead blotches along the leaf veins and sometimes distortion. Infected leaves and leaflets may drop from the tree, causing extensive defoliation in severe cases. Trees normally develop another crop of leaves following this defoliation.

Cause. Anthracnose of landscape trees is caused by several species of *Discula* (imperfect) or *Apiognomonia* (perfect) formerly referred to as *Gloeosporium* or *Gnomonia*. The fungus overwinters in fallen leaves and twigs as well as in branch cankers. In spring, the fungus produces spores, and in wet weather it infects leaves and twigs. Cool, wet weather favors twig and branch attack, whereas warmer wet weather favors leaf infection. The disease normally does not continue building up during the summer.

Control. Normally, little is done to control anthracnose, except to rake up and destroy fallen leaves and twigs to reduce fungal inoculum. Even that activity may not be effective unless combined with pruning all infected twigs and branches.

Fig. 12-10. Sycamore anthracnose: A, twig canker; B, leaf infection.

Fungicides such as Benlate, Mancozeb, Maneb, Zineb, or Basic Copper Sulfate can be used to reduce infections. They need to be applied first at budbreak in spring, again when first leaves are partially expanded, and finally two weeks later. Spraying might be warranted for young or particularly valuable specimens or following several successive severe anthracnose years. Fertilizing and watering trees to promote replacement foliage following defoliation may also help.

Verticillium Wilt

Verticillium wilt, one of the most common fungus diseases, is found on more than 300 kinds of plants, including shade and ornamental trees, food and fiber crops, and annual and perennial ornamentals. Among its more important hosts are the many species of maples planted along city streets. In fact, this disease has caused the death of more maples over the last half-century than any other disease. A survey in New Jersey by the senior author revealed that silver maple *(Acer saccharinum)* is the most susceptible of the streetside maples, with the Norway maple *(A. platanoides)*, red maple *(A. rubrum)*, and sugar maple *(A. saccharum)* decreasingly susceptible in that order. In other parts of the United States, however, it is the sugar maple that appears most susceptible.

The following additional trees are known to be susceptible to Verticillium wilt: hedge, black, Japanese, Japanese red, Schwedler's, sycamore, Drummond, tatarian, and trident maples; boxelder and California boxelder; horse-chestnut; tree-of-heaven; Spanish chestnut; southern and western catalpa; carob; redbud; camphor-tree; yellowwood; flowering dogwood; smoke-tree; quince; Japanese persimmon and Texas persimmon; persimmon; Russian-olive; white, green, black, and European ash; Kentucky coffee tree; China tree; goldenrain-tree; tuliptree; osage-orange; southern, saucer, and star magnolia; apple; sour-gum or tupelo; olive; tea olive; avocado; cork tree; pistachio; aspen; almond; sour, sweet, Mahaleb, or Morello cherry; garden and cherry plum; apricot and Japanese apricot; prune; pear; pin oak; black locust; sassafras; pepper-tree; Japanese pagoda-tree; American, littleleaf, and Crimean linden; American, English, slippery, Scotch, and Siberian elm.

Symptoms. Wilt symptoms vary somewhat with the kind of tree, its age, its location, and other factors, but there are some symptoms common to all situations. A sudden wilting of the leaves on one limb, on several limbs, or even on the entire tree is a general symptom (Fig. 12-11). The severity of this symptom depends less on species of tree than on the amount of infection in the below-ground parts. Long before wilting becomes visible, the fungus has been at work below ground. If most of the roots are infected, the tree will die quickly. On the other hand, if infections in the roots are confined to one side of the tree, only the parts above that side will show wilt symptoms. Periods of drought frequently accentuate the wilting symptoms and speed the death of the infected tree.

The wilted leaves may fall after they have dried completely, or they may hang on for the remainder of the season. Sometimes a few branches or even the entire top of the tree will die during the winter and hence fail to leaf out the following spring. Cases of this sort are usually attributed to winter injury.

Because many other agents cause similar symptoms, positive diagnosis of Verticillium wilt can be made only by culturing sapwood tissue in the laboratory. The cultures are taken from discolored sapwood, which occasionally ap-

Fig. 12-11. This maple is heavily infected by the fungus *Verticillium albo-atrum*.

pears in the branches but more commonly and more reliably near the base of the main trunk or in the main roots. When limb discoloration occurs, it frequently shows up on the undersides of infected branches nearest the trunk, reflecting direct movement of the fungus from the trunk to limbs via the xylem. Verticillium is a soil-inhabiting fungus primarily, and hence its most common avenue of entrance is through the roots. One would expect, therefore, that the discoloration of the sapwood would be more marked and more common closer to the point of entrance. This is actually the case, as we have discovered in our studies on the cause of landscape and streetside tree death over many years.

When the *Verticillium* fungus is present, the outer sapwood rings (those nearest the bark) show a discoloration. The color varies with the species of tree and the duration of the infection. In Norway maple it is a bright olive green; in American linden, dark gray; in tree-of-heaven, yellowish brown; in black locust, brown to black; in northern catalpa, purple to bluish brown; in American elm, brown (like the Dutch elm disease fungus); and in Japanese pagoda-tree, greenish black. But the presence of discolored sapwood tissue cannot be used as an exact diagnostic symptom. It merely indicates that a fungus may be involved. For example, the senior author has isolated at least six different kinds of fungi from sapwood of Norway, silver, and red maples which showed the bright green discoloration! Moreover, he reproduced the same discoloration in young, healthy maples by inoculating them with these fungi.

It is well to reiterate, therefore, that a positive diagnosis for Verticillium wilt, as for Dutch elm disease, can be made only by laboratory isolation tests.

Cause. Wilt is caused by the fungi *Verticillium dahliae* and *V. albo-atrum*. Many strains or segregates of this fungus are known to exist. The fungus can live in soil for some time without parasitizing roots. Although it is thought to enter roots and stems only through injuries, infection may occur in young, uninjured roots. Some research workers believe air currents may spread spores over long distances. When such spores lodge in tree wounds, aboveground infection may occur. Sapwood wounds are usually necessary for the start of an infection in the aerial portions of the tree. Contaminated pruning tools may promote aerial infections.

Once inside the water-conducting tissues, the *Verticillium* fungus grows upward longitudinally; that is to say, the infection spreads upward from the point of inoculation. The actual discoloration which results may be due to the presence of the fungus itself or to a chemical change resulting from the action of the fungus.

Control. Trees showing general and severe infection by the *Verticillium* fungus cannot be saved. Such trees should be cut down at once and destroyed quickly and completely. As many of the roots as possible should also be removed. A species of tree satisfactory for the particular location but not susceptible to wilt may be planted.

Narrow- and broad-leaved evergreens appear to be safe replacements for trees that died from the wilt disease. The following deciduous trees are not known to be susceptible to wilt and might be used as replacements: beech, birch, chestnut, crabapple, ginkgo, hackberry, hawthorn, hickory, holly, honey locust, hop-hornbeam, hornbeam, Katsura-tree, mountain-ash (European), mulberry, oak (white and bur), pawpaw, pear, pecan, planetree, serviceberry, sweetgum, sycamore, walnut, willow, and zelkova.

Where it is absolutely necessary to plant a Verticillium-susceptible tree in an infested site, soil fumigation with Vapam drenches, methyl bromide soil injections, and even formaldehyde drenches have been suggested. These materials are extremely toxic to the user, and only experienced, certified fumigators could be trusted with such a task. All previously existing roots would have to be removed from the soil and the soil would have to be well worked up beforehand to allow good fumigant penetration. Finally, a wait of several weeks between fumigation and planting is needed to allow the soil time to air out. Even with all this effort, the reliability of such a treatment would be questionable, especially if the dead tree was large.

We and other investigators have noted recovery of trees from mild cases of Verticillium wilt after liberal applications of a relatively soluble nitrogen fertilizer. On the other hand, it should be noted that Verticillium wilt is often fatal and fertilization can be a wasted effort, which could in some circumstances promote spread of the fungus in the tree. The material must be used early in the growing season on trees that show only one or two wilted branches. The heavy fertilization apparently stimulates leaf growth, which in turn enables the rapid formation of a thick layer of sapwood that seals in the infected parts beneath (Fig. 12-12).

We and other investigators have also tried injecting chemicals into infected trees or applying them to the soil around the roots as a curative measure. Such treatments are not reliably effective and hence must be considered as experimental at present.

Fig. 12-12. Greenish sapwood discoloration may be caused by Verticillium fungus. Several layers of healthy sapwood now cover the infected area in this branch.

Wood Decay

Living landscape trees are subject to internal decays of branches and trunk and of lower stem and roots. These decays are called heart rots, canker rots, and root and butt rots. They are sometimes hard to detect because the decaying tree may appear sound from the outside. The wood decay diseases are important to arborists because they afflict living trees, affecting the structural integrity and therefore the strength and relative hazard of the tree.

Decays of dead trees, of dead branches of living trees, or of wood products are sometimes called sapwood rots. Sapwood rots also represent a hazard, but an obvious one, because dead branches or dead trees are apparent. This discussion concentrates on decay in the living tree rather than in the dead one, though both are important.

Symptoms. Decay in living trees appears as a softening or weakening of the woody xylem tissues of the sapwood and the heartwood (Fig. 12-13). Although the extent of decay is not normally visible from the outside, the decay can sometimes be viewed through completely rotted branch stubs, wounds, and cracks in the bark.

If the tree were cut open, decayed wood might show brown rot or white rot symptoms. Brown rot fungi often produce a dry, crumbly, brown residue, which breaks up into cubical blocks, whereas white rot fungi usually produce a white, stringy decay. In either case, the wood is obviously weakened. A hollow where no wood is present can occur where decay has progressed extensively. An abnormal dark or light discoloration of the wood may be found in areas bordering the decay. The pattern of wood decay in the tree varies depending on the fungus, tree species, and growing site. Some fungi cause mainly canker rots, when wood and bark are both invaded, sometimes in sequence, so that large concentric callus ridges are formed in response to infection. Others cause root and butt rots, so that trees failing to generate roots faster than they are destroyed may show gradual decline or be easily uprooted. Some decays are primarily heart rots of the main trunk and branches of the tree.

Often fruiting bodies of the decay-causing fungus are present on the outside of the tree. These fruiting bodies, called conks, are indicators of advanced decay (Fig. 12-13). Conks may be leathery to woody and perennial or they may be soft, fleshy, and ephemeral. They range in size from an inch to several feet. The color is usually gray or brown, but it may be white, black, orange, or yellow. The fruiting body may protrude from the tree or develop flat against the tree surface. The downward-facing surface of the fruiting body, where the fungal spores are released, may be covered with tiny pores, slits, or teeth. Some conks produce no spores.

Cause. Wood decay in trees is caused by the growth of one of several fungi, mainly of the class Basidiomycetes, although some Ascomycetes and bacteria may also cause decay. The disease begins when a windblown fungal spore comes in contact with a tree wound and, given the right conditions (possibly involving other nondecaying microorganisms), germinates. Wounds such as branch stubs, lower trunk or root injuries, branch crotches, and pruning

Fig. 12-13. Top: Wood decay visible in old pruning wound. **Bottom:** Fruiting bodies at the base of an extensively decayed tree.

wounds are typical locations for this activity. The germinating spore produces a germ tube then branched hyphae and mycelia, which invade fiber, vessel, tracheid, and ray cells of the wood. The fungal hyphae release enzymes that break down cellulose (brown rot) or cellulose and lignin (white rot), thus providing food for the fungus and loss of rigidity for the tree.

However, trees are not passive victims; they are able to respond to wounding and fungal invasion. This response has been studied by Dr. Alex Shigo, formerly of the U.S. Forest Service, and co-workers. They developed the concept of compartmentalization of decay in trees (CODIT). There is considerable literature on this subject and the interested reader should consult Dr. Shigo's work for details.

When an injury occurs the invading wood-decay fungi encounter several tree defense mechanisms. Vertical movement of the decay fungus is impeded by plugging of xylem, tracheids, and vessels with resins and tyloses above and below the wound. The horizontal movement of decay fungi is impeded in an inward direction by the already existing growth rings, and in a radial direction by toxic substances produced by the ray cells. Decay is prevented from moving to new growth by barriers laid down by the vascular cambium after injury occurs.

Of these various defenses the strongest is that formed by the new growth. Consequently, decay will not proceed into subsequent yearly growth increments unless they are reinjured.

In circumstances where wounding is minimal and the tree is quick to respond to wounding and invasion, the tree suffers some internal discoloration and little or no decay, thus retaining its structural integrity. However, where wounding is extensive and when the tree is not capable of producing a quick, effective response, the decay fungus spreads and the tree is weakened.

The decay fungi gradually digest wood upward and downward as well as inward toward the center of the tree. Meanwhile, outside the original wound, callus cells and new wood may have formed. Decay fungi do not invade wood laid down after the wound has closed. Therefore, after a period of years, the outer cylinders of wood may be sound despite the internal decay. Canker rot fungi, able to invade both bark and wood tissues, cause internal decay while keeping the wound open and visible from the outside. Because most decay fungi primarily utilize the central portion of the tree, the term heart rot is used to describe this disease, even though both sapwood and heartwood (where present) are rotted.

The fungal mycelium continues to develop within the stem until, when conditions are right, the fungus begins to grow following a path of least resistance to the stem's outer surface. When it reaches the surface, the fungus produces a fruiting body, or conk. This often occurs near a dead branch stub or other injury. These fruiting bodies are capable of producing large number of spores, which then are available to start new infections of nearby wounded trees. The fruiting bodies of heart rot decay fungi produce spores in such great numbers that there are generally spores present at any wound site.

The presence of a fungal fruiting body signifies that internal decay has progressed extensively and, perhaps more significantly, that the tree has failed to limit decay and that additional decay is likely. Since the outer rings of sapwood, the cambium, and phloem carry out most of the tree's vital functions, the tree can appear very much alive despite the presence of extensive internal decay. There are hundreds of different wood decay fungi and each produces unique conks. The conks of some are perennial, remaining on the tree year after year, gradually increasing in size. Fungi such as *Fomes, Echinodontium, Phellinus, Perenniporia, Fomitopsis,* and *Ganoderma* produce such perennial shelflike or hooflike conks. Other important decay fungi such as *Armillaria, Collybia, Pleurotus,* and *Schizophyllum* produce short-lived mushrooms. *Phaeolus, Trametes, Stereum, Hericium, Inonotus, Heterobasidion,* and *Polyporus* produce annual shelf-like structures on decaying trees.

Control. Unnecessary wounding or injuries to the wood and bark should be avoided. Trees should be maintained in a vigorous condition and examined periodically for symptoms of decay and signs of the fungus-causing decay (conks). Structurally weakened trees may not be able to support their own weight and should be removed before they break and injure people or damage property.

Armillaria Root Rot

Some of our most valuable shade and ornamental trees are susceptible to Armillaria root rot disease. This disease is also called shoestring root rot disease because of the long, round, black strands of tissues, closely resembling shoestrings, produced by the fungus underneath infected bark, over infected roots, or in the soil. Among shade and ornamental trees, oaks and maples appear to be the most commonly infected, although the disease is occasionally destructive on apple, birch, black locust, cherry and other *Prunus* species, chestnut, empress-tree, eucalyptus, goldenrain-tree, Katsura-tree, larch, mountain-ash, pine, planetree, poplar, redbud, rhododendron, and spruce. In all probability the disease may occur on almost any tree or shrub grown, if the necessary conditions for infection are present.

Symptoms. The aboveground symptoms cannot be differentiated from those produced by many other diseases or agents that cause root or trunk injuries. Probably the most striking external symptom is a decline in vigor of a part or the entire top of the tree (Fig. 12-14). Where the progress of the disease is slow, branches die back from time to time over a period of several years.

The most positive signs of this disease, however, are found at the base of the trunk at or just below the soil line, or in the main roots in the vicinity of the root collar. Here, fan-shaped, white wefts of fungus tissue closely appressed to the sapwood are visible when the bark is cut away or lifted. Scraping or lifting the white wefts of mycelia, which have a strong mushroom odor, will reveal water-soaked sapwood. Where the entire top has wilted, the fungus tissue will be found completely around the trunk. Where a large branch has died back, or only one side of the tree shows poor vigor, the fungus will be found on one or two main roots or on one side of the trunk base. Where the tree has been dead for some time, the dark brown to black "shoestrings" may occur

Fig. 12-14. Branch dieback and general decline in this oak are symptoms of Armillaria root rot.

beneath the bark or in the soil near the infected parts. Clusters of light brown mushrooms, so-called honey mushrooms, may appear in the vicinity of the rotted wood in late autumn (Fig. 12-15). These rarely occur near infected street and shade trees, however, inasmuch as conditions are usually unfavorable for their development, or the infected trees are removed before the mushrooms have a chance to form.

Fig. 12-15. The mushroom of the Armillaria root rot fungus.

Cause. Armillaria root rot is caused by the fungus *Armillaria mellea.* Infection of living trees is thought to occur almost exclusively by means of root contact with the fungus shoestrings or by root grafts with infected trees. The fruiting stage, the honey mushroom, somewhat resembles the mushroom commonly sold in stores but is slightly larger, is yellowish brown, and the top of the cap is dotted with dark brown scales. The mushrooms usually grow in tight clusters with their stem bases pressed together. Spores released from the underside of the cap are blown by the wind to bark injuries at the base of living trees where infection can occur.

Because the disease is almost consistently associated with trees that were previously in poor vigor, several investigators believe that the fungus is only weakly parasitic and cannot attack vigorously growing trees. Others believe the causal fungus is capable of attacking living roots and can penetrate sound, healthy bark. The fact is that once the fungus becomes established in trees, it usually kills them in a relatively short while.

A similar and important root rot disease is caused by the fungus *Clitocybe tabescens.* The fungus kills the root collar cambium, producing mycelial fans under the bark. It also rots the roots, thus having a pathology similar to that of *Armillaria.* The main difference is that the mushroom lacks an annulus and furthermore *Clitocybe* does not produce the flat, black rhizomorphs or shoestrings of *Armillaria mellea.* Clitocybe root rot generally occurs in the South, whereas Armillaria occurs, often on the same tree species, in the North and West.

Control. Because much of the evidence at hand indicates that Armillaria root rot is associated with stressed trees, probably the best precautionary measures are fertilization, water, and soil improvement.

An infected tree whose entire root system or trunk is diseased cannot be saved. The large roots in the vicinity of the trunk as well as the trunk itself should be removed and destroyed. Soil in the immediate vicinity should also be removed.

Attempts to retard disease development on partly infected valuable trees are occasionally successful, especially where only one or two main roots or one side of the trunk is involved. In such cases, the soil within a radius of about 2 feet around the trunk should be removed in the spring or summer to expose the root collar and the large roots. Roots whose bark is completely rotted should be removed and the affected bark on partly diseased roots and on the trunk carefully cut away. All visible fungus tissue and water-soaked wood should be removed. The bark edges should be trimmed. Branches should be thinned out and small plants growing nearby removed to allow as much sunlight and dry air as possible in the vicinity of the treated parts.

Where the disease has killed one tree in a group, spread to nearby trees can occasionally be delayed or even completely checked by removing the affected and adjacent trees and their roots. All soil from the affected area should be carted away. This treatment is expensive and will not necessarily check the spread of the disease. The remaining trees should, of course, be fertilized, watered, and given all possible care to ensure continued vigorous growth.

The problem of planting another tree in the site on which one has previously died from this disease occasionally confronts the shade-tree commissioner, arborist, or property owner. The senior author has known of cases where three successively planted trees have died from Armillaria root rot under such conditions. Where a tree must be planted in the same place, the replacement should not be the same species as the one removed. Oaks, maples, and other highly susceptible species should be avoided.

Selected Bibliography

Blanchard, R. O., and T. A. Tattar. 1981. Field and laboratory guide to tree pathology. Academic Press, New York. 285 pp.

Carter, J. C. 1961. Illinois trees: Their diseases. Illinois Nat. Hist. Surv. Circ. 46:99 pp. (second printing with alterations)

Hepting, G. H. 1971. Diseases of forest and shade trees of the United States. U.S.D.A. Forest Service, Agriculture Handbook 386, Washington, D.C. 658 pp.

Himelick, E. B. 1969. Trees and shrub hosts of *Verticillium albo-atrum*. Illinois Nat. Hist. Surv. Biological Notes 66:8 pp.

Horst, R. K. 1979. Westcott's plant disease handbook, 4th ed. Van Nostrand Reinhold, New York. 803 pp.

Manion, P. D. 1981. Tree disease concepts. Prentice-Hall, Englewood Cliffs, N.J. 399 pp.

Pirone, P. P. 1978. Diseases and pests of ornamental plants, 5th ed. Wiley, New York. 566 pp.

Raabe, R. D. 1962. Host list of the root rot fungus *Armillaria mellea*. Hilgardia 33(2):25–88.

Shigo, A. L. 1986. A new tree biology. Shigo and Trees, Associates, Durham N.H. 595 pp.

Sinclair, W. A., H. H. Lyon, and W. T. Johnson. 1987. Diseases of trees and shrubs. Cornell University Press, Ithaca and London, 574 pp.

Stipes, R. J., and R. J. Campana (eds.). 1981. A compendium of elm diseases. American Phytopathological Society, St. Paul, Minn. 96 pp.

Tattar, T. A. 1978. Diseases of shade trees. Academic Press, New York. 361 pp.

Zwet, T. Vander, and H. L. Keil. 1979. Fire blight: A bacterial disease of rosaceous plants. U.S.D.A. Agricultural Handbook 510, Washington, D.C. 200 pp.

13

Coping with Tree Pests and Diseases

Once an actual or anticipated tree pest or disease problem has been identified the decision of what to do remains. There was a time when finding the right pesticide to kill the offending pest was the only decision required. Although arborists are still heavily dependent on chemical pesticides, their use in landscapes has been mitigated by laws, societal and environmental pressures, new scientific discoveries, and common sense. Several important steps should be followed in dealing with pest problems in the landscape:

1. Anticipate pest problems. The circumstances that promote their occurrence and their potential for damage should be well known. Key weather parameters should be monitored, as should the condition of the trees and the effect recent weather is having on them and their potential pests. The development of most fungal and bacterial pathogens is favored by specific temperature and moisture requirements. Many pest problems can be prevented if accurately anticipated.
2. Monitor trees in the landscape regularly. Discovery of pests before they cause serious problems and ability to locate the exact point of infection or infestation will make a rescue action, if needed, more successful.
3. Accurately determine the cause of the problem. If you do not know the answer, seek out competent help.
4. Determine a course of action. In some cases, the best decision will be to do nothing at the present time because the damage has already been done, there is nothing that can be done, or there will be little or no damage.
5. Carry out the control decision properly. Pest control results will not be satisfactory if improperly implemented.

A tree maintenance firm should have a pest control advisor who knows where to obtain good information about managing tree pests. Most states have county extension offices with staffs competent in pest identification and management or with access to this kind of information through land-grant university plant disease and insect clinics and university publications on tree pest control. Local, state, and federal foresters often have pest management information which relates to landscape trees. Meetings and workshops on tree care and tree pest management are held in many areas. Knowledge gained from textbooks is helpful, but it should be supplemented with real pest encounters. Practice and experience are essential.

Integrated Pest Management

An integrated pest management (IPM) program which incorporates every pest control method suited for a particular site can be used to manage landscape tree pests. All phases of tree maintenance need to be considered and be compatible with pest management objectives.

The landscape is too complex a place to look for every possible pest. Select the most important tree and pest combinations for monitoring, based on experience and advice. Monitor specifically for those pests regularly and systematically. In addition, it is very important to look for trees in trouble. If the tree does not look normal, seek out the cause of the problem! As mentioned, early detection allows timing and location of control measures to be more precise. In addition, any adverse effects on potential natural controls such as beneficial insects can be minimized.

Monitoring procedures vary from one landscape to another, but the following suggestions may be helpful:

1. Examine trees in the landscape routinely. Begin by seeking information about species, cultivar, age, and current growth rate. A knowledge of soil type, pH, fertility, drainage, and recent maintenance activities such as planting, pruning, and fertilization is important.
2. Identify the key potential pests. Sources of infestation or inoculum should be sought in the landscape or in the neighborhood. This will allow the destruction of pest insect egg masses and pupae as well as overwintering phases of fungal and bacterial pathogens. For Dutch elm disease and other devastating problems, the surrounding community should be checked so that a control strategy can be developed if necessary.
3. Use pheromone traps, if available, for the important insect pests which require precise timing of control measures. A trap in one landscape should be adequate for the needs of other landscapes within several miles. (This is discussed further in the next section.)
4. Keep track of degree-days. The rate of development of many insect pests can be predicted by keeping track of the accumulated degree-days above 40° F. The pest manager can thus determine when key stages in insect development will occur and time control activities more accurately.
5. Use detection methods such as shaking needle evergreen branches over white paper to detect mites, using two-sided tape to catch and count scale crawlers, and putting burlap bands on the tree trunk to detect gypsy moth larvae.

Deciding what to do about a pest problem will depend on one's idea of what the landscape should look like. A person expecting a formal, high-maintenance landscape will have different needs than one desiring an informal landscape where only the most fit trees are expected to survive. In addition, a person respecting all forms of life might seek a different approach to tree pest control than one with entomophobia. One IPM approach is to educate the public (or one's clients) to accept higher pest levels or damage thresholds than they otherwise might. Obviously one would consider this approach only when the perceived pest does not threaten the health or life of the tree.

Assuming that the overall approach has been decided or agreed upon, action should proceed. Cultural methods should be considered. Improvement of vigor and avoidance of plant stress by proper tree maintenance is almost always warranted. Sanitation measures should be applied routinely to destroy pest sources and to keep healthy trees from becoming contaminated with insects and parasitic microbes.

Modifying the tree habitat or environment to produce an unfavorable situation for pests is also appropriate where practical. For example, shade favors powdery mildew and many leaf spot organisms. Thinning a tree or removing overhanging branches from one nearby could reduce disease. Resistant cultivars should be considered wherever landscape trees are being replaced. Biological controls, although still largely undeveloped, may provide a safe and effective way to manage pest problems. Physical and mechanical controls may be considered as substitutes for chemical control strategies.

There are some inherent difficulties in relying on an IPM approach to managing landscape tree pests. One is the cost of monitoring. Our experience with a residential landscape IPM program implemented in Kentucky showed that most tree owners were unwilling to pay the cost of monitoring, relying instead on attempting pest control after the problem became obvious. Similarly, an IPM program run by a tree maintenance company in Maryland costs the customer more than a preventive spray schedule without monitoring. So although IPM may make biological, environmental, and sociological sense, some clients will not accept the increased cost. Pesticide use was reduced in the National Capitol Region, National Park Service following use of an IPM program, a benefit certainly worth the cost in this heavily used public area.

Accurate diagnosis is also a problem with an IPM approach, unless the program is backed by an unbiased pest diagnostic service. Too often, the tree care expert making the diagnosis is also the person who stands to gain monetarily from some kind of treatment, a situation that might make a tree owner uneasy. Little is known about pest threshold levels or aesthetic injury levels for trees. In many instances the biology of the pest is not well enough known to make an informed decision about its control.

Biological Control of Insects and Diseases

Biological control of insects includes the use of parasites, predators, and bacterial and viral organisms. More broadly, it also includes the use of microbial insecticides, sterilizing agents, and attractants or lures. Biological control of diseases include the use of microbial antagonists and disease-resistant cultivars.

Insects. One of the outstanding examples of biological control was the introduction of the vedalia ladybird beetle from Australia in 1887 into citrus orchards in California to control the cottony-cushion scale. Within two years of its release into the orchards, the larval and adult stages of this predator had annihilated the scale, a pest that had threatened to wipe out California's citrus industry.

In the urban landscape, parasitic wasps, natural enemies of shade tree aphids, have been used to reduce aphid problems of street trees in California. This approach may work for large, single-species plantings. Similarly, a parasite of elm leaf beetle eggs is being tried as a potential control of the elm pest in New Mexico.

The bacterium *Bacillus thuringiensis* produces a toxin which kills leaf-chewing larvae of butterflies and moths. When a caterpillar ingests 40,000 to 80,000 live spores of *B. thuringiensis*, it is poisoned by the toxic crystals produced by the bacteria during spore production. The *B. thuringiensis* dust is sold under such trade names as Agritol, Bakthane L 69, Biogard, Biotrol, Dipel, Larvatrol, and Thuricide.

Bacillus popillae, causal agent of milky disease of Japanese beetle grubs, is an effective biological control agent providing control of turf grass grubs for many years, once bacterial spores become distributed in the soil. If adult Japanese beetle control is to benefit trees, however, the bacterium would need to be used on a very wide scale over an entire community. This treatment is relatively expensive and effects may not be noticeable for several years.

Insect viruses are also being tested as a means of controlling certain caterpillars on trees and other plants. A virus obtained from diseased caterpillars has been used experimentally to control an infestation of the great basin tent caterpillar *(Malacosoma fragilis)*. Similarly, forest pines infested with European pine sawfly larvae were sprayed from an airplane with an infectious virus suspension, which greatly reduced defoliation compared to unsprayed trees. The gypsy moth nucleopolyhedrosis virus, formulated commercially as Gypchek, specifically affects gypsy moth, causing population reductions.

While some of these treatments may be useful in vast plantings, specific urban tree species are scattered throughout an area and spread of nonmotile insect parasites could be difficult in the diverse urban forest.

Sterilization of the males of certain harmful pests may have potential as a form of biological control. The insects are exposed to gamma radiation, rendering them sterile. Females mating with gamma ray-treated males produce no offspring; thus the insect population is greatly reduced. Several chemosterilants are also being tested. These affect the genetic material in the insects' reproductive organs, resulting in abortion of the fertilized egg.

Another important phase of biological control involves the use of sex attractants or pheromone traps (Fig. 13-1). These chemicals are placed in poisonous baits or on sticky surfaces to attract males, which then die. These traps are used to monitor pest populations so that insecticide treatments can be well timed. They are not used to reduce insect populations in a landscape. Entomologists in Kentucky found that traps containing a floral lure and Japanese

Fig. 13-1. Pheromone trap.

beetle sex attractant, designed to destroy Japanese beetles, actually increased the numbers of these pests in the landscape.

Pheromone traps are used to monitor populations of clear-winged moth borers such as dogwood borer, lilac borer, and ash borer. A synthetic sex attractant for the gypsy moth, a serious pest of fruit, forest, and shade trees in the northeastern United States, known as Disparlure, has been developed. Also available are pheromone traps for monitoring the population of codling moth, obliquebanded leafroller, peachtree borer, oriental fruit moth, lesser peach tree borer, and red-banded leafroller in orchards.

The insecticide Dimilin, a growth-regulating chemical, interrupts the growth process of insects such as gypsy moth caterpillars. Under normal conditions, before each molt (shedding the skin) the body begins to produce chitin, which forms the new outer skeleton. The Dimilin ingested by the caterpillar interferes with chitin production and results in death. Dimilin has been registered by the Environmental Protection Agency for use against several insects.

Eggs of some beneficial insects such as ladybird beetles, lacewings, and praying mantis are commercially available. Although these predators feed on some pest species such as aphids, mealybugs, and scales, they also feed on harmless insects, and one cannot assume that they will remain in the vicinity of their release. Homeowners also run the risk of losing their released predators to insecticides used by their neighbors. Additionally, populations of many predators do not build up to sufficient levels to affect the pest population until after the pest has seriously damaged the tree. Given the diversity of trees in the urban landscape and their uneven distribution, predators and parasites will not be widely used for landscape tree pest control without much more research and development.

Although it is true that in a few cases parasites and predators have controlled destructive insects successfully, it is unrealistic to expect biological means alone to control pests completely. During the twentieth century the U.S. Department of Agriculture has imported hundreds of natural enemies of pests. Some of these have survived and become established, but only a few have substantially controlled the pest they were imported to combat.

Diseases. Biological control of crown gall disease (Chapter 12) has been achieved commercially by using a related nonpathogenic strain of bacteria. *Agrobacterium radiobacter* strain 84 inhibits the pathogen, *A. tumefaciens,* by competing for infection sites around wounds. It is available commercially as Galltrol A. The product must be used as a preventive and will not cure already existing crown gall.

Chestnut blight disease control has been accomplished in European chestnut plantings by introducing hypovirulent strains of the pathogenic fungus, *Endothia parasitica,* into developing cankers. Hypovirulent strains of the fungus are not highly pathogenic to chestnuts, in contrast to the normal pathogenic strains. Hypovirulent strains are thought to be infected with a viruslike disease which weakens them. Thus normal strains exposed to hypovirulent strains may also become weakened and the chestnut tree then overcomes the canker. In the United States, research is being done to find a way to easily and rapidly transmit the "virus" to diseased trees.

Selection of disease-resistant tree cultivars is a form of biological control. The development of flowering crabapple cultivars has progressed to the point that cultivars are available that resist several important diseases: scab, fire blight, powdery mildew, rust, and black rot. New horticulturally pleasing disease-resistant cultivars are now available and should be requested when flowering crabapples are purchased for the landscape.

Chemical Pesticides

Chemical pesticides are convenient to use. Some are toxic to only one kind of pest, providing precise control as needed. Some are broad spectrum, thus allowing many kinds of pests to be controlled with one application. Some are systemic and can be applied directly to the tree, with little hazard to its surroundings. The decision to use a chemical for tree pest control should be based on sound biological, economic, aesthetic, and practical considerations. Preventive applications of chemicals for tree diseases and insect pests are warranted in the following circumstances:

1. The tree has very high value. Of course the control measure should not cost more than the replacement value of the tree!
2. There is a history of the problem on that individual tree or on that species

in the area. It is assumed that the problem will recur yearly, like scab disease of susceptible flowering crabapple. Even in this case, treatments may be unnecessary during a dry spring. Furthermore, if there is a history of the problem and cultural practices that readily reduce inoculum have not been attempted, spraying to overcome intense pest pressure could be futile.

3. The tree is already temporarily under stress from other problems, such as transplanting or defoliation.

4. The disease or insect pest is life-threatening, for example, a needle evergreen defoliation or a wilt disease having inoculum present only temporarily. If the pest has minimal impact on tree health, sprays are unnecessary, other than for aesthetic reasons.

5. The potentially serious disease or insect pest cannot be monitored or predicted. It is generally too late to attempt control when the symptoms are already showing.

Unanticipated pest problems should be dealt with on a case-by-case basis, with the decision to apply chemicals on a rescue basis lying with the tree expert.

There are obvious problems with pesticide dependence. Pest resurgence, a buildup of the pest population to even higher levels than before, or secondary pest outbreaks can occur, especially if pest predators are killed by the pesticide. Resistance to pesticides can also occur so that when a pesticide is really needed, it may not be effective. Misapplication of the pesticide can result in phytotoxicity and drift of the pesticide to nontarget locations. If misused, some chemical pesticides are hazardous to people, pets, and wildlife. Some pesticides present environment problems by contaminating soil and water supplies or by killing nontarget organisms. Despite these shortcomings, there are some kinds of serious tree pest problems that can be controlled only by using chemicals.

Chemical Application Methods and Equipment

Successful control of diseases and of insect pests is dependent on correct placement of the properly selected and prepared pesticide on the infested or susceptible parts of the tree. This is best accomplished with good spraying or injecting equipment and techniques.

Injections and Implants. Techniques and materials have been developed for controlling specific insects or diseases in certain trees by injecting insecticides or fungicides into the sapstream of the base of the trunk. The chemicals then move throughout the tree to provide protection and control. These systems are useful where sprays cannot be used. Much less chemical is needed and it is all delivered into the tree and not into the surrounding environment. Problems of uneven distribution of chemical in the tree crown do sometimes occur, however, and each injection creates a wound that is a potential colonization site for decay fungi. In this regard, the smaller the injection the better.

Trunk injections using the Mauget system of oxydemeton-methyl or Bidrin-containing injector units have provided many arborists with excellent results.

A small plastic cylinder containing the pesticide is attached to a short plastic tube inserted into a predrilled hole in the trunk. Installing the injector units requires knowledge and practice. Accordingly, these are used by arborists, nurserymen, and horticulturists who have participated in a special training course.

The Acecap implant technique uses Orthene-containing capsules which are placed in holes drilled in the trunk. The capsules dissolve in the sapstream liquid, releasing the pesticide to be distributed in the tree crown. Implants are simpler to use, but they require a larger trunk wound.

Apparatus is also available for injecting liquid insecticides and fungicides into the tree under high pressure. Small holes are first bored into the tree trunk to the depth of the outer layers of sapwood. A special injector screw is then inserted into the hole. A high-pressure hose is connected at one end to the screw and at the other end to a hydraulic sprayer capable of building pressures to 400 pounds per square inch. The insecticide and/or fungicide is then pumped into the tree at high pressure. Only experienced professional arborists should use this apparatus. The Elm Research Institute provides information on use of a low-pressure system for injection of fungicide into elm trees to control Dutch elm disease. The arrangement of a series of T-shaped nozzles connected to one another with tubing would be similar to the apparatus used for the high-pressure system described above.

Spraying Equipment. Small trees can be sprayed with hand-pumped (Fig. 13-2), bucket, hose-end, backpack, or small power sprayers. For low-growing trees, the "trombone" sprayer is very efficient and easy to operate. Large trees can be properly sprayed only with large spraying machines. Such machines are expensive and are owned principally by commercial arborists and by park and shade tree departments that do considerable tree work.

HYDRAULIC SPRAYERS. The modern spray machine (Fig. 13-3) consists of a large steel or Fiberglas tank, ranging in capacity from 100 to 1,000 gallons, for holding the spray material; an agitator inside the tank, which revolves to keep the spray material uniformly dispersed; and a gasoline or diesel-driven pump, which forces the material from the pump through the spray hose. These parts are usually mounted as a unit on skids to facilitate moving onto a truck or some other carrier, or they may be permanently mounted on a truck or trailer chassis. Certain accessories, such as adequate hose and spray guns with special nozzles, complete the equipment.

Spray machines are available with pumping capacities ranging from 5.5 gallons a minute at 300 pounds pressure to 55 gallons at 800 pounds. Specially made hose of ¾- or 1-inch size must be used to withstand some of the higher pressures. Several companies manufacture large spraying machines. Your local pesticide dealer has information on prices and specifications of spray equipment. The manufacturers whose names are listed in some of the trade magazines (e.g., *Arbor Age, American Nurseryman, Weeds, Trees, and Turf,* and *Grounds Maintenance*) also will provide information. Some of the magazines have special product directory issues. National and state tree industry trade shows are good places to see demonstrations of some of the equipment.

The stream of spray material and the height it will reach are governed by the

Fig. 13-2. Hand-held pump sprayer.

size and construction of the nozzle, the amount of material discharged, and the pressure behind it. There are five principal types of spray nozzles: the disk nozzle, which throws a hollow cone-shaped spray; the Vermorel nozzle, which throws a misty, more or less cone-shaped spray; the bordeaux nozzle, which throws a flat, fan-shaped driving spray; the spray gun, which throws a cone-type to solid stream spray; and the park nozzle, which throws a solid stream spray. Each type has its own particular advantages and uses.

The pressure of the spray material at the nozzle opening, as well as the type of nozzle, governs the success of a spraying operation. Other factors (e.g., the size of the nozzle opening) remaining the same, any increase in pressure breaks up the spray material into finer particles, forces them to a greater distance from the nozzle, and results in the application of a greater volume of the insecticide or the fungicide.

To spray a tree approximately 20 feet high, it is necessary to have a nozzle with a ⅛-inch opening in the disk, which will deliver 6 gallons of spray a minute at 180 pounds pressure. A tree 90 feet can be adequately sprayed with a nozzle having a 5⁄16-inch opening, releasing 44 gallons a minute at a pressure

Fig. 13-3. Hydraulic sprayer.

of 300 to 400 pounds. Because of friction as the spray moves through the hose, the pressure at the nozzle is much lower than that developed by the pump at the spray tank. For each 100-foot increase in length of 1-inch hose from the tank to the nozzle in the example cited above, the pressure in the tank must be increased by 33 pounds to provide sufficient pressure at the nozzle to enable coverage of the 90-foot tree.

Sprays are applied to tall trees in a so-called solid stream, that is, the material leaves the nozzle much as water issues from a fire hose. This stream, forced out under great pressure, soon reaches a height at which it breaks into a mist, which drifts onto the leaves and stems.

In contrast to solid-stream spraying, hydraulic mist spraying is used on trees less than 20 feet high. The spray breaks into a fine mist as soon as it leaves the nozzle, giving rapid and complete coverage. Most hand-held spray guns have nozzles that can be adjusted to accomplish both types of spraying.

Hydraulic sprayers must be handled properly to maintain peak efficiency. To

Fig. 13-4. Mist blower.

achieve this objective, lubricate daily, particularly all moving parts; pour chemical mixtures through a good strainer before they enter the tank; rinse out the tank, hose, and nozzles with clean water at the end of the day; and carry, rather than drag, the spray hose over rough terrain.

MIST BLOWERS. Mist blowers, or air sprayers, use blasts of air to propel the droplets of pesticide, in contrast to hydraulic sprayers, which use water as the vehicle for the pesticide. With mist blowers it is possible to cover more trees in shorter time and at far less cost. The use of highly concentrated materials not only speeds up the refilling time but sharply reduces runoff or drip waste, which is a major component of lost pesticide with hydraulic applications.

Some mist sprayers can disperse spray particles for relatively long distances, often up to 100 feet, against a mild wind. One should bear in mind, however, that if the pesticide vaporizes before it reaches its target, the dry chemical will not adhere to the leaf surface. In other words, to deposit a coating of pesticide properly, the liquid–pesticide mixture must reach the leaves before the liquid phase evaporates.

There are several sizes and types of mist blowers on the market. Some are more suitable for use on large shade trees (Fig. 13-4), and others are better adapted for watershed, park, and forest plantings. Small motor-powered backpack mist blowers (Fig. 13-5) are now available for low-growing plant materials.

Helicopters are now being used to apply insecticides and fungicides to large uninhabited areas such as woodlands and golf courses. The insecticide Sevin has been applied from helicopters to control gypsy moth caterpillars in woodland areas.

Fig. 13-5. Backpack mist blower.

Spraying Practices. Although correct spraying technique can be perfected only through practice, the following suggestions will help in its acquisition.

The intensity and direction of the wind must be considered before spraying operations begin. Poor coverage and considerable waste of material are bound to occur on windy days; therefore, pesticide application should never be attempted on windy days. Even in slight breezes, sprays will drift. Maximum coverage is obtained when the spray is directed toward the leeward sides of the leaves and twigs. As the spray is caught by the breeze, it is blown back in the opposite direction, thus ensuring complete coverage. The wind factor is of even greater importance when using mist blowers; that is to say, there should be little to no wind.

Another factor to be considered is the proximity of the trees to buildings and other structures. The operator should determine beforehand how to get best coverage of the tree while getting as little spray as possible on these structures. Where mist blowers are used near parked automobiles, the operator must take into account the possibility of coating the automobiles with the spray. The problem can be solved by enforcing regulations which prevent parking of automobiles under trees that are to be sprayed.

There is no ironclad rule as to the amount of spray that should be applied to any particular tree. The object of spraying is to cover every leaf, twig, and branch that is infested or that might later become infested or infected by an insect or parasite. Once complete coverage is obtained, the operator should move on to another part of the tree or to another tree. Usually it is not necessary to continue the spray once the material begins to drip from the leaves.

Thorough coverage is essential when using protectant pesticides. For protection against microbes that cause infectious diseases, both leaf surfaces usually need to be covered. Systemic pesticides are transported throughout the plant, hence complete coverage is less important.

Timing of pesticide applications is also important. To protect against infection by fungi or bacteria, the pesticide must be present on the plant before infection occurs. Once the pathogen has entered the plant, it is usually too late for the spray treatment to be of any value. Certain diseases such as scab of flowering crabapple can be controlled a few days after infection if the right systemic fungicide is used, but examples such as this are rare.

Pesticide Labels

The pesticide label is a legal document and all pesticide users are expected to understand the information printed on it. The label states the brand name of the product, the type of formulation, the common chemical name with its percentage as active ingredient indicated, and the weight or volume of the material as packaged. The signal words and symbols indicating relative toxicity; description of potential human, environmental, and physical and chemical hazards; antidotes to counteract accidental exposure; and a statement of whether it is a restricted or general use material are also included. The Environmental Protection Agency registration number and manufacturer's name and address are stated. Directions for use, label violation penalties, a reentry statement, if needed, and any restrictions for use of the product, if needed, are presented. Finally, storage and disposal directions are given.

Pesticide Toxicity. Pesticide labels indicate the pesticide's degree of toxicity to people. The important signal words to look for are *Danger/Poison,* with skull and crossbones, *Warning,* and *Caution* (Table 13-1). All pesticide labels state that the pesticide must be kept out of reach of children. Ill effects or injury resulting from pesticide exposure depends as much on how the pesticide is used as on the toxicity of the chemical. A high-hazard situation with even a low-toxicity pesticide can occur when the material is used constantly in a careless

Table 13-1. Relationship of pesticide label signal words to pesticide toxicity

Signal Words	Toxicity	Approximate Oral Dose of Pesticide Concentrate Needed to Kill the Average Person*
Danger/Poison	High	A taste to a teaspoonful
Warning	Moderate	A teaspoonful to an ounce (two tablespoons)
Caution	Low	More than an ounce (two tablespoons)

*Pesticides may also be toxic when inhaled or in contact with the skin. These kinds of exposures are more likely to occur than ingestion. A pesticide labeled Danger/Poison may be lethal if only small amounts are inhaled or contacted.

manner at a high concentration. A low-hazard situation with a pesticide can exist when a trained and knowledgeable applicator wearing protective clothing (Fig. 13-6) uses a dilute, infrequently needed material.

Fig. 13-6. Operator wearing appropriate safety clothing for applying hazardous pesticides, including hat, gloves, long sleeves and pants, and respirator.

Some additional safety precautions are:

- Always read the label before using pesticides. Respect warnings and precautions listed.
- Keep pesticides out of reach of children, pets, and irresponsible adults.
- Store pesticides in their original containers.
- Do not store pesticides near food.
- Do not smoke or eat while spraying.
- Avoid inhaling sprays.
- Do not spill pesticides on skin or clothing. Should some be accidentally spilled, remove contaminated clothing quickly and wash the skin thoroughly.
- Wash hands and face thoroughly with soap and water and change to clean clothing after spraying pesticides.
- Dispose of empty containers so they pose no hazard to any form of life.
- If symptoms of illness occur, call a physician or local poison control center.

Preparing Small Quantities of Pesticides. Pesticide labels frequently provide mixing instructions geared to the commercial user. Individuals wishing to use small quantities are sometimes put off by having to calculate teaspoons per gallon from pounds or pints per 100 gallons listed on the label. Table 13-2 can be used to make these calculations.

Specific Pesticides for Insect, Mite, and Disease Control

The materials mentioned in the following lists are generally available for tree pest control. Some pesticides can be used only by those people who have obtained certification for application of restricted use materials from their state regulatory agenices. In addition, people who are hired to apply pesticides need to obtain a license to do so from their state regulatory agency.

The pesticides that are discussed here are listed both by brand name and by common chemical name, and their toxicity signal words are stated. Different formulations of the same chemical may have different toxicities. An asterisk (*) denotes that the product mentioned is a trade name. Pesticide use patterns do change; by all means read the label before you buy, making sure that the

Table 13-2. Amount of wettable powder and liquid pesticide to use for various quantities of water*

	Quantity of Pesticide†	
Gallons of Spray	Wettable Powder	Liquid
100	1 lb	1 pt
50	8 oz	1 cup
5	3 tablespoons	5 teaspoons
1	2 teaspoons	1 teaspoon

*The density of wettable powders varies but is about half that of liquids.

†If the pesticide label calls for more pints or pounds per 100 gallons, simply multiply the number of cups, tablespoons, or teaspoons proportionally.

material you choose is currently legal and safe for the intended use. The miticides listed primarily control mites; however, many of the insecticides control both insects and mites.

Some materials may be safely combined with others to control fungus or bacterial diseases and insects in a single operation. Because improper mixing of these materials may cause injury to the sprayed plants or reduce the efficiency of one or several of the materials in the mixture, one should seek advice from state entomologists and plant pathologists as to the best and safest combinations. A spray compatibility chart is also available from the Meister Publishing Company, Willoughby, OH 44094, for a small charge.

Insecticides and Miticides

Abamectin. See Avid.*

*Acaraben** is a miticide chemically known as chlorobenzilate. Caution.

*Acecap** is a formulation of acephate contained in a capsule which is implanted into the tree trunk. Caution.

Acephate. See Acecap* and Orthene.*

Aldicarb. See Temik.*

Avermectin B1. See Avid.*

*Avid** is used to control leaf miners and spider mites of greenhouse ornamentals. The common chemical name is abamectin/avermectin B1. Warning.

*Azodrin** is a restricted-use chemical used for leaf miner, sawfly, thrips, aphid, leafhopper, plant bug, and mite control. The common chemical name is monocrotophos. Danger/Poison.

Azinphos-methyl. See Guthion.*

Bacillus thuringiensis is a microbe which as an insecticide is effective in the control of many kinds of caterpillars. It is harmless to humans and animal pets. The insecticide is sold under the trade names Bactospeine,* Bactur,* Bakthane,* Biotrol,* Dipel,* StanGuard,* and Thuricide.* Caution.

*Bactospeine.** See *Bacillus thuringiensis.*

*Bactur.** See *Bacillus thuringiensis.*

*Bakthane.** See *Bacillus thuringiensis.*

*Baytex** is also sold as Tiguvon* and is used for control of thrips, aphids, psyllids, and mites. The common chemical name is fenthion. Warning.

Bendiocarb. See Turcam.*

*Bidrin,** known as dicrotophos, is a systemic insecticide. It is an ingredient of Inject-A-Cide B* insecticide. Danger/Poison.

Bifenthrin. See Talstar.*

*Biotrol.** See *Bacillus thuringiensis.*

Carbaryl. See Sevin.*

Carbofuran. See Furadan.*

Chlorobenzilate. See Acaraben.*

Chlorpyrifos. See Dursban.*

*Cygon,** also sold as Rogor* (common chemical name is dimethoate), is effective in controlling hemlock fiorinia scale, honey locust mite, pine needle scale, and mealybug on yew. Some plants are injured by it, and it should therefore be used only on those recommended by the manufacturer. This is one of the least toxic of the systemic insecticides. Warning.

*Cythion.** See Malathion.

*Diazinon,** also sold as Spectracide,* controls soil insects such as wireworms and root worms as well as turf pests such as chinch bugs and many insects such as pear slug, bagworms, and leaf miners infesting ornamental plants. Warning.

*Dibrom,** known chemically as naled, is effective against many pests including bagworms, leaf miners, and aphids. Warning.

Dicofol. See Kelthane.*

Dicrotophos. See Bidrin.*

Diflubenzuron. See Dimilin.*

Dimethoate. See Cygon.*

*Dimilin** is used by state and federal pest control workers for control of gypsy moth. It is a growth-regulating insecticide, and the common name is diflubenzuron. Caution.

*Dipel.** See *Bacillus thuringiensis.*

*Dipterex,** also known under the common name trichlorfon and the trade name Dylox,* is a phosphate insecticide. It is very effective in the control of the omnivorous leafroller and many other caterpillars. Caution.

Disulfoton. See Di-Syston.*

*Di-Syston,** common name disulfoton, is a systemic insecticide used for control of leafhoppers, aphids, lace bugs, and mites. Danger/Poison.

Dormant oil sprays, or petroleum oils, properly prepared, are used to control many pests that live through the winter on the buds, twigs, and trunks of trees and other woody plants. These oils are applied when temperatures are above freezing just before budbreak for greatest effect. Dormant oil treatments physically injure the insect adults, nymphs, or eggs which are overwintering on the tree when the oil is applied. Dormant oil sprays are environmentally safe but occasionally phytotoxic.

The oils now recommended are referred to as ''superior'' oils and are highly refined, more paraffinic, petroleum oils having a high purity (unsulfonated residues more than 92%). These oils may have viscosities, or ''weights,'' of 100, 70, or 60. The most widely used, most effective, and least phytotoxic have a viscosity index of 70. Some oil sprays, those having a viscosity of 90 or more, are harmful to several plants. When using superior oils, be sure to have a recent label and consult it for precautions.

*Dursban** is used to control aphids, borers, scale crawlers, and several caterpillars that feed on shade trees. The common chemical name is chlorpyrifos. Warning.

*Dycarb.** See Turcam.*

*Dylox.** See Dipterex.*

Endosulfan. See Thiodan.*

Ethion is often combined with dormant oil to increase effectiveness against overwintering insects. it is sold as Nialate.* Warning.

Fenthion. See Baytex.*

*Ficam.** See Turcam.*

*Furadan,** also known by the common name carbofuran, is effective in controlling beetles, borers, and aphids of cottonwood and elm. This is a restricted-use systemic insecticide. Danger/Poison.

*Guthion** is used by some professional arborists, nurserymen, and foresters to control a wide variety of insects on shade trees and ornamentals. It is a restricted-use chemical not recommended for home gardeners' use. Azinphos methyl is the common chemical name. Danger/Poison.

Hexakis. See Vendex.*

*Gypchek,** the gypsy moth nucleopolyhedrosis virus, is a microbial insecticide.

*Imidan,** also sold as Prolate,* controls many chewing insects, including gypsy moth caterpillars and elm spanworms, and a wide variety of sucking insects. The common chemical name is phosmet. Caution.

*Inject-A-Cide** is a formulation of oxydemeton-methyl injected into trees using the Mauget microinjection feeder tube system.

*Inject-A-Cide B** is a formulation of Bidrin* injected into trees using the Mauget microinjection system.

*Kelthane** is a miticide used to control eriophyid and spider mites. The common chemical name is dicofol. Caution.

*Lannate** is a restricted-use chemical used to control foliar-feeding insects. Commonly called methomyl, this locally systemic chemical is also sold as Nudrin.* Danger/Poison.

*Lindane** is used as a trunk spray to control several types of borers of trees. Lindane,* common chemical name also lindane, is a restricted-use material. Warning.

Malathion, one of the most widely used organic phosphate insecticides, controls a wide variety of pests including aphids, scale crawlers, mealybugs, and many chewing insects. A premium grade of malathion, sold under the name Cythion,* has less odor than the regular product. Some formulations may be phytotoxic in hot weather. Consult the label. Warning.

*Marlate.** See methoxychlor.

Methidathion. See Supracide.*

Methomyl. See Lannate.*

Methoxychlor is an insecticide with long residual action used to control the Dutch elm disease vectors, elm bark beetles, and many foliar-feeding larvae. It is sold as Marlate.* Caution.

*Morestan** is used for control of mites, aphids, whiteflies, and psyllids. Oxythioquinox is the common chemical name. Caution.

Naled. See Dibrom.*

*Nudrin.** See Lannate.*

*Omite** is an effective spider mite control chemical. Consult the label for specific uses. The common chemical name is propargite. Danger/Poison.

*Orthene** is effective in controlling aphids, bagworms, cankerworms, gypsy moths, and many other insect pests. The common chemical name is acephate, an ingredient of Acecap* tree trunk capsule implants. Caution.

Oxamyl. See Vydate L.*

Oxydemeton-methyl. See Inject-A-Cide.*

Oxythioquinox. See Morestan.*

Phosmet. See Imidan.*

*Prolate.** See Imidan.*

Propargite. See Omite.*

*Rogor.** See Cygon.*

*Sevimol** is a formulation of Sevin* in molasses, which adheres to sprayed foliage better than Sevin* used alone.

*Sevin,** known under the common name carbaryl, is used to control many insects infesting trees. Repeated use will result in an increase in spider mites. It is toxic to bees and will defoliate Boston ivy and Virginia creeper vines. Caution.

*Sevin 4 Oil,** a combination of Sevin* and oil, is used to control the spruce budworm and western spruce budworm. It can be applied to large forest areas from an airplane. Ground applications can be made to high-value trees with a mist blower. Caution.

*Spectracide.** See Diazion.*

*StanGuard.** See *Bacillus thuringiensis*.

*Supracide** is a restricted-use insecticide–miticide that is effective against certain pests infesting nursery-grown trees. The common chemical name is methidathion. Danger/Poison.

*Talstar** is the trade name for the synthetic pyrethroid insecticide called bifenthrin. It is used for control of aphids, mites, and leaf rollers. Danger/ Poison.

*Tedion** is a long residual miticide effective against the egg stage. The common chemical name is tetradifon. Caution.

*Temik** is a highly toxic restricted-use systemic insecticide used to control aphids and leaf miners. It is formulated as a granular material which is incorporated into the soil. The common chemical name is aldicarb. Danger/Poison.

Tetradifon. See Tedion.*

*Thiodan** is used to control aphids and some borers as well as other insects. The common chemical name of this restricted-use chemical is endosulfan. Warning.

*Thuricide.** See *Bacillus thuringiensis*.

Trichlorfon. See Dipterex.*

*Turcam,** also sold as Vicam,* Ficam,* and Dycarb,* is used to control a wide range of foliar-feeding insects such as bagworms, mealybugs, and tent caterpillars. The common chemical name is bendiocarb. Warning.

*Vendex** is a miticide having the common chemical name hexakis. It is not injurious to beneficial mite species. Danger/Poison.

*Vicam.** See Turcam.*

*Vydate L** is a highly toxic insecticide which controls a wide range of foliar feeding insects. Oxamyl is the common name of this restricted-use chemical. Danger/Poison.

Nematicides

Many nematicides are also insecticides and are found in the previous list. Some nematicides, used as soil fumigants, are also general biocides and kill soilborne insects, fungi, and weed seeds. These soil-applied biocides are toxic to trees and can only be used prior to planting.

Aldicarb. See Temik.*

*Bromo-O-Gas.** See Chloropicrin and Methyl bromide.

*Brozone.** See Chloropicrin and Methyl bromide.

Carbofuran. See Furadan.*

Chlorinated hydrocarbons are the major ingredients of Telone* and D-D Soil Fumigant* (100%), Vidden-D* (99%), Terrocide 15D* and Shell D-D + Chloropicrin* (85%), Vorlex* (80%), and Nemex* (50%). These general fumigants are injected into the soil and sealed with plastic, water, or compaction. Danger/Poison.

*Chlor-O-Pic.** See Chloropicrin.

Chloropicrin is used both as a general soil fumigant and, when used in small amounts, as an odiferous warning agent. Chloropicrin is a component of Chlor-O-Pic* and Picfume* (99%), Nemex* (50%), Larvabrome* and Dowfume MC-33* (33%), Terrocide 15D* and Shell D-D + Chloropicrin (15%),* Brom-O-Gas* and Dowfume MC-2* (2%), and Brozone* and Nemaster* (1.4%). It must be sealed in with plastic as it is applied. Danger/Poison.

*Dasanit** may be used to control nematodes of established boxwood, holly, juniper, and yew. It is incorporated into the soil as a drench or as granules. Common chemical name is fensulfothion. Danger/Poison.

*D-D Soil Fumigant.** See Chlorinated hydrocarbons.

*Dowfume MC-2.** See Chloropicrin and Methyl bromide.

*Dowfume MC-33.** See Chloropicrin and Methyl bromide.

Ethoprop. See Mocap.*

Fensulfothion. See Dasanit.*

*Furadan** is a restricted-use systemic nematicide and insecticide. The common chemical name is carbofuran. Danger/Poison.

*Larvabrom.** See Chloropicrin and Methyl bromide.

Methyl bromide is used as a general fumigant because it is odorless and also highly toxic; it is mixed with at least small amounts of chloropicrin as a warning agent. It is applied under plastic. Methyl bromide is a component of Brom-O-Gas* and Dowfume MC-2* (98%), Brozone* and Nemaster* (68.6%), Dowfume MC-33* (67%) and Larvabrom* (33%). Danger/Poison.

*Mocap** can be used for nematode control either before or after planting of boxwood, holly, and yew. The common chemical name of this nematicide is ethoprop. Danger/Poison.

*Nemacur** is a nematicide that can be used after planting. It controls nematodes of boxwood, crabapple, holly, juniper, pine, spruce, and yew. The common chemical name is phenamifos. Danger/Poison.

*Nemaster.** See Chloropicrin and Methyl bromide.

*Nemex.** See Chlorinated hydrocarbons and Chloropicrin.

Phenamifos. See Nemacur.*

*Picfume.** See Chloropicrin.

Oxamyl. See Vydate.*

*Shell D-D + Chloropicrin.** See Chlorinated hydrocarbons and Chloropicrin.

*Telone.** See Chlorinated hydrocarbons.

*Temik,** used as a nematicide and insecticide, is highly toxic. It is incorporated into the soil beneath the tree, taken up by the roots, and redistributed within the host. The common name is aldicarb. Danger/Poison.

*Terrocide 15D.** See Chlorinated hydrocarbons and Chloropicrin.

*Vapam** is a general soil fumigant used to kill nematodes before planting. It is applied as a drench, or by soil injection, and is also used to sever tree root grafts for Dutch elm disease control. Caution.

*Vidden-D.** See Chlorinated hydrocarbons.

*Vorlex** is a general fumigant used to treat soil infested with weed seeds, insects, nematodes, and fungi. Vorlex contains, in addition to chlorinated hydrocarbons, 20 percent methyl isothiocyanate. Danger/Poison.

*Vydate** is a systemic insecticide and nematicide used for nematode control on apple, arborvitae, boxwood, buckeye, cherry, Chinese elm, cypress, Douglas-fir, fir, hemlock, holly, juniper, magnolia, maple, mulberry, pine, spruce, and yew. It can be used after planting. The common chemical name is oxamyl. Danger/Poison.

*Zinophos** is a nematicide and insecticide that can be used either before or after planting. Soil drench or root dip treatments of boxwood, cherry, crabapple, dogwood, holly, pear, pine, redbud, red cedar, Russian-olive, willow, and yew are used. Danger/Poison.

Fungicides and Bactericides

Many fungicides are available for controlling tree diseases, but there are only a few bactericides available for control of bacterial diseases of trees. Following are some of the materials approved in most parts of the United States:

*Acme Bordeaux Mixture.** See Bordeaux mixture.

*Agrimycin.** See Streptomycin.

*Agri-Strep.** See Streptomycin.

*Aliette** is a systemic fungicide used for control of *Phytophthora* and *Pythium* root and crown rot diseases. The common chemical name is fosetyl-Al. Caution.

*Arasan.** See Thiram.*

*Arbotect 20-S** is a benzimidazole fungicide with the common chemical name of thiabendazole. It is used for elm injection treatments for Dutch elm disease. Caution.

*Banrot** is used as a soil drench treatment for damping-off. It is a combination of etridiazole for control of water molds and thiophanate-methyl for *Rhizoctonia* control. Banrot* can be used in the production of many kinds of trees in containers. Warning.

*Basi-Cop.** See Basic Copper Sulfate.*

*Basic Copper Sulfate** is one of several fixed copper compounds and is also sold as Basi-Cop,* Copper 53 Fungicide,* Microcop,* T-B-C-S 53,* and Tribasic Copper Sulfate.* These materials are moderately effective fungicides and are fairly effective bactericides, some being used for fire blight control. Sometimes these materials are phytotoxic to new growth. Danger/Poison.

*Bayleton** is a systemic fungicide used for control of rusts and powdery mildews of landscape trees. The common chemical name is triadimefon. Warning.

*Benlate,** common chemical name benomyl, is sold under the names of Benlate Benomyl Fungicide,* Bonide Benomyl,* Miller's Systemic Fungicide,* and Science Systemic Fungicide.* This fungicide controls many fungus leaf spots and blotches, blights, rots, scabs, and powdery mildews. Caution.

Benomyl. See Benlate.*

*Black Leaf Bordeaux Powder.** See Bordeaux mixture.

*Blitex 80 MN.** See Maneb.

*Bonide Benomyl.** See Benlate.

Bordeaux mixture is one of the oldest and still one of the most useful fungicides. The killing principle in the spray, after it is made up and ready for use, is a mixture of certain salts of copper. It is prepared by mixing copper sulfate (bluestone, blue vitriol) and hydrated lime in various proportions. The formula 4-4-50, for instance, means that 4 pounds of copper sulfate and 4 pounds of hydrated lime are used with 50 gallons of water. These proportions are so generally useful that this may be considered a standard mixture. It has been found, however, that the amount of lime may be considerably reduced, resulting in less foliage spotting residue.

To prepare small quantities of bordeaux, the following procedure is suggested. Dissolve 4 ounces of copper sulfate crystals in 1 gallon of water. Then dissolve 2 ounces of hydrated lime in 2 gallons of water. Finally, while stirring, add the copper sulfate solution to the lime water. This makes 3 gallons of approximately a 4-2-50 bordeaux.

Mixing the two solutions together forms an insoluble or "fixed" gelatinlike copper precipitate. When it dries on the leaves, it forms a membranous coating which clings. Since the adhesiveness is lost if the mixture stands too long, bordeaux mixture should always be fresh when used.

Ready-made bordeaux mixture is sold under the names Acme Bordeaux Mixture,* Black Leaf Bordeaux Powder,* Bordo,* Bor-dox,* Copper Bordo,* Copper Hydro Bordo,* Miller 658,* Ortho Copper Fungicide,* and Pratt Bordeaux Mix.* Bordeaux mixture, like the other copper-containing materials in this list, has similar uses in control of bacterial and some fungal diseases. For other fixed coppers, see Basic Copper Sulfate,* C-O-C-S,* Copper oxychloride and Kocide 101.*

One objectionable feature of bordeaux mixture and other copper compounds is that they may cause some injury to plants when used on new growth or during cool wet weather. Danger/Poison.

*Bordo.** See Bordeaux mixture.

*Bor-dox.** See Bordeaux mixture.

*Bravo.** See Daconil 2787.*

Calcium polysulfide. See Lime sulfur.

*Captan,** also sold under the name Orthocide,* is effective in preventing scab on crabapple leaves and fruits and many other leaf spot diseases of trees. Caution.

*Carbamate.** See Ferbam.*

*Chem-Neb.** See Maneb.

*Chem-Zineb.** See Zineb.

*Chipco 26019** is used to control Botrytis blight of flowering plum and cherry, and Rhizoctonia damping-off of dogwood. The common chemical name is iprodione. It is also sold as Rovral.* Caution.

Chlorothalonil. See Daconil 2787.*

*C-O-C-S** is also called copper oxychloride sulfate. It is used for bacterial and fungal disease control. Danger/Poison.

*Copper Bordo.** See Bordeaux mixture.

*Copper 53 Fungicide.** See Basic Copper Sulfate.*

*Copper Hydro Bordo.** See Bordeaux mixture.

Copper hydroxide. See Kocide 101.*

Copper oxychloride is a fixed copper compound used for disease control. It is sold as Coprantol.* Danger/Poison.

Copper oxychloride sulfate. See C-O-C-S.*

*Coprantol.** See Copper oxychloride.

*Coromate.** See Ferbam.*

*Cyprex,** commonly known as dodine, is effective against the leaf blight of sycamores and black walnuts. Danger/Poison.

*Daconil 2787** is a broad-spectrum protectant fungicide used for control of fungal leaf spot diseases of trees. The common chemical name is chlorothalonil, which is also the ingredient in Bravo.* Warning.

Dinocap. See Karathane.*

*Dithane M-22.** See Maneb.

*Dithane M-45.** See Mancozeb.

*Dithane Z-78.** See Zineb.

Dodine. See Cyprex.*

Etridiazole. See Banrot* and Terrazole.*

Fenarimol. See Rubigan.*

*Ferbam,** also sold under the names Carbamate,* Coromate,* Fermate,* Fermate Ferbam Fungicide,* Karbam Black,* and Vancide FE,* was one of the first so-called carbamates to be marketed. It controls the cedar rust disease of crabapples, many leaf spots, blights, and scab. Caution.

*Fermate.** See Ferbam.*

*Fermate Ferbam Fungicide.** See Ferbam.*

*Folpet.** See Phaltan.*

*Fore.** See Mancozeb.

Fosetyl-Al. See Aliette.*

*Funginex** is a sterol inhibitor fungicide having good activity against rust and powdery mildew diseases of trees. It is also known as triforine. Danger/ Poison.

Iprodione. See Chipco 26019.*

*Karathane** is a fungicide specifically active against powdery mildew and mites. The common chemical name is dinocap. Caution.

*Karbam Black.** See Ferbam.

*Kocide 101** is copper hydroxide, used for bacterial and fungal disease control such as Bordeaux mixture is used. Caution.

*Lignasan BLP** is used as an aid in the control of Dutch elm disease. It is injected into the trunk of elms in late spring. It is chemically referred to as MBC phosphate. Caution.

Lime sulfur is effective in controlling powdery mildews on many trees. It also helps combat infestations of spider mites. In concentrated form it is used during the dormant period to destroy overwintering stages of aphids, mites, and scale insects. As a dormant spray 1 part of the concentrated lime sulfur is diluted in 8 or 10 parts of water. Lime sulfur should not be used near buildings, walls, or trellises because it will stain them. It can be diluted 1 part in 50 parts of water and used as a spray during the growing season. Lime sulfur is also available as a dust. Neither the dust nor the spray should be used when the air temperature is above 85° F (29° C). It is also called calcium polysulfide. Caution.

Mancozeb is a broad-spectrum fungicide for prevention of leaf spot diseases of trees. It is sold under the trade names Dithane M-45,* Fore,* Manzate 200,* and Zyban.* Caution.

Maneb, also sold as Blitex 80 MN,* Chem Neb,* Dithane M-22,* Manex,* and Manzate 200,* is a protectant fungicide effective in combating leaf spot diseases of trees. Caution.

*Manex.** See Maneb.

*Manzate 200.** See Maneb.

MBC phosphate. See Lignasan BLP.*

Metalaxyl. See Subdue.*

*Microcop.** See Basic Copper Sulfate.*

*Miller 658.** See Bordeaux mixture.

*Miller's Systemic Fungicide.** See Benlate.

*Ornalin** is used to control Botrytis and Sclerotinia blights of several landscape trees. The common chemical name is vinclozolin, also the ingredient of Ronilan* and Vorlan*. Caution.

*Orthocide.** See Captan.*

*Ortho Copper Fungicide.** See Bordeaux mixture.

Oxytetracycline. See Terramycin.*

*Parzate.** See Zineb.

PCNB. See Terraclor.*

*Phaltan,** also sold under the name Folpet,* is used to control many fungus diseases of trees. It is one of the carbamate fungicides presently available that will control powdery mildew. The chemical name is folpet. Caution.

*Phytomycin.** See Streptomycin.

Piperalin. See Pipron.*

*Pipron,** also known by the common name piperalin, is effective in combating powdery mildew of catalpa and many ornamental plants. Caution.

*Pratt Bordeaux Mix.** See Bordeaux mixture.

*Ronilan.** See Ornalin.*

*Rovral.** See Chipco 26019.*

*Rubigan** is a locally systemic fungicide used for crapemyrtle powdery mildew control. Fenarimol is the chemical name. Caution.

*Science Systemic Fungicide.** See Benlate.

Streptomycin, also known as Agrimycin,* Agri-Strep,* and Phytomycin,* is effective in controlling diseases such as fire blight of apple, crabapple, pear, and walnut. Caution.

*Subdue** is a systemic fungicide used to control water molds such as *Pythium* and *Phytophthora.* The common name is metalaxyl. Warning.

Sulfur has long been used to control many plant diseases such as powdery mildew. It also controls spider mites. See Lime sulfur for details. Caution.

*T-B-C-S 53.** See Basic Copper Sulfate.*

*Terraclor** is an excellent soil-applied fungicide for control of Rhizoctonia. PCNB (pentachloronitrobenzene) is the common chemical name. Caution.

*Terramycin** is used to control mycoplasma-caused diseases such as palm lethal yellowing. The common chemical name of this antibiotic is oxytetracyclene.

*Terrazole,** also sold as Truban,* is effective against soilborne fungi such as *Pythium.* The common chemical name is etridiazole. Warning.

Thiabendazole. See Arbotect 20-S.*

Thiophanate-methyl. See Banrot,* Zyban,* and Topsin-M.*

*Thiram,** also sold as Arasan* and Thylate,* is used to control rust and scab of crabapples and other diseases. Caution.

*Thylate.** See Thiram.*

*Topsin-M** is a systemic fungicide having activity similar to benomyl fungicide. The common chemical name is thiophanate-methyl. Caution.

Triadimefon. See Bayleton.*

*Tribasic Copper Sulfate.** See Basic Copper Sulfate and Bordeaux mixture.

Triforine. See Funginex.*

*Truban.** See Terrazole.*

*Vancide FE.** See Ferbam.*

Vinclozolin. See Ornalin.*

*Vorlan.** See Ornalin.*

Zineb, also sold as Chem-Zineb,* Dithane Z-78,* and Parzate,* is used as a preventive fungicide for anthracnose of planetrees and other fungus diseases. Caution.

*Zyban** is a combination of thiophanate-methyl and mancozeb used to control scab, powdery mildew, anthracnose, twig blight, and rust of several trees. Caution.

Spreading and Sticking Agents

Any material that helps spread spray in an even, unbroken film over the surface of leaves, bark, or insects' bodies is known as a spreader; this action is usually accomplished by lowering the interfacial tension of the spray. Materials that help the spray adhere firmly to the leaves are known as stickers.

Formerly soaps, household detergents, and some oils were recommended as spreaders. We now know that some, particularly soaps, are alklaine and may completely detoxify certain insecticides. Others, such as oils, may not be compatible with fungicides containing sulfur.

Where a spreader–sticker is required it is best to depend on specially prepared products.

Commercial spreaders and stickers widely available include du Pont Spreader–Sticker (SS₃), Triton B1956, Filmfast, Nu-Film, Ortho Dry Spreader, and Spray Stay.

Selected Bibliography

Anonymous. 1975. Apply pesticides correctly: A guide for commercial applicators. U.S.D.A. and U.S.E.P.A., Washington, D.C. 45 pp.

Hartman, J. R., J. L. Gerstle, M. Timmons, and H. Raney. 1986. Urban landscape IPM in Kentucky: A case history. *J. Environ. Hortic.* 4:120–24.

Jones, R. K., and R. C. Lamke (eds.). 1982. Diseases of woody ornamental plants and their control in nurseries. North Carolina Agriculture Extension Service, Raleigh. 130 pp.

Meister, R. T. (ed.). 1987. Farm chemicals handbook. Meister Publishing Co. Willoughby, Ohio.

McGrath, H., J. Feldmesser, and L. Young (eds.). 1986. Guidelines for the control of plant diseases and nematodes. U.S.D.A. Agriculture Handbook Number 656 Washington, D.C. 274 pp.

Olkowski, H., and W. Olkowski. 1976. Entomophobia in the urban ecosystem, some observations and suggestions. Bull. Entomol. Soc. Am. 22:313–17.

Potter, D. A. 1986. Urban landscape pest management, in Advances in urban pest management, G. W. Bennett and J. M. Owens (eds.). Van Nostrand Reinhold, New York, pp. 219–51.

Quist, J. A. 1980. Urban insect pest management for deciduous trees, shrubs, and fruit. Pioneer Science Publications, Greeley, Co. 169 pp.

Raup, M. J., and R. M. Noland. 1984. Implementing landscape plant management programs in institutional and residential settings. J. Arboric. 10:161–69.

Thomson, W. T. 1977. Pesticide guide: Tree turf and ornamental. Thompson, Fresno, Calif. 134 pp.

III

Abnormalities of Specific Trees

Abnormalities of Specific Trees

This section deals primarily with the diseases and arthropod pests of trees. The trees are listed alphabetically by common name. The cultural requirements of some of the trees are discussed briefly before the common diseases and pests and their control are described.

Control of many pests and problems of trees depends on maintenance of good growing conditions so that the tree has energy reserves to respond to and ward off pest problems. Failure to recognize the importance of integrating tree maintenance practices into pest control operations can lead to unsatisfactory pest control.

The chemical control measures suggested here should be used as a guide. Since pesticide registrations are constantly changing, it is important for the user to read the pesticide label before purchasing and using the product. In addition, most county extension offices have information relating to control of landscape tree problems, with control measures that are suited for local conditions. For further information about cultural, biological, and chemical control and application methods, refer to Chapter 13. More detailed information about important selected abiotic diseases, insect pests, and parasitic diseases is found in Chapters 10, 11, and 12. Approaches to problem diagnosis are presented in Chapter 9.

ACACIA (Acacia)

Acacias typically produce bright yellow flowers and are grown outdoors in the Southwest and in greenhouses elsewhere. This legume is a relatively fast grower.

Diseases

The fungus diseases of acacia are not important. A twig canker caused by *Nectria ditissima* and *Fusarium lateritium* (*Gibberella baccata*) has been reported from California. Leaf spotting by the fungus *Physalospora fusca*, a species of *Cercospora*, and the alga *Cephaleuros virescens* occasionally develops in the South. In California powdery mildew caused by the fungus *Erysiphe polygoni* is common. Root and trunk rots caused by the fungi *Phymatotrichum omnivorum*, *Armillaria mellea*, *Clitocybe tabescens*, and *Ganoderma applanatum* have been reported from the South and the West.

Insects

COTTONY-CUSHION SCALE. *Icerya purchasi.** This ribbed-scale bark louse, which has been a serious pest of acacia and many other trees in California and throughout the southern United States, has been troublesome in greenhouses in the eastern states for a number of years. The mature insect has a ribbed, cottony covering through which may be seen its soft body, sometimes ¼ inch long. The white, fluted mass that extends from the scale is composed largely of eggs covered by a protecting waxy mass. The scale itself is much smaller and is usually inconspicuous. At its earliest stage the scale is greenish and not contained in a woolly excretion.

Several other species of scale occasionally infest acacia. These are California red, which is round and reddish in color; dictyosperm; greedy, a small gray species; latania, similar to greedy; lesser snow; oleander, a pale yellow kind; and San Jose, which is small and gray with a nipple in its center.

Control. Malathion sprays or aerosols are effective against the crawler stage of scale insects.

CATERPILLARS. *Argyrotaenia citrana* and *Sabulodes caberata.* The former, known as the orange-tortrix, a dirty-white, brown-headed caterpillar, webs and rolls the leaves of many trees and shrubs on the West Coast. The latter, known as the omnivorous looper, a yellow to pale pink or green caterpillar, with yellow, brown, or green stripes on the sides and back, also feeds on a wide variety of plants in the same region.

Control. Spray with Sevin when the caterpillars are young.

FULLER ROSE BEETLE. See under Camellia.

ALBEZZIA PSYLLID. *Psylla uncatoides.* This is an important pest of some species of acacia in California. The feeding causes yellowing and enough honeydew may be produced to make this pest bothersome.

OTHER INSECTS. Acacia is host to a thornbug, *Umbonia crassicarnis;* a leafhopper, *Kunzeana kunzii;* and a species of spittle bug, *Clastoptera arizonana.*

*The scientific name of the insect is given after the common name.

AILANTHUS. See Tree-of-Heaven.

ALDER *(Alnus)*

Most species of alder grow best in moist soils. Those grown in ornamental plantings are less subject to diseases than to insect pests. The few diseases that occasionally appear produce little permanent damage. It is a relatively fast growing and short-lived tree.

Diseases

CANKER. A number of fungi cause cankers and dieback of the branches. Among the most common are *Nectria galligena, Solenia anomala,* and *Botryosphaeria obtusa.* Trunk cankers may also be caused by *Diatrypella oregonensis, Hymenochaete agglutinans,* and *Didymosphaeria oregonensis.*

Control. No simple control measures are known. Cankered branches should be pruned to sound wood and the trees fertilized and watered to increase their vigor.

LEAF CURL. *Taphrina macrophylla.* The leaves of red alders grow to several times their normal size, are curled and distorted, and turn a decided purple when affected by this disease. Infection is apparent on the leaves as soon as they appear in the spring.

Three other species, *T. amentorium, T. occidentalis,* and *T. robinsoniana,* are known to cause enlargement and distortion of the scales of female catkins. These project as curled, reddish tongues, which are soon covered with a white glistening layer of fungus tissue.

Control. Spray the trees with Ferbam in late fall or while they are still dormant in the spring. This will destroy most of the spores, which overwinter in the bud scales and on the twigs and which are largely responsible for early infections. Sprays are suggested only for valuable specimens.

POWDERY MILDEW. Several species of powdery mildew fungi attack the female catkins of alder. The most common, *Erysiphe aggregata,* develops as a white powdery coating over the catkin. Two other species, *Microsphaera penicillata* and *Phyllactinia guttata* occur less frequently.

Control. These fungi rarely cause enough damage to warrant control measures. Where justified, Benlate or sulfur sprays will give control.

LEAF RUST. *Melampsoridium alni.* Some damage is done to alder leaves by this rust, which breaks out in small yellowish pustules on the leaves during the summer. The leaves later turn dark brown. It is suspected that the alternate host of this rust is a conifer.

Control. The disease is never severe enough to justify control measures.

Insects

WOOLLY ALDER APHID. *Prociphilus tessellatus.* The downward folding of leaves, in which are found large woolly masses (Fig. III-1) covering bluish black aphids, is typical of this pest. Eggs are deposited in bark crevices and pass the winter in these locations. Maples are also infested by this species.

Control. Malathion sprays when the young begin to develop in spring give excellent control.

ALDER FLEA BEETLE. *Altica ambiens.* The leaves are chewed during July and August by dark brown larvae with black heads. The adult, a greenish blue beetle ⅕ inch long, deposits orange eggs on the leaves in spring.

Control. Sevin sprays when the larvae begin to feed will control this heavy feeder.

ALDER LACE BUG. *Corythucha pergandei.* This species infests not only alder but occasionally birch, elm, and crabapple.

Control. Spray with malathion or Sevin in late spring.

Fig. III-1. Woolly alder aphid on alder.

ALDER PSYLLID. *Psylla floccosa.* This sucking insect is common on alders in the northeastern United States. The nymphal stage produces large amounts of wax. When massed on the stems, the psyllids resemble piles of cotton.

Control. Spray with malathion when the nymphs or adults are seen.

ALDER BORER. *Saperda obliqua.* This borer attacks alder.

OTHER INSECTS. Alder is a preferred host of gypsy moth (see under Elm) and host to fall webworm (see under Ash); a leafroller, *Archips negundanus;* European alder leaf miner, *Fenusa dohrnii;* elm calligrapha (see under Linden); mottled willow borer (see under Willow); adult bronze birch borer (see under Birch); birch aphid, *Euceraphis betulae;* and an eriophyid mite.

ALMOND, FLOWERING *(Prunus triloba).* See under Cherry.

ARBORVITAE *(Thuja)*

In the wild, the American arborvitae, also called white cedar, inhabits moist, sometimes swampy areas along banks of streams. When under cultivation, it prefers a moist but well-drained soil. The unthrifty condition of many trees in ornamental plantings is often associated with waterlogged soils.

Many horticultural varieties of American arborvitae are now available from nurserymen. Most of these are subject to the same troubles that affect the parent tree.

Diseases

LEAF BLIGHT. Although most destructive on the western red cedar of the Northwest, leaf blight occasionally appears on arborvitae. The disease is most prevalent on trees under 4 years old, but it may attack trees of all ages.

Symptoms. From one to four irregularly circular, brown to black cushions appear on the tiny leaves in late spring. The leaves then turn brown, and the affected areas appear as though scorched by fire. Toward fall the leaves drop, leaving the branches bare.

Cause. The fungus *Fabrella thujina* causes

leaf blight. The brown to black cushions on the leaves are the fruiting bodies of this fungus. Spores are discharged into the air from these bodies during rainy periods in summer, to initiate new infections.

Control. Small trees or nursery stock can be protected by several applications of bordeaux mixture or other copper sprays in midsummer and early autumn.

TIP BLIGHT. This disease resembles leaf blight but occurs primarily on varieties of *Thuja orientalis,* the golden biota or golden arborvitae being most susceptible. The leaves near the branch tips turn brown in late spring or early summer. Tiny black bodies, which are visible through a magnifying lens, occur on infected leaves.

Cause. The fungus *Coryneum thujinum* causes tip blight. Several other fungi, including *Cercospora thujina, Pestalotiopsis funerea,* and *Phacidium infestans,* are also associated with tip and twig blights.

Control. Spray with a copper fungicide two or three times at weekly intervals starting in late May.

JUNIPER BLIGHT. The fungus *Phomopsis juniperovora,* which is the cause of the blight of red cedar and other species of *Juniperus,* also attacks arborivitae.

Control. See Phomopsis twig blight, under Juniper.

Abiotic Diseases.

BROWNING AND SHEDDING OF LEAVES. The older, inner leaves of arborvitae turn brown and drop in fall. When this condition develops within a few days or a week, as it does in some seasons, many persons feel that a destructive disease is involved. Actually it is a natural phenomenon similar to the dropping of leaves of deciduous trees. When the previous growing season has been favorable for growth, or when pests such as red spider have not been abundant, the shedding occurs over a relatively long period through fall and consequently is not so noticeable.

WINTER BROWNING. Rapid changes in temperature, rather than drying, in late winer and early spring are responsible for browning of arborvitae leaves.

Insects and Related Pests

ARBORVITAE APHID. *Cinara thujafilina.* This reddish brown aphid with a white bloom infests roots, stems, and leaves of arborvitae, Italian cypress, and retinospora.

Control. Treat soil with Cygon to control the root-infesting stage, and spray with malathion or Diazinon to control pests above ground.

BAGWORM. See under Juniper.

CEDAR TREE BORER. See under Redwood.

LEAF MINER. *Argyresthia thuiella.* The leaf tips turn brown as a result of the feeding within the leaves of the small leaf miner maggot. The adult stage is a tiny gray moth with a wingspread of 1/3 inch. The maggots overwinter in the leaves. This pest also feeds on juniper.

Control. Trim and destroy infested leaves. Spray with Diazinon in mid-June. Soil drenches in May and August with Cygon will also give some control.

ARBORVITAE WEEVIL. *Phyllobius intrusus.* White to pink larvae with brown heads feed on the roots of arborvitae and junipers. The adult, covered with greenish scales, feeds on the upper parts of the plants from May to July.

Control. Spray with Sevin in late May and repeat in mid-June to control the adult stage, if needed.

MEALYBUG. *Pseudococcus ryani.* This pest also attacks incense cedar, Norfolk Island pine, and redwood.

Control. Spray with malathion or Cygon

SCALE. *Lecanium fletcheri.* This dark brown, flat, more or less hemispherical scale belongs to the so-called naked or unarmored scales. Newly hatched young have an amber color, which changes to pale orange-yellow. The insect overwinters as a partly grown scale on the stems, branches, and leaves. The pest is relatively unimportant on this host but is rather serious on yews.

Several other species of scales also attack arborvitae: European fruit lecanium, Glover, a juniper scale relative, and San Jose.

Control. Malathion or Sevin sprays are very effective if applied during midsummer when the young are crawling about.

SPIDER MITE. See spruce spider mite, under Spruce.

OTHER INSECTS. Arborvitae is also host to a bark-feeding caterpillar, *Dioryctria abietella;* the cypress webber, *Epinotia subvirdis;* a weakened tree-attacking bark beetle, *Phloeosinus canadensis;* and gypsy moth (see under Elm).

ASH (Fraxinus)

Ash trees are popular landscape plants. Cultivars of native ash are available. Some, such as blue ash, can survive several centuries and attain large size. White, green, and black ash are unusually free from attacks by fungus parasites but are frequently attacked by several insect pests. Other species of ash native to Europe and Asia Minor are also available.

Diseases

DIEBACK. In the northeastern United States, white ash has been affected by a branch dieback and tree decline disease, and since 1940 occasional death of trees has been reported. The primary cause of the disease has not been clearly established, although tobacco ringspot and tobacco mosaic viruses have been associated with it. The virus was transmitted both by grafting and by nemas. The ash decline may be a complex problem also involving drought and opportunistic canker-causing fungi such as *Cytophoma pruinosa.*

Control. Control measures have not been developed.

WITCHES' BROOM. This is a mycoplasmalike organism. This disease, also called ash yellows, is fairly widespread in the northern United States. Bunching, proliferating shoot growth and gradual decline characterize the disease. It is most damaging to white ash.

Control. No controls are known.

ANTHRACNOSE. *Discula sp. (Gloeosporium aridum).* Brown spots occur over large areas of the leaves, especially among the veins (Fig. III-2). In wet seasons, infected leaflets drop prematurely.

Control. Valuable specimens should be sprayed with Captan, zineb, Benlate, Topsin-M, Zyban, Daconil 2787, or mancozeb three times at 2-week intervals, starting when the buds break open.

Fig. III-2. Anthracnose of white ash caused by the fungus *Gloeosporium aridum*.

LEAF SPOTS. Several leaf-spotting fungi occur on ash, including *Cercospora fraxinites, C. lumbricoides, C. Texensis, Septoria besseyi, S. leucostoma,* and *S. submaculata*. Mycosphaerella leaf spots of ash involve several stages of the causal fungi: *Mycosphaerella effigurata (Marssonina fraxini, Asteromella fraxini); Mycosphaerella fraxinicola (Cylindrosporium fraxini, Phyllosticta virdis)*.

Control. Gather and destroy fallen leaves. This practice is usually sufficient to keep leaf spot diseases at a minimum. In cases where the disease was severe the previous year and spring conditions remain wet, spray with Daconil 2787 or mancozeb two or three times at 10-day intervals, beginning when the buds begin to open.

CANKERS. *Cytospora annulata, Diplodia infuscans, Dothiorella fraxinicola, Botryosphaeria dothidea, Nectria cinnabarina, N. coccinea,* and *Sphaeropsis* sp. At least six fungi cause branch and trunk cankers on ash. A species of *Dothiorella* was associated with the death of many white ash trees in the metropolitan New York area in 1977.

Control. Prune out infected branches. Maintain trees in good condition by proper feeding, watering, and spraying.

RUST. *Puccinia sparganioides.* The leaves of green and red ash are conspicuously distorted and the twigs are swollen by this fungus. The spores of the fungus, yellow powder in minute cups, appear over the swollen areas. The spores produced on ash are incapable of reinfecting ash but infect the so-called marsh and cord grasses.

Control. The disease is rarely destructive enough to warrant special control measures. Sulfur sprays can be used where the disease is serious on valuable specimens.

POWDERY MILDEW. *Microsphaera fraxini* and *Phyllactinia guttata.* These fungi cause powdery mildew of ash.

Control. Bayleton fungicide applied when mildew first appears controls this disease.

HEART ROT. Some aggressive heart rot fungi attack older ash trees, causing decay and hollowing. These include species of *Fomes, Pleurotus, Polyporus, Perennipora,* and *Phellinus.* See Chapter 12.

Abiotic Diseases

AIR POLLUTION. Ash is relatively sensitive to ozone injury. See Chapter 10.

Insects and Related Pests

ASH BORER. *Podesesia syringae fraxini.* This borer attacks ash and mountain-ash in the Prairie states. It burrows into the lower part of the tree trunk.

Control. Cut out and destroy severely infested trees. Lindane or Dursban sprays applied around the trunk when the adult moths are depositing eggs may be helpful.

CARPENTER WORM. *Prionoxystus robiniae.* Large scars along the trunk, especially in crotches, and irregularly circular galleries about ½ inch in diameter, principally in the wood, are produced by a 2- to 3-inch pinkish white caterpillar, the carpenterworm. The adult moth, with a wingspread of nearly 3 inches, deposits eggs in crevices or rough spots on the bark during June and early July. A period of 1 to 4 years is necessary to complete the life cycle.

Control. Spray or paint bark of trunk and main branches with methoxychlor in late June and repeat twice at 2-week intervals. Inject commercial borer paste such as Bortox into burrows and seal openings.

RED-HEADED ASH BORER. *Neoclytus acuminatus.* This pest can be a problem of weakened and newly set trees in the Midwest.

LILAC BORER. *Podosesia syringae syringae.* Rough, knotlike swellings on the trunk and limbs and the breaking of small branches at the point of injury are indications of the presence of the lilac borer, a brown-headed, white-bodied larva ¾ inch long. The adult female appearing in late spring is a moth with clear wings having a spread of 1½ inches, the front pair of which is deep brown. The larvae pass the winter underneath the bark.

Control. Spray the trunk and main branches with methoxychlor or Dursban in early May and repeat twice at 3-week intervals.

OTHER BORERS. Many other borers infest ash. Among the more common are brown wood, California prionus, flatheaded apple tree, the Pacific flatheaded, and leopard moth (see under Maple).

Control. State entomologists will supply spraying dates and other information necessary to control these pests.

LILAC LEAF MINER. *Caloptila syringella.* Light yellow, ¼-inch larvae first mine the leaves of ash, deutzia, privet, and lilac and then roll and skeletonize the leaves. Small moths emerge from overwintering cocoons in the soil in May to deposit eggs in the undersides of the leaves. A second brood emerges in July.

Control. Spray with Diazinon or malathion in spring before larvae curl the leaves. Repeat in mid-July.

FALL WEBWORM. *Hyphantria cunea.* In fall webworm infestations, the leaves are chewed in August and September and the branches are covered with webs or nests, which enclose skeletonized leaves and pale yellow or green caterpillars 1 inch long. The adult month has white- to brown-spotted wings with a spread of 1½ inches.

Control. Cut out or remove nests, or spray thoroughly with *Bacillus thuringiensis,* Diazinon, Dylox, Dursban, Orthene, methoxychlor, or Sevin when the webs are first visible.

BROWN-HEADED ASH SAWFLY. *Tomostethus multicinctus.* Trees may be completely defoliated in May or early June by yellow sawfly larvae. The adult is a beelike insect that lays eggs in the outer leaf margins. Winter is passed in the pupal stage in the ground.

Control. Spray with Sevin about the middle of May.

ASH FLOWER GALL. *Eriophyes fraxinoflora.* This disease is caused by small mites which attack the staminate flowers of white ash. The flowers develop abnormally and form very irregular galls up to ½ inch in diameter (Fig. III-3). These galls dry out, forming clusters which are conspicuous on the trees during winter.

Control. Superior oil applied during the dor-

Fig. III-3. Ash flower gall caused by mites.

mant period or Sevin or Kelthane applied after the buds swell and before the new growth emerges in spring will provide control.

OYSTERSHELL SCALE. *Lepidosaphes ulmi.* Masses of brown bodies shaped like an oyster-shell and about 1/10 inch long, covering twigs and branches, are characteristic of this insect. The pests overwinter in the egg stage under the scales. The young crawling stage appears in late May. The pit-making pittosporum scale, *Asterolecanium arabidis,* and walnut scale (see under Walnut) also feed on ash.

Control. Spray in late spring, before the buds open, with 1 part concentrated lime sulfur in 10 parts of water, or with dormant oil plus ethion according to the directions of the manufacturer. Malathion or Sevin sprays applied when the young are crawling about in May and June also give control.

OTHER INSECTS. Ash is subject to attack by the foliar-feeding spiny io moth caterpillar (see io moth, under Sycamore); forest, eastern, and California tent caterpillars (see under Maple, Willow, and Strawberry-Tree); a southern Cal-

ifornia bagworm, *Oiketus townsendi;* Japanese weevil (see under Holly); a western woolly aphid, *Prociphilus californicus;* ash plant bug, *Tropidosteptes* sp.; cambium miner (see under Holly); buffalo treehopper, *Stictocephala bubalus;* and ash midrib gall midge, *Contarinia candensis.*

ASPEN. *See Poplar.*

ATHEL TAMARISK *(Tamarix)*

This Mediterranean native is a medium-sized tree, tolerant of salt but not tolerant of cold. It can be used in arid regions, and is relatively tolerant of pest problems.

AUSTRALIAN-PINE *(Casuarina)*

This tree has been widely planted as an ornamental and as windbreaks in southern Florida. It is fast growing, pinelike, with spreading, drooping branches. *Casuarina* does not tolerate freezing weather.

Diseases

In Florida the Australian-pine is subject to a root rot caused by *Clitocybe tabescens* and to a species of root-knot nema. In California it is subject to the root-rot fungus *Armillaria mellea*.

Control. Effective control measures are not available.

Insects

AUSTRALIAN-PINE BORER. *Chrysobothris tranquebarica.* This flat-headed borer also attacks red mangrove trees. The adult females are greenish bronze beetles, ½ to ¾ inch long, which deposit eggs on the bark in April.

Control. Cut out beetle-infested branches or trees in fall or winter. Valuable ornamental specimens can be protected by methoxychlor sprays applied in April.

OTHER INSECTS. Two species of mealybugs—citrus and long-tailed—and seven species of scales—barnacle, brown soft, cottony-cushion, Dictyospermum, latania, long soft, and mining—also may infest Australian-pine.

Control. Control measures are rarely used.

AVOCADO (Persea)

Avocado is sometimes considered a highly desirable ornamental for subtropical areas.

Diseases

ANTHRACNOSE. *Colletotrichum gloeosporioides.* Greenhouse plants in the northern states and garden plants in the southern are subject to attack by this fungus. It causes a general wilting of the ends of branches and the development of cankers on the stem and spots on the leaves and flowers.

Control. If necessary, spray with bordeaux mixture or some other copper fungicide.

ROOT ROT. *Phytophthora cinnamomi.* Root rot, also known as decline, is the most destructive disease of avocado in California. The causal fungus is soilborne and seedborne and thrives under conditions of poor soil drainage.

Control. Treat seed in hot water at 120° to 125° F (49° to 50° C) for 30 minutes to eliminate seedborne infections. The avocado rootstock variety Duke is said to be rather resistant to attack by the causal fungus.

SCAB. *Sphaceloma perseae.* This is a serious disease of avocado in Florida and Texas.

Control. Copper sprays will control this disease.

OTHER DISEASES. Other important diseases of avocado are canker, caused by *Botryosphaeria dothidea,* root rot by *Armillaria mellea,* and wilt by *Verticillium albo-atrum.*

Control. Measures for the control of the latter two are discussed in Chapter 12.

Insects

Golden mealybug, *Nipaecoccus nipae;* mealybugs, *Pseudococcus* sp.; thrips, *Heliothrips haemorrhoidalis;* and tuliptree, ivy, latania, and dictyospermum scales attack avocado.

BALD CYPRESS (Taxodium)

This genus, native to the eastern United States, has both evergreen and deciduous members. Bald cypress thrives on wet sites and also tolerates dry soil, forming knees in wet locations. It is relatively free of fungus parasites and insect pests.

Diseases

TWIG BLIGHT. This disease, also present on a number of other conifers, including aborvitae, junipers, cypress, and yews, is rarely serious. It causes a spotting of leaves, cones, and bark and, in very wet seasons, a twig blight.

Cause. The fungus *Pestalotiopsis funerea* is frequently associated with twig blight. It is not considered a vigorous parasite but becomes mildly pathogenic on trees weakened by mites, dry weather, sunscald, or low temperatures.

Control. Control measures are rarely applied, although a copper fungicide would probably be effective.

WOOD DECAY. A number of fungi belonging to the genera *Echinodontium, Fomes, Lenzites, Poria,* and *Polyporus* are associated with wood decay of bald cypress. The last is sometimes found on living trees.

Control. Avoid wounding the trunk base of valuable trees. Keep trees in good vigor by watering and, if needed, by feeding.

OTHER FUNGUS DISEASES. Canker caused by species of *Septobasidium* and heart rot caused by *Fomes geotropus, F. extensus,* and *Ganoderma applanatum* have been reported from the Gulf Coast.

Control. No control measures are known for these diseases.

Insects

SOUTHERN CYPRESS BARK BEETLE. *Phloeosinus taxodii.* This beetle attacks bald cypress stems.

CYPRESS MOTH. *Recurvaria apicitripunctella.* The larval stage of this moth mines the leaves of bald cypress and hemlock, then webs them together in late summer. The female adult is a small yellow moth with black markings and fringed wings.

Control. Spray with Sevin when the larvae begin to mine the leaves in late spring and before they web the leaves together.

BANANA *(Musa)*

Plants of the cultivated banana, *Musa paradisiaca* var. *sapientum,* are subject to a number of serious bacterial and fungus diseases as well as many insects. Such plants grown in northern greenhouses for display and educational purposes are subject to mealybugs, whiteflies, and scales. These can be controlled with malathion sprays.

In Florida the dwarf banana, *M. nana,* is subject to anthracnose caused by the fungus *Gloeosporium musarum,* leaf blight by the bacterium *Pseudomonas solanacearum,* and the southern root-knot nema, *Meloidogyne incognita.* The burrowing nema, *Radopholus similis,* occurs in Louisiana on the roots of ornamental banana trees.

Control. Plant pathologists at southern agricultural experiment stations will provide control measures.

BANYAN *(Ficus benghalensis)*

The only pests reported on this tree are insects: the long-tailed mealybug and several species of scales—black, Chinese obscure, California red, lesser snow, green shield, and mining.

Control. Malathion sprays are suggested for these insects attacking valuable ornamental specimens. Summer oil emulsions are also effective.

BASSWOOD. See Linden.

BEECH *(Fagus)*

Beeches are among our most beautiful trees; their dormant shapes and winter bark contribute much to any landscape.

The American beech *(F. grandifolia)* is a handsome, low-branched, slow-growing tree. Its dense foliage and shallow rooting habit, however, make difficult the maintenance of a lawn under it. It grows best in cool, moist soils and can withstand neither constant trampling over the soil nor common city conditions very well. It is generally extremely difficult to transplant.

The European beech *(F. sylvatica)* is similar to the American species except that its bark is darker gray and its foliage more glossy. Its many horticultural varieties include purple beech, variety *atropunicea,* with deep purple leaves; weeping beech, variety *pendula,* one of the world's most beautiful weeping trees; weeping purple beech, variety *purpureopendula;* and cutleaf, variety *laciniata,* with very beautiful, narrow, almost fernlike leaves. Other new or interesting kinds are Darwyck beech, *F. sylvatica fastigiata;* golden-leaved beech, variety *aurea;* a red fern-leaved beech, variety *Rohanni;* and the tricolor beech with pink, white, and green leaves that turn a coppery bronze.

Diseases

The dying of a large number of American beeches is believed by some investigators to be related to drought and subsequent death of the roots. Associated with the progressive weakening of some of the trees, however, is a trouble known as leaf mottle or scorch.

LEAF MOTTLE. This disease has appeared somewhat sporadically in the eastern United States since the late 1940s.

Symptoms. In spring, small translucent spots surrounded by yellowish green to white areas appear on the young unfurling leaves. These spots turn brown and dry, and by the first of June the mottling is very prominent, especially between the veins near the midrib and along the outer edge of the leaf. Within a few weeks the brown areas increase in number until the entire leaf presents a scorched appearance. A considerable part or, in some instances, all of the leaves then drop prematurely. Where complete defoliation occurs, new leaves begin to develop in July. The second set of leaves in such cases appears quite normal and drops from the trees at the normal time in the fall.

The extensive loss of foliage early in the season exposes the branches to the direct rays of the sun. Because beeches are sensitive to such rays, considerable scalding of the bark occurs. Such injury predisposes the tree to attacks by the two-lined chestnut borer and possibly by other insects. Heavy infestations of borers will kill large branches or even the entire tree.

Cause. The cause of leaf mottle is not known. Investigations by the senior author many years ago failed to reveal definite clues. Site, age of tree, water supply, or atmospheric conditions do not appear to be correlated with the incidence or severity of the trouble. For example, trees growing in open lawns appear to be as subject as trees in the woods; young trees seem to contract the trouble as readily as old ones; trees growing in light, well-drained soils apparently are as susceptible as those in heavy, moist soils; and the disease appears to be as prevalent in wet seasons as in dry ones. The symptoms are not typical of any known nutrient deficiency, and injections of magnesium nitrate and ferrous sulfate into the trunks have failed to produce any improvement.

It is thought that leaf mottle is actually a form of leaf scorch. Trees that exhibit mottling have a different genetic background from those that are not mottled. Hence one tree may respond differently from the other to the same soil conditions.

Control. Until more is known about the cause, the only recommendation that can be made is to provide good growing conditions. The bark of particularly valuable specimens that are completely defoliated might well be protected from the midsummer sun with burlap or some other material until the second set of leaves has developed sufficiently to provide the necessary shade.

BLEEDING CANKER. *Phytophthora cactorum.* This disease, described under maple, occurs occasionally on beech. The name suggests the most common symptom visible on beech, maple, and elm, an oozing of a watery light brown or thick reddish brown liquid from the bark.

The fungus also causes crown canker on dogwood and collar rot of fruit trees.

Control. No effective preventive measures are yet known. Infected specimens should be cut down and destroyed to prevent spread to nearby trees. Mildly affected trees have been known to recover. Avoid bark wounds near the base of the tree.

BEECH BARK DISEASE. The natural stands of beech in eastern Canada and northeastern United States have been severely damaged by this disease since the early 1930s. Although it is primarily a disease of forest beech, it is potentially dangerous to ornamental beech.

The attack of the woolly beech scale, *Cryptococcus fagisuga* and possibly other scales, combined with the fungus *Nectria coccinea* var. *faginata,* and possibly other *Nectria* fungi, causes this disease.

Infestations of the scale on the bark always precede those of the fungus. During August and September, countless numbers of minute, yellow, crawling larvae appear over the bark. By late autumn they settle down and secrete a white fluffy material over their bodies. This substance is very conspicuous, and the trunks and branches appear to be coated with snow. Through the feeding punctures of the insect, the *Nectria* fungus then penetrates the bark and kills it. The insects soon die because of the disappearance of their source of food.

Death of the infected bark is followed by drying; infected areas on the trunk thus are depressed and cracked. Eventually, deeply sunken cankers are formed, which assume a more or less circular or oval shape. The destruction of the bark and other tissues of the tree leads to a progressive dying of the top. Late in the season the leaves usually curl and turn brown, the twigs die, and new buds fail to form. The dead leaves remain attached throughout the winter. In spring, affected branches fail to produce foliage, and other branches, lacking reserve food materials, produce small yellow leaves that usually die during summer. Eventually the entire tree dies.

Control. Because infestations of the woolly beech scale must precede fungus penetration, the eradiation of the insect pest will prevent the start of the disease. Malathion sprays applied to the trunks and branches of valuable ornamental trees growing in the infested areas in early August and in September should control the young scales. A dormant lime sulfur spray applied to trunks and branches will control overwintering adult scales. Oil sprays are also effective but are not reliably safe on beech trees.

CANKERS. *Asterosporium hoffmanni, Cytospora sp., Strumella coryneoidea, Endothia gyrosa, Nectria galligena,* and *N. cinnabarina.* Canker and branch dieback of beech may be caused by any one of the five fungi listed.

Control. Prune and destroy infected branches.

LEAF SPOTS. *Gloeosporium fagi* and *Phyllosticta faginea.* Leaf spots develop late in the growing season.

Control. Control can be achieved by spraying with zineb, maneb, or copper fungicides.

POWDERY MILDEW. *Microsphaera erineophila* and *Phyllactinia guttata.* Two species of powdery mildew fungi occasionally develop on beech leaves in late summer.

Control. Benlate or wettable sulfur sprays will control powdery mildews.

WOOD DECAY. *Hericium erinaceus, Fomes fomentarius,* and *Phellinus igniarius.* Three species of fungi have been observed in decayed wood. See Chapter 12.

Insects

BEECH BLIGHT APHID. *Prociphilus imbricator.* The bark is punctured and juices are extracted by the beech blight aphid, a blue insect covered with a white cottony substance. The pests also feed on the leaves.

Control. Spray with Sevin or malathion when the pests first appear in spring. Repeat 2 weeks later if necessary.

WOOLLY BEECH APHID. *Phyllaphis fagi.* Leaves are curled and blighted by the woolly beech aphid, a cottony-covered insect, the cast skins of which adhere to the lower leaf surface. The purple beech is more commonly infested than the American beech. The giant bark aphid (see aphids, under Sycamore) also feeds on beech.

Control. Before the leaves are curled, spray with the same formula as that recommended for the beech blight aphid.

BEECH SCALE. *Cryptococcus fagisuga.* White masses of tiny circular scales, each $1/40$ inch in diameter, on the bark of the trunk and lower branches are the common signs of beech scale infestation. The eggs are deposited on the bark in late June and early July, and the young crawling stage appears in August and September. The pest overwinters as a partly grown adult scale. Beech scale is primarily a pest of forest trees and is mentioned here only because of its association with the beech bark disease. It is present in the northeastern United States.

Control. See under beech bark disease.

CATERPILLARS. Many other caterpillars chew the leaves of beech, including the hemlock looper, saddled prominent, walnut, fall and spring cankerworms, elm spanworm, and green mapleworm. The larval stage of the following moths also infests the leaves of this host: imperial, io, luna, and the rusty tussock.

Control. Valuable ornamental specimens can be protected from any of these pests with methoxychlor or Sevin sprays. These are most effective if applied when the caterpillars are small.

BROWN WOOD BORER. *Parandra brunnea.* Winding galleries in the wood, made by the white-bodied, black-headed borers, $1 1/4$ inches long, and tiny holes in the bark, made by emerging shiny brown beetles $3/4$ inch long, are typical signs of brown wood borer infestation. The eggs are deposited in bark crevices or in decayed wood. Leopard moth larvae (see under Maple) also invade beech wood.

Control. Infestations can be prevented to a large extent by avoiding mechanical injuries to bark and wood and by treating open wounds.

TWO-LINED CHESTNUT BORER. See under Oak.

ASIATIC OAK WEEVIL. See under Oak.

SCALES. In addition to the *Cryptococcus* scale mentioned earlier, beech trees are subject to the following scales: black, cottony-cushion, cottony maple, hickory lecanium, European fruit lecanium, oystershell, Putnam, and San Jose.

Control. Spray with malathion when the young scales are crawling about and with dormant lime sulfur in early spring.

EASTERN TENT CATERPILLAR. See under Willow.

DATANA CATERPILLAR. See yellow-necked caterpillar, under Oak.

GYPSY MOTH. See under Elm.

OTHER INSECTS. Beech leaf miner, *Brachys aeruginosus;* maple trumpet skeletonizer; and maple leaf cutter (see under Maple) attack beech. Mites, including oak mite (see under Oak), and eriophyids infest beech.

BIRCH *(Betula)*

The birches, graceful as well as beautiful, are highly susceptible to ice and snow breakage. Extremely particular about soil conditions, they are unable to adapt themselves as street trees and are used principally on lawns.

The canoe birch *(B. papyrifera)* has attractive bark and a single trunk. Gray birch *(B. populifolia)* is gray-barked, and many specimens have several trunks.

Ornamental birches are susceptible to several fungus parasites and insect pests. Of these, the bronze birch borer is mainly responsible for the death of trees used in ornamental plantings.

River birch *(B. nigra)* with its shaggy reddish brown bark likes moist sites and is resistant to bronze birch borers.

The Monarch birch *(B. maximowicziana),* introduced into the United States from Japan in 1893 by Professor Sargent of the Arnold Arboreteum, is highly resistant to the bronze birch borer. Its leaves are larger than those of native species and more distinctly heart-shaped. Monarch birch does better in urban environments than the other birches. It should be more widely planted as a lawn or park tree.

Most native birches, except river birch, are best grown in cold climates.

European white birch *(B. pendula)* is also widely planted but very susceptible to borers.

Diseases

LEAF BLISTER. Two fungi cause leaf blister on many species of birch. *Taphrina carnea* produces red blisters and curling of the leaves on many of the birch species, and *T. flava* forms yellow blisters on gray and canoe birches.

Control. Gather and destroy all fallen leaves. Spray with Ferbam or Zineb just before buds open in spring.

LEAF RUST. *Melampsoridium betulinum.* Leaves of seedlings and of mature trees are sometimes attacked by a rust which causes spotting and defoliation. The rust pustules are bright reddish yellow. The spores from these pustules carry the infection from leaf to leaf. The alternate or sexual stage causes a blister rust on larch. In mixed forest plantings both hosts may be seriously injured.

Control. In ornamental plantings the disease rarely becomes destructive enough to warrant special control measures.

LEAF SPOT. Several leaf spot fungi attack birch. The fungus *Colletotrichum gloeosporoides* (*= Glomerella cingulata*) produces brown spots with a dark brown to black margin. The fungus *Cylindrosporium betulae* forms smaller spots with no definite margin. Both fungi may become sufficiently prevalent to cause some premature defoliation. *Marssonina betulae* causes a leaf blotch.

Control. Gathering and destroying fallen leaves usually suffices for practical control. Bordeaux mixture or Benlate may be used in late spring as a preventive.

CANKER. Black, paper, sweet, and yellow birches are particularly susceptible.

Symptoms. Young cankers seldom are readily visible. Upon close examination, however, they are seen to be darker in color than the adjacent healthy tissue and appear water-soaked. Later the edge of the diseased area cracks, thus exposing the canker (Fig. III-4). Callus tissue forms over the cracked area but becomes infected and dies. The process is repeated annually, forming concentric rings of dead callus. The bark within the cankered zone falls away, leaving the wood exposed. When the canker completely girdles the stem or trunk, the distal portion dies.

Fig. III-4. Canker on black birch.

Cause. Canker is caused by the fungus *Nectria galligena.* Small, globose, dark red fruit bodies of the fungus are barely visible to the unaided eye on the dead bark.

Control. In thick stands it is advisable to remove trees with trunk infections. Trees having cankers or galls on the branches may be saved by pruning out and destroying the cankers. Since the cankers originate on young growth, inspection of the plantings and early destruction of the cankered young trees are advisable. Trees in ornamental plantings should be fertilized and watered to maintain good vigor.

DIEBACK. *Melanconium betulinum.* Trees weakened by drought may be attacked by this fungus, which causes a progressive dieback of the upper branches. Infestations of the bronze birch borer mentioned below may cause similar symptoms, however.

Control. Prune affected branches to sound wood, and fertilize and water heavily to help revitalize the tree.

WOOD DECAY. Several wood-decay fungi attack birches. One, *Piptoporus betulinus,* attacks dying or dead birches and produces shelf- or hoof-shaped, gray, smooth, fungus bodies along the trunk. Others, such as *Torula ligniperda, Fomes fomentarius, Inonotus obliquus, Phellinus laevigatus,* and *Ganoderma applanatum,* are associated with decay of living trees.

Control. Wood decays cannot be checked once they have become extensive. Avoidance of wounds and maintenance of the trees in good vigor by fertilization are the best preventive practices.

OTHER FUNGUS DISEASES. Birches are subject to several other fungus diseases: powdery mildews caused by *Microsphaera ornata* and *Phyllactinia guttata,* leaf spots by *Discula betulina* and *Septoria betulicola,* and stem cankers by *Botryosphaeria obtusa* and *Diaporthe alleghaniensis. Gloeosporium* and *Diaporthe* cause sunken, black stem cankers on young branches; distal leaves turn brown.

Control. Control measures are rarely necessary for these diseases.

VIRUS DISEASE: LINE PATTERN MOSAIC; APPLE MOSAIC VIRUS. Decline, dieback, and death of both white and yellow birches in forest plantings may be virus-induced. White birch in ornamental plantings may also be affected but to a lesser extent. Yellow to golden line and ringspot patterns on the leaves are the most striking symptoms.

Control. Control measures have not been developed for this virus disease.

Abiotic Disease

AIR POLLUTION. Birch is relatively sensitive to sulfur dioxide. See Chapter 10.

Insects and Related Pests

APHIDS. *Euceraphis betulae* and *Calaphis betulaecolens.* The former, known as the European birch aphid, is yellow and infests cut-leaved and other birch varieties. The latter, the common birch aphid, is a large green species which produces copious quantities of honeydew followed by sooty mold. Giant bark aphid (see aphids, under Sycamore) is also found on birch.

Control. Spray thoroughly with Diazinon, dimethoate, or malathion as soon as these aphids appear.

WITCH-HAZEL LEAF GALL APHID. *Hamamelistes*

spinosus. This insect, which causes the formation of cone galls on witch-hazel, migrates from this host to birches in summer. It feeds on the undersides of the leaves and resembles nymphs of whiteflies.

Control. Spray with malathion when aphids are present.

CASE BEARER. *Coleophora fuscedinella.* Leaves are mined and shriveled and small cases are formed under the leaves by the case bearer, a light yellow to green caterpillar, ⅕ inch long with a black head. The adult is a brown moth with a wingspread of ⅖ inch. The pest overwinters as a larva in a case attached to the bark.

Control. Spray in late spring, before growth begins, with 1 part concentrated lime sulfur in 8 parts of water, or with standard malathion solution in late July or early August.

SCALES. Several types, including hickory le-

canium, terrapin, frosted, and walnut, infest birch.

BIRCH LEAF MINER. *Fenusa pusilla.* Gray, canoe, and cut-leaf birches are especially susceptible to attacks by the leaf miner, a small white worm which causes leaves to turn brown in late spring or early summer (Fig. III-5). The adult is a small black sawfly which overwinters in the soil as a pupa. The first brood begins to feed anytime from very early to late May, depending on the season and the location of the trees. The first brood causes most damage because it attacks the tender spring foliage. Other broods hatch during the summer, but these cause less damage because they do not attack mature foliage but confine their feeding to leaves on sucker growths and to newly developing leaves in the crowns of the trees.

Control. Malathion or Sevin sprays will con-

Fig. III-5. The leaf miner causes brown blotches on birch leaves. The tiny adult flies and the larvae are also shown.

trol the birch leaf miner. The first spray should be applied about May 1. If the spring is a cold one, the first application can be delayed a week or so. For best control, an additional application should be made 10 days later. To control the second brood of leaf miners, spray again about mid-June. If mines are already present in early summer, apply Cygon or Orthene.

BIRCH SKELETONIZER. *Bucculatrix canadensisella.* The lower leaf surface is chewed and the leaf is skeletonized and may turn brown as a result of feeding by this skeletonizer, a yellowish green larva ¼ inch long. The adult moth has white-lined, brown wings with a spread of ⅜ inch.

Control. Spray the upper and lower sides of the leaves with Diazinon or Sevin about mid-July.

BRONZE BIRCH BORER. *Agrilus anxius.* Varieties of birch grown in parks as ornamental shade trees, especially where the soil is poor, and trees grown elsewhere under adverse con-

ditions become prey to this borer. The grub is from ½ to 1 inch long, flat-headed, and light-colored. The adult stage is a beetle ½ inch long. The beetles, which feed on foliage for a time, deposit their eggs in slits in the bark. The borers make flat, irregular, winding galleries just beneath the bark of the main trunk. (Fig. III-6). Heavy infestations usually kill the trees (Fig. III-7). Poplar and willow borers and dogwood borers also attack birch.

Control. Spray the trunk and branches thoroughly and the leaves lightly in early June and twice more at 2-week intervals with Lindane or Guthion. Keep the trees in good vigor by fertilizing and watering when needed.

BIRCH LACE BUG. *Corythuca pallipes.* Nymphs feed on birch, causing a stippling symptom on the leaf surface. Controls are not normally needed.

SEED MITE GALL. *Eriophyes betulae.* Another conspicuous gall on paper birch and other species is caused by the seed mite. The galls are

Fig. III-6. Bronze birch borer exit holes.

Fig. III-7. These birches are dying back because of infestations of the bronze birch borer.

about 1 inch in diameter, made up of many adventitious branches and deformed buds. They may resemble witches' brooms.

Control. This pest is never serious enough to require control measures.

OTHER INSECTS. Dusky birch sawfly, *Croesus latitarsus;* yellow-necked caterpillar (see under Oak); cambium miner (see under Poplar); white-humped caterpillar (see under Poplar); white-marked tussock moth; mourning-cloak butterfly (see under Elm); elm spanworm (see under Elm); cecropia moth, *Hyalophora cecropia;* eastern tent caterpillar (see under Willow); forest tent caterpillar (see under Maple); and leaf roller, *Archips* sp. feed on birch leaves. Japanese beetle (see under Linden); potato leafhopper, *Empoasca fabae;* planthopper (see under Cherry); spider mites; and giant hornet wasp (see under Franklin-Tree) also infest birch.

BLACK GUM. See Tupelo.

BOXWOOD *(Buxus)*

Boxwood requires a well-drained, neutral soil with an occasional light application of ground limestone. In poorly drained soil, it is susceptible to winter injury and fungus diseases. A few kinds, particularly the tree box, grow into small trees and hence are discussed here.

Diseases

CANKER. One of the most destructive diseases of boxwood is canker.

Symptoms. The first noticeable symptom is that certain branches or certain plants in a group do not start new growth as early in spring as do others, and the new growth is less vigorous than that on healthy specimens. The leaves turn from normal to light green and then to various shades of tan. Infected leaves turn upward and lie close to the stem instead of spreading out like the leaves on healthy stems. The diseased leaves and branches show small, rose-colored, waxy pustules, the fruiting bodies of the fungus. The bark at the base of an infected branch is loose and peels readily from the gray to black discolored wood beneath. Infection is frequently found to take place at the bases of small dead

shoots or in crotches where leaves have been allowed to accumulate.

Cause. Canker is caused by the fungus *Pseudonectria rousselliana*, which is found on the stem cankers. Another stage of the same fungus, known as *Volutella buxi*, attacks the leaves and twigs (Fig. III-8), producing pale rose-colored spore masses on yellowed leaves. A species of *Verticillium* is occasionally associated with the dieback of twigs. The fungus *Nectria desmazierii*, whose imperfect stage is known as *Fusarium buxicola*, is also capable of causing canker and dieback of boxwood.

Control. Dead branches should be removed as soon as they are noticeable, and cankers on the larger limbs should be treated by surgical methods. The annual removal and destruction of all leaves that have lodged in crotches is recommended. Four applications of fixed copper bordeaux mixture 3-3-50 or lime sulfur 1-50 have been shown to be very effective in preventing canker. The first should be made after the dead leaves and dying branches have been removed and before growth starts in the spring; the second, when the new growth is half completed; the third, after spring growth has been completed; and the fourth, after the fall growth has been completed. The boxwood should be fertilized with occasional applications to the soil of well-rotted cow manure, or commercial fertilizers, and ground limestone.

BLIGHT. Several fungi are associated with the blighting of boxwood leaves. The most common are *Phoma conidiogena* and *Hyponectria buxi*. The exact role of these fungi in this disease complex is not clear.

Control. The fungicides recommended for canker control will also control blight.

LEAF SPOTS. *Macrophoma candolei, Phyllosticta auerswaldii, Fusarium buxicola,* and *Collectotrichum* sp. Leaves turn straw-yellow and are thickly dotted with small black bodies, the fruiting structures of the first fungus listed above. The others also cause leaf spotting. All are apparently limited in their attacks to foliage weakened by various causes.

Control. Leaf spots may be controlled by shaking out all fallen and diseased leaves from the center of the bush and destroying them. All dead branches in the center of specimen plants

Fig. III-8. Volutella canker of boxwood.

or hedges should be removed to allow better aeration. An application of bordeaux mixture or any ready-made copper fungicide before growth starts in spring may be beneficial. This spray will discolor the foliage, but the unsightly effect is soon hidden by new growth.

ROOT ROT. *Phytophthora cinnamomi.* "Off-color" foliage followed by sudden wilting and death of the entire plant is characteristic of this disease. Yews and a large number of other woody ornamental plants are also subject to this disease. Another species, *P. parasitica,* also causes a root rot and blight of boxwood. Root infection by the fungus *Paecilomyces buxi* was found to be associated with boxwood decline in Virginia.

Control. Infected plants cannot be saved. Pasteurize infested soil and replant with healthy specimens.

Nonparasitic Diseases

WINTER INJURY AND SUN SCALD. Most boxwood troubles in the northeastern United States are due to freezing and sunscalding, which primarily injure the cambium of unripened wood. Several distinct types of symptoms are exhibited by winter injury. Young leaves and twigs may be injured when growth extends far into fall or begins too early in spring. Leaves may turn rusty brown to red as a result of exposure to cold, dry winds during winter. A dieback of leaves, twigs, and even the entire plant may occur on warm winter days when the above-ground tissues thaw rapidly and lose more water than can be replaced through the frozen soil and roots. Another type of winter injury is evinced by the splitting and peeling of the bark. The bark becomes loosened and the stems are entirely girdled, resulting in death of the distal portions.

Control. Fertilizers should be applied in late fall, preferably, or very early in spring. Adequate windbreaks should be provided during winter, especially in the more northern latitudes. In New York City spraying with an antitranspirant on a mild day in December, and again on a mild day in the following February, will provide as much protection from winter winds as do burlap windbreaks. A heavy mulch consisting of equal parts of leaf mold and cow manure should be applied to prevent deep freezing and to aid in supplying water continuously.

Insects and Related Pests.

BOXWOOD LEAF MINER. *Monarthropalpus buxi.* Oval, water-soaked swellings on the lower leaf surface result from the feeding inside the leaves by the leaf miner, a yellowish white maggot, ⅛ inch long. The adult is a tiny midge, ⅒ inch long, which appears in May.

Control. When the adult midges are seen in May and June (about the time the weigelas are in full bloom), spray the boxwood leaves with Diazinon, Cygon, malathion, or Sevin.

BOXWOOD PSYLLID. *Psylla buxi.* Terminal leaves are cupped and young twig growth is checked by the boxwood psyllid, a small, gray, sucking insect covered with a cottony or white waxy material (Fig. III-9). The adult is a small green fly with transparent wings having a spread of ⅛ inch.

Control. Spray in mid-May and again 2 weeks later with Diazinon, Lindane, malathion, or Sevin.

BOXWOOD WEBWORM. *Galasa nigrinodis.*

Fig. III-9. Cupping of boxwood leaves (left and right) caused by the boxwood psyllid *(Psylla buxi).* Leaves in the center are normal.

This pest chews leaves and forms webs on boxwood.

Control. Spray with malathion or Sevin when the young larvae begin to feed.

GIANT HORNET WASP. See giant hornet wasp, under Franklin-Tree.

YELLOW-NECKED CATERPILLAR. See under Oak.

MEALYBUGS. Comstock mealybug is discussed under Catalpa. The ground mealybug, *Rhizoecus falcifer,* feeds on the roots of many shrubs and trees.

Control. Cygon, malathion, or Sevin sprays applied to the leaves and stems will control the Comstock mealybug. A dilute solution of Diazinon applied to the soil around the base of the boxwood should control the ground mealybug.

SCALES. Many species of scales, California red, cottony maple, cottony-cushion, lesser snow, oystershell, Japanese wax, ivy, Glover, and European fruit lecanium, may infest boxwood.

Control. Spray with a dormant oil to control overwintering scales. Where infestations are heavy, follow with a malathion, Orthene, or Sevin spray when the young are crawling about in May, June, and July.

Other Pests

BOXWOOD MITE. *Eurytetranychus buxi.* A light mottling followed by brownish discoloration of the leaves is caused by infestation of eight-legged mites, about 1/64 inch long when full grown. Winter is passed in the egg stage. The eggs hatch in April, and the young mites begin to suck out the leaf juices. By June or July, considerable injury may occur on infested plants. As many as six generations of mites may develop in a single season.

Control. The dormant oil spray recommended for scales will destroy many overwintering mites. Cygon, Diazinon, or Kelthane applied during May and June will also control mites.

NEMAS. *Pratylenchus pratensis.* Leaf-bronzing, stunted growth, and general decline of boxwood may result from invasion by meadow nemas. They enter the roots, usually near the tips, and move through the cortical tissue. Invaded portions soon die and the plant forms lateral roots above the invaded area. These

laterals in turn are infested. Repeated infestations and lateral root production result in a stunted root system resembling a witches' broom. Even heavy rains may fail to wet such densely woven root bundles. Boxwoods are also subject to several other parasitic nemas, including the southern root-knot nema, *Meloidogyne incognita;* various ring nemas, *Criconema, Criconemoides,* and *Procriconema;* and the spiral nema *Helicotylenchus.*

Control. Several chemicals are available for controlling ectoparasitic nemas such as *Pratylenchus pratensis,* which live outside the roots. Zinophos, Vydate, or Mocap can be used as a soil drench around old, infested boxwood. The treatment must be repeated each year for several years before good control is achieved. It must be supplemented by shearing and fertilization to stimulate root activity.

Endoparasitic nemas such as the root-knot nema *Meloidogyne incognita* which live inside the roots of boxwood, have been controlled with Dasanit or Mocap.

The life of infested but untreated plants may be prolonged by providing good care and by soaking the soil thoroughly during dry spells. Before boxwood is replaced in infested soil, the planting site should be fumigated with any one of several materials available for that purpose. Plant pathologists at state experiment stations will advise on selection and use of fumigants.

BUCKEYE. See Horsechestnut.

BUCKTHORN *(Rhamnus)*

This small tree or tall shrub is sometimes used in the landscape.

Insects

Buckthorn may be attacked by black, gloomy, and ivy scales.

CAJEPUT *(Melaleuca)*

This native of Australia grows to medium size in Florida and California where it is grown as a street tree and in landscapes. It is not troubled by serious diseases in the United States.

Insects

At least fourteen species of scale as well as the citrus mealybug have been found on this host in Florida.

Control. Malathion or Sevin sprays will control the crawler stage of the scales and the mealybug. Summer oil emulsion is also effective.

CALIFORNIA LAUREL *(Umbellularia)*

This California and Oregon native has high aesthetic value in many of its native situations and is widely used in home and park landscaping. It is also called Oregon myrtle and pepperwood, the latter because of a strong, pungent, sneeze-inducing camphorlike odor emitted from crushed green bark and foliage.

Leaf Diseases

A bacterium, *Pseudomonas lauracearum,* and fungi *Kabatiella phoradendri* f. sp. *umbellulariae* and *Colletotrichum gloeosporioides* occasionally cause serious leaf blight in California. *Mycosphaerella arbuticola* and several sooty mold fungi also appear on leaves. Control measures have not been worked out.

CANKER. *Nectria galligena.* Canker also attacks this tree. Unnecessary wounds should be avoided.

Insect Pests

California laurel aphid, *Euthoracaphis umbellulariae,* sometimes mistaken for immature whitefly, is found on this host. See Chapter 11.

CAMELLIA *(Camellia japonica and C. sasanqua)*

Commonly a garden shrub, camellia can grow into a small, broad tree. Camellias are grown outdoors in the warmer parts of the country and indoors in the colder parts. The *sasanqua* varieties are said to be more winter hardy than the *japonica* varieties in the northeastern part of the country along the Atlantic Coast.

Fungus and Algal Diseases

BLACK MOLD. *Meliola camelliae.* The abundant black fungus growth of the *Fumago* stage covers the leaves and twigs of this host. The ascospores are brown, each provided with several crosswalls.

Control. Spray with malathion to control insects such as aphids and scales which secrete the substance on which this fungus grows. Promptly pick off and destroy infected leaves and discard all debris from infected plants.

CANKER. *Glomerella cingulata.* A canker and dieback of camellias is widespread and frequently destructive in the southern states. It also occurs on greenhouse-grown plants in the North. The fungus enters only through wounds. In nature, the usual entrance points are scars left by the abscission of leaves in spring.

In Florida a species of *Phomopsis* causes somewhat similar symptoms.

Control. Prune and destroy cankered twigs. Where the cankers occur on the main stem of large plants, surgical removal of the diseased portions should be attempted, followed by application of a fungicide. Copper fungicides applied periodically to the leaves and stems may help prevent new infections.

FLOWER BLIGHT. *Ciborinia camelliae.* This blight is confined to the flowers, which turn brown and drop. It occurs in the Pacific Coast states and in Gulf and other southern states from Texas to Virginia. All species and varieties of camellias appear equally susceptible to the blight.

Another flower-blighting fungus, *Sclerotinia sclerotiorum,* has been reported from North Carolina. A bud and flower blight is occasionally caused by *Botrytis cinerea,* particularly after the plants have been subjected to frost.

Control. To control the *Ciborinia camelliae* blight, pick off and discard all old camellia blossoms before they fall. Benlate, Ferbam, sulfur, or mancozeb sprays help prevent infection. Infections can also be prevented by placing a 3-inch mulch of wood chips or other suitable material around the base of each plant. Such a barrier will prevent the fungus bodies in the soil beneath from ejecting their spores into the atmosphere and onto the leaves. Soils heavily infested with sclerotia (which later produce ascocarps) may be treated with Ferbam or Captan. PCNB (Terraclor) provides even more effective control but must be used in soils free of plants.

No special controls have been developed for the *Botrytis* bud and flower blight or for *Sclerotinia sclerotiorum*.

LEAF BLIGHT. *Cephaleuros virescens.* The epidermal cells are attacked by this alga, which spreads rapidly over the leaf and causes it to blacken and die.

Control. Remove diseased leaves. Badly infested specimens may be sprayed with a copper fungicide.

LEAF GALL. *Exobasidium camelliae.* The leaves and stems of new shoots are thickened and distorted by this fungus.

Control. Spray once before the leaves unfurl with either Ferbam or zineb.

LEAF SPOT. *Cercospora theae.* This leaf spot, first reported from Louisiana, develops under conditions of overcrowding, partial shade, and high humidity of a lath or shade house.

Control. No controls have been developed.

ROOT ROT. *Phytophthora cinnamomi.* This disease is common not only on camellia but also on avocado, maple, pine, and rhododendron, as well as many other woody plants. Excessive moisture and poor soil drainage favor its development.

Control. Improve drainage. Drenching soil around living plants with zineb may help.

SPOT DISEASE. *Pestalotiopsis maculans.* More or less irregular round blotches run together, causing a silvery appearance of the upper surface of the leaves. The diseased area is sharply marked off from the healthy portion. The pycnidia or fruiting bodies of the fungus are visible as black dots. Leaf fall sometimes results. Several other fungi produce leaf spotting: *Phyllosticta camelliae, P. camelliaecola,* and *Sporonema camelliae.* A species of *Sphaceloma* causes scabby spots on the leaves.

Control. Collect and destroy all diseased leaves. Spray larger plantings with some fungicide, such as bordeaux mixture, if spotting of the foliage is not objectionable; otherwise spray the plants with wettable sulfur.

Fig. III-10. Upper: Yellow mottle virus symptoms. **Lower:** Bud drop of camellia.

Virus Diseases

The virus diseases of camellia are not well understood. Leaf and flower variegation is presumed to be caused by a virus, inasmuch as the condition was transmitted by graftage from variegated *Camellia japonica* to uniformly green varieties of *C. japonica* and *C. sasanqua.* Some yellow variegation, however, may be due to genetic changes rather than virus infection. Such variegations usually follow a uniform and rather typical pattern which is more or less similar on all leaves. The senior author has observed many greenhouse-grown camellias in the North with typical ringspot patterns in the leaves (Fig. III-10).

Control. Plants suspected of harboring a virus should be discarded, or at least isolated from healthy plants.

Abiotic Diseases

BUD DROP. Camellias grown in homes, in greenhouses, and even outdoors frequently lose their buds before opening, or the tips of the young buds and edges of young petals turn brown and decay (Fig. III-10). Bud drop from indoor-grown plants usually is due to overwatering of the soil or to some other faulty environmental condition such as insufficient light, excessively high temperatures, or a potbound condition of the roots. Bud drop in the Pacific Northwest may result from a severe frost in September or October, severe freezing during the winter, or an irregular water supply. In California it may result from lack of adequate moisture.

CHLOROSIS. Deficiency of some elements in the soil may result in chlorosis.

EDEMA. Frequently brown, corky, roughened swellings develop on camellia leaves grown in greenhouses. The condition is associated with overwatering of the soil during extended periods of cloudy weather.

SUNBURN. This condition appears on leaves as faded green to brown areas with indefinite margins. It occurs on the upper exposed sides of bushes, particularly those transplanted from shaded to very sunny areas.

SALT INJURY. Camellias cannot tolerate high soil salinity even though they grow best in the acid soils and temperate climate of our eastern and Gulf Coast areas. Salt levels above 1800 parts per million in the soil solution were fatal to camellias in greenhouse tests. Azaleas were found to be equally susceptible to high salt concentrations.

Insects and Other Animal Pests

FLORIDA RED SCALE. *Chrysomphalus aonidum.* This scale insect, common on citrus and other plants in the greenhouse, has been found to live on camellia leaves. The scales are dark brown and more or less circular. It is easy to remove the scale with a needle; this exposes the very light yellow body of the insect, firmly attached to the leaf by its sucking organ. The leaf shown in Figure III-11 was photographed after several of the scales had been thus removed; the insects appear as white spots.

Many other species of scales infest camellias. They include black, California red, camellia, chaff, cottony taxus, degenerate, dictyosperm,

Fig. III-11. Florida red scale on camellia.

euonymus, Florida wax, glover, greedy, hemispherical, Japanese wax, latania, Mexican wax, oleander, oystershell, peony, and soft scale. Two other scales known as the camellia parlatoria and the olive parlatoria also attack this host.

Control. Spray with malathion or Sevin to control the young crawling insects. Repeat the treatment at 2-week intervals for heavily infested plants. In the South, Florida Volck and similar oil emulsions are effective on outdoor plants. Oil sprays should not be applied when the temperature is above 85° F (29° C), but malathion and Sevin may be applied at any time.

TEA SCALE. *Fiorinia theae.* The most serious pest of outdoor camellias in the South, this scale also infests greenhouse-grown camellias in the North as well as ferns, palms, orchids, figs, and several other plants. It can be distinguished superficially by its oblong shape and the ridge down the center parallel to the sides.

Control. The same as for Florida red scale.

MEALYBUGS. *Planococcus citri* and *Pseudococcus adonidum.* These two white insects are usually found in the leaf axils and shoot buds.

Control. Spray with malathion.

FULLER ROSE BEETLE. *Pantomorus cervinos.* This snout-beetle occasionally infests camellias, roses, palms, and many other plants.

A number of other beetles also infest this host. The most common are rhabdopterus, flea beetle, and the grape colaspis.

Control. Sevin sprays will control the various kinds of beetles.

SPOTTED CUTWORM. *Amathes c-nigrum.* This

cutworn has been found feeding on the blasted buds of camellias in greenhouses. Apparently it is able to climb up among the branches, leaves, and flower buds.

Control. Spray with Sevin.

THRIPS. The browning of the tips of buds, followed by decay and dropping, has frequently been found to be due to the attacks of a species of thrips. This should be clearly distinguished from the bud drop caused by overwatering (see above).

Control. Spray with malathion.

WEEVILS. *Otiorhynchus sulcatus* and *O. ovatus.* The black vine weevil and the strawberry root weevil feed on the leaves; the larval stages feed on the roots and the base of the stem. The Japanese weevil, *Pseudocneorhinus bifasciatus,* feeds on leaves like other weevils.

Control. Spray the lower parts of the plants and the soil surface with Diazinon.

OTHER INSECTS. A great number of other insects infest camellias in greenhouses, homes, and outdoors. These include the following aphids: black citrus, melon, green peach, and ornate. The following caterpillars chew the leaves: omnivorus looper, orange-tortrix, and western parsley. The fruit tree leafroller and the greenhouse leaf tier also chew the leaves and roll or tie them together. The greenhouse whitefly is common on greenhouse-grown plants.

Control. Most of these pests can be easily controlled with malathion. The caterpillars can also be controlled with Sevin.

ROOT NEMA. *Meloidogyne incognita.* Camellias are unusually resistant to root-knot nemas, although the pests have been recorded on these plants in Texas. Another species of nema, *Hemicriconemoides gaddi,* has been reported on the roots of camellia in Louisiana.

Control. Drenching the soil around infested trees with Mocap or Dasanit may provide control.

MITES. Southern red mite (see under Holly) attacks camellia as does *Cosetacus camelliae,* an eriophyid mite.

CAMPHOR-TREE *(Cinnamomum)*

An Asian native, this tree is popular in the southern, southwestern and western United States

but does poorly in alkaline, clay soils. This tree has escaped from cultivation from Florida to Louisiana and has become naturalized in Florida.

Diseases

WILT. Verticillium wilt has been found on this tree in the San Francisco Bay area of California. See Chapter 12.

ROOT ROT. Clitocybe and Armillaria root rots have also been reported for this host. See Chapter 12.

ANTHRACNOSE. *Glomerella cingulata (Gloeosporium).* Leaf spot, canker, and shoot blight are symptoms of this disease.

POWDERY MILDEW. This disease is caused by *Microsphaera alni.* See Chapter 12.

THREAD BLIGHT. *Ceratobasidium stevensii.* This occurs in Louisiana.

Insects

SCALES. Sixteen species of scale insects infest camphor-trees in Florida. The most destructive one, which can be fatal to this host, is the camphor scale, *Pseudaonidia duplex.* It is circular in shape, convex, dark blackish brown, and $\frac{1}{10}$ inch across.

MITES. Other pests of the camphor-tree are three species of mites—avocado red, plantanus, and southern red—and camphor thrips, *Liriothrips floridensis.*

Control. Summer oil emulsion, applied in February in the Deep South and repeated in 30 days, will control the scale insects. Mites are controlled with Kelthane. Malathion will control the camphor thrips.

CAROB *(Ceratonia)*

This tree grows in Mediterranean climates and its pods are made into a chocolate substitute. Used as a screen or shrubby hedge, carob is somewhat messy and its male flowers are foulsmelling. Carob is susceptible to Verticillium wilt and attacked by oleander scale.

CATALPA *(Catalpa)*

Two species of catalpa, common and western (northern), are used in ornamental plantings.

Common catalpa (*C. bignonioides*) is a messy tree and should be planted only where its flowers and fruits are not objectionable. Western catalpa (*C. speciosa*) is hardy and grows rapidly. Several leaf spots and powdery mildews, in addition to the Verticillium wilt disease (Chapter 12), occur on these hosts. Insects are usually more of a problem than diseases.

Diseases

LEAF SPOT. A common disease of catalpa, leaf spot generally appears during rainy seasons.

Symptoms. Tiny water-soaked spots, scattered over the leaf, appear in May. The spots turn brown and increase in size until they attain a diameter of about ¼ inch. Holes in the leaves are common as a result of dropping out of the infected tissue. Where the spotting is unusually heavy, the leaves may drop prematurely.

Cause. Three fungi, *Phyllosticta catalpae, Gloeosporium catalpae,* and *Cercospora catalpae,* are frequently associated with these spots (Fig. III-12). Injury by the catalpa midge, discussed below, and infection by bacteria are believed to increase the susceptibility to leaf spots. *Alternaria catalpae* may be a secondary invader.

Control. Gather and destroy the leaves in fall. Spray valuable susceptible trees three times with bordeaux mixture or any other copper fungicide, first as the leaves unfurl, then when the leaves are half-grown, and again when they are full grown.

POWDERY MILDEW. Two species of mildew fungi, *Microsphaera elevata* and *Phyllactinia guttata,* attack catalpas.

Control. Spray valuable trees with Benlate or wettable sulfur.

WILT. Verticillium wilt is an important problem. See Chapter 12.

WOOD DECAY. The western (northern) catalpa is extremely susceptible to heartwood decay caused by the fungus *Trametes versicolor.* The heartwood becomes straw-yellow, light, and spongy. Another fungus, *Polyporus catalpae,* causes decay of the trunk near the soil line. The wood becomes brown, tough, brittle, and full of cracks.

Control. Avoid wounds, inasmuch as the

Fig. III-12. Leaf spot caused by the fungus *Phyllosticta catalpae.*

spores of both fungi enter through bark injuries. Keep the trees in good vigor by fertilizing and watering.

OTHER DISEASES. Among other diseases of catalpa are twig dieback, caused by *Botryosphaeria dothidea,* and root rot by *Armillaria mellea* and *Phymatotrichum omnivorum.*

Control. Control measures have not been developed.

Abiotic Disease

AIR POLLUTION. Catalpa is relatively sensitive to sulfur dioxide. See Chapter 10.

Insects

COMSTOCK MEALYBUG. *Pseudococcus comstocki.* This small, elliptical, waxy-covered insect attacks catalpa primarily, but it is also found on apple, boxwood, holly, horsechestnut, magnolia, maples, osage-orange, poplar, and Monterey pine. After hatching in late May, the young crawl up the trunk to the leaves, where they suck out the juices and devitalize the tree. Twigs, leaves, and trunks may be distorted as a result of heavy infestations. The eggs winter in bark crevices or in large masses hanging to the twigs.

Control. Before the buds open, spray with 1 part concentrated lime sulfur in 10 parts of water, or with winter-strength dormant oil. A malathion or Sevin spray applied to the trunk and branches in late spring and again in early summer will provide additional control.

CATALPA MIDGE. *Cecidomyia catalpae.* Leaves are distorted, and circular areas inside the leaves are chewed, leaving a papery epidermis, as a result of infestation by tiny yellow maggots. The adult, a tiny fly with a wingspread of $\frac{1}{16}$ inch, appears in late May or early June to lay eggs on the leaves. Winter is passed in the pupal stage in the soil.

Control. Cultivate the soil beneath the trees to destroy the pupae, and spray in late May with malathion.

CATALPA SPHINX. *Ceratomia catalpae.* Leaves may be completely stripped from a tree by a large yellow and black caterpillar, the sphinx, which attains a length of 3 inches. The adult female is a grayish brown moth with a 3-inch wingspread. The winter is passed as the pupal stage in the ground.

Control. Spray the foliage with Orthene or Sevin early in June when hornworms are small and again in mid-August.

WHITEFLY. *Tetraleurodes* sp. Infestation is often followed by the appearance of sooty mold.

SCALE. See white peach scale, under Cherry.

CEDAR *(Cedrus)*

This genus contains several very beautiful evergreen trees like the Atlas cedar (*C. atlantica*) and the cedar of Lebanon (*C. libani*). Cedars are relatively free of pests and diseases. These "true cedars" produce cones, unlike juniper, false cypress (also called white cedar), and arborvitae, which are sometimes called cedars.

Diseases

TIP BLIGHT. The fungus *Sphaeropsis sapinea,* formerly called *Diplodia pini,* occasionally causes canker and dieback of branch tips in the South.

Control. The same as for tip blight of pine.

ROOT ROT. Several fungi, including *Armillaria mellea, Clitocybe tabescens,* and *Phymatotrichum omnivorum,* are associated with root and trunk decay. The last attacks a great variety of trees, shrubs, and ornamental and food plants in the South.

Control. No effective, practicable control measures are known.

Insects

BLACK SCALE. *Saissetia oleae.* This dark brown to black scale is primarily a pest of citrus on the West Coast. It attacks a wide variety of trees and shrubs in the South and West, including the Deodar cedar (*C. deodara*). Besides extracting juice from the plant, it secretes on the leaves and stems a substance on which the sooty mold fungus grows. Cottony-cushion, greedy, and latania scales have also been reported on cedar.

Control. Sevin or malathion sprays when the young are crawling about in May and June are effective.

DEODAR WEEVIL. *Pissodes nemorensis.* This brownish snouted weevil feeds on the cambium of leader and side branches of Deodar, Atlas, and Lebanon cedars. It deposits eggs in the bark, and the $\frac{1}{3}$-inch-long white grubs which hatch from the eggs burrow into the wood. Eventually the leaders and terminal twigs turn brown and die. Small trees may be killed by this pest.

Control. A methoxychlor spray applied in April, when the beetles are feeding, will control this insect.

OTHER INSECTS. Red-headed pine sawfly, *Neodiprion lecontei;* bagworm (see under Juniper); and mealybug, *Pseudococcus longispinus,* attack cedar.

CEDAR, INCENSE *(Libocedrus)*

This handsome columnar or narrow, pyramidal tree grows mainly in the northwestern United

States, although it can be grown in warmer parts of the country. Useful as a tall hedge or windbreak, the foliage of this tree is aromatic when crushed.

Diseases

The bacterial crown gall caused by *Agrobacterium tumefaciens*, blight by the fungus *Herpotrichia nigra*, branch canker by *Coryneum cardinale*, needle cast by *Lophodermium juniperinum*, root rot by *Phymatotrichum omnivorum*, and rust by *Gymnosporangium libocedri* are recorded on incense cedar. The latter forms conspicuous witches' brooms, killing small sprays of foliage.

Control. Control measures are rarely needed.

Insects

The cypress bark beetle, cypress tip moth, cypress mealybug, and four species of scales—cypress, juniper, pine needle, and Putnam—infest incense cedar.

Control. Effective controls for the cypress bark beetle have not been developed. Mealybug and scales can be controlled with Sevin or malathion sprays.

Other Pests

The mistletoe *Phoradendron juniperinum* f. *libocedri* infests incense cedar in the western states.

Control. No control is known.

CEDAR, WHITE. See False Cypress.

CHAMAECYPARIS. See False Cypress.

CHASTE-TREE *(Vitex)*

This low-growing tree or shrub produces showy purple clusters of blooms in late summer in the North and earlier in the South.

Diseases

LEAF SPOT. *Cercospora viticis* This disease occurs on chaste-tree along the Gulf Coast.

Control. The disease is rarely severe enough to warrant control measures.

ROOT ROT. *Phymatotrichum omnivorum.* Root rot is present on chaste-tree in Texas.

Control. Control measures are not practicable.

Insects

SCALES. Pit-making pittosporum and oleander scales attack chaste-tree.

CHERRY, JAPANESE FLOWERING, BLACK, AND CHOKE *(Prunus serrulata, P. yedoensis, P. serotina, P. virginiana, and related species); also FLOWERING ALMOND (P. dulcis), FLOWERING PEACH (P. persica) and FLOWERING PLUM (P. cerasifera).*

Flowering *Prunus* species are used extensively as ornamentals. In latitudes north of southern New Jersey they are occasionally damaged by low winter temperatures, as is evidenced by longitudinal cracks (Fig. III-13) on the south or west side of the trunk or in the branch crotches. Flowering *Prunus* species are best planted in fall. Native black cherries, which develop into large trees, are adapted throughout the eastern United States.

Landscape cherries, peaches, plums, and almonds are susceptible to some of the fungi and insect pests that attack the *Prunus* species grown for fruit. These parasites, however, have not been studied in detail on landscape varieties, and any treatment must be based on the control measures suggested for fruit crops. Information on the diseases and insects of fruiting *Prunus* is readily available from the state agricultural experiment stations.

Bacterial and Fungus Diseases

SHOT-HOLE. *Xanthomonas pruni.* The bacterium that attacks peaches and cherries in orchards is known to attack Japanese cherries also, causing a familiar "shot-hole" appearance. The infected tissue dries up and falls out, leaving a hole about ⅛ inch in diameter. This bacterium is also capable of causing stem canker

Fig. III-13. Winter injury to flowering cherry. This species is particularly susceptible in eastern areas north of Philadelphia.

and gummosis, although other bacteria such as *Pseudomonas syringae* are more often involved. Shot-holes in cherry leaves may also be caused by the fungus *Coccomyces hiemalis*, discussed below, and by virus infection.

Control. Where the *Xanthomonas* bacterium causes serious damage to flowering cherries, spray with Cyprex, mentioned below under leaf spot.

LEAF SPOT. *Blumeriella jaapii,* formerly called *Coccomyces hiemalis.* During rainy springs this disease is widespread. The reddish spots on the leaves drop out, leaving circular holes. Complete defoliation may follow.

Control. Benlate, Cyprex, Captan, or Ferbam sprays will control this rather prevalent disease. The first application should be made when the flower petals fall, followed by two more applications at 2-week intervals.

POWDERY MILDEW. *Podosphaera clandestina and Sphaerotheca pannosa.* The Japanese cherry is subject to the same powdery mildew that attacks edible cherries. The leaves and twigs become coated with a mat of fungus

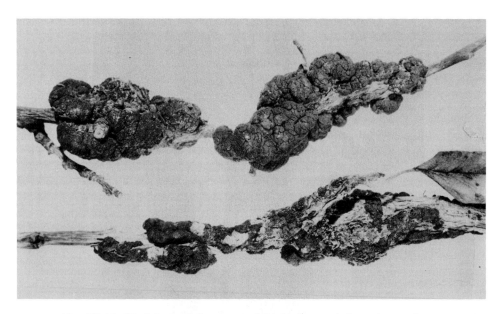

Fig. III-14. Black knot of cherry caused by the fungus *Apiosporina morbosa*.

growth, which causes dwarfing and death of these branches. The disease is uncommon.

Control. Benlate sprays will control this disease.

BLACK KNOT. *Apiosporina morbosa.* Black, rough cylindrical-shaped galls (Fig. III-14) develop on the twigs of apricots, cherries, and plums. Neglected trees appear to be especially subject to this disease. Wild black cherries are susceptible to black knot, and when growing near the landscape, they may be an important source of inoculum for disease of domestic *Prunus.*

Control. Prune knotted twigs and excise knots on large branches during the winter. Then spray with zineb or Ferbam when the trees are dormant, at pink bud stage, at full-bloom stage, and 3 weeks later. Lime sulfur and Tribasic Copper Sulfate are registered for use.

WITCHES' BROOM. *Taphrina wiesneri.* Japanese flowering cherry seems to be susceptible to this disease. Large branches will sometimes become deformed by development of many irregular dwarfed branches to form a witches' broom. Blossoms develop and leaves come out on the brooms earlier than on the normal branches. Sometimes large numbers of very small brooms develop all over the tree, killing the end branches and eventually the whole tree. *T. deformans* causes peach leaf curl, and *T. communis,* plum pockets.

Control. Cut off and destroy the brooms. Spray Ferbam in fall or in early spring.

CROWN GALL. *Prunus* species are susceptible. See Chapter 12.

SHOOT AND TWIG BLIGHT. *Monilinia fructicola and M. laxa.* The fungi cause flowers to turn brown and rot in moist weather and also cause leaf blight and fruit brown rot.

Control. Prune and destroy infected twigs. To prevent disease, spray with Benlate or Captan as flowers open, and repeat in 10 days.

WILT. Verticillium wilt is found in this genus. See Chapter 12.

PERENNIAL CANKER. *Cytospora cincta and C. leucostoma.* These fungi cause serious damage to *Prunus.*

Control. Prune out cankered twigs and branches.

BACTERIAL LEAF SPOT AND TWIG CANKER. *Pseudomonas syringae* pv. *syringae.* This common epiphyte infects *Prunus* species and is also involved in ice nucleation.

ROOT ROT. *Prunus* species are susceptible to

attack by *Armillaria mellea.* See Chapter 12. *Xylaria mali,* the dead man's finger fungus, also decays cherry roots.

WOOD DECAY. *Phellinus tuberculosus.* See Chapter 12.

Virus Diseases

Prunus species are susceptible to mosaic, ringspot, green ring mottle, and stem pitting viruses. Virus infection of fruit trees can be prevented by careful sanitary practices during propagation. Once landscape trees are infected, there is no control.

Mycoplasma Diseases

Peach yellows and X-disease are important mycoplasmal diseases of *Prunus.* Yellowing, tufted growth, leaf roll and shothole, and decline are some symptoms of these diseases. Control measures are unknown.

Abiotic Disease

YELLOWING. Very often yellowing and premature defoliation of flowering cherry occur without previous spotting of the leaves. These symptoms are associated with excessively wet or dry soils or with low-temperature injuries to the crown and roots. As a rule, a second set of leaves is formed after the premature defoliation in spring or early summer. The new leaves are normal to all appearances.

Insects and Other Animal Pests

PEACH TREE BORER. *Sanninoidea exitiosa.* The grubs of the peach borer cause a great amount of damage to flowering peach as well as to related forms, frequently causing death of the trees. The damage is marked by profuse gummosis at the crown and on the main roots just below the surface of the soil. The trees fail to grow properly and the leaves turn yellowish. The frass or borings of the grubs becomes mixed with the gum. If one follows down the burrows with a chisel or a penknife, white flat grubs from ½ to 1 inch long, with brown heads, may be found.

Control. The grubs must be removed from young trees by hand grubbing; that is, the ground should be pulled back from the base of the tree and the grubs dug out with the aid of a grubbing chisel or other implement with a curved blade. The burrows should be followed down until the grubs are found. Young trees up to 3 years old should be examined twice a year. In trees over 3 years old, the grubs can be controlled by the use of paradichlorobenzene (PDB). The soil should be dug away from the base of the tree to a depth of 3 or 4 inches and the cavity leveled off. For trees 3 years old, apply ½ to ¾ of an ounce of the fumigant in a circle around the base. For older trees, use 1 to 1½ ounces. The soil is then replaced in the hole and mounded around the base of the tree. While paradichlorobenzene may be applied during the summer, it is safer to use it in late September of October. If it is applied in the summer, the soil should be removed about 1 month or 6 weeks later. If it is allowed to remain too long in the soil during hot weather, injury is likely, especially to young trees.

Spraying or painting the trunk with methoxychlor in early July and repeating three times at 2-week intervals is a good preventive treatment.

LESSER PEACH TREE BORER. *Synanthedon pictipes.* This borer attacks growing tissue anywhere in the trunk from the ground to the main branches. The adult, a metallic, blue-black, yellow-marked moth, emerges in June in the vicinity of New York City and in May in the South to deposit eggs on the bark higher up in the tree than the peach borer.

Control. The adult female of this borer begins to deposit eggs earlier than does the peach tree borer. Hence the Lindane applications should be started by mid-June. The insecticide should be applied to the main branches as well as to the main trunk. The other applications should follow at about 3-week intervals.

BORERS. Shot-hole and peach bark beetle borers, *Scolytus rugulosus* and *Phloeotribus liminaris,* attack weakened trees; leopard moth larvae (see leopard moth borer, under Maple) also tunnel into wood. American plum borer (see under Planetree) also attacks *Prunus.* Another clear-winged moth, the dogwood borer, also attacks *Prunus* (see borers, under Dogwood).

ORIENTAL FRUIT WORM. *Graphiolitha molesta.* Wilting of the tips of twigs may be due to boring by the oriental fruit worm, a small pinkish white larva, about ½ inch long. The adult female is about ½ inch long and is gray with chocolate brown markings on the wings. Larvae overwinter in the soil. Oak twig pruner and twig girdler each attack *Prunus*.

Control. Where only a few trees are involved, removal and destruction of wilted tips as they appear is usually sufficient. Large numbers of trees can be protected by periodic applications of methoxychlor sprays to which a compatible mite killer such as Kelthane has been added. Applications should be started as soon as the leaves begin to emerge and should be repeated twice at 10- to 12-day intervals.

PEAR SLUG. *Caliroa cerasi.* These so-called slugs, olive green, semitransparent, and slimy, are the larvae of a sawfly. They are about ½ inch long, swollen at the front, and shaped somewhat like a tadpole (Fig. III-15). They occasionally infest *Prunus* and may completely skeletonize the leaves. There are two generations a year in the northern states and three in the southern.

Control. Spray Sevin or Diazinon twice at 2-week intervals starting 2 weeks after bloom.

PLANTHOPPERS. *Metcalfa pruinosa* and *Ormensis septentrionalis.* These planthoppers injure shrubs and trees by sucking the juices of

Fig. III-15. Pear slug sawfly larvae skeletonizing a cherry leaf.

Fig. III-16. Planthoppers *(Ormensis septentrionalis)* on a cherry branch.

the more tender branches, which they cover with a woolly substance. The former, also called mealy flata, is ¼ inch long with purple-brown wings and covered with white woolly matter. It also attacks mulberry. The latter is a beautiful blue-green covered with white powder; it is illustrated in Fig. III-16. These insects are about ½ inch long, are very narrow, and also feed on hawthorn. The rose and other leafhoppers and buffalo tree hoppers also injure *Prunus* foliage.

Control. The problem is rarely serious enough to warrant control measures.

EASTERN TENT CATERPILLAR. See under Willow. Wild cherries are the natural hosts of the tent caterpillar. See Fig. III-17.

OTHER CATERPILLARS. *Dichomeris ligulella,* a juniper webworm relative; yellow-necked caterpillar (see under Oak); red-humped caterpillar (see under Poplar); fall webworm (see under Ash); forest and California tent caterpillars (see under Maple and Strawberry-Tree); and ugly nest caterpillar, *Archips cerasivoranus,* are all *Prunus* foliage feeders.

CAMBIUM MINER. See under Holly.

LEAF MINERS. California casebearer, *Coleo-*

Fig. III-17. Eastern tent caterpillars on wild cherry. Egg masses on twigs.

phora sacramenta; and cherry leaf miner, *Paraleucoptera heinrichi,* a holly leaf cherry pest, have been reported.

MITES. Spindle gall and marginal fold-forming eriophyid mites are found on cherry leaves.

SAN JOSE SCALE. *Aspidiotus perniciosus.*

Gray, closely appressed masses of circular scales, $\frac{1}{10}$ inch in diameter, with a raised nipple in the center, are characteristic of the San Jose scale. The winter is passed in the immature stages on the bark.

Control. Dormant oil sprays or lime sulfur

1-10 on the trunk, branches, and twigs before the buds open in spring is effective. A more complete control can be obtained by following with Sevin, Diazinon, or Orthene sprays in May and June.

WHITE PEACH SCALE. *Pseudaulacaspis pentagona.* The presence of this scale is indicated by white incrustations on the bark, masses of circular white scales $\frac{1}{10}$ inch in diameter, which suck juices from below the bark. Other *Prunus* scales include cottony-cushion, cottony maple, black, hickory lecanium, terrapin, European peach, globose, greedy, frosted, oystershell, oleander, walnut, euonymus, and Comstock mealybug.

Control. While the tree is dormant, spray with 1 part concentrated lime sulfur in 10 parts of water, or with dormant strengths oils. Diazinon, methoxychlor, or Sevin sprays applied when the young are crawling about in early summer will also control these pests.

GREEN PEACH APHID. *Myzus persicae.* This insect has been found to be very injurious during the summer to a number of varieties of flowering peach and related ornamentals.

Control. Spray with malathion, or Sevin when insects first appear, and repeat a week later if necessary.

BLACK PEACH APHID. *Brachycaudus persicae.* This aphid is common on commercial plantings of peaches.

Control. The same as for green peach aphid.

WATERLILY APHID. *Rhopalosiphum nymphaeae.* Ornamental species of *Prunus* grown near waterlily ponds are seriously attacked by the waterlily aphid. The insects migrate to the trees in May and June and in autumn. Apple aphid, *Aphis pomi,* also attacks *Prunus.*

Control. Spray with malathion when the insects appear on the leaves.

ROOT NEMA. *Pratylenchus penetrans.* Research at the New York State Experiment Station revealed that this nema attacks the roots of edible cherry trees. It is possible that the roots of Japanese flowering cherries are also susceptible to the same nema.

ASIATIC GARDEN BEETLE. *Maladera castanea.* The leaves are chewed during the night by a brown beetle $\frac{1}{4}$ inch long. During the day the insect hides just below the soil surface.

Control. Sevin sprays will keep this pest in check.

JAPANESE BEETLE. See under Linden.

OTHER BEETLES. Fuller rose beetles (see under Camellia) also attack *Prunus.*

CHESTNUT *(Castanea)*

Although the majestic American chestnut is no longer a part of the natural landscape in eastern United States, substitutes such as Chinese chestnuts are occasionally planted as landscape and backyard nut trees.

Diseases

BLIGHT. The rapid disappearance of one of our best forest, ornamental, and nut tree, the American chestnut, as a result of infection by one of the most virulent tree parasites, is too well known to warrant much discussion in this book. Despite tireless efforts and tremendous monetary expenditures, dead and diseased chestnut trees are all that remain of the losing battle man has waged to check this invader. Although there are few optimistic signs, some say it is too early to predict the ultimate fate of this valuable tree. They point out that cankers on diseased trees appear to develop less rapidly than in former years and that more and more partly healed cankers are visible. Other hopeful signs, they assert, are the fact that sprouts continue to develop at the base of dead stumps and from the roots, some attaining a height of more than 30 feet and a trunk diameter of 8 or more inches, and the fact that thousands of other sprouts, though not attaining so large a size, are able to produce nuts before they succumb.

No one will dispute the statement that chestnut blight disease has done more than any other single factor in American history to make the public tree-conscious. Within the span of 80 years many people witnessed the passing of this irreplaceable tree. Believed to be of minor importance when first reported by the late Hermann Merkel, who found a few infected trees in Bronx Park, New York City, in 1904, the disease proceeded to wipe out the chestnut stands in New England forests and along the eastern slopes of the Allegheny and Blue Ridge mountains, the principal range of this host. Today some chestnuts still stand at the fringes or outside of this tree's natural range: in Michigan, Oregon, and Tennessee. It is safe to say, however, that they too will soon suffer the same fate as their eastern kin, for blight has been reported in these states.

Though blight is essentially a disease of the American chestnut, it commonly occurs on the chinquapin (*Castanea pumila*) as well. The causal fungus has been found growing on red maple, shagbark hickory, live oak, and staghorn sumac (*Rhus typhina*), and on dead and dying white, black, chestnut, and post oaks.

Symptoms. The only evidences of the once magnificent chestnut trees are the tremendous barkless trunks that still stand among living trees of other species, or the rotting stumps around whose edges vigorous living sprouts and small trees continue to appear. Many young trees seem to escape the disease for the first dozen years of their lives. Whether this is owing to their smooth bark, which is relatively free from injury during this period, or whether they have some degree of resistance in infancy is not definitely known. Once the destructive fungus gains a strong foothold, however, the leaves on one branch, or several branches, or even on the entire young tree suddenly wilt, turn brown, and hang dry on the branches (Fig. III-18). The death of the leaves and of the stems to which they are attached results from the death and girdling of some portion below. When the infection occurs high in the tree, only the distal

Fig. III-18. Chestnut blight on young sprouts. The leaves hang dry and brown on girdled stems. Notice the stump of the original tree.

parts are affected. When it develops at the base near the soil line, the entire top of the tree dies.

Close inspection of the lower parts of dying twigs, branches, or trunks reveals the presence of cankers. These are discolored, slightly sunken areas in the bark. Layers of flat, fan-shaped, buff-yellow wefts of fungus tissue are usually revealed between the bark and the sapwood when the bark in the cankered area is lifted or peeled carefully.

Cause. Blight is caused by the fungus *Endothia parasitica.* It is sometimes referred to as *Cryphonectria parasitica.* When a considerable bark area has been invaded, tiny, pinpoint fruiting bodies, the tips of which barely protrude from the bark, are developed. In damp weather, sticky yellowish orange masses of spores ooze from the openings in these bodies. The spores are splashed by rain or are carried by birds and by crawling and flying insects to wounds in the bark below or to nearby trees, where they cause new infections. A second type of fruiting structure is also developed. This is a small, black, flask-shaped body embedded in the bark with a relatively long beak protruding above the surface. The spores from this type are shot into the air and are blown for many miles by the wind. In fact, spore traps placed on high buildings in New York City, when the epidemic was at its peak years ago, caught many spores of this type. The windblown spores lodge in open wounds on chestnut trees and produce new infections when conditions are favorable. The mycelium, the vegetative stage of the fungus, penetrates the bark, cambium, and sapwood with considerable rapidity and soon completely girdles the infected twig, branch, or trunk. Chestnut blight has been unchecked because it is impossible to prevent the dissemination of spores by wind, birds, and insects.

Control. Although many suggestions and recommendations have been published, none has proved effective in controlling chestnut blight. Persistent efforts have been made to find some chemical that when injected into the tree would check the development of cankers, but none has been found.

Recent research by plant pathologists at several experiment stations indicates that cankers can be restricted by the introduction of a hy-

povirulent (weak) parasitic strain of *Endothia parasitica* into the canker. Ways are still being sought to transmit the hypovirulent strains to cankered trees so that the fungus causing the disease can become weakened too. Other lines of research suggest that a moist soil compress prepared from soil beneath the tree and affixed to cankers on stems and branches will cause canker remission. The extensive labor involved and possible retreatment needed makes this method impractical on a large scale.

After 70 years of research in this country, no American chestnut has been found with sufficient resistance to be of practical value.

When research scientists finally realized that the chestnut blight fungus was uncontrollable, substitutes were sought. The answer seemed to be the introduction of Asiatic chestnuts, which were known to be resistant to blight. Although a few of these chestnuts were introduced earlier, the greatest numbers were introduced in the late 1920s. But their introduction resulted in the appearance of other diseases, such as the blossom-end rot of the nuts caused by the fungus *Glomerella cingulata* and twig canker, discussed below.

Asiatic chestnuts (*C. japonica* and *C. mollissima*) are now widely available from many nurseries. Most of them are best suited for ornamental plantings and nut production rather than as forest trees. Many have a shrublike growth habit, with multiple trunks arising near the ground level. They do not grow as tall and straight as the American chestnut and cannot compete on wooded areas when interplanted with other trees. The varieties of *C. mollissima* that are best for orchard cultivation are Abundance, Kuling, Meiling, and Nanking.

TWIG CANKER OF ASIATIC CHESTNUTS. The increased use of Asiatic species of chestnuts, because of their resistance to blight, has revealed that these species are susceptible to a less destructive disease, twig canker.

Symptoms. Cankers on trunks, limbs, or twigs result from fungus invasion of the bark, cambium, and sapwood. Cankers on the branches may cause complete girdling, followed by death within a single season. When complete girdling does not take place in one season, the canker may callus over temporarily, and the infection then may continue the following season, to cause death of the parts above the canker.

Cause. Several fungi have been associated with this disease, the most common and virulent one being *Cryptodiaporthe castanea.* It has been found to kill young trees in nurseries and older trees in permanent plantings, but more commonly it kills individual branches, thus decreasing growth and deforming the trees.

Control. Twig canker is most prevalent on trees in poor vigor. Maintaining good vigor will do much to ward off attacks. Planting sites should be carefully selected and fertilization and watering practiced to ensure vigorous growth. All unnecessary injuries to the trees should be avoided, since the causal organisms penetrate and infect most readily through bark wounds. Trees should be carefully inspected in early summer, when symptoms are most apparent, and affected twigs should be pruned to sound wood and destroyed. Large cankers on the trunk or larger branches should be removed by surgical methods.

OTHER DISEASES. The fungus *Monochaetia kansensis* causes a leaf spot of *Castanea mollissima* in Kansas and Mississippi. Powdery mildews, *Microsphaera americana* and *Phyllactinia guttata,* also attack chestnut.

Insects

The insect pests which attack American chestnut are of no importance because of the destruction of the trees by blight. Following, however, are some insects that can attack living American and Asiatic chestnuts.

WEEVILS. *Curculio auriger* and *C. proboscideus.* Two species of weevils native to the United States may seriously damage the nuts of the Asiatic species of chestnuts. Asiatic oak weevil (see under Oak) feeds on chestnut.

Control. Spray the trees with methoxychlor before the weevils lay eggs in August. Treat the soil beneath Asiatic chestnut trees with ethylene dibromide in spring to kill the larval stage. Nuts harvested from infested trees which have not been treated as described should be enclosed with methyl bromide in a tight container for 3 hours.

CATERPILLARS. Yellow-necked caterpillar (see under Oak) and spring cankerworm and white-marked tussock moth (see under Elm) feed on chestnut foliage.

BORERS. Brown wood borer (see under Beech), leopard moth borer (see under Maple), and two-lined chestnut borer (see under Oak) all attack chestnut.

JAPANESE BEETLE. See under Linden.

TWIG PRUNER. See oak twig pruner, under Hickory.

SCALES. Obscure and lecanium scales (see under Oak) attack chestnut.

GIANT BARK APHID. See aphids, under Sycamore.

SPIDER MITES. See oak mite, under Oak.

CHINABERRY *(Melia)*

This southeast Asia native, which thrives in difficult growing conditions, is planted in the southeastern United States for quick, dense shade.

Diseases

Chinaberry is susceptible to *Peronospora parasitica* downy mildew, *Phyllactinia guttata* powdery mildew, and *Cercospora meliae* and *C. subsessilis* leaf spots. Controls for these diseases are normally not needed.

Insects

Two scales, greedy and white peach, feed on this host.

CHINESE JUJUBE *(Ziziphus)*

This low-growing tree is adapted to dry, alkaline, salty soils.

CHINESE PISTACHIO *(Pistacia)*

This tree is widely adapted in the United States and tolerates summer heat. Chinese pistachio has no serious disease or insect problems.

CHINESE TALLOW TREE *(Sapium)*

This medium-sized tree, adapted to mild climates, grows best in moist acid soils. It grows rapidly and is relatively free of diseases and insect pests.

CITRUS *(Citrus)*

Citrus trees are grown in backyard landscapes as well as in commercial orchards in the frost-free areas of the southern and western United States.

Diseases

BACTERIAL DISEASES. Citrus foliage is susceptible to two foliar bacterial diseases: Citrus blast, *Pseudomonas syringae,* in California; and the dreaded citrus canker, *Xanthomonas citri.* In the latter case, even patio citrus trees are subject to intense scrutiny from plant industry officials because of citrus's commercial importance.

LEAF SPOTS. Foliar diseases caused by fungi include scab, *Elsinoe fawcetii;* melanose, *Diaporthe citri;* leaf spot, *Septoria citri* or *S. limonum;* grease spot, *Cercospora citri-grisea;* tar spot, *C. gigantea;* and anthracnose, *Glomerella cingulata.*

TRUNK AND ROOT DISEASES. Stems and roots are affected by brown rot gummosis or foot rot caused by several *Phytophthora* species. *Clitocybe tabescens* and *Armillaria mellea* also attack citrus roots.

OTHER DISEASES. Citrus is also susceptible to a spiroplasma, several virus diseases, and to deficiencies of the mineral nutrients copper, zinc, manganese, magnesium, boron, and iron.

Insects

BAGWORM. *Oeketus abbotii.* Bagworm feeds on citrus.

WEEVILS. Fuller rose beetle (see under Camellia) and citrus root weevil attack citrus.

APHIDS. Several types including black citrus, spirea, potato, and cowpea feed on citrus, often producing sufficient honeydew to feed sooty mold fungi.

WHITEFLIES. Greenhouse whitefly, *Trialeurodes vaporariorum,* and woolly whitefly, *Aleurothrixus floccosus,* withdraw sap from leaves and produce honeydew.

SCALES. The following attack citrus: long-tailed and Comstock mealybug, cottony-cushion, black,

hemispherical, ivy, California red, Florida red, dictyospermum, citrus, snow, and tea.

OTHER INSECTS. A treehopper, leaf-footed bug, and greenhouse and Cuban laurel thrips attack citrus.

CORK TREE *(Phellodendron)*

Cork trees grow well in cities and are unusually free of pests. The only insects recorded on this host are the lesser snow scale, *Pinnaspis strachani,* and the pustule scale, *Asterolecanium pustulans.*

Malathion or Sevin sprays applied in late spring or early summer will control the crawler stages of these pests.

COTTONWOOD. See Poplar.

CRABAPPLE, FLOWERING *(Malus)*

Over 600 species and varieties of flowering crabapple are grown in North America. Flowering crabapple, needing some winter chilling, is widely adapted except in southern California and other mild weather areas. Flowering crabapple trees used for ornamental purposes are subject to attack by many of the fungi and insects that occur on commercial fruiting apples.

Some species and cultivars that are unusually resistant to the five most common diseases—scab, fireblight, cedar–apple rust, powdery mildew, and black rot—include 'Adams', 'Albright', 'Baskatong', 'Christmas Holly', 'Cotton Candy', 'David', 'Dolgo', 'Henry Kohankie', 'Jewelberry', 'Prof Sprenger', *Malus* × *robusta* c. *Persicifolia,* 'Robinson', *sargentii* cv. 'Tina', 'Sentinel', and 'Sugartime'. There are several other excellent crabapples such as 'Beverly', 'Coralburst', 'Liset', *rocki,* and 'Snowdrift', which are susceptible only to one of the important diseases. They would be good choices in areas where the disease in question is rare. Many county extension offices have information on the best cultivars to grow locally.

Bacterial and Fungus Diseases

FIRE BLIGHT. *Erwinia amylovora.* This disease will not be serious most seasons on the flowering crab if commercial orchards of pears and apples are not nearby.

Control. See under fire blight, in Chapter 12. Use disease-resistant crabapples listed above.

BLACK ROT. *Botryosphaeria obtusa (imperfect stage, Sphaeropsis malorum).* This fungus causes cankers on the trunks of crabapples. It often gains entrance through wounds made by lawn mowers and other maintenance equipment. This fungus also causes a leaf spot called "frogeye" and a fruit rot disease.

Control. Avoid wounding trees. Increase the tree's vigor by fertilizing and watering. Fungicide sprays will control leaf spot.

RUST. *Gymnosporangium juniperi-virginianae.* When common red cedar with cedar-apple rust galls is grown near the landscape, a number of brown or orange spots, bearing the aecial stage of the rust, may later be very conspicuous on the leaves of flowering crabapples nearby. Much defoliation may follow heavy infection.

Control. If practicable, eliminate any red cedars *(Juniperus)* within as great a distance as possible (up to a mile) or spray with a fungicide such as Bayleton, Ferbam, Daconil 2787, mancozeb, Zyban, or Thiram. Four to five applications at 7- to 10-day intervals are needed, starting in spring when the orange fungus masses appear on junipers. For most effective results, the sprays should be applied before rainy periods.

SCAB. *Venturia inaequalis.* Olive-drab spots ¼ inch in diameter appear on the leaves, which drop prematurely, and the fruits are disfigured. The imperfect stage of the fungus, *Fusicladium dendriticum,* overwinters on the twigs and fallen leaves. Many of the old-fashioned flowering crabapples such as 'Almey' and 'Hopa' are so susceptible to scab that in wet seasons, they are nearly defoliated by midsummer.

Control. Plant disease-resistant crabapples. Spray with Benlate, Daconil 2787, Captan, or mancozeb four or five times at 10-day intervals starting at budbreak. We found that three sprays of Benlate at 3-week intervals starting at budbreak provided good control. We also learned that susceptible trees provided with good scab control bloomed more profusely the following season than those allowed to become defoliated.

POWDERY MILDEW. *Podosphaera leucotricha.* Leaves and terminal shoots of many crabapple cultivars are attacked by this fungus.

Control. See Chapter 12.

OTHER DISEASES. Flowering crabapples are also subject to thread blight (Fig. III-19), *Ceratobasidium stevensii;* canker, *Botryosphaeria dothidea;* collar rot, *Phytophthora cactorum;* and crown gall, *Agrobacterium tumefaciens.* Virus diseases causing leaf mosaic and stem

Fig. III-19. Flowering crabapple thread blight.

and limb abnormalities are also found on flowering crabapple.

Abiotic Disease

AIR POLLUTION. Crabapples are sensitive to sulfur dioxide. See Chapter 10.

Insects and Related Pests

Most insects that infest the leaves and twigs, such as aphids, alder lace bug, leafhoppers, and several kinds of caterpillars, are easily controlled with malathion if applied when the young are feeding. Oystershell, San Jose, and other scales can be controlled with a dormant oil spray, or with malathion or Sevin when the young scales are crawling about in late spring. Malathion can be safely combined in the Ferbam–sulfur spray recommended for rust control.

PERIODICAL CICADA. *Magicicada septendecim.* This pest (Fig. III-20), also known as the 17-year locust, damages branches of crabapple, apple, oak, and many other trees by making deep slits in the bark during the egg-laying period. Such branches are easily broken during windy weather. Following hatch, nymphs drop to the ground, enter the soil, and begin feeding on roots. Tree decline may be observed after many years.

Control. A spray containing 1 ounce of Sevin in 5 gallons of water gives excellent control if applied in spring and summer at the time the cicadas are in the trees.

FRUIT TREE LEAF ROLLER. *Archips argyrospilus.* This pest feeds and produces webbing on flowering crabapple in the spring. It also feeds on cherry, poplar, oak, hawthorn, and elm.

Control. Apply Imidan or Guthion sprays in late May and early June.

FOLIAR-FEEDING CATERPILLARS. Crabapple leaves are food for yellow-necked caterpillar, tussock moth, eastern tent caterpillar, and many other lepidopterous larvae.

JAPANESE BEETLE. See under Linden.

BORERS. Leopard moth and flatheaded borers (see under Maple), dogwood borer (see borers, under Dogwood) as well as several other insects feed on wood of flowering crabapple.

APHIDS. Apple aphid, *Aphis pomi,* feeds primarily on crabapple leaves; *Eriosoma rileyi* is a bark-feeding species.

SCALES. Cottony-cushion, cottony maple, black, European fruit lecanium, calico, frosted, oystershell, greedy, California red, and San Jose scales infest crabapples.

EUROPEAN RED MITE. *Panonychus ulmi.* This species also infests the leaves of black locust, elm, mountain-ash, and rose, in addition to many kinds of fruit and nut trees.

Control. A superior dormant oil applied in early spring just before new growth begins will provide control. When infestations are particularly heavy, applications of Acaraben, Kelthane, or malathion in late May, and again two weeks later, may be necessary.

PEAR SLUG. See under Cherry.

CRAPEMYRTLE (Lagerstroemia)

This beautiful shrubby tree is reliably hardy only in the South, although fine specimens are growing in protected places in the latitude of

Fig. III-20. Female periodical cicadas lay their eggs in young twigs and branches.

New York City. Crapemyrtle grows best in a hot, dry climate and in a fertile, well-drained loam.

Diseases

POWDERY MILDEW. *Erysiphe lagerstroemiae.* This mildew is most serious in the spring and fall months. The leaves and shoots are distorted and stunted. Inflorescences are also attacked, the flower buds failing to open. The shoots and leaves may be coated with white growth, the leaves tending to assume a reddish color beneath the mildew. The whole plant may sometimes be affected.

Two other species, *Phyllactinia corylea* and *Uncinula australiana,* occasionally infect this host.

Control. Bayleton or Rubigan are very effective in controlling the powdery mildew on this host. Wettable sulfur and Benlate sprays also control it. A dormant lime sulfur spray just before buds open in spring is recommended, in addition to the other sprays, where mildew is unusually prevalent. Japanese crapemyrtle has more mildew resistance.

OTHER FUNGUS DISEASES. Black spot caused by a species of *Cercospora,* tip blight by *Phyllosticta lagerstroemia,* leaf spot by *Cercospora lythracearum* and *Pestalotiopsis maculans,* and root rot by *Clitocybe tabescens* are other diseases of this host. Thread blight, caused by *Ceratobasidium stevensii,* also affects crapemyrtle.

Control. Leaf spots and tip blight can be controlled with several applications of a copper fungicide at 3-week intervals, starting when the new growth appears in spring. Root rot cannot be controlled.

Insects

CRAPEMYRTLE APHID. *Myzocallis kahawaluokalani.* This species attacks only crapemyrtle. It exudes a great amount of honeydew, on which the sooty mold fungus, *Capnodium* sp., thrives.

Control. Spray with malathion or Sevin early in the growing season.

FLORIDA WAX SCALE. *Ceroplastes floridensis.* This reddish or purplish brown scale covered with a thick, white, waxy coating tinted with pink attacks a wide variety of shrubs in the south.

Control. Spray with Diazinon, malathion, or Sevin when the crawling stage is about.

JAPANESE BEETLE. This pest does extensive damage to crapemyrtle. See under Linden.

LEAFFOOTED BUG. *Leptoglossus* sp. This insect feeds on crapemyrtle flowers.

CRYPTOMERIA *(Cryptomeria)*

Japanese cedar *(C. japonica),* an Asiatic native, is a good park tree, becoming tall and pyramidal. This tree is practically free of fungus parasites and has no important insect pests.

Diseases

LEAF BLIGHT. Leaves and twigs may be blighted by a species of *Phomopsis* during rainy seasons.

Control. Pruning out affected leaves and twigs is usually sufficient. Copper fungicides can be used on valuable specimens. Dormant oil sprays, which may cause serious damage, should never be used on this host.

LEAF SPOT. Two fungi, *Pestalotiopsis cryptomeriae* and *P. funerea,* are frequently associated with a leaf spotting of this host. Infection probably follows winter injury or some other agent.

Control. Same as for leaf blight.

CUCUMBERTREE *(Magnolia)*

This tree is widely but not frequently distributed in cool, moist forests south of New York and Michigan. It makes a handsome ornamental having a similarity to tuliptree, a close relative.

Diseases

Nectria galligena causes target cankers on stressed trees and *Phyllosticta cookei* is a leaf spot pathogen of cucumbertree.

CYPRESS* *(Cupressus)*

Arizona cypress *(C. glabra)* and Monterey cypress *(C. macrocarpa)* are well adapted to their

*Certain species of *Chamaecyparis* and *Taxodium* are also called cypress.

native regions. The former is grown as a Christmas tree in the south. Italian cypress *(C. sempervirens)* is widely planted in the western United States, where it tolerates dry soils.

Diseases

CANKER. Twigs and branches may be girdled by cankers, and the entire tree may be killed. Cypress, particularly the Monterey cypress, junipers, and Oriental arborvitae, are all subject to this disease.

Cause. The fungus *Seiridium cardinale* forms spores in branch and twig cankers. The sexual stage of the fungus is said to be a species of *Leptosphaeria. Botryosphaeria dothidea* is another canker-causing pathogen of cypress.

Control. Because the fungus spores are spread by wind-splashed rain, by pruning tools, and perhaps by insects and birds, control is difficult. Remove and destroy severely infected trees. Drastically prune mildly infected ones and spray periodically with a copper fungicide, starting at the beginning of the rainy period.

CYTOSPORA CANKER. The columnar form of Italian cypress, *C. sempervirens,* along the California coast is most subject to this disease. Occasionally it also affects the horizontal form of Italian cypress and the smooth cypress.

Symptoms. Smooth reddish brown cankers, from which resin flows, develop on young branches. Diseased bark on older branches becomes cracked and distorted with a more abundant flow of resin.

Cause. The fungus *Cytospora cenisia* f. *littoralis* causes this canker.

Control. Prune and destroy dead or dying branches. Trees with cankers on the trunks should be removed and destroyed.

OTHER DISEASES. Cypresses are subject to several other diseases: needle blight caused by *Cercospora thujina (C. sequoiae),* crown gall by the bacterium *Agrobacterium tumefaciens,* and monochaetia canker by *Monochaetia unicornis.*

Control. Control measures are rarely adopted.

TWIG BLIGHT. See Phomopsis twig blight, under Juniper.

ROOT ROT. *Clitocybe tabescens* and *Armil-*

laria mellea. These two fungi attack cypress. See Chapter 12.

Insects

CYPRESS APHID. *Siphonatrophia cupressi.* This large green aphid infests blue and Monterey cypress.

Control. Spray with malathion or Sevin when the young begin to feed.

CYPRESS MEALYBUG. *Pseudococcus ryani.* Primarily a pest of Monterey cypress in California, this species also infests arborvitae, redwood, and other species of cypress. *Ehrhornia cupressi,* another mealybug, also infests cypress.

Control. Control measures have not been developed.

CATERPILLARS. The caterpillars of the following moths feed on cypress: tip moth, *Argyresthia cupressella;* webber, *Epinotia subviridis;* imperial, *Eacles imperialis;* and white-marked tussock moth (see under Elm).

Control. Spray with Sevin when the caterpillars are small.

BARK SCALE. *Ehrhornia cupressi.* This pink scale covered with loose white wax infests Monterey cypress primarily. It occasionally attacks Guadalupe and Arizona cypress and incense cedar. Leaves of heavily infested trees turn yellow, then red or brown.

Control. Malathion sprays applied to the trunk and branches are said to be effective in combating this scale.

OTHER SCALES. Several other scales, including the cottony-cushion and the juniper, occasionally infest cypresses. These are controlled with malathion sprays.

CYPRESS MOTH. *Recurvaria stanfordi.* This moth mines leaves of Monterey cypress.

CYPRESS TIP MOTH. *Argyresthia* and *Recurvaria* spp. Both species attack foliage tips.

BAGWORM. See under Juniper.

DESERT WILLOW *(Chilopsis)*

A catalpa relative, desert willow is a small tree native to the arid southwestern United States. Few pathogens or insect pests are important on this host. *Phyllosticta erysiphoides* causes a leaf spot disease.

DOGWOOD *(Cornus)*

The dogwoods, among our best ornamental low-growing trees, are subject to several important fungus diseases and insect pests. Flowering dogwood *(C. florida)* is native to the eastern United States and cultivars adapted to various parts of the range are available. Pagoda dogwood *(C. alternifolia),* a little hardier, and Kousa dogwood *(C. kousa),* which blooms later, are also good landscape trees.

Diseases

CROWN CANKER. The most serious disease of flowering dogwood *(C. florida),* crown canker, is prevalent in several northeastern states.

Symptoms. An unthrifty appearance is the first general symptom. The leaves are smaller and lighter green than normal and turn prematurely red in late summer. At times, especially during dry spells, they may curl and shrivel. Later, twigs and even large branches die. At first the diseased parts occur principally on one side of the tree, but within a year or two they may appear over the entire tree.

The most significant symptoms and the cause for the weak top growth is the slowly developing canker on the lower trunk or roots, at or near the soil level. Although the canker is not readily discernible in the early stages, it can be located by careful examination. Cutting into it will reveal that the inner bark, cambium, and sapwood are discolored. Later, the cankered area becomes sunken, and the bark dries and falls away, leaving the wood exposed. When the canker extends completely around the trunk base or the root collar, the tree dies.

Cause. Crown canker is caused by the fungus *Phytophthora cactorum.* The same parasite apparently causes the so-called bleeding canker disease of maples and canker of American beech. It also parasitizes a large number of other plants. *P. cactorum* can live over in the soil in partly decayed organic matter, and its spores may be washed to nearby uninfested areas. It appears

to gain entrance primarily through wounds, and then invades the tissues in all directions. Thus far this basal canker is found only on transplanted trees in ornamental plantings. It is our observation in greenhouse tests that dogwoods subjected to periodic flooding and *P. cactorum* sustained root and collar rot damage whether or not they were wounded. Dogwoods grown in infested soil saturated with water for 48 hours every 2 weeks were damaged.

Control. Crown canker cannot be controlled after the fungus has invaded most of the trunk base or root collar. Control is possible if the infection is confined to a relatively small area at the trunk base. After the cankered area is outlined, a strip of healthy bark approximately 1 inch wide should be removed all around the edge of the canker. The bark removal should be deep enough to uncover the cambium. Within the area thus left bare, all discolored bark and sapwood should be carefully cut out with a gouge. The edge of the wound should be disinfested with 70 percent alcohol or 10 percent bleach and allowed to dry out. Success with this procedure will depend on early detection of the infection and thorough and complete surgical treatment.

Because the disease has been found only on transplanted trees, and because the fungus appears to enter the host plants more readily through wounds, some observers have suggested that infection occurs primarily through unavoidable injuries inflicted during transplanting. Another avenue of entrance suggested, especially on long-established trees, is wounds made by lawn machinery and cultivating tools. Consequently, all wounds should be avoided. The possibility of injury to valuable lawn specimens can be decreased by providing some protection around the trunk base, such as organic mulch to eliminate the need for mowing near the trunk.

Areas where dogwoods have died from *Phytophthora* infections should not be replanted with dogwoods for several years unless the soil is fumigated. Where losses have occurred in nurseries, drench the soil with Subdue or Terrazole to kill any of the fungus still present in the soil. The chemical treatments will not cure already infected trees.

Fig. III-21. Dogwood canker.

DIEBACK. *Botryosphaeria dothidea.* Dieback of dogwood branches, particularly the pink-flowering kinds, is frequently caused by this species of *Botryosphaeria.* The dieback can erroneously be attributed to dogwood borers.

Control. No effective control measures have been developed.

CANKER. A canker disease of unknown cause affects the trunk and lower branches (Fig. III-21) of flowering dogwoods in landscapes, nurseries, and in the wild. The disease causes a one-sided flattening of the stem and scaly, rough bark. These cankers are preferred sites for dogwood borer egg laying.

FLOWER AND TWIG BLIGHT. *Botrytis cinerea.* In rainy seasons in the eastern United States, the white flower bracts fade and rot and infect leaves on which they fall (Fig. III-22). In some cases twigs are also blighted.

Control. Spray the entire tree lightly with Benlate or zineb early in the flowering period.

ANTHRACNOSE. A leaf spotting, blighting, and twig dieback disease caused by the fungus *Discula* has been devastating to trees in the northeastern United States during recent years. This disease also occurs on dogwood in the Pacific Northwest.

Control. Provide good growing conditions, including adequate water, light, and borer controls. Prune out dead and dying branches, and limb and trunk sprouts. Sprays of Daconil 2787 or mancozeb, made at 10-day intervals during leaf expansion and again during wet periods, will prevent anthracnose.

LEAF SPOTS. *Ascochyta cornicola, Cercospora cornicola, Colletotrichum gloeosporioides, Elsinoë corni* (Fig. III-23), *E. floridae, Phyllosticta cornicola, Ramularia gracilipes, Septoria cornicola* (Fig. III-24), *and S. floridae.* Many species of fungi cause leaf spots on this host.

Control. Among the fungicides that are effective in combating leaf spots are Fore, Zyban, Daconil 2787, and Captan. Applications should be made once a month, starting in April when the flower buds are in the cup stage and continuing until the flower buds for the following year are formed in late summer.

POWDERY MILDEW. *Microsphaera pulchra* and *Phyllactinia guttata.* Powdery mildews may attack dogwood, entirely covering the leaves with a thin white coating of the fungus.

Control. Spray with Benlate or Bayleton.

TWIG BLIGHTS. *Myxosporium everhartii, Cryptostictis sp., and Sphaeropsis sp.* Cankering and blighting of dogwood twigs may be caused by three species of fungi.

Control. Prune and destroy infected twigs. Fertilize and water to increase vigor of the tree.

OTHER DISEASES. Other diseases of dogwood include root rots caused by *Armillaria mellea* in the North, its counterpart *Clitocybe tabescens* in the South, and *Phymatotrichum omnivorum.* The Pacific dogwood *(Cornus Nuttallii)* is sub-

Fig. III-22. Blight of flowering dogwood leaves caused by the fungus *Botrytis cinerea.*

ject to canker caused by *Nectria gallignea,* collar rot by *Phytophthora cactorum,* and leaf disease by *Placosphaeria cornicola.*

Control. Control measures are not available.

The native dogwood, Cornus florida, is especially sensitive to dry soils. In prolonged dry spells its leaves wilt at the margins and turn brown and appear as though scorched by fire. Applying water to specimens growing in lawns during dry spells will help to prevent this condition.

Insects

BORERS. At least seven kinds of borers attack dogwoods. The most serious are the flatheaded borer (see under Maple) (Fig. III-25) and the dogwood borer, *Synanthedon scitula.* The dogwood borer adult (Fig. III-26) is a clear-winged moth which resembles a wasp. Eggs are laid on the bark, and borer larvae enter through wounds and scars. They feed in the cambium and produce sawdustlike frass, which may appear on the bark surface. Infested trees eventually become weakened. This insect is attracted to injured trees and trees growing in full sun. Dogwood twig borers, *Oberea tripunctata* (see under Elm) and *O. ulmicola,* also attack dogwood.

Control. In the latitude of New York, paint or spray the trunks and branches with Dursban, Lindane, or Thiodan in early May and twice more at 2-week intervals to get control.

DOGWOOD CLUB GALL. *Mycodiplosis clavula.* Club-shaped galls or swellings, ½ to 1 inch long, on twigs of flowering dogwood are caused by a reddish brown midge which attacks the twigs in late May. Small orange larvae develop in the galls, which drop to the ground in early fall (Fig. III-27).

Control. Pruning and destroying twigs with

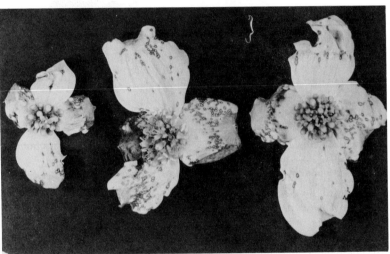

Fig. III-23. Small, slightly raised reddish gray spots on dogwood leaves are caused by the fungus *Elsinoë corni*. Spots on the white bracts are also caused by *E. corni*.

the club galls during summer will provide control. Spraying weekly six to seven times with Sevin starting in late May will also control this pest.

LEAF MINER. *Xenochalepus dorsalis.* The flat yellowish white larvae, up to ¼ inch long, attack black locust leaves, making blisterlike mines on the underside. The adult beetles skeletonize dogwood leaves by feeding on the underside.

Control. Spray with malathion when the adult

beetles are feeding on the leaves or before the larvae enter the leaves.

SCALES. Several species of scales may infest dogwood: dogwood, cottony maple (Fig. III-28), obscure, oystershell, San Jose, ivy, mealybug, calico, white peach, and tea.

Control. Spray with a lime sulfur or miscible oil, dormant strength, in late fall or early spring. Follow with malathion or Sevin in mid-May and again in June.

WHITEFLY. *Tetraleurodes mori.* Leaves of dog-

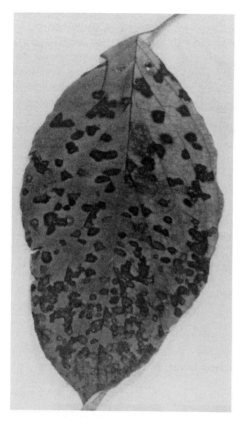

Fig. III-24. Septoria leaf spot of dogwood.

Fig. III-25. Flatheaded apple tree borer and galleries in dogwood branch.

Fig. III-26. Adult stage of the dogwood borer *Synanthedon scitula*.

Fig. III-27. Dogwood club gall opened to show larvae, *Mycodiplosis alternata*.

Fig. III-28. Cottony maple scale on dogwood leaves.

wood, mulberry, holly, maple, and planetree infested by whitefly are usually sticky from honeydew secreted by black, oval, scalelike young, with a prominent white border. The adults are tiny white flies, which dart away when the leaves are disturbed.

Control. Spray with malathion in midsummer and again in late summer.

OTHER INSECTS. Among other insects which occasionally infest dogwood are melon and potato aphid, psyllid, pitted ambrosia beetle, redhumped caterpillar, green maple worm, giant hornet, leafhoppers, greenhouse thrips, leafroller, and the dogwood sawfly. The latter is especially damaging to gray dogwood. The control for most of these is achieved with Sevin sprays.

JAPANESE WEEVIL. See under Holly.

TWIG GIRDLER. See under Hackberry.

DOUGLAS-FIR *(Pseudotsuga menziesii)*

Although native to western North America from Alaska to Mexico, this tree is fairly well adapted elsewhere and is used as a landscape tree in eastern North America.

Diseases

GALL DISEASE. Causes unknown, bacteria suspected. Galls are formed on the twigs of Douglas-fir and big-cone spruce in California. Trees up to 15 years in age are most susceptible. When galls develop on the main stem, girdling and death of the upper portion follow.

Control. Control measures have not been developed.

CANKERS. *Valsa abietis, Leucostoma kunzei, Dasyscypha ellisiana, D. pseudotsugae, Phacidiopycnis pseudotsugae,* and *Phomopsis lokoyae.* A number of fungi cause cankers on this host, most of them being prevalent in the Pacific Northwest.

Control. Control measures are rarely practiced except on important ornamental specimens.

LEAF CAST. *Rhabdocline pseudotsugae.* Yellow spots first appear near the needle tips in fall. The spots enlarge in spring, then turn reddish brown, contrasting sharply with adja-

cent green tissues. With continued moist weather, the discoloration spreads until the entire needle turns brown. When many groups of needles are so affected, the trees have a brown, scorched aspect when viewed from a distance.

Two other fungi produce leaf cast of Douglas-fir: *Phaeocryptopus gaeumannii* and *Rhabdogloeum hypophyllum.* The former produces symptoms closely resembling those of *Rhabdocline pseudotsugae,* and both may be present in the same tree.

Control. No control measures for large trees have been reported. Severe outbreaks in nurseries or on small trees can probably be prevented by spraying with Daconil 2787, Mancozeb, or copper fungicides at the time the spores are being discharged.

OTHER DISEASES. Douglas-fir is susceptible to a leaf and twig blight caused by the fungus *Botrytis cinerea,* which is serious in wet springs; rust by the fungus *Melampsora medusae,* the alternate stage of which occurs on poplar; needle blight by *Rosellinia herpotrichioides;* and witches' broom by the mistletoe *Arceuthobium.* The fungus *Dermea pseudotsugae* causes a dieback and death of young Douglas-firs in northern California. Controls have not been developed.

Insects

APHID. *Cinera pseudotsugae. Cinara* is one of a complex of three aphids which seriously damage young Douglas-fir.

Control. The aphid can be controlled with Sevin or malathion sprays in April or May.

COOLEY SPRUCE GALL APHID. See Cooley spruce gall adelgid, under Spruce.

TUSSOCK MOTH. *Orgyia pseudotsugata.* Douglas-fir tussock moth may occasionally completely defoliate landscape Douglas-fir in the western United States.

SCALES. Two scale insects, black pineleaf and pine needle, infest Douglas-fir. Malathion or Sevin sprays are very effective against the young crawling stage of these pests.

OTHER INSECTS. This host is also subject to the spruce budworm, Douglas-fir beetle, white pine weevil, pitch moth, spruce and eriophyid mites, and cambium miners. The controls for these pests are discussed under more common hosts such as pine, spruce, and yew.

ELM (Ulmus)

Elms are important landscape trees worldwide. In 1981 it was estimated that the 136 million landscape elms throughout the world had a value of at least 13.6 billion dollars (Compendium of Elm Diseases). Almost three dozen species of elms are thought to exist worldwide and many of them are important landscape trees. Elms grow fast and are tolerant of poor soils, heat, and drought. Unfortunately, Dutch elm disease has decreased the value of American elm (U. americana) and Siberian elm (U. pumila), an undesirable tree, has tarnished Asiatic elm reputations. Chinese elm (U. parvifolia) is actually an excellent disease-resistant tree with good form, although not the vaselike form of American elm.

Diseases

WETWOOD. A term applied to certain forms of wilt, branch dieback, and internal and external fluxing of elms. It occurs in American, Moline, Littleford, English, Siberian, and slippery elms.

Symptoms. Elms affected by wetwood have dark, water-soaked malodorous wood. The condition is usually confined to the inner sapwood and heartwood in trunks and large branches. Little or no streaking occurs in the outer sapwood, and no discoloration is seen in the cambial region or phloem.

Cause and Control. See Chapter 12.

DUTCH ELM DISEASE. Much has been written both in the United States and in Europe on the Dutch elm disease, the most destructive fungus disease attacking elms. The misleading name given the disease merely refers to the place where it was first identified in 1919, the Netherlands. The disease is believed to have entered the United States in the late 1920s on burled elm logs from Europe. After killing 40 million of the 70 million landscape elms throughout the United States, it now is known to be present in almost every state where elms are grown.

Symptoms. Wilting leaves on one or more branches followed by yellowing, curling, and

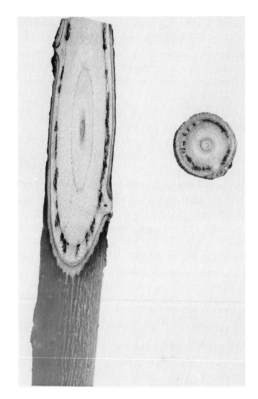

Fig. III-29. Discoloration in twig resulting from infection by the Dutch elm disease fungus, *Ophiostoma ulmi*.

dropping of all but a few of the leaves at the branch tips about midsummer are the first outward symptoms. The disease is often recognizable in winter by the tuft of dead brown leaves adhering to the tips of curled twigs. When diseased twigs or branches are cut, spots or flecks are visible in the sapwood near the bark (Fig. III-29). A longitudinal section of the diseased twig show long brown streaks following the grain of the wood.

Discoloration in the wood also occurs in several other less destructive elm diseases. Consequently, a positive diagnosis can be made only after the fungus responsible for the discoloration has been isolated. Plant pathologists at state agricultural experiment stations are equipped to make a diagnosis, provided adequate twig specimens showing internal discolorations are submitted to them.

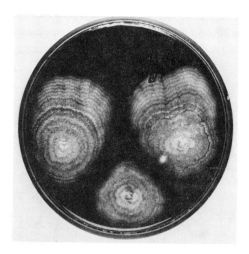

Fig. III-30. The Dutch elm disease fungus as it appears in pure culture when isolated from infected elm twigs.

Cause. The Dutch elm disease is caused by the fungus *Ophiostoma ulmi (Ceratocystis ulmi),* the asexual stages (Fig. III-30 and 31) of which are *Pesotum* and *Sporothrix.* An aggressive, highly pathogenic strain of the fungus now exists in many parts of Europe and North America.

Several species of insects are largely responsible for the spread of the fungus from diseased to healthy trees. Among the most common vectors in this country are the smaller European elm bark beetle, *Scolytus multistriatus,* and the native elm bark beetle, *Hylurgopinus rufipes.* A number of other boring insects are also known to transmit the causal fungus, or are suspected of transmitting it. The fungus may occasionally be spread by rainwater and through grafting of roots of a diseased with a healthy elm. Other suspected means of spread are windblown spores, birds, and pruning tools.

The fungus penetrates the tree only through wounds. Once inside the wood, it can spread rapidly either by spores developed in the wood vessels and carried in the sap stream or by growth of the fungus hyphae. Death of affected branches is believed to be due to toxins produced by the fungus or to lack of water as a result of the plugging of the vessels by materials formed in the process of fungus invasion. It has been proposed that compartmentalization of the fungus by the tree also results in the walling off of starch reserves at a time when energy reserves may already be low, leading to tree death by starvation.

Control. After almost 50 years of research, no simple positive control has been developed. The most effective controls are indirect, that is,

Fig. III-31. Spore-bearing coremia of the *Pesotum ulmi* asexual stage of the Dutch elm disease fungus.

against the insects (bark beetles) which are the principal disseminators of the fungus and against sources of fungal inoculum. Sanitation—removing and destroying dead and dying elm trees—is extremely important in control of Dutch elm disease. Sources of fungal inoculum are reduced when infected trees are destroyed. In addition, removing dead and dying elms eliminates breeding sites for the bark beetle vectors. Thorough sanitation requires removal of all dead, dying, or devitalized elm material such as sick trees, hurricane-damaged ones, broken limbs, elmwood piles, and elm fenceposts. Such material must be removed from a relatively large area because the smaller European elm bark beetle, the principal disseminator in this country, can fly more than 3 miles in search of suitable breeding places, but these same beetles may also feed on living trees. Such beetles have been found to carry viable spores of the Dutch elm disease fungus for more than 2 miles. The elm bark beetles breed and feed in all species of elms that grow in this country.

Injection of the herbicide cacodylic acid into diseased elm trees can also be used to reduce beetle populations. Treated trees are quickly colonized by bark beetles, but herbicide-induced drying of the bark prevents maturation of the larvae. This "trapping" of bark beetles reduces the probability of Dutch elm disease spread in a community.

Another way to prevent the movement of these beetles is to use a residual-type contact insecticide on elm trees in March or April. Methoxychlor is the insecticide of choice for this use.

Pruning diseased branches has often been suggested as a means of checking this disease. Although costly, therapeutic pruning done when symptoms first begin to occur will preserve most infected trees. The infected tree has a better chance of survival if the infection in the excised branch has not penetrated too far down in the branch. A minimum of 10 feet of healthy wood must be removed along with the discolored wood. The extent of the infection can be determined by peeling back the bark and seeing how far down the tree the xylem discoloration has gone. Chain saw blades and lubricating oil should be disinfested regularly. It has been observed that healthy trees that are pruned in late July, August, and September are more likely to contract the disease than those pruned at other times of the year.

The disease may be spread by root grafts between a diseased and a healthy tree. Soil treatments with Vapam, a general purpose soil fumigant, will kill a narrow zone of roots between trees, thus breaking grafts. The Vapam label provides instructions for use. Digging a narrow trench between trees will also break the grafts between healthy and diseased trees.

Frequent reports appear in the press and in scientific literature about a chemical which will prevent the Dutch elm disease or cure mildly infected trees. Lignasan BLP, a water-soluble form of benomyl, is recommended by the Elm Research Institute and has been widely tested. Trunk injections are made under pressure with special apparatus any time during the growing season but preferably in spring when the trees reach full leaf. Details on dosage and procedures are available from the Elm Research Institute.

Arbotect 20-S, is also a systemic fungicide for use as a Dutch elm disease preventive and as a therapeutic treatment for mildly infected trees. The active ingredient is thiabendazole. Arborists are having some success with Arbotect.

The Dutch elm disease fungus is highly variable. Some strains of the fungus are apparently highly resistant to benomyl and presumably to chemically related fungicides such as Lignasan and Arbotect.

Another product, Phyton 27, has also been cleared for Dutch elm disease control. Not yet widely tested, the chemical seems to work in some situations. Fungisol, another benomyl relative, is injected via Mauget feeder tubes.

None of the chemicals or injection systems is 100 percent effective. All of the injection systems cause injury to the tree; the larger the hole, the greater the possibility of discoloration and decay.

At the moment and for all practical purposes, as thorough a sanitation program as possible is perhaps the most important control practice. This involves a community-wide program of dead tree and elm firewood removal. One needs to use some judgment before engaging in a Dutch elm disease control program. The surrounding neighborhood in an area of several square blocks must be scouted thoroughly for sources of disease inoculum, the already infected dead and dying trees. If a real threat exists, then for valuable trees, extraordinary measures can be taken.

The greatest hope for eventual control lies in the discovery of American elms that have natural immunity to the disease. A hybrid elm, 'Urban Elm', is resistant to Dutch elm disease. This elm is a cross between an elm from the Netherlands and a Siberian elm *(Ulmus hollandica* var. *vegeta* × *U. carpinifolia)*. The new tree will grow to moderate size, making it more suitable for urban planting than the American elm. It has an upright branching form as compared with the very familiar umbrella shape of the American elm. Several commercial nurseries are now propagating 'Urban Elm'. The cultivars 'Dynasty', 'Thompson', 'Jacan', 'Regal', 'Pioneer', and 'Homestead' are reportedly resistant.

Another disease-resistant hybrid clone, 'Sapporo Autumn Gold', was developed by tree breeders at the University of Wisconsin. Several commercial nurseries are now propagating it. 'American Liberty' is a Dutch elm disease-resistant American elm selection.

The 'Christine Buisman' elm, formerly thought to be highly resistant to the disease, has been found to be susceptible to both Dutch elm disease and the mycoplasma disease phloem necrosis. Other Dutch selections such as 'Dodoens', 'Lobel', 'Groenveld', and 'Plantyn' are reportedly resistant. The Hanson Manchurian elm, like most Asiatic elms, is decidedly resistant to the disease. A close relative of the American elm, Japanese keaki *(Zelkova serrata)*, is much more resistant to the disease than is the American elm. It is vase-shaped and has bark resembling beech and elmlike foliage that turns red in the fall.

Occasionally one reads that well-fertilized elms are less likely to contract Dutch elm disease. Actually, the reverse is true. Fertilizers increase the vessel group size, which makes the trees more susceptible to the disease. Unless the trees show a severe nutrient deficiency, they should not be fertilized more often then once every 3 or 4 years.

WILT. *Deuterophoma ulmi* Early symptoms are the drooping and yellowing of the leaves, which are more or less mottled and which later become brownish and rolled. The foliage on trees whose trunks are infected is very dwarfed. Much dieback of twigs and branches occurs. This fungus is spread by wind, rain, insects, and birds. The parasite enters through wounds of leaves and tender shoots and develops in the water-conducting system. Branches having dieback develop cankers with associated pycnidia.

Control. Severly infected trees should be removed and destroyed. Mildly infected ones should be pruned heavily to remove as much of the diseased wood as possible. Because the fungus may develop internally well beyond the area of the external symptoms, pruning does not always produce the desired results. Heavy fertilization may help mildly diseased trees to recover. The number of leaf infections can be reduced by applying combination sprays containing a fungicide and an insecticide. Fertilization is suggested as a general precautionary measure, despite the fact that there seems to be no correlation between the vigor of the tree and its susceptibility to the disease.

VERTICILLIUM WILT. *Verticillium albo-atrum.* The symptoms of this disease are so much like those of the Dutch elm disease and Cephalosporium wilt that culturing of the fungus is necessary to distinguish them. This wilt may in time cause the death of large elms but usually does not become epidemic. See Chapter 12.

CANKERS. *Botryosphaeria dothidea, Coniothyrium* sp., *Cytospora ambiens, Nectria galligena, N. cinnabarina, Phoma* sp., *Phomopsis* sp., *Phytophthora inflata, Schizoxylon microsporum,* and *Sphaeropsis ulmicola.* Many species of fungi cause cankers and dieback of twigs and branches of elms. *Botryodiplodia hypodermia* and *Tubercularia ulmea* cause Siberian elm cankers in the Great Plains.

Control. Many small cankers can be eradicated by surgical means. The cuts should extend well beyond the visibly infected area to ensure complete removal of fungus-infected tissue. If the canker has completely girdled the stem, prune well below the affected area and destroy the prunings.

BLEEDING CANKER. *Phytophthora inflata.* This disease, also called pit canker, is characterized by seepage of reddish brown fluid from the margins of perennial cankers produced on the trunk or limbs of trees growing under adverse conditions.

Control. See under Maple.

LEAF BLISTER. *Taphrina ulmi.* Small blisters which lead to abnormal leaf development follow an attack by this fungus. Infection usually takes place soon after the leaves unfold.

Control. Valuable specimens subject to this disease should be sprayed with concentrated lime sulfur or bordeaux mixture in spring just before growth starts.

LEAF SPOTS. *Cercospora sphaeriaeformis, Cylindrosporium tenuisporium, Coryneum tumoricola, Gloeosporium inconspicuum, G. ulmicolum, Monochaetia desmazierii, Phyllosticta confertissima, P. melaleuca, Mycosphaerella ulmi, Septogloeum profusum, and Coniothyrium ulmi.* There are so many fungi that cause leaf spots of elms that only an expert can distinguish one from another. Probably the most prevalent leaf spot, however, is that caused by *Stegophora ulmea,* the first stympom of which appers early in spring as small white or yellow flecks on the upper leaf surface. The flecks then increase in size, and their centers turn black (Fig. III-32). If infections occur early and are heavy, the leaves may drop prema-turely. Usually, however, the disease becomes prevalent in late fall about the time the leaves drop normally, and consequently little damage to the tree occurs.

Control. Gather the fallen leaves in autumn and place them in trash cans to be carted away. Ferbam, fixed copper, Benlate, or zineb sprays applied three times at 10- to 12-day intervals, starting when the leaves are half-grown, will give good control of leaf spots.

POWDERY MILDEWS. *Microsphaera neglecta, Phyllactinia guttata and Uncinula macrospora.* These species of powdery mildew fungi develop their mycelia on both sides of the leaves and cause a yellowish spotting.

Control. Damage is so slight that spraying is usually unnecessary.

WOOD DECAY. A number of fungi are associated with decay of elm wood. They cannot be checked once they have invaded large areas of the trunk. Many can be prevented from gaining access to the interior of the tree, however, by avoiding bark injuries, by properly treating injuries that do occur, and by keeping the tree in good vigor by fertilizing and watering.

PHLOEM NECROSIS. This disease is also called elm yellows and is caused by a mycoplasmalike organism. It is even more deadly than Dutch elm disease. Thousands of elms in the Middle West have died from its effects since the early 1940s. The disease occurs from the Great Plains to the East Coast.

Symptoms. The earliest symptoms appear in

Fig. III-32. Leaf spot of elm caused by the fungus *Stegophora ulmea.*

the extreme top of the tree at the outer tips of the branches. Here the foliage becomes sparse, and the leaves droop because of downward curvature of the leaf stalks. Individual leaves curl upward at the margin, producing a trough-like effect that makes them appear narrow and grayish green. They are often stiff and brittle. The leaves then turn yellow and fall prematurely. These symptoms appear throughout the tree and are not confined to one or several branches, as are wilt disease infections.

After initial infection and before top symptoms appear, the small fibrous roots are killed. Eventually the larger roots die. One of the most typical signs is the discoloration of the phloem tissue that precedes the death of the larger roots. The discoloration frequently extends into the trunk and branches. The cambial region becomes light to deep yellow, and the adjacent phloem tissue turns yellow, then brown, with small black flecks scattered throughout. After this, the phloem tissues are browned and killed. Moderately discolored phloem has an odor resembling wintergreen. In chronic cases of phloem necrosis, there is a gradual decline over a 12- to 18-month period before the tree dies. In acute cases, an apparently healthy and vigorous tree may wilt and die within 3 to 4 weeks.

Cause. The phloem necrosis organism can be transmitted experimentally by grafting patches of diseased bark, scions, or roots on healthy trees. In nature the mycoplasma is transmitted by insects such as the white-banded leafhopper, *Scaphoideus luteolus.* The disease sometimes appears in nearby trees through root grafts.

Control. Control of elm yellows in susceptible trees is almost impossible. Tetracyclene, shown to temporarily suppress other mycoplasma diseases, has been reported to suppress phloem necrosis. Asiatic and European elms are resistant.

MOSAIC. This virus disease, which causes yellow mottling of leaves, is rare and relatively harmless. It can be transmitted from diseased to healthy trees through both pollen and pistils. Cherry leafroll virus causes elm mosaic.

SCORCH. This bacterial disease occurs in the vicinity of Washington, D.C. Foliar necrosis and a gradual crown deterioration and eventual death are the principal symptoms. See Chapter 12.

Control. No control is known for mosaic or scorch.

Abiotic Disease

AIR POLLUTION. Elms are sensitive to sulfur dioxide. See Chapter 10.

Insects and Other Animal Pests

APHIDS, WOOLLY APPLE APHID. *Eriosoma lanigerum.* Infestations result in stunting and curling of the terminal leaves. The growing tip may be killed back for several inches. This bluish white aphid spends two generations on elm, migrates to rosaceous host roots, then goes back to elm, overwintering in the egg stage in bark crevices. Woolly elm aphid *(E. americana)* has a similar life history, having serviceberry as the alternate host. Giant bark aphid (see aphids, under Sycamore), elm leaf aphid *(Tinocallis ulmifolii),* woolly elm bark aphid *(E. rielyi),* woolly hawthorn aphid *(E. crategi),* and woolly pear aphid *(E. pyricola)* also feed on elm.

Control. In spring just after the buds burst, spray with Diazinon, malathion, or Dursban. Repeat in 10 days to 2 weeks.

ELM COCKSCOMB GALL. *Colopha ulmicola.* Elongated galls which resemble the comb of a rooster are formed on the leaves as a result of feeding and irritation by wingless yellow-green aphids. Eggs are deposited in bark crevices in fall. Two other species, *C. ulmisaccula* and *Eriosoma langninosa,* behave similarly.

Control. A malathion spray applied as the buds open in spring will destroy the so-called stem-mother stage.

ELM CASE BEARER. *Coleophora ulmifoliella.* When elms are infested by the case bearer, small holes are chewed in the leaves and angular spots mined between the leaf veins by a tiny larva. The adult is a small moth with a ½-inch wingspread. The pest overwinters in the larval stage in small cigar-shaped cases made from leaf tissue.

Control. Malathion or Sevin sprays in May will control this pest.

ELM LACE BUG. *Corythucha ulmi.* In eastern states this insect may do considerable damage to elms. It first infests the tender foliage in spring, later causing a characteristic spotting of the leaves, which turn brown and die. Black specks of the excreta on the underside of the leaves are also characteristic.

Control. Spray with malathion or Sevin when the young bugs appear in spring.

ELM LEAF BEETLE. *Pyrrhalta luteola.* Two distinct types of injury are produced by this pest. Soon after the leaves unfurl in spring, rectangular areas are chewed in them by the adult beetles, brownish yellow insects ¼ inch long (Fig. III-33). Later in the season the leaves are skeletonized and curl and dry up as a result of the feeding on the lower surface by the larvae, black grubs with yellow markings. The eggs are deposited on the lower leaf surface by the beetles.

Control. Spray with methoxychlor, Orthene, or Sevin in May and June. Mites may increase on leaves sprayed too often with Sevin. The masses of grubs or pupae around the base of the tree can be destroyed by wetting the soil with a dilute solution of malathion prepared from the emulsifiable concentrate.

ELM LEAF MINER. *Fenusa ulmi.* Leaves of American, English, Scotch, and Camperdown elms are mined and blotched in May and June by white, legless larvae, sometimes as many as

Fig. III-34. Engraving of the wood made by the larvae of the smaller European elm bark beetle.

20 in a single leaf. Adult females are shining black sawflies, which deposit eggs in slits in the upper leaf surfaces.

Control. Spray the leaves in late May and early June with Diazinon, malathion, or Sevin.

SMALLER EUROPEAN ELM BARK BEETLE. *Scolytus multistriatus.* Trees in weakened condition are most subject to infestation by the smaller European elm bark beetle. The adult female, a reddish black beetle ¹⁄₁₀ inch long, deposits eggs along a gallery (Fig. III-34) in the sapwood. The small white larvae that hatch from the eggs tunnel out at right angles to the main gallery. Tiny holes are visible in the bark when the adult beetles finally emerge. The beetle is of interest because it is one of the principal vectors of the Dutch elm disease fungus. Adult beetles will feed to a slight extent on buds and bark of twigs during summer. Native bark beetle, *Hylurgopinus rufipes,* having a similar life history, is also a Dutch elm disease vector.

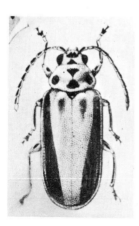

Fig. III-33. Adult elm leaf beetle, which occasionally becomes a household pest.

Control. Remove and destroy severely infested branches or trees, and fertilize and water weakened trees. Methoxychlor should be applied in April to control this pest and thus prevent it from spreading the Dutch elm disease fungus.

DOGWOOD TWIG BORER. *Oberea tripunctata.* This girdler causes the dropping of many small twigs of the elm in May and June. The female partially girdles branches up to ½ inch in diameter. When the eggs hatch, the grubs bore 4 to 5 inches down the center of the twig, which may have broken off at the girdled point. The grubs are dull yellow, ¾ inch long; they winter in the twigs. The adult beetles appear in the spring.

Control. To destroy the larvae, gather and discard all fallen twigs as soon as they are noticed; to destroy the adults, spray the trees with Sevin or methoxychlor in June.

ELM BORER. *Eutetrapha tridentata.* Weakened trees are also subject to attack by the elm borer, a white grub 1 inch long which burrows into the bark and sapwood and pushes sawdust out through the bark crevices. The adult is a grayish brown beetle ½ inch long, with brick red bands and black spots. Eggs are deposited on the bark in June. The larvae overwinter in tunnels beneath the bark. Similarly, red-headed ash borer (see under Ash) also attacks elm.

Control. Spray the bark of the trunk and branches in late June and mid-July with methoxychlor.

DOGWOOD BORER. See borers, under Dogwood.

TWIG GIRDLER. See twig girdler, under Hackberry.

LEOPARD MOTH BORER. See under Maple.

SPRING CANKERWORM. *Paleacrita vernata.* Spring cankerworm, also called inchworms or measuring worms, are looping worms, of various colors and about 1 inch in length, that chew the leaves. The adult female moth, which is ½ inch long and is wingless, climbs up the trunk to deposit eggs in early spring.

Control. Spray the leaves with *Bacillus thuringiensis* (Dipel, Thuricide), Orthene, methoxychlor, Imidan, or Sevin in May while the worms are small. In spring the trunk may be banded with a sticky material such as Tree Tanglefoot to trap the adult females as they crawl up the tree to deposit their eggs.

FALL CANKERWORM. *Alsophila pometaria.* The leaves are chewed by the fall cankerworm, a black and green worm about 1 inch in length. The adult female is a wingless moth that deposits eggs in late fall on the twigs and branches. This insect also attacks oak and maple.

Control. The spray recommended for spring cankerworm will control this pest. Bands of sticky material should be applied by late September to trap the adult females as they climb the trees.

ELM SPANWORM. *Ennomos subsignarius.* This pest, whose adult stage is known as the snow-white linden moth, caused heavy defoliation of deciduous trees in some parts of the northeastern United States in the summer of 1970. It feeds mostly on beech, elm, hickory, horsechestnut, linden, maple, oak, pecan, and yellow birch. The larvae are about 1½ to 2 inches long, brownish black with bright red head and anal segments. Eggs are laid in midsummer in groups on branches and hatch the following spring. The moths appear in late July in such great numbers that they resemble a snow shower. Io moth, *Automeris io,* larvae feed on elm.

Control. Sevin or Imidan sprays applied when the trees come in full leaf will control the spanworm.

GYPSY MOTH. *Lymantria dispar.* The leaves of a large number of forest, shade, and ornamental trees are chewed by the gypsy moth larva, a hairy, dark gray caterpillar with pairs of blue and red dots down its back, ranging up to 3 inches in length. Among the most susceptible trees are apple; speckled alder; gray, paper, and red birches; hawthorn; linden; oaks; poplars; and willows. Trees that are also favored as food include ash, balsam fir, butternut, black walnut, catalpa, red cedar, flowering dogwood, sycamore, and tuliptree. The average annual damage caused by feeding of this insect amounts to nearly 3 million dollars. More than a million acres of woodlands have been 25 to 100 percent defoliated in a single year. See Chapter 11.

Control. Bacillus thuringiensis, Imidan, methoxychlor, Sevin, Dimilin, Orthene, and

Fig. III-35. Gypsy moth adult and egg mass.

Sevimol sprays all control this pest. They are most effective on young caterpillars. Gypchek, formulated from a gypsy moth virus, may also be used.

Destroying the egg masses (Fig. III-35) during winter or early spring also helps to protect valuable ornamental trees.

A synthetic sex lure, Disparlure, was developed by United States Department of Agriculture scientists. Male gypsy moths are lured into special traps by this material, which enables federal and state officials to determine the presence and density of gypsy moths in any particular area.

MOURNING-CLOAK BUTTERFLY. *Nymphalis antiopa.* The leaves are chewed by the caterpillar stage of this butterfly, which is 2 inches long,

spiny, with a row of red spots on its back. The adult has yellow-bordered, purplish brown wings. It overwinters in bark cavities and other protected places and deposits masses of eggs around small twigs in May.

Control. Spray with *Bacillus thuringiensis* or Sevin when the caterpillars are small. Prune and destroy infested twigs and small branches.

WHITE-MARKED TUSSOCK MOTH. *Orgyia leucostigma.* The leaves are chewed by the tussock moth larva, a hairy caterpillar 1½ inches long, with a red head, longitudinal black and yellow stripes along the body, and a tussock of hair on the head in the form of a Y (Fig. III-36). The adult female is a wingless, gray, hairy moth which deposits white egg masses on the trunk and branches. Larvae emerge from eggs in May

Fig. III-36. White-marked tussock moth larva.

and can devour large amounts of foliage before being discovered. Red-humped caterpillar (see under Poplar) and *Schizura ipomaeae* caterpillar also feed on elm.

On the West Coast, the larvae of the western tussock moth, *Hemerocampa vetusta,* feed on almond, apricot, cherry, hawthorn, oaks, pear, plum, prune, walnut, and willows in addition to elms.

Control. Spray the foliage with Sevin, malathion, or methoxychlor when the young cat-

erpillars begin to feed in the spring, and again in August.

SCALES. Many species of scales infest elms: brown elm, calico, camphor, citricola, cottony maple, elm scurfy, European elm, European fruit lecanium, walnut, gloomy, hickory lecanium, San Jose, obscure, European peach, frosted, Comstock mealybug, oystershell, Putnam and scurfy. The European elm scale, *Gossyparia spuria,* (Fig. III-37), is a soft scale not protected by a waxy covering.

Fig. III-37. European elm scale, *Gossyparia spuria.*

Control. Spray all surfaces with a superior miscible oil before the buds open in spring, or with Orthene or Sevin in late May and again in mid-June.

MITES. The four-spotted mite, *Tetranychus canadensis;* Schoene spider mite, *Tetranychus schoene;* and elm spider mite, *Eotetranychus matthyssei,* infest elms, causing the leaves to turn yellow prematurely.

Control. The inclusion of a miticide such as Kelthane or Acaraben in the regular insecticide spray will curb these pests. Dormant oil applications can be used against the egg stage.

OTHER INSECTS. Elm is attacked by elm calligrapha, elm sawfly, Japanese beetle, and linden looper, each discussed under Linden. Carpenterworm, fall webworm (see under Ash), and forest tent caterpillar (see under Maple) infest elm.

EMPRESS-TREE *(Paulownia)*

The empress-tree, native of China, is prized most for its showy clusters of violet flowers in early spring. It is relatively free of pests, but it is sensitive to extremes in the weather. No important insects are known to attack the empress-tree.

Diseases

LEAF SPOTS. *Ascochyta paulowniae* and *Phyllosticta paulowniae.* Two species of fungi cause leaf spots on this host in very rainy seasons.

Control. Gather and destroy fallen leaves. Preventive sprays containing copper or dithiocarbamate fungicide can be used on extremely valuable specimens in rainy seasons.

MILDEW. *Phyllactinia guttata* and *Uncinula clintonii.* Two species of powdery mildew fungi occasionally infect empress-tree leaves.

Control. See Chapter 12.

WOOD DECAY. *Polyporus spraguei* and *Trametes versicolor* These fungi are constantly associated with a wood decay of this host.

Control. As with other wood-decaying fungi, control is difficult. Avoiding wounds near the trunk base and keeping the tree in good vigor

by fertilizing and watering when necessary are suggested.

TWIG CANKER. *Phomopsis imperialis.* Occasionally twigs and small branches are affected by this disease.

Control. Prune and destroy infected branches.

EUCALYPTUS. See Gum-Tree.

FALSE CYPRESS *(Chamaecyparis)*

Some, such as Port Orford cedar *(C. lawsoniana),* also called Lawson cypress, a handsome ornamental, and Alaska cedar *(C. nootkatensis)* are native to the Pacific Northwest, while others *(C. thyoides)* are East Coast or Japanese natives. False cypress is also known as white cedar and is best grown in moist circumstances.

Diseases

BLIGHT. *Phomopsis juniperovora.* The leaves of false cypress may be attacked by this fungus. (See Phomopsis twig blight under Juniper.) *Didymascella chamaecyparissi* also causes a destructive leaf and tip blight.

WITCHES' BROOM. *Gymnosporangium ellissii.* This fungus enters the leaves and travels down into the living bark of the twigs. The presence of the fungus stimulates the formation of a large number of buds which develop to form characteristic witches' brooms, shown in Figure III-38. Eventually the branch with its broom dies. During the early spring (in April) brown telial horns grow out from infected branches; they are about 1/4 inch long and threadlike. Spores from these horns are carried in the wind to the bayberry, *Myrica,* which in turn becomes infected; a light orange-colored rust appears on the leaves and does considerable damage. The rust occasionally affects sweetfern, *Comptonia.* When the seedlings of false cypress are attacked, the trees become dwarfed; trees 15 to 20 years old, if they live, may not be over a foot or two high. When young trees are infected at the growing point, the main trunk is prevented from developing normally and the tree is stunted. The fungus is deep-seated; sometimes it is even found in the pith region. Heavily broomed trees may die.

Fig. III-38. Witches' broom on false cypress.

Control. No effective control has been proposed other than separating the two hosts. The fungus acts slowly, and thus removal of the brooms prevents the spread of the parasite to the bayberry, which otherwise endangers the evergreens.

SPINDLE BURL GALL. *Gymnosporangium biseptatum.* This rust fungus is more or less local in its infection, probably first infecting the leaves and then penetrating into the young branches. It stimulates an excessive growth of wood into long burls, which may be several inches in diameter. The branches that bear these burls eventually die. When infection occurs at the base of a young tree, the burl may continue to grow with the tree without doing particular damage. Burls a foot across have been seen at the bases of trees of about the same diameter. Occasionally two species of rust attack the trees at the same point, and a combination of burl and witches' broom results. The alternate host of the rust which causes the latter abnormality

is the serviceberry *(Amelanchier).* Ornamental cedars grown individually are rarely infected.

Another species of rust, *Gymnosporangium fraternum,* attacks the leaves of false cypress. It is of little consequence on this host but does some damage to chokeberry *(Aronia),* its alternate host.

Control. Control measures are unavailable.

ROOT ROT. *Phytophthora lateralis.* This highly destructive disease affects native species in the Pacific Northwest. It is especially serious on Lawson cypress. The fungus infects leaves, stems, and trunk in addition to the roots. Another species, *P. cinnamomi,* causes root rot of Lawson cypress seedlings in Louisiana.

Control. No satisfactory controls have been developed.

Insects

Among the pests that attack false cypress are the arborvitae weevil (see under Arborvitae),

Fig. III-39. Bags of the bagworm on false cypress twigs.

the spruce spider mite (see under Spruce), the larvae of the imperial moth *Eacles imperialis,* the bagworm (see under Juniper) (Fig. III-39),and the juniper scale *Diaspis carueli.* Details on control of the bagworm are given under Juniper.

FIG *(Ficus)*

Of the many species of fig, some—such as Moreton Bay and weeping figs—are being used as street and landscape trees in California and Florida. Figs are subject to a number of fungal leaf spots, crown gall, twig and branch cankers, root rots, and nematode damage. We have observed that Benjamin figs grown indoors under low light intensities are subject to a Phomopsis twig blight. The causal fungus is *Diaporthe cinerescens,* with a *Phomopsis* conidial state. Moving the affected plants nearer to windows alleviates the problem.

FIR *(Abies)*

Several parasitic microorganisms and insects attack firs but rarely produce extensive damage in ornamental plantings. Lack of vigor in many

trees is more likely caused by an unfavorable environment. They do not do well in hot, dry climates.

Firs thrive best in light, porous, acid soils which are well drained and yet are continually moist. They also require full sunlight but appear to grow more vigorously in valleys or protected places than near hilltops or in exposed situations. Firs are sensitive to air pollution injury.

Diseases

NEEDLE AND TWIG BLIGHT. In the northeastern United States, needle and twig blight occurs commonly on balsam fir and to a lesser extent on Colorado and Alpine firs. Noble and Fraser firs are also susceptible but are attacked infrequently.

Symptoms. The needles of the current season's growth turn red and shrivel, and the new twigs are blackened and stunted. Severely infected trees appear as though scorched by fire or damaged by frost. The lower branches are most heavily infected. Needles infected in previous years remain attached to the twigs.

Cause. The fungus *Rehmiellopsis balsameae* causes needle and twig blight. It overwinters on diseased needles and twigs. In spring, fruiting bodies mature on these parts and release spores, which infect the newly developing needles. The fungus *Cenangium abietis* occasionally attacks firs but is more common on pine.

Control. In ornamental plantings, control is possible by pruning and destroying infected twigs and by applying a copper fungicide three times at 12-day intervals, starting when the new growth begins to emerge from the buds.

LEAF CAST. *Bifusella abietis, B. faullii, Hypodermella mirabilis, H. nervata, Lophodermium autumnale,* and *L. lacerum.* When attacked by any one of these fungi, needles turn yellow, then brown, and drop prematurely. Elongated black bodies appear along the middle vein of the lower leaf surface. Spores shot from black fruiting bodies in summer to young leaves germinate and penetrate the new growth.

Control. Copper sprays applied as for needle and twig blight will control leaf cast.

CANKERS. *Cylindrocarpon* sp., *Cytospora pini* and *C. abietis, Cryptosporium macrospermum, Scoleconectria balsamea,* and *S. scolecospora.* Occasionally sunken dead areas on the trunk and branches of firs in ornamental plantings result from infection by one of the several fungi listed. On balsam fir, the fungus *Aleurodiscus amorphus* forms narrowly elliptical cankers with a raised border on the main trunk of young trees and centering around a dead branch.

Control. Sanitation, avoidance of bark injuries, and fertilization to maintain the trees in good vigor are suggested.

ROOT ROT. *Phytophthora cinnamomi.* This fungus has been destructive to Fraser fir seedlings growing in southern Appalachian nurseries. The fungicide Subdue reportedly suppresses this disease.

ARMILLARIA ROOT ROT AND WOOD DECAY. These problems are caused, respectively, by the fungus *Armillaria mellea* and by fungi belonging to the genera *Phellinus, Odontia, Coniophora, Polyporus, Lenzites,* and *Haematostereum.*

Control. No completely effective control is known. Avoid bark injuries and keep the trees in good vigor by watering and fertilization when needed.

RUSTS. *Milesia fructuosa, Hyalospora aspidiotus, Uredinopsis mirabilis, U. osmundae, U. phegopteridis, Pucciniastrum pustulatum, P. goeppertianum, Melampsora abieti-capraearum, Caeoma faulliana, Peridermium ornamentale,* and *Melampsora cerastii.* Most of these rust fungi attack forest firs but are seldom found on ornamental specimens. *Melampsorella caryophyllacearum* infection causes a common yellow witches' broom of balsam fir. Broomed shoots are upright and dwarfed with short yellow needles that drop in less than a year, leaving bare shoots. The rust is perennial in the broomed shoots; it also causes trunk and branch swellings.

Control. Measures to control rusts are rarely adopted because the fungi do not cause much damage. Most of the rusts listed above have alternate hosts which are needed to complete the life cycles of the fungi. The elimination of the alternate host will result in nearly complete disappearance of the fungus. A rust expert must be consulted for the name of the alternate host before any eradicatory steps are taken. Periodic applications of sulfur sprays during summer will also control rusts, but these are not practical for large trees.

SOOTY MOLD. Needles of firs are covered more frequently than those of most other species of evergreens by black sootlike material. This substance consists of fungus tissues that exist on the secretions of aphids and other insects. The black mold will usually disappear if the insects are kept under control.

Insects

BALSAM TWIG APHID. *Mindarus abietinus.* The leaves and shoots of white and balsam firs, as well as spruce, are attacked by this green aphid, which is covered with white waxy secretions. Affected shoots are roughened and curled.

Control. Spray with malathion in late April and repeat in mid-May if necessary.

BALSAM GALL MIDGE. *Paradiplosis fumifex.* Small, subglobular swellings at the base of the leaves are caused by this midge.

Control. Spray the newly developing leaves in late April with malathion.

BARK BEETLE. *Pityokteines sparsus.* Seepage of balsam from the trunk, reddening of the needles, and death of the upper parts of the tree result from infestations of the balsam bark beetle, a pest ⅒ inch long. Vigorous trees are attacked.

Control. Prune and destroy infested parts.

CATERPILLARS. Several caterpillars feed on the needles of firs. These are the larval stage of the hemlock looper moth, the spotted tussock moth, the balsam fir sawfly, the Zimmerman pine moth, and the pine butterfly.

Control. Spray with Sevin when the caterpillars are young.

SPRUCE SPIDER MITE. See spider mite, under Spruce.

SCALE. Several kinds of scale insects, including the cottony-cushion, oystershell, and pine needle scale, infest firs.

Control. Spray with lime sulfur as for spider mite, or with malathion or Sevin during May and June to control the young crawling stages.

BALSAM WOOLLY APHID. *Adelges piceae.* This introduced adelgid is becoming increasingly prevalent in the Northeast. It attacks twigs and buds and causes dieback of twigs and tree-tops.

Control. Spray with malathion when adelgids become numerous.

BAGWORM. See under Juniper.

SPRUCE BUDWORM. See under Spruce.

CEDAR TREE BORER. See under Redwood.

CAMBIUM MINER. See under Holly.

Other Pests

DWARF MISTLETOE. *Arceuthobium campylopodum.* In California this mistletoe is a widespread serious pest of red fir *(Abies magnifica)* and white fir *(A. concolor).*

Control. In valuable trees, remove dwarf mistletoe by pruning.

FIREWHEEL TREE *(Stenocarpus)*

This Australian native grows in mild climates, preferring rich, slightly acid soil. It is relatively free of diseases and pests.

FLAME-TREE. See *Poinciana.*

FRANKLIN-TREE *(Franklinia alatamaha)*

Franklinia, native to Georgia, is a small to medium-sized tree which blooms in the fall. It prefers partial shade, wind protection, and rich, acid soil.

Diseases

LEAF SPOT. *Phyllosticta gordoniae.* Spots on the leaves of this host occasionally occur in rainy seasons.

Control. Valuable specimens can be sprayed with copper or dithiocarbamate fungicides.

BLACK MILDEW. *Meliola cryptocarpa.* In the Deep South, leaves of the Franklin-tree may be covered by this black fungus.

ROOT ROT. *Phymatotrichum omnivorum.*

Control. Measures are rarely adopted to control black mildew or root rot.

WILT. *Phytophthora sp.* This is a highly destructive wilt disease in container-grown Franklin-trees. Affected trees wilt suddenly during hot weather and then drop their leaves. Infected roots are brownish black in color, and black cankers occur along the stems.

Control. Provide good soil drainage. A soil drench of Subdue or Aliette may help suppress the fungus.

Insects

SCALES. Three species of scale insects are known to attack the Franklin-tree: red bay, *Chrysomphalus perseae;* walnut, *Aspidiotus juglansregiae;* and *Lecanium* sp. The last-named was found by the senior author on specimens submitted from New York State.

Control. Spray with malathion or Sevin when the young are crawling in late spring.

GIANT HORNET WASP. *Vespa crabro germana.* This wasp tears the bark not only from the Franklin-tree but also from birch, boxwood, poplar, willow, and other trees and shrubs. The wasp is dark reddish brown with orange markings on the abdomen. It is the largest wasp in the United States—1 inch long. The pest nests in trees, buildings, or underground.

Control. Apply a Sevin spray to the trunk

when the pest begins to tear the bark in July. Blow Diazinon dust into wasp nests.

FRINGE-TREE (Chionanthus)

Native to the southeastern United States, fringe tree is small and usually multitrunked.

Diseases

LEAF SPOTS. *Cercospora chionanthi, Phyllosticta chionanthi, Septoria chionanthi,* and *S. eleospora.* These four species of fungi are known to cause leaf spotting on this host.

Control. Spray with a copper fungicide or with one of the dithiocarbamates.

POWDERY MILDEW. *Phyllactinia guttata.* The leaves of fringe-tree are occasionally affected by this disease.

Control. Where mildew is severe, spray with Benlate or wettable sulfur.

OTHER DISEASES. Fringe-tree is occasionally subject to several canker diseases caused by *Botryosphaeria dothidea, Phomopsis diatrypea,* and *Valsa chionanthi.*

Control. Prune diseased branches.

Insects

SCALES. The rose scale *Aulacaspis rosea* and the white peach scale (see under Cherry) infest fringe-tree.

Control. A dormant oil spray applied in early spring will control these pests. Methoxychlor or Sevin applied in mid-May and again in mid-June will control the crawler stage.

GINKGO (Ginkgo)

Ginkgo is successfully being grown as a shade tree throughout the United States, except in the coldest northern states. It withstands urban conditions and is easy to transplant. There are several cultivars available, and planting of male ginkgo is usually recommended because a disagreeable odor is associated with female fruit hulls.

Ginkgo, also known as maidenhair-tree, is unusually resistant to fungus and insect attack. A fungitoxic substance, α-hexenal, is believed to be responsible for its resistance to fungus diseases, although a waxy cuticle has also been implicated. Leaf spots have been attributed to three fungi: *Glomerella cingulata, Phyllosticta ginkgo,* and *Epicoccum purpurascens.* The damage by these fungi is negligible. In 1966 the senior author isolated a bacterium from another leaf spot (Fig. III-40) but was unable to prove definitely that it was responsible for the spotting. In Czechoslovakia a similar disease is attributed to a virus.

Several wood-decaying fungi, including *Irpex lacteus, Trametes versicolor,* and *Fomes meliae,* have also been reported, but these are of rare occurrence.

Control. No effective control is known.

Insects and Other Pests

Few insects attack this tree. Among those occasionally found are the omnivorous looper, *Sabulodes caberata;* the grape mealybug, *Pseudococcus maritimus;* the white-marked tussock moth, *Orgyia leucostigma;* the American plum borer (see under Planetree); and the fruit tree leafroller, *Archips argyrospilus.*

The southern root-knot nema *Meloidogyne incognita* has been reported on ginkgo in Mississippi.

Control. Control measures are rarely applied.

GOLDEN-CHAIN (Laburnum)

This medium-sized tree does best in moist limestone soils and is hardy in the South.

Diseases

LEAF SPOT. *Phyllosticta cytisii.* Leaves are subject to spotting when infected by this fungus. The spot, at first light-gray, later turning brown, has no definite margin. The black fruiting bodies of this fungus dot the central part of the spot. Another fungus, *Cercospora laburni,* also causes leaf spots.

Control. In areas where these leaf spots are troublesome, apply a copper fungicide several times at 2-week intervals during rainy springs.

TWIG BLIGHT. *Fusarium lateritium.* Brown le-

Fig. III-40. Leaf spot of *Ginkgo biloba*, the cause of which is unknown.

sions on the twigs followed by blighting of the leaves above the affected area in very wet springs is characteristic of this disease. The sexual stage of this fungus is *Gibberella baccata*.

Control. Prune and destroy infected twigs and spray as for leaf spot.

LABURNUM VEIN MOSAIC. Conspicuous veinbanding of *Laburnum alpinum* in Maryland is said to be due to a virus. Although tobacco ringspot virus was isolated from infected plants, there was no evidence that it caused the disease.

Control. Remove and destroy infected plants.

Insects and Related Pests

APHIDS. *Aphis craccivora.* The cowpea aphid, black with white legs, clusters at the tips of the branches. The bean aphid, *A. fabae,* also infests golden-chain.

Control. Spray with malathion or Sevin when the aphids appear in late spring.

GRAPE MEALYBUG. *Pseudococcus maritimus.*

The grape mealybug occasionally infests goldenchain both above and below ground.

Control. Mealybugs infesting the branches and twigs can be controlled with malathion sprays. Those infesting the roots can be curbed by wetting the soil with Diazinon.

NEMA. *Meloidogyne hapla.*

Control. Control measures are rarely practiced for this pest.

GOLDENRAIN-TREE *(Koelreuteria)*

This native of Asia tolerates a wide range of growing conditions.

Diseases

CORAL SPOT CANKER. *Nectria cinnabarina.* Small, depressed, dead areas in the bark near wounds or branch stubs are caused by this fungus. Tiny coral-pink bodies are formed on the dead bark.

Control. Prune infected branches back to

sound wood. Fertilize and water to maintain vigor.

OTHER FUNGUS DISEASES. The only other fungus diseases reported as affecting goldenrain-tree are a leaf spot caused by a species of *Cerocospora,* wilt by *Verticillium albo-atrum,* and root rot by *Phymatotrichum omnivorum.*

Control. For wilt, see Chapter 12. The other diseases are never destructive enough to warrant control measures.

Insects

SCALES. Three species of scales infest goldenrain-tree: lesser snow (see under Poinciana); mining, *Howardia biclavis;* and white peach (see under Cherry).

Control. Dormant oil sprays applied in early spring will control these scales. Malathion or Sevin sprays applied in late May and in mid-June will control the crawler stage of the pests.

GUAVA *(Psidium guajava)*

Diseases

In Florida this host is subject to a leaf and fruit spot caused by the fungus *Glomerella cingulata,* a thread blight by *Ceratobasidium stevensii,* a leaf spot by *Cercospora psidii,* and a root rot by *Clitocybe tabescens.*

Control. Effective control measures are unavailable.

Insects and Other Animal Pests

SCALES. Nine species of scale insects infest guavas: barnacle, black, chaff, Florida red, Florida wax, greedy, green shield, hemispherical, and soft.

Control. Spray with malathion or Sevin to control the crawling stage of these pests.

OTHER INSECTS. Occasionally guava is attacked by the Mexican fruit fly, *Anastrepha ludens;* the long-tailed mealybug, *Pseudococcus adonidum;* and the eastern subterranean termite, *Reticulitermes flaviceps.*

Control. Control measures have not been developed.

SOUTHERN ROOT-KNOT NEMA. *Meloidogyne in-*

cognita. This nema infests guava roots in Florida.

Control. This pest is difficult to control on plants growing outdoors.

GUMBO-LIMBO *(Bursera)*

Diseases

SOOTY MOLD. *Fumago vagans.* A heavy growth of this sooty mold develops on the plant as a result of infestation with the brown soft scale (see below). The results are sometimes very serious.

Control. Spray with malathion or Sevin to control the scale insect which secretes the substance on which the sooty mold fungus lives.

OTHER DISEASE. In the Deep South, branches of this host may be killed by the fungus *Physalospora fusca.*

Control. Prune infected branches.

Insects

BROWN SOFT SCALE. *Coccus hesperidum.* This scale insect, which often remains inconspicuous because of its tendency to assume the color of the twigs or leaves upon which it is feeding, does great damage to greenhouse plants. The lower sides of the leaves are sometimes heavily infested with light green scales, which often choose the petioles or twigs and the upper parts of the trunk on which to develop their mature stages. The scales are arranged longitudinally along the smaller branches and main stem. They are very flat and thin, more or less transparent. Though the young are usually sluggish, not migrating far from the mother scale, they do travel to the upper leaves. This scale is said to be a general feeder, attacking many plants in the greenhouse and also tropical fruit outdoors. Eight other species of scales may infest gumbo-limbo.

Control. Spray the plants with malathion or Sevin every 2 weeks until all the young scales are destroyed.

GUM-TREE *(Eucalyptus)*

Gum-trees, also called eucalyptus, are largely natives of Australia. They are widely grown in

California and some parts of the southwestern United States. They are fast growing and require a mild climate. When killed by frost in parts of northern California, gum-trees became a fire hazard the following season.

Diseases

LEAF SPOTS. *Hendersonia sp., Monochaetia monochaeta, Mycosphaerella moelleriana,* and *Phyllosticta extensa.* Several species of fungi cause leaf spots of gum-tree. They are rarely serious enough to justify control measures.

CROWN GALL. The bacterium *Agrobacterium tumefaciens* and many wood-decaying fungi also occur on this host.

OTHER FUNGUS DISEASES. Among other diseases of this host are canker and twig blight caused by *Botryosphaeria dothidea,* basal canker by *Cryphonectria cubensis,* root rot by *Armillaria mellea* in California, root rot by *Clitocybe tabescens* in Florida and by *Phymatotrichum omnivorum* in Texas.

Control. Control measures for crown gall and Armillaria root rot are given in Chapter 12. The other diseases are not serious enough to warrant control practices.

EDEMA. Several species of *Eucalyptus* grown in greenhouses as ornamentals are subject to a physiological disease which is manifested by intumescences or blisterlike galls on the leaves. Sections of these galls show several layers of cells formed one above the other. These growths usually crack open and become rust-colored. The disease is difficult to diagnose because it looks so much like the work of a blister mite or a rust fungus. It is not caused by a parasite but results from the accumulation of too much water through poor ventilation of the greenhouse or through overwatering of the plants.

Insects and Related Pests

A goodly number of insects attack gum-trees. Among the more common are the cowpea aphid; three species of borers—California prionus, nautical, and Pacific flatheaded; the lygus bug; three kinds of caterpillars—California oakworm, omnivorous looper, and orange-tortrix; long-tailed mealybug; greenhouse thrips; and

three species of mites—avocado red, platanus, and southern red.

Control. Aphids can be controlled with malathion, borers with methoxychlor, caterpillars with Sevin, and mites with Kelthane.

SCALES. Many species, including black, California red, greedy, dictyospermum, and ivy scales, infest gum-trees.

Control. Use malathion.

EUCALYPTUS LONGHORN BORER. *Phoracantha semipunctata.* This potentially destructive pest was discovered in southern California in 1984. It kills trees when the larvae tunnel into the tree and riddle the inner bark and cambium with frass-filled galleries. Adult beetles lay eggs under loose bark of moisture-stressed trees, generally avoiding healthy trees. Larvae attempting to penetrate healthy trees are smothered by copious quantities of gum produced by the tree. Stressed trees lack this defensive ability.

Control. No practical control measures have been developed for large trees.

HACKBERRY (Celtis)

Common hackberry, despite witches' broom and nipple gall problems, is a tough tree, adapted to harsh growing conditions, as in the Great Plains. Sugar hackberry, grown in the South, and European hackberry are also tolerant of a wide range of soil conditions

Diseases

LEAF SPOTS. *Cercospora spegazzini, Cercosporella celtidis, Cylindrosporium defoliatum, Phleospora celtidis, Phyllosticta celtidis, Pseudoperonospora celtidis,* and *Septogloeum celtidis.* Many fungi cause leaf spots on hackberry in rainy seasons, or late in the season on senescing leaves.

Control. Leaf spots are rarely serious enough to warrant control, but a copper or dithiocarbamate fungicide would probably be effective.

POWDERY MILDEW. *Pleochaeta polychaeta.* Both sides of the leaves are attacked, the mildew being visible either as a thin layer over the entire surface or in irregular patches. The small black fruiting bodies, ascocarps, develop mostly on the side opposite the mildew.

Control. Spray valuable specimens with

Benlate when mildew begins to become prevalent.

WITCHES' BROOM. In the eastern and central states the American hackberry is extremely susceptible to the witches' broom disease.

Symptoms. The early symptoms are visible on the buds during the winter. Affected buds are larger, more open, and hairier than normal ones. Branches that develop from such buds are bunched together, producing a broomlike effect, which is most evident in late fall or winter. The witches' brooms are more unsightly than harmful. Their presence, however, causes branches to break off more readily during windstorms, and the exposed wood is then subject to decay.

Cause. The exact cause of the witches' broom disease is not definitely known. An eriophyid mite, *Eriophyes celtis,* and the powdery mildew fungus *Sphaerotheca phytoptophila* are almost constantly associated with the trouble and are believed to be responsible for the deformation of the buds that results in the bunching of the twigs.

Control. No effective control measures are known. Pruning back all infected twigs to sound wood and possibly spraying with 1 part of lime sulfur in 10 parts of water in early spring might help. This should be followed with two applications of Kelthane at 2-week intervals starting in mid-May. The Chinese hackberry is less susceptible than the common hackberry and should be substituted in areas where the disease is prevalent. The southern hackberry *(C. mississippiensis)* is also less subject to witches' broom.

GANODERMA ROT. *Ganoderma lucidum.* This fungus is capable of attacking living trees, causing extensive decay of the roots and trunk bases. See under Maple.

WOOD DECAY. Hackberries are subject to decay by several different heart rot fungi. See Chapter 12.

Insects and Related Pests

HACKBERRY NIPPLE-GALL MAKER. *Pachypsylla celtidis-mamma.* Small round galls opening on the lower leaf surfaces and resembling nipples (Fig. III-41) are caused by a small jumping louse or psyllid. Another species of psyllid,

Fig. III-41. Hackberry nipple galls.

Pachypsylla celtidisvesicula, produces blister galls.

Control. Spray with Diazinon, Sevin, malathion, or Orthene in May when the leaves are one-quarter grown.

SCALES. Several species of scales including camphor, cottony-cushion, cottony maple, gloomy, hickory lecanium, obscure, oystershell, Putnam, San Jose, and walnut occasionally infest this host.

Control. Malathion, Orthene, or Sevin sprays when the young are crawling about in spring will provide control. Where infestations are heavy, a dormant oil or lime sulfur spray before leaves emerge in spring should also be applied.

MOURNING-CLOAK BUTTERFLY. See under Elm.

PAINTED HICKORY BORER. See under Hickory.

TWIG GIRDLER. *Oncideres cingulata.* This beetle chews a continuous notch around the twig proximal to the point of egg deposition in the twig. Girdled twigs die and break off; the larva completes its development there.

Control. No chemical control is available. Pick up and destroy fallen branches.

RED-HEADED ASH BORER. See under Ash.

HARDY RUBBER TREE *(Eucommia)*

This tree, adapted to temperate regions, is unusually free of insects and diseases.

HAWTHORN *(Crataegus)*

Hawthorns, sometimes called thorns, are members of the rose family, to which the apple and pear belong. As such, they are frequently attacked by the same fungi, bacteria, and insects that attack those trees. There are many species and cultivars of hawthorn, each having different susceptibilities to pests.

Diseases

LEAF BLIGHT. In late summer English hawthorn *(Crataegus oxyacantha)* and Paul's scarlet thorn *(C. oxyacantha pauli)* may be completely defoliated by the leaf blight disease. Cockspur thorn and Washington thorn appear to be much more resistant to the trouble.

Symptoms. Early in spring small, angular, reddish brown spots appear on the upper leaf surface (Fig. III-42). As the season advances, the spots enlarge, then coalesce, and the leaves finally drop.

Cause. The fungus *Diplocarpon mespili (Entomosporium mespili)* causes leaf blight. It is primarily a parasite of hawthorns, although some investigators believe it is also responsible for a similar disease of pear and quince.

Numerous black, flattened, orbicular bodies are visible, which the aid of a hand lens, in the discolored spots on both leaf surfaces. These bodies contain spores that initiate numerous infections during rainy springs.

Control. Gather and destroy all the leaves in the fall to reduce infections the following spring, inasmuch as the fungus survives from one season to the next primarily in diseased leaves.

There is some evidence that the fungus also overwinters as mycelium in small cankers on the bark. Because such sources of inoculum cannot be completely eliminated, sprays containing Benlate, Fore, or zineb are needed, These fungicides should be applied three times at 10-day intervals. The first application should be made immediately after the leaves have unfolded.

Trees that have been subjected to several consecutive years of severe leaf blight infections may be considerably weakened. A good tree fertilizer, applied in fall or spring, will help such trees to regain their vigor.

FIRE BLIGHT. Hawthorns, especially the English hawthorn, are subject to the bacterial disease fire blight, discussed in Chapter 12.

RUSTS. At least twelve rust fungi attack hawthorns. Of these the cedar quince and cedar hawthorn rusts, *Gymnosporangium clavipes* and *G. globosum,* are the most common on ornamental species. The former (Fig. III-43) attacks both the stems and the fruits of hawthorn, causing deformation of fruit. The twigs are also attacked and deformed, developing abnormal, antlerlike branches. The fungus breaks out in little cuplike structures called cluster-cups, from which quantities of bright orange spores are shed. These spores are borne by the wind to nearby red cedars and infect their leaves and young twigs. The fungus is perennial in the cedars but annual in the hawthorns. *G. glo-*

Fig. III-42. Leaf blight of hawthorn caused by the fungus *Diplocarpon mespili.*

bosum produces on the leaves spots (Fig. III-44) which vary from light gray to brown. The cluster-cups are long, slender, tubelike. This fungus also attacks apple trees; it seldom does much damage to hawthorn. Another stage of the fungus lives for two or three years on the red cedar before killing the small branches which it attacks. See cedar rusts, under Juniper.

The Washington thorn *(Crataegus phaenopyrum)* and the cockspur thorn *(C. crus-galli)* are resistant to rusts.

Control. If practicable, eliminate susceptible junipers within one mile of hawthorns. If not practical, spray the hawthorns with Daconil 2787, Bayleton, Zyban, Fore, Ferbam, or a mixture of Ferbam and wettable sulfur four to five times at 7- to 10-day intervals when orange masses appear on junipers.

LEAF SPOTS. A large number of fungi are known to cause leaf spots on hawthorns. Among the more common ones are *Cercospora confluens, C. apiifoliae, Cercosporella mirabilis, Cylindrosporium brevispina, C. crataegi, Gloeosporium crataegi, Hendersonia crataegicola, Septoria crataegi,* and *Monilinia johnsonii.* The last also causes spots on the fruits, as does the scab fungus *Venturia inaequalis.*

Control. Zineb, mancozeb, or fixed copper sprays will control most of the fungi listed.

POWDERY MILDEW. Two species of powdery mildew fungi, *Phyllactinia guttata* and *Podosphaeria clandestina,* attack hawthorns.

Fig. III-43. Rust of hawthorn caused by the fungus *Gymnosporangium clavipes*.

Fig. III-44. Cedar rust symptoms on hawthorn leaves.

Control. The dithiocarbamates, which control leaf blight and rust, are ineffective against powdery mildew. For these fungi use Benlate, Bayleton, Zyban, or sulfur.

SCAB. *Venturia inaequalis.* Olive-drab spots ¼ inch in diameter appear on the leaves, and smaller spots appear on the fruits. Leaves drop prematurely, and the fruits are disfigured. The fungus may also overwinter on the twigs.

Control. Spray with Captan, dodine, or Fore, four of five times at 10-day intervals when the leaves are half-grown.

Insects and Other Pests

APHIDS. Many species of aphids infest hawthorns: apple, apple grain, four-spotted hawthorn, rosy apple, woolly apple, and woolly hawthorn.

Control. Spray with malathion, Orthene, or Diazinon before the pests become numerous.

APPLE LEAF BLOTCH MINER. *Profenusa collaris.* This wasplike sawfly leaf miner can be very destructive in hawthorn nurseries and ornamental plantings, especially to *Crataegus crus-*

galli and certain other species and cultivars. The principal damage is done during May or June. The first symptom is a small channel in the leaf, which widens to a blisterlike area light brown in color. The miner begins work at one edge of the leaf near the stalk and continues on that side toward the point of the leaf. The inner parts of the leaf are usually completely consumed, only the epidermis and veins remaining. Only the leaves that are unfolding are attacked. Much defoliation follows the work of these insects. The same species attacks many fruit trees, including apple, cherry, plum, quince, and sweet-scented crabapple.

Control. Spray in early May and repeat twice at 2-week intervals with malathion or Sevin.

APPLE AND THORN SKELETONIZER. *Anthophila pariana.* This pest feeds on leaves of apple, pear, and hawthorn in the northeastern United States. The adult moth is dark gray to reddish brown with a ½-inch wing expanse. The fully grown larva is ½ inch long, with a yellowish green body and a pale brown head. Hawthorn leaves may also be skeletonized by the pear slug, a sawfly larva (see under Cherry).

Control. Spray with Sevin when the larvae begin to feed.

WESTERN TENT CATERPILLAR. *Malacosoma pluviale.* Tawny or brown caterpillars with a dorsal row of blue spots feed primarily on hawthorns as well as wild cherry and alder, making tents at the same time as the fall webworm. The moths are smaller and somewhat lighter than the eastern tent caterpillar adults.

Other caterpillars that may feed on hawthorns include the eastern tent, the forest tent, the red-humped, the variable oak leaf, and the walnut caterpillar. The caterpillar stages of the gypsy moth, the cecropia moth, and the western tussock moth also occur on this host.

Control. Spray with *Bacillus thuringiensis,* Dylox, methoxychlor, or Sevin when the caterpillars begin to chew leaves.

BORERS. Four borers infest hawthorns—flat-headed apple tree, roundheaded (see under Mountain-Ash), pear, and shot-hole.

Control. Maintain trees in good vigor by fertilizing when necessary and watering during dry spells. Paint or spray the trunk and branches with methoxychlor at periodic intervals during the growing season. State entomologists will supply the proper dates for each locality.

LACE BUGS. *Corythucha cydoniae and C. bellula.* Lace bugs are not common on hawthorn but occasionally appear to be serious. They feed on the undersides of the leaves, depositing small, brown, sticky spots of excreta. Once the infestation is started, the insects breed during summer and become numerous. Lace bugs have sucking mouthparts. Adults are about ⅛ inch long and have lacy wings. Nymphs may be spiny.

Control. Spray with Lindane, Orthene, malathion, or Sevin in May and July.

PLANTHOPPERS. See under Cherry. Rose leaf-hopper, *Edwardsiana roseae,* also feeds on hawthorn.

Control. Spray the trees forcefully with pyrethrum or rotenone sprays, or a combination of both. The insects must be thoroughly wetted by the sprays.

SCALES. The following kinds of scales infest hawthorn: azalea bark, barnacle, cottony maple, European fruit lecanium, Florida wax, frosted, lecanium, Putnam, oystershell, scurfy, soft, San Jose, and walnut.

Control. Spray with lime sulfur or miscible oil, dormant strength, just before plant growth begins in spring. Follow in mid-May and in June with Diazinon, malathion, or Sevin sprays to control the crawling stages.

TWO-SPOTTED MITE. *Tetranychus urticae.* During the summer months, especially in dry weather, these mites often become sufficiently numerous to injure foliage. Infested leaves take on a gray or yellow cast and may be covered with fine silky threads (see Chapter 11). Eriophyid mites, such as leaf blister mites, also attack hawthorns.

Control. If the hawthorn is given a dormant clean-up spray in spring with commercial lime sulfur, the eggs of the mites are killed. If not, spray with Acaraben, hexakis, or Kelthane in early June and repeat in a few weeks, if necessary.

HEMLOCK *(Tsuga)*

Hemlocks are native to the United States and are among our most graceful and highly prized

Fig. III-45. Fruiting bodies of *Dermatea* canker of hemlock.

evergreens. They grow best in a fairly damp soil where their roots may be cool. The soil must be well drained, however, and moderately acid. Hemlocks do not tolerate heat and drought. Hemlocks are among the few evergreens that thrive near the trunks of large deciduous trees. Like many other conifers, they do not adapt to city conditions and rarely prosper in small front yards of suburban homes.

Hemlocks are less susceptible to diseases and most insect pests than are other conifers, such as firs, pines, and spruces.

Diseases

BLISTER RUST. *Pucciniastrum vaccinii* and *P. hydrangeae*. Young hemlocks and the lower leaves of older trees have yellowish blisters or pustules, from which the spores sift out during June and July. Rhododendron is an alternate host of *P. vaccinii*, which causes a rust-brown leaf spot, more or less injurious in nurseries. Wild and cultivated hydrangeas are alternate hosts of *P. hydrangeae*.

Control. The only control known for these rusts is to be sure that neither of the alternate hosts is thereafter planted in any given region. Spraying with Ferbam may help.

CANKERS. At least six species of fungi, *Botryosphaeria tsugae*, *Cytospora* spp., *Leucocytospora kunzei*, *Dermatea balsamea* (Fig. III-45), *Hymenochaete agglutinans*, and *Phacidiopycnis pseudotsugae*, are known to cause cankers on hemlocks.

Control. Prune affected branches and spray with a copper fungicide if affected trees are particularly valuable.

LEAF BLIGHT. *Fabrella tsugae*. In late summer, leaves of Eastern hemlock turn brown and drop prematurely when attacked by this fungus. Small black fruiting bodies of the fungus occur on the fallen leaves. These produce spores the following spring, which initiate new infections.

Control. Leaf blight rarely damages the trees sufficiently to necessitate measures other than gathering and destroying fallen infected leaves in autumn.

SIROCOCCUS BLIGHT. *Sirococcus conigenus*. Western hemlock is susceptible to this shoot tip killing disease. Controls have not been developed.

NEEDLE RUST. *Melampsora farlowii* and *M. abietis-canadensis*. Eastern hemlock and, to a lesser extent, Carolina hemlock are attacked by these fungi. In late May or early June some of the new leaves turn yellow. Within 2 weeks the shoots to which these leaves are attached turn yellow, become flaccid, and droop. Most of the needles then drop from the affected shoots. Severely rusted trees appear as though their branch tips had been scorched by fire. Red, waxy, linear fungus bodies occur on the lower leaf surfaces, on the shoots, and on the cones.

Control. A spray consisting of 4 pounds of dry lime sulfur in 50 gallons of water applied at weekly intervals in May has given good control of the disease on small trees in nursery plantings. The use of this material for large trees may be justified in unusual circumstances.

HEARTWOOD ROT. *Ganoderma lucidum, Echinodontium tinctorum, and Coniophora puteana.* These fungi cause heart rot or decay of the tissues immediately beneath the bark at the base of the trunk; this results in the death of the tree.

Control. No effective control measures are known. Avoid wounding the bark, and fertilize and water trees to keep them in good vigor.

Abiotic Diseases

SUNSCORCH. Ornamental hemlocks are frequently subject to severe burning or scorching when the temperature reaches 95° F (35° C). The ends of the branches may be killed for several inches back.

DROUGHT INJURY. Hemlocks are more sensitive to prolonged periods of drought than most other narrow-leaved evergreens. The damage is most severe on sites with southern exposures or on rocky slopes where the roots cannot penetrate deeply into the soil. Thousands of hemlocks died in the northeastern United States as a result of severe droughts in the years 1960 to 1966.

AIR POLLUTION. Hemlocks are sensitive to ozone. See Chapter 10.

Insects and Related Pests

HEMLOCK WOOLLY APHID. *Adelges tsugae.* This pest appears as white tufts on the bark and needles. It is capable of killing young ornamental hemlocks.

Control. When the pest becomes prevalent, spray with malathion, Dylox, or Diazinon.

HEMLOCK BORER. *Melanophila fulvoguttata.*

Wide, shallow galleries in the inner bark and sapwood result from boring by a white larva ½ inch long. The adult, a flat, metallic-colored beetle with three circular reddish yellow spots on each wing cover, deposits eggs in bark crevices.

Control. Prune and destroy severely infested branches. Keep the tree in good vigor by fertilizing and watering.

HEMLOCK LOOPER. *Lambdina fiscellaria.* Hemlocks may be completely defoliated by a pale yellow caterpillar with a double row of small black dots along the body, which is more than an inch long at maturity. The adult moth has tan to gray wings, which expand to more than 1 inch. Two species, *L. athasaria athasaria* and *L. fiscellaria lugubrosa,* also defoliate hemlocks.

Control. Spray with Sevin or methoxychlor when the larvae are small.

HEMLOCK FIORINIA SCALE. *Fiorinia externa.* This scale may infest hemlock leaves, and occasionally those of spruce, causing them to turn yellow and drop prematurely. Both male and female scales are elongated. The females are pale yellow to brown and are almost completely covered with their own cast skins. There are two generations a year in the New England states.

Control. Cygon sprays are very effective against this scale. In the northeastern United States, a foliar spray in mid-May gives adequate control of this pest. An additional application in July is necessary for complete control. Dormant oil sprays are also effective.

HEMLOCK SCALE. *Abgrallaspis ithacae.* The adult female, circular and nearly black, infests the lower surfaces of hemlock leaves, causing premature leaf fall. In heavy infestations it may move to the twigs and branches. Another hemlock scale, *A. pini,* also circular and black, infests hemlock, Douglas-fir, and many species of pine.

Yet another species of scale, *Tsugaspidiotus tsugae,* closely resembles *A. ithacae.* It is also circular, somewhat darker brown-black, and nippled at the center. The heaviest infestations of this scale have been in Fairfield County, Connecticut.

Control. Spray with a superior dormant oil,

or a combination of the oil and ethion, in April before new growth emerges. A Cygon spray during the growing season is also effective.

GRAPE SCALE. *Aspidiotus uvae.* Hemlock hedges can be destroyed by infestations of this small, dingy-white scale, which has yellowish nipples or exuviae. Japanese wax scale also attacks hemlock.

Control. Malathion or Sevin sprays applied from mid- to late June provide good control.

SPRUCE LEAF MINER. *Taniva albolineana.* This species, more common on spruce, occasionally mines the leaves of hemlock.

Control. Spray with malathion when the pests begin to feed in late May.

HEMLOCK ERIOPHYID MITE. *Nalepella tsugifoliae.* The unthrifty look of some hemlocks may be due to infestations of this mite.

Control. Kelthane sprays applied in early April will control this pest. Superior oil or oil plus ethion in early April is also effective.

SPIDER MITES. Hemlocks in ornamental plantings are extremely susceptible to several other species of mites. The spruce spider mite (see under Spruce) is prevalent on hemlock. The two-spotted mite, *Tetranychus urticae,* feeds on the undersides of the needles, sucking the juice from the cells; the needles turn pale and become spotted. Eggs and mites are usually covered with delicate webs.

Control. Mites can be controlled on this host by spraying with Acaraben, or Kelthane. These should not be applied to tender foliage during the hottest part of the day, for the susceptibility of hemlock foliage to intense heat is aggravated by spraying.

OTHER PESTS. Hemlocks are also subject to the bagworm, cypress moth, blackheaded budworm, Japanese weevil (see under Holly), black vine weevil (see under Yew), fir flatheaded borer, spruce budworm, gypsy moth, and hemlock sawfly.

Control. Control measures are rarely used.

HICKORY (Carya)

Hickories are not commonly planted as shade trees, but wooded lot homesites frequently include hickories. Pignut hickory grows on dry upland sites in the eastern United States; most other hickories grow in moist sites having deep soils. Most hickories have naturally deep taproots, perhaps making transplanting difficult, thus limiting their use as landscape trees.

Diseases

CANKER. Several canker diseases occasionally occur on hickory. These are associated with the fungi *Strumella coryneoidea, Nectria galligena,* and *Rosellinia caryae. Poria spiculosa* causes a serious canker rot of hickory stems. These cankers, with thick deep callus folds, appear as rough, circular trunk swellings having depressed centers with an old decayed branch stub.

Control. Prune dead or weak branches. Avoid bark injuries and keep the trees in good vigor by fertilizing and by watering during dry spells. Keep borer and other insect infestations under control.

CROWN GALL. *Agrobacterium tumefaciens.* This bacterial disease occurs occasionally on hickory. Heavily galled limbs and branches resembling crown gall are actually caused by *Phomopsis* sp. infections.

Control. Prune and destroy infected twigs or branches.

LEAF SPOTS. Several leaf-spotting fungi occur on hickory. Of these, *Gnomonia caryae* is the most destructive. It produces large, irregularly circular spots, which are reddish brown on the upper leaf surface and brown on the lower. The margins of the spots are not sharply defined, as are those of many other leaf spots. The minute brown pustules on the lower surface are the summer spore-producing bodies. Another spore stage develops on dead leaves and releases spores the following spring to initiate new infections. The fungus *Monochaetia monochaeta* occasionally produces a leaf spot on hickory but is more prevalent on oaks. The fungus *Marssoniella juglandis* also attacks hickory but is more destructive to black walnut. Two species of *Septoria, S. caryae* and *S. hicoriae,* also cause spots on hickory.

Control. Gather and destroy leaves in fall to kill fungi they harbor. To protect valuable specimens, spray the leaves with maneb or zineb when the leaves unfurl, when half-grown, and again when full-grown.

POWDERY MILDEWS. *Phyllactinia guttata and Microsphaera caryae.* Two fungi cause mildewing of leaves.

Control. Control measures are rarely adopted.

WITCHES' BROOM. *Microstroma juglandis.* This fungus, which causes a leaf spot of butternut and black walnut, is capable of causing a witches' broom disease on shagbark hickory. The brooms, best seen when the trees are dormant, are composed of a compact cluster of branches. Early in the growing season the leaves on these branches are undersized and curled with white, moldy growth on the lower leaf surface; later they turn black and fall. Another witches' broom disease of hickory and pecan, called bunch disease, is caused by a mycoplasma.

Control. No effective control measures have been developed, but a dormant lime sulfur spray, followed by Ferbam sprays during the growing season, is suggested.

TRUNK ROTS. *Ganoderma applanatum, Oxyporus populinus, Phellinus igniarius, Climacodon septentrionalis, and Inonotus andersonii.* These fungi cause hickory trunk rots.

Insects

HICKORY LEAF STEM GALL APHID (ADELGID). *Phylloxera caryaecaulis.* Hollow green galls in June, which turn black in July, on leaves, stems, and small twigs of hickory are caused by the sucking of this small louse. In June the insides of the galls are lined with minute shiny lice of varying sizes. Galls range from the size of a small pea to more than ½ inch in diameter.

Control. A dormant oil spray in early spring just as growth begins should destroy many overwintering lice. Diazinon or malathion sprays early in the growing season at budbreak and repeated 2 weeks later also give control. Sprays are not effective once the galls begin to develop.

APHIDS. Black pecan and giant bark aphids attack hickory.

HICKORY BARK BEETLE. *Scolytus quadrispinosus.* Young twigs wilt as a result of boring by the bark beetle, a dark brown insect ⅕ inch in length. The bark and sapwood are mined, and the tree may be girdled by the fleshy, legless larvae ¼ inch long. The larvae overwinter under the bark.

Control. Spray the foliage with Sevin when the beetles appear in July. Remove and destroy severely infested trees, and peel the bark from the stump. Increase the vigor of weak trees by fertilization and watering.

CATERPILLARS. The leaves of park trees may be chewed by one of the following caterpillars: elm spanworm, hickory-horned devil, red-humped, walnut, yellow-necked, white-marked tussock moth, fall cankerworm, hickory tussock moth, fall webworm, fruit tree leafroller, and green mapleworm.

Control. All caterpillars are readily controlled with *Bacillus thuringiensis* or Sevin sprays.

JUNE BUGS. *Phyllophaga sp.* The leaves may be chewed during the night by light to dark brown beetles, which vary from ½ to ⅞ inch in length. The beetles rest in nearby fields during the day. The larva is ¾ to 1 inch long, white, and soft-bodied with a brown head. Three or more years are required for completion of the life cycle of most species.

Control. Spray the leaves with Sevin during late May or early June, depending on the locality. The larval stage, which feeds on grasses in lawns and golf courses, can be controlled by treating the lawns with Diazinon or Dursban.

PECAN CIGAR CASEBEARER. *Coleophora caryaefoliella.* The leaves are mined, turn brown, and fall when infested by the pecan cigar casebearer, a larva ⅓ inch long, with a black head. The adult female is a moth with brown wings that have fringed hairs along the edge and a spread of ⅖ inch. The larvae overwinter on twigs and branches in cigar-shaped cases ⅛ inch long.

Control. Spray with malathion as soon as the leaves are fully developed.

PAINTED HICKORY BORER. *Megacyllene caryae.* The sapwood of recently killed trees is soon riddled by painted hickory borers, creamy white larvae that attain a length of ¾ inch. The adult beetle is dark brown, has zigzag lines on the back, and is ¾ inch long. Eggs are deposited in late May or early June. Dogwood and red-headed ash borers also attack hickory (see under Dogwood and Ash).

Control. Remove and destroy dead trees immediately. Inject a nicotine paste such as Bortox into the tunnels of live trees and then seal the openings with chewing gum, grafting wax, or putty. Spraying the bark with Lindane in June, July or August might also help. Thiodan will also control this borer but its use is restricted to professional arborists and nurserymen in some states.

OAK TWIG PRUNER. *Elaphinoides villosus.* Twigs weakened by tunneling activity of this pest, a larva ½ inch long, are then broken off by the wind and fall to the ground. The adult is a reddish brown beetle ¾ inch long. The larvae overwinter inside the twigs on the ground. Another insect, the twig girdler, also attacks hickory (see under Hackberry).

Control. Gather and destroy severed branches and twigs in autumn or early spring.

SCALES. Several species of scales—grape, obscure, walnut, and Putnam—occasionally infest hickories.

Control. Spray with Orthene, Diazinon, or Sevin during late spring and early summer, or with lime sulfur just before growth starts in spring.

MITE. *Eotetranychus hicoriae.* This pest occasionally infests hickory leaves, causing them to turn yellow and drop prematurely. An eriophyid gall mite also attacks hickory.

Control. Spray with malathion or a miticide such as Kelthane.

OTHER INSECTS. Butternut woolly worm (see under Walnut), Asiatic oak weevil (see under Oak), maple leafhopper (see leafhoppers, under Maple), and planthoppers (see under Cherry) also feed on hickory.

HOLLY *(Ilex)*

The American holly, *Ilex opaca,* native to the eastern United States, is not subject to large numbers of pests. Most of the troubles experienced with trees in ornamental plantings result from improper transplanting practices or unfavorable soil conditions. In localities where it can withstand the winters, holly will thrive in almost any type of soil that is well drained and contains considerable amounts of humus. Incorporating several bushels of well-rotted oak leaf mold into the soil at transplanting time will help to provide the conditions favored by this tree. When the tree is established in its new site, cottonseed meal should be occasionally worked into the soil to supply nitrogen.

Many other species of holly, both of native and Asiatic origin, are also grown. Unless specifically noted, the diseases and insects described here affect primarily the American holly. American holly leaves remain attached for 3 years and are shed in the spring, thus giving foliar pests plenty of time to attack.

Diseases

BACTERIAL BLIGHT. *Corynebacterium ilicis.* This disease was first found in a holly orchard on Nantucket Island in 1957. Leaves and shoots of the primary growth appear scorched in June and July. Diseased shoots wilt, droop, and dry but persist. The infection progresses into the woody shoots of the previous year's growth, where the leaves turn black (Fig. III-46).

Control. Copper fungicides will probably control this disease. Excessive use of nitrogenous fertilizers and cultivation of soil beneath the trees increase the trees' susceptibility to bacterial blight.

CANKER. *Botryosphaeria dothidea, Diaporthe eres, Nectria coccinea, Physalospora ilicis, Phomopsis (Diaporthe) crustosa,* and *Diplodia* sp. Sunken areas on the twigs and stems may be caused by these fungi.

Control. Prune diseased branches and spray with copper fungicides several times in late spring.

ANTHRACNOSE. *Gloeosporium ilicis.* Native holly *(Ilex opaca)* leaves develop dead blotches which look like winter scorch symptoms but with a prominent black line margin.

Control. Pick off and destroy affected leaves. New foliage can be protected with either a copper fungicide or Ferbam spray applied twice at 2-week intervals, starting when the leaves reach full size.

LEAF ROT, DROP. *Pellicularia filamentosa.* As a result of invasion by the *Rhizoctonia* stage of the causal fungus, the leaves of American holly

Fig. III-46. Blight of American holly caused by the bacterium *Corynebacterium ilicis*.

cuttings may decay and drop about 2 weeks after the cuttings are inserted into the rooting medium. The disease first appears as a cob-webby coating, a combination of the fungus threads and grains of sand adhering to the undersides of the leaves which touch the sand.

Control. Insert cuttings in clean fresh sand or in pasteurized old sand. Do not use cuttings taken from holly branches that touch the ground.

LEAF SPOTS. *Cercospora ilicis, C. ilicicola, C. pulvinula, Cylindrocladium sp., Englerulaster orbicularis, Gloeosporium aquifolii, Macrophoma phacidiella, Microthyriella cuticulosa, Phyllosticta concomitans, P. terminalis, Rhytisma ilicinicola, R. velatum, Sclerophoma sp., and Septoria ilicifolia.* Many fungi cause brown spots of the leaves.

Control. Infected leaves should be picked off and destroyed, and the vigor of the trees improved by incorporating oak leaf mold or cottonseed meal into the soil. In addition, water should be provided during dry spells. Applications of bordeaux mixture or any other copper spray in late summer and in early fall will largely prevent the formation of the spots. More lasting results are obtained, however, by improving the soil conditions. Copper fungicides may cause some injury, especially to leaves that

have been punctured by the holly leaf miner. They also leave an unsightly residue which persists for some time. Benlate, Ferbam, or Fore sprays should be used where copper sprays are likely to cause injury.

POWDERY MILDEWS. *Microsphaera nemopanthis and Phyllactinia guttata.* In the South, holly leaves may be affected by these mildew fungi.

Control. Where the disease becomes prevalent, spray with Benlate, Bayleton, or wettable sulfur.

SPOT ANTHRACNOSE. *Elsinoë ilicis.* Leaves of Chinese holly *(I. cornuta)* in the South are occasionally affected by this disease. Two types of lesions occur on the leaves: numerous tiny black spots and a large leaf-distorting spot more than an inch in length, which is confined to half of the leaf blade. Lesions on the shoots and berries may also occur.

Control. Periodic applications of copper fungicides will provide control.

TAR SPOT. Native holly and English holly *(I. aquifolium)* are subject to the tar spot disease.

Symptoms. Yellow spots appear on the leaves during late May. These turn reddish brown and finally black by fall. A narrow border of yellow tissue remains around the darkened spots. The

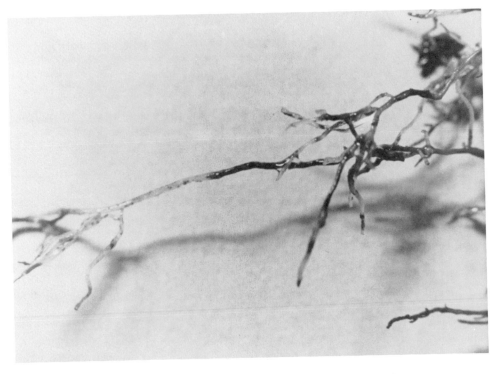

Fig. III-47. Holly roots showing black lesions characteristic of black root rot.

efficiency of the leaf is reduced when a large number of spots develop. Premature defoliation seldom occurs.

Cause. Tar spot is caused by the fungus *Phacidium curtisii.* Spores produced in the blackened areas initiate new infections in early spring.

Control. Gather and destroy badly spotted leaves. Spray with a copper fungicide several times at 2-week intervals starting in late spring. Such sprays may cause slight injury if the season is cool.

BLIGHT. *Botrytis cinerea.* Blossoms and new growth are blighted in wet seasons.

TWIG DIEBACK. Black stem cankers and black spots on the leaves of English holly grown in the Pacific Northwest are typical of this relatively new fungus disease. During cool rainy weather, complete defoliation and severe twig blighting of the lowermost branches are common.

Cause. The fungus *Phytophthora ilicis* causes this disease.

Control. Although no control measures have been developed, it is highly probable that periodic applications of copper fungicides will provide control.

BLACK ROOT ROT. *Thielaviopsis basicola,* (also referred to as *Chalara elegans*). This fungus, the cause of Japanese and blue holly decline and death, also infects American holly roots (Fig. III-47). We have observed this fungus associated with declining American holly in Kentucky. During drought seasons, whole sections of trees having infected roots will wilt and die.

Control. Provide good growing conditions, including adequate water.

OTHER DISEASES. A virus disease called yellow leaf spot affects holly; a red leaf spot is caused by a species of *Sclerophoma;* and several species of nemas cause root-knot.

Control. Control measures are usually unnecessary.

Nonparasitic Diseases

SPINE SPOT. Small gray spots with purple halos are caused by the puncturing of the leaves by the spines of adjacent holly leaves. A careful examination with a hand lens will reveal tiny circular holes or an irregular tear at the center of each spot.

Spine spot is often confused with slits made by the holly leaf miner. The latter have neither a gray center nor a purple halo.

LEAF SCORCH. A browning or scorching of the leaves, common on holly in late winter or early spring, is of nonparasitic origin. Occasionally it is caused by the presence of water or ice on the leaves at the time the sun is shining brightly. This causes a scalding, followed by invasion by secondary organisms and finally by scorching.

Hollies planted in wind-swept areas are also more susceptible to so-called winter drying. The leaves in late winter or early spring lose water faster than it can be replaced through the roots. As a result the leaf edges wilt and turn brown. In exposed situations, newly transplanted holly should be protected with some sort of windbreak or sprayed with an antidesiccant such as Wilt-Pruf NCF.

Insects

BEETLES. The black blister beetle, the Japanese beetle, and the potato flea beetle occasionally infest hollies.

Control. Spray with a Sevin–Kelthane mixture when the beetles first appear.

BERRY MIDGE. *Asphondylia ilicicola.* The larvae of this pest infest holly berries and prevent them from turning red in fall.

Control. Where only a few trees are involved, hand picking and destruction of infested berries should keep this pest under control. Diazinon spray applied in early June will control the midge where many trees are being grown.

CAMBIUM MINER. *Marmara* and other spp. Thin-barked trees such as young holly, fir, pine, birch, and cherry show serpentine mines in the bark. These larvae feed primarily on phloem and generally do little damage. Controls are not normally used.

BUD MOTH. *Rhopobota naevana ilicifoliana.* Holly in the Pacific Northwest is subject to this pest, the larval stage of which feeds on the buds and terminal growth inside a web.

Control. Spray with methoxychlor or Sevin between the opening of the leaf bud and the time of blossoming.

HOLLY LEAF MINER. *Phytomyza ilicis.* Yellow or brown serpentine mines or blotches in leaves are produced by the leaf miner, a small yellowish white maggot, 1/6 inch long, that feeds between the leaf surfaces (Fig. III-48). The adult is a small black fly that emerges about May 1 and makes slits in the lower leaf surfaces, where it deposits eggs.

Another species, the native leaf miner, *P. ilicicola,* produces very slender mines and may occur on the same tree. Holly leaf miners are the most damaging pest of landscape hollies.

Fig. III-48. Blotch on holly leaf caused by the leaf miner *Phytomyza ilicis.*

Control. Diazinon, Sevin, or Dylox sprays applied when the adults are first observed in mid-May and again in early June provide effective control. Orthene or Cygon can be used in June to control the larvae in the mines.

HOLLY SCALE. *Dynaspidiotus britannicus.* Circular, flat, $\frac{1}{16}$-inch scales infest the berries, leaves, and twigs of holly in the West.

Control. Malathion, methoxychlor, or Sevin sprays applied in July to the young crawling stage give excellent control. Superior oil or oil plus ethion applied in April is also helpful.

PIT-MAKING SCALE. *Asterolecanium puteanum.* This scale is becoming more prevalent on American holly and Yaupon (*I. vomitoria*) in the southern states. Oval, $\frac{1}{16}$ inch in diameter when mature, and pale yellow in color, this insect embeds itself in the bark and causes a pitted and swollen condition of the stems somewhat resembling that produced by the golden oak scale. Branches are distorted, the leaves take on an abnormal color, and at times there is considerable dieback from the branch tips.

Control. The same as for holly scale.

OTHER SCALES. Many other species of scale insects attack hollies: black, cottony maple leaf, cottony taxus, Japanese wax, nigra, latania, purple, dictyospermum, gloomy, walnut, euonymus, California red, greedy, lecanium, oleander, oystershell, peach, soft, and tea. Comstock mealybug also feeds on holly.

Control. Most of these scales can be controlled with malathion or Sevin sprays applied when the young are crawling about in spring, the application being repeated several times at 10-day intervals.

JAPANESE WEEVIL. *Callirhopalus bifasciatus.* The leaves of ash, elm, dogwood, hemlock, and oak, as well as of holly, are occasionally attacked by this pest. The beetles are about $\frac{1}{4}$ inch in length, varying from light to dark brown, with striations on the wing covers. Weevil feeding results in notches cut into the leaf margin. This pest, found in the Northeast and Midwest, will sometimes lie motionless on the ground.

Control. Orthene or Sevin sprays applied in late summer will control this pest.

WHITEFLY. See under Dogwood.

APHIDS. *Toxoptera aurantii.* The citrus aphid feeds on holly.

PSYLLID. *Metaphalaria ilicis.* The yaupon psyllid forms a reddish gall on yaupon holly.

Other Pests

SOUTHERN RED MITE. *Oligonychus ilicis.* This mite has become a serious pest of holly.

Control. Apply a dormant oil spray in March or April just before new growth starts, then spray with Vendex, Diazinon, or Kelthane about mid-May and repeat in 10 days.

HONEYLOCUST. See Locust, Honey

HOP-HORNBEAM (Ostrya)

Hop-hornbeam, also known as ironwood, is a slow-growing tree with a rounded crown and slender, pendulous, often contorted branches. This American native is tolerant of a wide range of soils.

Diseases

Hop-hornbeam is susceptible to cankers caused by *Aleurodiscus* sp., *Nectria* sp., and *Strumella coryneoidea;* to leaf spots by *Cylindrosporium dearnessi* and *Septoria ostryae;* to powdery mildews by *Microsphaera ellisii, Phyllactinia guttata,* and *Uncinula macrospora;* to root rots by *Armillaria mellea* and *Clitocybe tabescens;* to a leaf blister by *Taphrina virginica;* and to a rust caused by *Melampsoridium carpini. Gnomoniella carpinea* (imperfect, *Monostichella robergei*) causes twig cankers and leaf spots of ironwood.

Control. The controls for these diseases are discussed under more seriously affected hosts.

Insects

Among the insects that attack hop-hornbeam are the birch lace bug, the melon aphid, pitted ambrosia beetle, two-lined chestnut borer, and two species of scales, cottony-cushion and latania.

Control. These insects are rarely serious enough to require control measures.

HOP-TREE *(Ptelea)*

Diseases

LEAF SPOTS. *Cercospora afflata, C. pteleae, Phleospora pteleae, Phyllosticta pteleicola,* and *Septoria pteleae.* These five species of fungi cause leaf spots on hop-tree.

Control. Pick off and destroy spotted leaves. Other control measures are rarely necessary.

RUST. *Puccinia windsoriae.* This fungus occasionaly occurs on hop-tree. The alternate stage of the fungus is found on grasses.

Control. No controls are required.

ROOT ROT. *Phymatotrichum omnivorum.*

Control. Control measures have not been developed.

Insects

TWO-MARKED TREEHOPPER. *Enchenopa binotata.* These little sucking insects, ¼ inch in length, with a long, proboscislike head portion, resemble miniature quail or partridges; they are dark brown in color with two white spots. When disturbed, the insects jump very rapidly from place to place. They secrete honeydew. The egg masses are covered by a snow-white frothy substance (Fig. III-49) like that secreted by spittle insects but much firmer in consistency. From a distance, infested branches resemble those infested with woolly aphids or cottony-cushion scale. Walnut, redbud, hickory, locust, and sycamore are also infested.

Control. Spray with malathion or with a pyrethrum–rotenone compound when the insects are young.

SCALE. *Pseudaulacaspis pentagona.* This pest, also known as West Indian peach scale, infests a wide variety of trees in the warmer parts of the country.

Control. Malathion or Sevin sprays when the crawling stage is moving about in the early spring will provide control.

HORNBEAM *(Carpinus)*

European hornbeam has an attractive pyramidal shape. It is relatively tolerant of air pollution.

Fig. III-49. Treehopper *(Enchenopa binotata)* egg masses on hop-tree.

Diseases

LEAF SPOTS. *Clasterosporium cornigerum, Monostichella robergei, Gnomoniella fimbriella, Phyllosticta* sp., and *Septoria carpinea.* These five fungi cause leaf spotting of hornbeam.

Control. Leaf spots are rarely serious enough to warrant control measures. Copper or Ferbam sprays are effective in preventing heavy outbreaks.

CANKERS. *Pezicula carpinea* and *Nectria galligena.* These two species of fungi frequently cause bark cankers, sometimes leading to severe dieback of branches.

Control. Badly cankered trees cannot be saved. Prune out twigs and branches of mildly infected ones. The use of copper sprays may also be justified in some situations.

TWIG BLIGHT. *Fusarium lateritium.* In the South this fungus causes a twig blight on hornbeam. The sexual stage of this fungus is *Gibberella baccata.*

Control. Prune and destroy infected twigs on valuable ornamental specimens.

OTHER DISEASE. The felt fungus, *Septobasidium curtisii*, occurs on a number of trees in the South. The felt is purplish black and covers the branch and the scale insect the fungus parasitizes.

Control. Control measures are not needed.

Insects

SOURGUM SCALE. *Phenacaspis nyssae.* This scale is nearly triangular, flat, and snow-white.

Control. Malathion or Sevin sprays directed to the lower leaf surfaces in late spring are effective.

TWO-LINED CHESTNUT BORER. See under Oak. This pest attacks European hornbeam following severe winters, killing branches in the upper crown, and sometimes killing the entire tree.

LEAFHOPPER. See under Locust, Honey.

MAPLE PHENACOCCUS. See under Maple.

HORSECHESTNUT *(Aesculus)*

Valued for its shape and palmate leaf, horsechestnut tends to be messy and troublesome because of leaf blotch and scorch. Other species in this genus are called buckeye and some are good landscape trees. Buckeye and horsechestnut are subject to many of the same diseases, with buckeye generally less damaged. Unlike buckeye, horsechestnut is not native to the United States.

Diseases

ANTHRACNOSE. *Glomerella cingulata.* Leaf petioles, midribs, and veins turn brown, distinguishing this disease from leaf blotch. In addition, terminal shoots become blighted down to several inches below the buds. Diseased tissue is shrunken, and the epidermis and young bark are ruptured; pustules are formed containing the pink spores of *Colletotrichum*, the imperfect stage of the anthracnose fungus.

Control. Spray as described below for leaf blotch.

CANKER. *Nectria cinnabarina.* This disease is said to attack the branches and to cause much defoliaton of old trees; however, the fungus is apparently mostly saprophytic.

LEAF BLOTCH. *Guignardia aesculi.* This fungus disease is very serious in nurseries, where it often causes complete defoliation of the stock. The spots may be small or so large they include nearly all the leaf. At first they are merely discolored and water-soaked in appearance; later they turn a light reddish brown with a very bright yellow marginal zone (Figs. III-50 and 51). When the whole leaf is infected it becomes

Fig. III-50. Early stage of the horsechestnut leaf blotch disease.

Fig. III-51. Advanced stage of the disease.

dry and brittle and usually falls. The small black specks seen in the center of the spot are the fruiting bodies of *Phyllosticta,* the imperfect stage of the fungus. The leaf stalks are also attacked. This leaf blotch is very similar to scorch, often seen on shade trees along streets and in city parks. The two diseases can be distinguished by the small, black, pimplelike fruiting bodies on the leaf blotch caused by the fungus. The fungus overwinters on the old leaves where it has produced its perfect stage. The ascospores are the means by which the disease is spread in the spring. The first signs of infection may not appear until some time in July.

Control. Old leaves under diseased trees should be raked up and carted away. Spray with dodine, Fore, or zineb, two to four times at 10-day intervals, starting after the buds open. The total number of applications is governed by the weather. The disease is always more severe during very wet springs.

LEAF SPOT. *Septoria hippocastani.* Small brown circular spots occasionally develop on the leaves of this host. The slender spores may be seen when the fruiting structures are observed under the microscope.

Control. The sprays recommended for leaf blotch will also control this fungus.

POWDERY MILDEW. *Uncinula flexuosa.* This disease is prevalent in the Middle West, where the undersides of leaves frequently are covered with white mold. The fruiting bodies of the winter stage of the fungus appear as small black dots over the mold.

Control. Spray trees with Bayleton, Benlate, or wettable sulfur a few times at weekly intervals, starting when the mildew appears.

WOUND ROT. *Collybia velutipes.* The fungus enters through wounds and destroys the wood, later forming clusters of mushroomlike fruiting bodies. It is one of the fungi which may be found during the winter months still attached to the trees. The fruiting bodies have dark brown, velvety stalks. Other fungi such as *Polyporus, Stereum, Hypoxylon,* and *Trametes* may also decay living horsechestnut.

Control. Remove limbs that have been killed by this rot and cover the cut surface to prevent the entrance of fungus spores into the wound.

OTHER DISEASES. Horsechestnut is susceptible to three stem diseases: wilt caused by *Verticillium albo-atrum,* canker by *Diaporthe ambigua,* and bleeding canker by *Phytophthora cactorum.* The former is discussed in Chapter 12 and the latter under crown canker in dogwood.

Some horsechestnut trees are susceptible to nonparasitic leaf scorch. The scorching usually becomes evident in July or August. First the margins of the leaves become brown and curled. Within 2 or 3 weeks the scorch may extend over the entire leaf. Some observers report that scorch is more prevalent in hot, dry seasons having polluted air, but serious injury also has been noted in wet seasons. Trees that are prone to scorch will show symptoms every year regardless of the kind of weather, whereas others nearby may show none.

Control. Prune susceptible trees and provide them with good growing conditions. Fertilize and water when necessary.

Insect Pests

COMSTOCK MEALYBUG. See under Catalpa.

WHITE-MARKED TUSSOCK MOTH. See under Elm.

JAPANESE BEETLE. See under Linden.

ELM SPANWORM. See under Elm.

APHIDS. *Drepanaphis* and *Periphyllus* spp. These aphids infest horsechestnut.

LEAFHOPPERS. *Empoasca* sp. This pest causes premature defoliation of California buckeye.

WALNUT SCALE. See under Walnut.

BAGWORM. See under Juniper.

FLATHEADED BORER. See under Maple.

JACARANDA *(Jacaranda)*

This native of Brazil produces showy lavender flowers in spring. Leaves are double compound, giving the tree a feathery look. Jacaranda, planted as a street and landscape tree in southern California, tends to be weak wooded with narrow branch crotch angles. Pest problems are minimal.

JAPANESE LILAC TREE *(Syringa)*

This is a hardy, small tree that is drought-resistant.

JAPANESE PAGODA TREE. See Pagoda-Tree.

JAPANESE SNOWBELL *(Styrax)*

This small tree has pendent flowers and grows best in well-drained, rich soil. It suffers few problems.

JUNEBERRY See Serviceberry

JUNIPER *(Juniperus)*

Because of their evergreen habit, diversity of form, and other desirable qualities the species and varieties in the genus *Juniperus,* known commonly as cedars and junipers, are extensively used in ornamental plantings.

Diseases

CEDAR RUSTS. The several species of rust fungi that attack various species of *Juniperus* occasionally produce material damage. These rusts require hosts other than cedar for the completion of their life cycles, inasmuch as the spores produced on one host are unable to reinfect the plant upon which they are formed. The fungi cannot survive, therefore if either of the alternate hosts is removed. For this reason, several important apple-growing states prohibit the planting of susceptible cedars in the vicinity of apple orchards.

The rust fungi all belong to the genus *Gymnosporangium.* The cedar hosts are *Juniperus virginiana* and many of its varieties, *J. sabina, J. scopulorum, J. communis,* and *J. sibirica.* The alternate hosts for these rusts belong to the order *Rosales,* including the cultivated apple, pear, and ornamental crab; *Crataegus; Amelanchier; Cydonia; Sorbus; Aronia;* and others of less importance.

Although at least seven different species of rust fungi attack *Juniperus,* only three are important in the East. These are known as the cedar–apple rust, the cedar–quince rust, and the cedar–hawthorn rust.

The cedar–apple rust, caused by the fungus *Gymnosporangium juniperi-virginianae,* attacks junipers and apples. On juniper twigs it produces large, globose galls, ranging up to 1½ inches in diameter. Long gelatinous orange tendrils or spore horns protrude from these galls during rainy weather in spring (Fig. III-52). The spores formed on the orange tendrils are carried by wind and insects to developing apple leaves and fruits. Yellow or orange lesions are then produced on the leaves or fruits as a result of infection by these spores. This stage of the

Fig. III-52. Orange galls of the cedar–apple rust fungus *Gymnosporangium juniperi-virginianae* showing gelatinous ''horns.''

disease is most destructive, inasmuch as severe defoliation may occur, especially on apple. Late in the season, fungal spores from apple are blown to juniper, causing foliage infection there. Galls having orange telial horns are produced some 18 to 20 months later, thus repeating the cycle. Galls produce spores in only one season, drying up after spring.

Control of Cedar–Apple Rust. On apple, hawthorn, and other rosaceous hosts, as many as six applications, at 10-day intervals, of a wettable sulfur, mancozeb, or a mixture of wettable sulfur and Ferbam may be needed for good control. The initial application should be made just before an expected rainy spell and as soon as the leaves emerge in spring.

Cedar–apple and cedar–hawthorn rust-resistant junipers include 'Foemina' and *sargentii* Chinese juniper; 'Aureospica' *depressa,* 'Repanda', *saxatilis,* and 'suecica' common junipers; Savin juniper; *fargesii* red cedar; and 'Tripartita' eastern red cedar.

The cedar–quince rust, caused by the fungus *G. clavipes,* attacks junipers, quince, and the fruit of the hawthorn, the juneberry, and the apple. It appears in the spring and differs somewhat in appearance from the cedar–apple rust. On juniper, it produces slight fusiform swellings on the twigs (Fig. III-53) branches, and occasionally the trunk. During wet weather, reddish orange spore masses break through the bark. Of the three rusts discussed, quince rust is the most destructive on the evergreen hosts. On the fruits of the rosaceous hosts, quince rust produces slightly swollen areas covered with tiny deep red dots. Long, white tubelike structures eventually emerge, open, and expose orange-brown spore masses. These spores then reinfect the *Juniperus* hosts. Cedar–quince galls are perennial on juniper, producing spores in spring for several seasons.

Control of Cedar–Quince Rust. On quince, hawthorn, and other rosaceous hosts, three applications, at 10-day intervals, of the sprays recommended for cedar–apple rust will give control. The first application should be made when the blossom buds are opening.

The cedar–hawthorn rust, caused by the fun-

Fig. III-53. Cedar–quince rust of juniper.

Fig. III-54. Phomopsis twig blight of juniper.

gus *G. globosum,* attacks junipers, hawthorn, and to a lesser extent apple. On junipers, it produces small irregularly shaped galls, less than an inch in diameter, with wedge-shaped, gelatinous, orange spore masses. The symptoms on hawthorn and apple are similar to those produced by the cedar–apple rust.

Control of Cedar–Hawthorn Rust. The spray schedule suggested for the control of cedar–apple rust is also recommended for the control of cedar–hawthorn rust.

A fourth rust, juniper broom rust occurs throughout North America. The fungus, *Gymnosporangium nidus-avis* attacks serviceberry, apple, and quince as well as juniper. On the latter host, symptoms include fusiform branch swellings with longitudinal cracks, and foliar witches' brooms.

PHOMOPSIS TWIG BLIGHT. Although primarily a disease of seedlings and nursery stock, twig blight may appear on 8- to 10-foot trees in ornamental plantings and on larger native red cedars in some parts of the country. The disease becomes progressively less serious, however, as the trees become older, and little damage occurs on trees over 5 years old.

Though primarily a disease of the common red cedar *(J. virginiana)* and its horticultural varieties, twig blight has been found on more than a dozen other groups, among which are aborvitae, cypresses, retinosporas, white cedar, and other species of *Juniperus.*

Symptoms. The tips of branches first turn brown (Fig. III-54), followed by progressive dying back until an entire branch or even the entire young tree is killed.

Cause. Twig blight is caused by the fungus *Phomopsis juniperovora.* Older trees in the vicinity of an evergreen nursery may harbor the fungus on diseased twigs. The fungus may also be carried on small, diseased branch tips that accompany purchased seed.

In spring and summer, large numbers of spores are produced in fruiting bodies on infected twigs and branches. During rainy periods the spores ooze out from the bodies and are spread to other plants by wind, rain, and laborers.

Control. Where practicable, prune out and destroy infected branch tips. Spray with Benlate or Zyban plus a spreader–sticker once in the fall and three or four times at 2-week intervals in spring, starting when warm weather begins.

The use of resistant varieties offers considerable promise as a means of avoiding twig blight. Cultivars and varieties reportedly resistant are Chinese juniper—'Foemina', 'Iowa', 'Keteleeri', 'Pfitzeriana Aurea', 'Robusta', *sargentii* and cv. 'Glauca', and 'Shoo smith'; common juniper—'Ashfordii', 'Aureo-spica', *depressa*, 'Hulkjaerhus', 'Prostrata Aurea', 'Repanda', *saxatilis*, 'Suecica'; creeping juniper —'Depressa', 'Procumbens'; savin—'Broadmoor', 'Knap Hill', 'Skandia'; western red cedar—'Silver King', 'Campbellii', *fargesii*, 'Prostrata', 'Pumila'; and red cedar—'Tripartita'.

The following cultivars, all of *Chamaecyparis pisifera*, are reported to be resistant: *filifera aureovariegata, plumosa aurea, plumososa argentea, plumosa lutescens,* and *squarrosa sulfurea.*

KABATINA TWIG BLIGHT. *Kabatina juniperi.* This fungus causes a serious twig blight of juniper similar in symptoms to Phomopsis tip blight. Tips turn brown and die in spring as new growth begins, suggesting that infections may occur during the previous season. The fungus enters through wounds; infection may be associated with insect feeding.

Control. This disease is not easily controlled. We have found no fungicides that control the disease well, although mancozeb fungicides may help. Juniper cultivars have not been evaluated for resistance to Kabatina, but Phomopsis resistant-types may be resistant to both diseases.

CERCOSPORA BLIGHT. *Cercospora sequoiae* and *C. sequoiae* var. *juniperi.* Cercospora blight can be distinguished from Phomopsis and Kabatina blights because Cercospora blight leaves trees with healthy tips and blighted foliage away from the tip. Bordeaux mixture applied in late spring and early summer will prevent Cercospora blight.

ROOT ROT. *Phytophthora cinnamomi.* Although many junipers are considered to be resistant to this fungus, some cultivars are susceptible.

Control. Infected plants cannot be saved. Pasteurize infested soil and replant with healthy specimens. Subdue or Aliette fungicides can be used to prevent root rot.

WOOD DECAY. Several species of wood-decay fungi, such as *Heterobasidion annosus* and *Antrodia juniperina,* are occasionally found on junipers but rarely become destructive enough to warrant special treatments. The fungi usually enter through injuries at the base of the tree. Prevention and treatment of wounds and fertilization to maintain the tree in good vigor are recommended.

Abiotic Diseases

Hybrid junipers are among evergreens that may be seriously injured and even killed by coatings of ice which last for several days. Individual branches or whole trees may die from the aftereffects.

Insects and Related Pests

ROCKY MOUNTAIN JUNIPER APHID. *Cinara sabinae.* Twig growth may be checked and the entire tree weakened by heavy infestations of the Rocky Mountain aphid, a reddish brown insect ⅛ inch long. The so-called honeydew that it secretes is a good medium for the sooty mold fungus, which may completely coat the leaves and further weaken the tree.

Control. Spray with malathion or Sevin when the aphids are visible.

RED CEDAR BARK BEETLE. *Phloeosinus dentatus.* The adult beetle is about ¹⁄₁₆ inch long. It lays its eggs in narrow excavations about 1 or 2 inches long. As the young grubs hatch, they bore out sidewise, making galleries of a characteristic pattern which resembles the markings made in elms by the elm bark beetle.

Control. The cedar bark beetle is more likely to attack trees recently transplanted or those that are suffering from lack of water. Spraying with methoxychlor will probably help.

BAGWORM. *Thyridopteryx ephemeraeformis.* Red cedars are among the most susceptible of ornamentals to the attacks of bagworms. The caterpillar builds around itself an elongated sack (Fig. III-55) which grows to 2 to 3 inches in length as the insect grows. The presence of these feeding insects is likely to be overlooked because of the protecting bags made of green leaves. The adult stage is a moth and the pest

Fig. III-55. Juniper bagworms.

overwinters as eggs in the old female bags. Bagworms have a wide host range, which includes needled and broad-leaved trees.

Control. Pick off the bags by hand or cut them with a pruning pole in late summer or in fall or winter and destroy them. If this is not done, spray with *Bacillus thuringiensis,* Baytex, Dycarb, Dylox, malathion, Orthene, Sevin, or Diazinon in late spring when the young caterpillars begin to feed.

JUNIPER MIDGE. *Contarinia juniperina.* Blisters at the base of the needles and death of leaf tips are produced by small yellow maggots, the adult stage of which is a small fly.

Control. Spray the leaves in mid- to late April with dimethoate or malathion.

JUNIPER SCALE. *Carulaspis carueli.* The needles, particularly of the Pfitzer juniper, turn yellow as a result of sucking by tiny circular scales, which are at first snow-white then turn gray or black (Fig. III-56). The pest overwinters in the female adult stage. Greedy and black scales also attack juniper.

Fig. III-56. Juniper scale, *Carulaspis carueli.*

Control. Spray with 1 part of concentrated lime sulfur solution in 10 parts of water, or a dormant oil plus ethion, before growth starts in spring, or Sevin, methoxychlor, or malathion spray should be applied about the middle of June and again in July.

JUNIPER WEBWORM. *Dichomeris marginella.* The twigs and needles are webbed together, and some turn brown and die when infested by the juniper webworm, a ½-inch-long brown larva with longitudinal reddish brown stripes. The adult female, a moth with a wingspread of ⅗ inch, appears in June and deposits eggs that hatch in 2 weeks. The winter is passed in the immature larval stage. This pest also attacks the creeping juniper *(J. horizontalis).*

Control. Spray with Diazinon, Dylox, or Sevin plus Kelthane in late July and again in mid-August.

CYPRESS TIP MOTH. *Argyresthia* sp. This moth attacks foliage tips.

JUNIPER MEALYBUG. *Pseudococcus juniperi.* This dark red mealybug infests junipers in the Middle West.

Control. Spray with Cygon, malathion, or Sevin when the pests appear.

TAXUS MEALYBUG. See under Yew.

TWO-SPOTTED MITE. See under Hawthorn.

SPRUCE SPIDER MITE. See under Spruce.

OTHER INSECTS. *Periploca nigra,* juniper twig girdler, is a pest in California. *Clastoptera juniperina* is a western spittlebug of juniper. Arborvitae weevil and leaf miner also feed on juniper (see under Arborvitae).

KATSURA-TREE *(Cercidiphyllum)*

Katsura is tolerant of shade and moist soil, growing well in the eastern and northern United States. The only diseases recorded on this host are cankers caused by a species of *Phomopsis* and a species of *Dothiorella,* and root rot by *Armillaria.* Pruning cankered branches below the infected area should keep the canker disease under control. See Chapter 12 for the control of Armillaria root rot.

KENTUCKY COFFEE TREE *(Gymnocladus)*

Kentucky coffee tree has bipinnate compound leaves and produces large, coarse, heavy pods.

It grows best in rich, light soils. This species is remarkably free of fungus parasites and insect pests.

Diseases

LEAF SPOT. Three fungi, *Cerospora gymnocladi, Phyllosticta gymnocladi,* and a species of *Marssonia,* have been reported on this host.

Control. Special control measures are rarely required, but periodic applications of copper or dithiocarbamate fungicides will protect valuable specimens.

OTHER DISEASES. A root rot caused by the fungus *Phymatotrichum omnivorum* and a root and butt rot by *Ganoderma lucidum* are the only other fungus diseases known on *Gymnocladus.*

Control. Control measures have not been developed

Insects

OLIVE SCALE. *Parlatoria oleae.* This insect is purplish brown, and its female shell is ovate, circular, dirty gray, and very small. The female begins laying eggs in spring. Walnut scale also attacks Kentucky coffee tree.

Control. Spray in late spring with either malathion or Sevin to control the crawler stage.

LARCH *(Larix)*

Larches grow in almost any type of soil, including clay and limestone. They do best, however, in moist, loamy soils and in full sunshine. They do not tolerate dry soils or sandy hillsides in climates where the summers are hot.

The varieties most commonly used in ornamental plantings are the American larch or tamarack, the European larch, the Japanese larch, and the Western larch.

Diseases

EUROPEAN LARCH CANKER. This disease has been particularly destructive on larches in Europe for a long time. It is discussed here to acquaint aborists with its symptoms so that suspicious cases can be reported to the proper authorities immediately.

A few localized outbreaks arising from the importation of diseased plants from Europe occurred in the New England states from 1927 to 1935. Fortunately, these were stamped out as soon as discovered. Because of the similarity of symptoms to several less destructive canker diseases, larch canker has been erroneously reported as existing on various trees.

European and American larches are known to be very susceptible to canker, whereas the Japanese larch is relatively resistant.

Symptoms. Stems of young trees and branches of small diameter die suddenly as the result of girdling by cankers. On trunks and branches of large diameter, cankers that increase in size each year are formed. A heavy flow of resin is evident from the cankers. The slowly developing cankers on the trunk or large branches result in considerable distortion and swelling of the affected members. White, hairy, cup-shaped fruiting bodies, about ¼ inch in diameter, develop in the cankered tissues.

Cause. The fungus *Lachnellula willkommii* causes larch canker. It enters through wounds and destroys the inner bark and cambium. Frost injury may also be associated with canker.

Control. European investigators have found that selection of favorable planting sites and maintenance of trees in good vigor discourage severe outbreaks of the disease. It is hoped that any future outbreaks in the United States will be discovered in time and eradicated so as to prevent extensive spread of the disease.

OTHER CANKERS. Four other species of fungi cause cankers: *Dasyscypha ellisiana, Aleurodiscus amorphus, Leucostoma (Valsa) kunzei,* and *Phomopsis* species, some of which are associated with senescing tissues.

Control. Canker diseases are difficult to control. Keeping trees in good vigor by fertilizing and watering when needed will help to reduce the severity of infection.

LEAF CAST. *Hypodermella laricis.* The needles of the American and Western larches and the spur shoots that bear them may be killed by this disease. Early symptoms are yellowing, followed by browning of the needles. Very small, elliptical, black fruiting bodies of the fungus appear on the dead leaves during the winter.

Several other fungi, including *Cladosporium*

sp., *Lophodermium laricis, L. laricinum,* and *Meria laricis,* also cause leaf cast or leaf blights of larch. The leaf cast diseases are most common on ornamental larches in the western United States.

Control. Gather and destroy the needles in late fall or winter to eliminate the most important source of inoculum. This usually ensures satisfactory control. Spraying trees of ornamental value with dilute lime sulfur solution or with bordeaux mixture or any other copper fungicide may be advisable.

NEEDLE RUSTS. Three rust fungi attack larch needles. They develop principally on the needles nearest the branch tips. Affected needles turn yellow and have pale yellow fungus pustules on the lower surfaces.

The fungus *Melampsora paradoxa* occurs on American, European, Western, and Alpine larches. Its alternate hosts are several species of willows. Larches can be infected only by spores developing on willows, but the spores on willows can reinfect the willow.

The fungus *Melampsora medusae* attacks American larch and its alternate host, poplar. As with *M. paradoxa,* spores on larch cannot reinfect larch but must come from poplars.

Melampsoridium betulinum affects American larch and several species of birches. The spore stage on birch can infect birch as well as larch.

Control. Where larches are the more valuable specimens in an ornamental planting, the removal of the alternate host will prevent infection. Infected needles should be submitted to a rust specialist for determination of the exact species involved before attempts to eradicate the alternate hosts are made. Spraying trees of ornamental value with a dilute lime sulfur solution or with a copper fungicide may be advisable.

WOOD DECAY. Several species of fungi are constantly associated with the various types of wood decay of larch. Those most common in the eastern United States are *Heterobasidion annosum, Phellinus pini,* and *Phaeolus schweinitzii.* They are found mainly on older, neglected trees.

Control. Little can be done to check wood decays by the time they are discovered. Maintaining valuable specimens in good vigor by periodic fertilization and by watering during dry spells will do much to prevent initial infections. Wounds on the trunks of such trees should be treated promptly.

Abiotic Disease

AIR POLLUTION. Larch is sensitive to sulfur dioxide and ozone. See Chapter 10.

Insects

LARCH CASEBEARER. *Coleophora laricella.* In May and June larches infested with this caterpillar suffer from an extensive browning of the leaves. The leaves are mined by a small caterpillar which used pieces of the needles to form a cigar-shaped case ¼ inch long. The black-headed caterpillar eats a hole in the leaf either at the end or in the middle and feeds as a miner in both directions as far as it can without leaving the case. The miners winter in the cases, which are attached to the twigs. The moths emerge in late June or July.

Control. To kill the casebearers that have survived the winter, spray in early spring with lime sulfur, dormant strength, or with a miscible oil. If this is not done, spray with Sevin as soon as the insects begin to feed in early spring, and again later in the summer when the next generation begins to feed.

LARCH SAWFLY. *Pristiphora erichsonii.* In sawfly infestations, the needles are chewed by ¾- to 1-inch-long, olive green larvae covered with small brown spines. The adult is a wasplike fly with a wingspread of ⅘ inch. Eggs are deposited in incisions on twigs in late May and June. The larvae overwinter in brown cocoons on the ground.

Control. Gather and destroy fallen needles beneath the tree. If necessary, spray with methoxychlor or Sevin in May or early June as soon as the insect begins to feed.

WOOLLY LARCH APHIDS. *Adelges laricis and A. lariciatus.* White woolly patches adhering to the needles are typical signs of this pest. The adult adelgids (not really aphids) are hidden beneath the woolly masses. The winged adults migrate to pines, and another generation returns to the larch the following season. Eggs are deposited at the bases of the needles in the spring. Young adelgids overwinter in bark crevices.

Control. Spray with 1 part concentrated lime sulfur in 10 parts of water before growth starts, or with Thiodan or malathion when the young are hatching in May.

OTHER INSECTS. The larvae of the gypsy moth and the white-marked tussock moth (see under Elm), the Japanese beetle (see under Linden) and *Orgyia pseudotsugata,* Douglas-fir tussock moth, also chew larch leaves.

Control. Spray with *Bacillus thuringiensis* or Sevin when the pests begin to feed.

LINDEN (Tilia)

Linden is also called basswood. The larger leaved American basswood and the smaller leaved European linden are both used in landscapes. They grow best in fertile, moist soils, but they can tolerate variable soils, heat, and drought associated with urban circumstances.

Diseases

CANKER. *Nectria cinnabarina.* Twigs and larger branches bear cinnabar-colored fruiting bodies of the fungus, each body about the size of a pinhead. These ascocarps break through the bark and are readily seen without a hand lens. The same or similar fungi attack apples, oaks, and other trees.

Other cankers on linden are caused by *Aleurodiscus griseo-cana* and *Strumella coryneoidea.*

Control. Cut out and destroy all cankered branches and remove and destroy twigs and branches that have fallen to the ground.

LEAF BLIGHT. *Cercospora microsora.* Circular brown spots with dark borders characterize this disease. The spots are very numerous, sometimes causing the entire leaf to turn brown and fall off. Young trees are most seriously affected. The sexual stage of the causal fungus is *Mycosphaerella microsora.*

Control. The same as for leaf blotch, below.

LEAF BLOTCH. European lindens are occasionally affected.

Symptoms. Splotchy brown spots with irregular margins occur in various parts of the leaf.

Cause. Leaf blotch is caused by the fungus *Asteroma tiliae,* which is thought to overwinter on diseased fallen leaves.

Control. Gather and destroy fallen leaves.

LEAF SPOTS. *Phlyctaena tiliae* and *Phyllosticta tiliae.* These leaf spots are relatively rare and hence control measures are unnecessary.

POWDERY MILDEWS. *Microsphaera alni, Phyllactinia guttata,* and *Uncinula clintonii.* Lindens are susceptible to powdery mildew fungi. They rarely cause enough damage to require control measures.

Control. Valuable specimens may be sprayed with Benlate or wettable sulfur when the mildew appears.

OTHER DISEASES. Other diseases reported on lindens include canker caused by *Botryosphaeria dothidea* and wilt by *Verticillium alboatrum.*

Control. See under Planetree and Chapter 12.

Insects and Related Pests

LINDEN APHID. *Myzocallus tiliae.* Sap is sucked from the leaves and a sticky substance is exuded by a yellow and black aphid with clouded wings.

Control. Spray with Orthene or malathion when the young aphids appear on the leaves in spring.

JAPANESE BEETLE. *Popillia japonica.* This metallic green-bronze beetle skeletonizes the leaves of its host in midsummer. It is attracted to certain trees in a planting, feeding high up on a sunny side (Figs. III-57, 58, and 59). The host range of Japanese beetle is very wide, but linden, sassafras, and horsechestnut are favored. The larval stage is a grub which feeds on roots of turf grasses in late summer.

Control. Trees should be sprayed with Sevin, methoxychlor, Orthene, or Diazinon in early July and again 10 days later. Milky spore disease, a bacterial disease of Japanese beetle, can reduce beetle populations if it is applied throughout a community.

ELM CALLIGRAPHA. *Calligrapha scalaris.* Ragged holes remain in leaves chewed by creamy-white larvae with yellow heads. The adult beetle is ⅜ inch long, oval, yellow, with green spots on the wing covers and a broad, irregular, coppery green stripe down the back. Lemon yellow eggs are deposited on the lower leaf surface in late June or early July. The beetles hibernate in the ground.

Control. Spray the foliage with Sevin as the larvae appear.

CATERPILLARS. Among the many caterpillars that chew the leaves of this host are cankerworms, yellow-necked caterpillar, elm spanworm, the variable oak leaf caterpillar, and the larvae of the gypsy, cynthia, cecropia, mourning-cloak, and white-marked and hickory tussock moths.

Control. Spray with Sevin when the caterpillars are young.

BASSWOOD LEAF MINER. *Baliosus ruber.* The beetles feed on the underside of the leaves, eating out all tissues exept the veins. The larvae also work on the underside, making large, blisterlike mines. The foliage turns brown, withers, and falls off.

Control. See locust leaf miner under Locust, Black.

ELM SAWFLY. *Cimbex americana.* These smooth caterpillars, pale green with a black stripe down the middle of the back, are about 1 inch long. They curl up tightly when at rest. Elm, willow, maple, and poplar are other hosts frequented.

Control. The standard methoxychlor sprays used to control this pest on elms will also work on linden. Straight methoxychlor sprays may bring on an outbreak of mites, and it is therefore wise to include a mite-killer in the mixture. Sevin may also be used.

LINDEN LOOPER. *Erannis tiliaria.* This pest, also known as the basswood looper, infests apple, birch, elm, hickory, and maple. The caterpillars are 1½ inches long at maturity, bright yellow, with 10 longitudinal wavy black lines down the back (Fig. III-60). The moth, buff colored with a 1¾-inch wingspread, deposits its eggs from October to November.

Control. Bacillus thuringiensis, malathion, Sevin, or methoxychlor sprays applied in spring when this or other young caterpillars begin to feed will provide control.

LINDEN MITE. *Eriophyes tiliae.* In midsummer, especially in dry weather, the leaves become

Fig. III-57. Japanese beetles on linden and the characteristic appearance of the chewed leaves. **Fig. III-58. Upper inset:** Japanese beetle. **Fig. III-59. Lower inset:** Larval stage of Japanese beetle.

infested with this mite, which causes them to turn brown and dry up.

Light brown woolly patches along the veins on the lower leaf surface are produced by the mite *Eriophyes tiliae*.

Control. Lindens are especially susceptible to mite damage when straight methoxychlor sprays are used. Hence valuable trees should be sprayed with a good miticide such as Kelthane. A good preventive is to use this material with the methoxyclor from the start.

EUROPEAN LINDEN BARK BORDER. *Chrysoclista linneela.* This whitish larva with a light brown head bores into the bark of lindens. It does not affect the cambial area but honeycombs the bark to such an extent that decay-producing organisms have easy access.

Control. No control measures have been developed.

LINDEN BORER. *Saperda vestita.* Broad tunnels beneath the bark near the trunk base or in roots are made by the linden borer, a slender

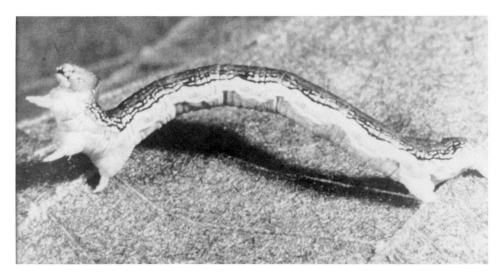

Fig. III-60. Linden looper.

white larva 1 inch long. The adult, a yellowish-brown beetle ¾ inch long, with three dark spots on each wing cover, feeds on green bark. Eggs are deposited in small bark crevices made by the beetle.

Other borers that attack lindens are the flat-headed (see under Maple), the American plum (see under Planetree), red-headed ash (see under Ash), and the brown wood.

Control. Dig out the borers with a flexible wire, or inject a nicotine paste such as Bortox into the tunnels and seal the openings. Spraying the trunk base with methoxychlor may control many of the young larvae after they hatch out of their eggs.

SCALES. Nine species of scale insects infest this host: cottony maple, European fruit lecanium, oystershell, Putnam, San Jose, terrapin, tuliptree, walnut, and willow.

Control. Valuable specimens should be sprayed from time to time with Diazinon, malathion, or Sevin.

WALNUT LACE BUG. See under Walnut.

TWIG GIRDLER. See under Hackberry.

LOCUST, BLACK *(Robinia)*

This tree, a native of the eastern United States, is a quick-growing, short-lived tree. Black lo-cust tolerates heat, drought, and neglect. Globe-shaped thornless types are potentially useful landscape trees.

Diseases

CANKER. *Aglaospora anomala, Nectria galligena,* and *Diaporthe oncostoma.* Cankers on twigs and death of the distal portions may be due to any one of these three fungi.

Control. Prune and destroy infected twigs.

DAMPING-OFF. *Phytophthora parasitica.* In nurseries, serious damage to seedings from 1 to 3 weeks old may be caused by this fungus. The young plants droop and their cotyledons curl. This is followed by wilting and the collapse of the entire seedling, which decays within a few days.

Control. Use clean soil or steam-pasteurize old soil for setting out seedlings.

LEAF SPOTS. *Cladosporium epiphyllum, Cylindrosporium solitarium, Gloeosporium revolutum, Phleospora robiniae,* and *Phyllosticta robiniae.* Many fungi cause leaf spots of black locust.

Control. Control measures are seldom practiced.

POWDERY MILDEWS. *Erysiphe polygoni, Microsphaera diffusa,* and *Phyllactinia guttata.*

White coating of the leaves occurs only occasionally on black locust.

Control. These mildews are never serious enough to warrant control measures.

WOOD DECAY. Nearly all the older black locusts growing along roadsides and in groves in the eastern United States harbor one of several wood-decay fungi. Nearly all these decays have followed infestation of the locust borer, *Megacyllene robiniae*.

The fungus *Phellinus robiniae* causes a spongy, yellow rot of the heartwood. It infects the trunk through tunnels made by the locust borer or through dead older branches. After extensive decay of the woody tissues, the fungus grows toward the bark surface where it produces hard, woody, bracket- or hoof-shaped fruiting structures nearly 1 foot wide. The upper surface of the structure is brown or black and is cracked; the lower surface is reddish brown.

Control. Because wood decays of black locust become established mainly through locust borer tunnels, their prevention rests primarily on freedom from the insect pests. Effective control of the latter, however, cannot easily be attained, and for this reason heartwood decays will continue to be prevalent. Cavity treatments should never be attempted on black locusts.

WITCHES' BROOM. Black locust and honey locust are subject to this condition. The disease, though common on the sprouts, rarely occurs on the large trees. It is characterized by production in late summer of dense clusters or bunches of twigs from an enlarged axis. The bunched portions ordinarily die during the following winter. The disease was long considered to be virus-induced, but recent evidence educed by Japanese and American scientists indicates that a mycoplasmalike organism is the cause of many witches' broom diseases.

One witches' broom of black locust is reportedly caused by tomato spotted wilt virus.

Control. Infected trees appear to recover naturally. No definite control measures are known.

Insects

LOCUST BORER. *Megacyllene robiniae.* Galleries formed by this borer may extend in all directions into the wood, which is discolored or blackened. The trees become badly disfigured. Young plantings may be entirely destroyed (Fig. III-61). The adult is a black beetle about ¾ inch long, spotted with bright yellow, transverse, zigzag lines (Fig. III-62). The young grubs first bore into the inner bark and sapwood.

Control. Cut and destroy badly infested trees. Kill young borers in trunk and branches by inserting Bortox into the burrows and sealing the openings with chewing gum or putty. Spraying with methoxychlor, Thiodan, or Lindane in late August or early September gives effective control. Maintain trees in good vigor by proper watering, pruning, and fertilization.

LOCUST LEAF MINER. *Odontota dorsalis.* The beetles live through the winter and attack the young leaves in early May. They skeletonize the upper surface and lay their eggs on the under surface. The larvae enter the leaf and make irregular mines in the green tissue. A second generation of beetles emerges in September.

Control. Spray with Diazinon or Sevin early in July to kill the young larvae as they begin to mine the leaves.

LOCUST TWIG BORER. *Ecdytolopha insiticiana.* Elongated, gall-like swellings 1 to 3 inches long on the twigs are caused by the feeding and irritation of the pale yellow larvae of the locust twig borer. The adult female is a grayish brown moth with a wing expanse of ¾ inch. The pest overwinters in the pupal stage among fallen leaves. The carpenterworm (see under Ash) also attacks locust.

Control. Prune and destroy infested twigs in August, and gather and destroy fallen leaves in autumn.

SCALES. Many species of scale insects infest black locust: black, cottony-cushion, cottony maple, frosted, greedy, oystershell, Putnam, San Jose, soft, and walnut.

Control. Spray with malathion or Sevin when the young scales are crawling about in spring. Repeat the application in about 2 weeks.

OTHER INSECTS. Yellow-necked caterpillar, Asiatic oak weevil, honey locust plant bug, silver-spotted skipper caterpillar, greenhouse whitefly, bagworm, and treehoppers infest locust trees.

Fig. III-61. Branch dieback of black locust caused by locust borer infestations.

Fig. III-62. Adult stage of the locust borer, *Megacyllene robiniae*.

Other Pests

DODDER. *Cuscuta sp.* This well-known flowering vine, which grows as a parasite on various plants, causes considerable damage to seedlings. It may also be a factor in transmission of certain virus diseases in border plantings.

Control. Dacthal herbicide spray applied to the soil in early spring will prevent germination of dodder seed.

LOCUST, HONEY *(Gleditsia)*

Honey locust is native to the eastern and central United States. It is tolerant to a wide range of growing conditions. Only the thornless types should be used in the landscape.

Diseases

TAR SPOT. *Plagiosphaeria gleditschiae.* This is a serious blight in the southern states. Numerous black fruiting bodies (acervuli) of the fungus develop on the lower side of the leaves. The ascocarpic stage develops throughout the summer and lives through the winter. In the Middle West other leaf spots are caused by *Cercospora condensata* and *C. olivacea.*

Control. Gathering and destroying of all fallen leaves should provide practical control.

CANKERS. *Thyronectria austro-americana* (imperfect, *Gyrostroma*), *T. denigrata, Nectria cinnabarina* (imperfect, *Tubercularia vulgaris*), *Cytospora gleditschiae,* and *Dothiorella sp.* These five fungi are known to cause stem cankers on honey locust. The senior author

found that *Thyronectria* killed a number of honey locusts at the United Nations gardens in New York several years ago (Fig. III-63). The last-named fungus causes extensive areas of necrosis, cracking, and peeling of the trunk bark along with a brown discoloration.

Control. Effective control measures have not been developed.

POWDERY MILDEW. *Microsphaera ravenelii.* This mildew fungus is widespread on honey locusts.

Control. Control measures are rarely used for this disease on this host.

RUST. *Ravenelia opaca.* One rust disease is known to occur on honey locust.

Control. No control measures are necessary.

WOOD DECAY. Like most trees, honey locusts are subject to wood-decaying fungi. Among the more prevalent ones are a species of *Fomes, Laetiporus sulphureus, Daedalea ambigua, D. elegans, Ganoderma lucidum,* and *Xylaria mali. G. lucidum* occurs on living trees, which suggests that it is a vigorous parasite, as it is on species of maple. Dead man's fingers, *Xylaria* fruiting bodies, have been observed in lawns near declining honey locust trees. *Xylaria* causes root rot.

Control. No effective controls are known. Avoid mechanical injuries around the base of the tree, provide good growing conditions, and fertilize and water the tree to maintain good vigor.

Abiotic Disease

AIR POLLUTION. Honey locust is very sensitive to ozone. See Chapter 10.

Insects and Related Pests

HONEY LOCUST BORER. *Agrilus difficilis.* A flat-headed borer burrows beneath the bark of honey locust and eventually may girdle the tree. Large quantities of gum exude from the bark near the infested nodes. The adult beetles, which emerge in June, are elongate, ½ inch long, black with a metallic luster. The borer preferentially attacks stressed trees.

Control. No effective controls are known, although it is likely that methoxychlor sprays

Fig. III-63. The honey locust *(Gleditsia triacanthos)* at left in the United Nations gardens died as a result of infection by the fungus *Thyronectria austro-americana.*

applied to the trunks during the egg-laying period in mid-June and early July may prove effective.

POD GALL MIDGE. *Dasineura gleditschiae.* This midge causes globular galls ⅛ inch in diameter at the growing tips (Fig. III-64). The pest seems to prefer some of the newer thornless varieties such as 'Moraine' and 'Shademaster' to the ordinary honey locust. The adult midge appears in April when the leaves begin to emerge. It deposits eggs singly or in clusters among the young leaflets. The larvae hatch within a few

days and begin to feed on the inner surface of the leaflets.

Control. Diazinon or Orthene sprays applied in May control this pest. Because the pest has many natural enemies, outbreaks may be brief.

WEBWORM. Damage (Fig. III-65) is similar to that observed on silk-tree, (see under Silk-Tree). Most tolerant to least tolerant honey locust cultivars are 'Moraine', 'Skyline', 'Shademaster', 'Imperial', and 'Sunburst'. Winter survival of pupae in the North is thought to be enhanced when relatively warm overwintering sites such

Fig. III-64. Pod gall of honey locust caused by the insect *Dasineura gleditschiae*. Uninfested leaves are seen on the lower right side.

as buildings are located within a few yards of infested honey locusts.

HONEY LOCUST PLANT BUG. *Diaphnocoris chlorionis.* Discoloration of leaves and stunting of new growth is caused by this widely distributed pest (Figs. III-66 and 67). Complete defoliation may occur during heavy infestations. Adults are $^{3}/_{16}$ inch long and pale green. Both adults and nymphs are difficult to detect because their color blends with foliage and growing tips on which they feed. Yellow-leaved strains of honey locust, such as 'Sunburst,' are more susceptible to this bug than are green-leaved ones like 'Shademaster.'

Control. Spray susceptible varieties with Orthene or Sevin a week or 10 days after buds burst.

LEAFHOPPER. *Marcropsis fumipennis.* These $^{1}/_{8}$-inch-long, wedge-shaped, pale active sucking insects cause stippled, pale leaves.

Control. Spray in June, July, or August, when leafhoppers are first noticed, with malathion, Sevin, Orthene, or Diazinon.

SPIDER MITE. *Platytetranychus multidigituli.* This mite causes yellow stippling of the leaves, which drop prematurely. Defoliated trees usu-

Fig. III-65. Mimosa webworm damage on honey locust.

Fig. III-66. Honey locust plant bug injury and discoloration of foliage.

ally leaf out again in late summer but are considerably weakened.

Control. Spray with Kelthane early in July and repeat every 2 weeks as needed.

BAGWORM. See under Juniper.

SPRING AND FALL CANKERWORMS. See under Elm.

WALNUT CATERPILLAR. See under Walnut.

TWIG GIRDLER. See under Hackberry.

SCALES. Cottony maple (see under Maple), San Jose (see under Cherry), Walnut (see under Walnut), black, and hickory lecanium scales feed on honey locusts.

OTHER INSECTS. Cowpea aphid (see Chapter 11) and whitefly attack honey locust.

MADRONE. See Strawberry-Tree.

MAGNOLIA *(Magnolia)*

Many species of magnolia are native to the southeastern and eastern United States. The three major groups include evergreen, late-blooming deciduous, and early-blooming decid-

Fig. III-67. Adult of honey locust plant bug.

uous types. Early-blooming magnolias should not be planted in protected, sunny locations, but in areas where bloom would be delayed and frost damage to flowers avoided.

Magnolias are subject to a number of fungus diseases and insect pests. Only the more important are treated here.

Diseases

BLACK MILDEWS. *Irene araliae, Meliola amphitrichia, M. magnoliae,* and *Trichodothis comata.* A black, mildewy growth covers the leaves of magnolias in the Deep South.

Control. Magnolias may be sprayed with wettable sulfur.

LEAF BLIGHT. *Ceratobasidium stevensii.* The leaves of *Magnolia grandiflora* may be blighted by this fungus, which also affects apple, citrus, dogwood, Japanese persimmon, pecan, quince, and many shrubs in the South.

Control. Valuable specimens can be protected with two or three applications of copper fungicides.

LEAF SPOTS. *Alternaria tenuis, Cladosporium fasciculatum, Mycosphaerella milleri, Colletotrichum* sp., *Coniothyrium fuckelii, Epicoccum nigrum, Exophoma magnoliae, Glomerella cingulata, Hendersonia magnoliae,*

Micropeltis alabamensis, Phyllosticta cookei, P. glauca, P. magnoliae, Septoria magnoliae, and *S. niphostoma.* These fifteen species of fungi cause leaf spots on magnolias.

Control. Fungus leaf spots on valuable specimens can be controlled by periodic applications, early in the growing season, of copper or dithiocarbamate fungicides.

LEAF SCAB. *Elsinoë magnoliae.* In the Deep South *Magnolia grandiflora* leaves may be spotted by this fungus.

Control. The disease is not serious enough to warrant control measures.

DIEBACK. *Phomopsis sp.* Cankers with longitudinal cracks in the bark are formed on the larger limbs and trunks. The wood is discolored a blue-gray. The bark is dark brown over the affected areas. Apparently healthy branches also may be discolored.

Control. No suggestion for control has been made.

NECTRIA CANKER. *Nectria magnoliae.* This fungus produces symptoms similar to those of *N. galligena* (see canker, under Walnut), but it infects only magnolias and tuliptrees.

Control. Prune and destroy cankered branches. Keep trees in good vigor by watering, spraying, and fertilizing when necessary.

WOOD DECAY. A heart rot, associated with the fungi *Fomes geotropus* and *F. fasciatus,* has been reported on magnolia. Affected trees show sparse foliage and dieback of the branches. In early stages the rot is grayish black, with conspicuous black zone lines near the advancing edge of the decayed area. The mature rot is brown. The causal fungi gain entrance through wounds.

Control. No control measures are effective once the rot has become extensive. Avoid trunk wounds and maintain good vigor by fertilization and watering.

ALGAL SPOT. *Cephaleuros virescens.* Leaves and twigs infested by this alga have velvety, reddish brown patches.

Control. Control measures are usually unnecessary.

Wilt caused by *Verticillium albo-atrum,* angular leaf spot by *Mycosphaerella milleri,* and leaf scab by *Sphaceloma magnoliae* are among other diseases of magnolia.

Insects

MAGNOLIA SCALE. *Neolecanium cornuparvum.* Underdeveloped leaves and generally weak trees may result from heavy infestations of the magnolia scale, a brown, varnishlike hemispherical scale, ½ inch in diameter, with a white, waxy covering (Fig. III-68). The young scales appear in August and overwinter in that stage.

Control. A dormant oil–ethion spray in early spring just before new growth emerges will control the adult, overwintering scales. The crawler stage of both this and tuliptree scale, unlike most scales, appears in late summer. Hence the use of Orthene, Sevin, or a mixture of malathion and methoxychlor or Sevin for the crawler stage should be delayed until late August or early September.

TULIPTREE SCALE. *Toumeyella liriodendri.* This scale also infests magnolias and lindens at times.

Control. The same as for magnolia scale.

OTHER PESTS. The following scales also occur on magnolias: black, California, chaff, cottonycushion, European fruit lecanium, Florida wax, glover, greedy, oleander, purple, and soft. Spray the young, crawling stages of these scales in May and June with malathion or Sevin.

The Comstock mealybug, the omnivorous looper caterpillar, and the citrus whitefly also infest magnolias. Malathion sprays, applied when the pests appear, give good control.

A species of eriophyid mite infests *Magnolia grandiflora* in the South. Kelthane or Tedion sprays will control this pest.

The sassafras weevil (see under Sassafras) occasionally infests magnolias.

Planthoppers (see under Cherry) and greenhouse thrips (see under Maple) also attack magnolia.

MAIDENHAIR. See Ginkgo.

MANGO (Mangifera indica)

Mango produces an important fruit in the tropics. This tree with its wide-spreading crown is grown in some landscapes in southern Florida and California.

Fig. III-68. Magnolia scale, *Neolecanium cornuparvum.*

Diseases

ANTHRACNOSE. *Glomerella cingulata.* This is perhaps the most prevalent disease of mango in the South. Leaves are spotted, flowers and twigs blighted, and fruits rotted by this fungus.

Control. Spray valuable specimens periodically with a copper or dithiocarbamate fungicide.

POWDERY MILDEW. *Oidium mangiferae.* Flower panicles and foliage may be injured when infected by this powdery mildew fungus.

Control. Spray with Benlate, or wettable sulfur when mildew appears.

OTHER FUNGUS DISEASES. In Florida mango is subject to twig blight caused by a species of *Phomopsis;* leaf spots by *Pestalotiopsis mangiferae, Phyllosticta mortoni,* and *Septoria* sp.; scab by *Elsinoë mangiferae;* and sooty mold by species of *Capnodium* and *Meliola.*

Control. Control measures are rarely necessary.

Insects

SCALES. Many species of scale insects including California red, tea, and armored *(lindingaspis)* infest mango.

Control. Spray with malathion or Sevin.

OTHER INSECTS. The long-tailed mealybug and greenhouse thrips also infest mango.

Control. Same as for scales.

MAPLE *(Acer)*

The many species of maple in the United States range from shrub-sized patio trees to forest giants. Their leaves are palmately lobed, easily recognized (as on the Canadian flag) and set in opposite pairs. Leaf size ranges from a few inches to a foot; color is variable in summer and often spectacular in fall. Distinctive winged seeds (samaras) consist of paired seeds set in flat membranous wings to facilitate dispersal. Some maples are desired landscape trees; others are prohibited as street trees. Some, like box-elder, are adapted to harsh growing conditions and may be desirable in the Great Plains while being undesirable in the East. Norway maple is said to be adapted to urban environments.

Fungus and Bacterial Diseases

BASAL CANKER. The senior author observed this disease on Norway maples in New Jersey in the early 1940s.

Symptoms. An early symptom is a thin crown resulting from a decrease in the number and size of the leaves. Trees die within a year or two following this period of weak vegetative growth. A more striking symptom is the pres-ence of cankers at the base of the trunk near the soil line. The inner bark, the cambium, and in many instances the sapwood are reddish brown in the cankered area (Fig. III-69). Death occurs when the entire root system decays or when the cankers completely girdle the trunk.

Cause. Basal canker is caused by the fungus *Phytophthora*. A similar disease in Wisconsin was found to be caused by *P. citricola*. The fungus appears to be most destructive on trees growing in poorly drained or shallow soils. Phloem and young xylem are killed by this disease.

Control. No effective control measures are known. Diseased trees should be removed and destroyed. New plantings of Norway maples should be made in well-drained soils, high in organic matter. Frost cracks and mechanical injuries near the base of the trunk should be properly treated and trees fertilized and watered to maintain good vigor.

BLEEDING CANKER. This highly destructive dis-

Fig. III-69. Bark cut away to show the reddish brown discoloration of the sapwood produced by the fungus *Phytophthora*.

ease was first reported from the New England states centering around Rhode Island. The senior author found this disease in New Jersey in October 1940. Norway, red, sycamore, and sugar maples as well as oaks, elms, and American beech are susceptible to this disease. The severity of the disease is associated in some way with hurricane damage.

Symptoms. The disease is named for its most characteristic primary symptom, the oozing of sap from fissures overlying cankers in the bark. Infected inner bark, cambium, and sapwood develop a reddish brown necrotic lesion, which commonly exhibits an olive green margin. These symptoms differ markedly from those produced in trees affected with basal canker. A secondary symptom, the wilting of the leaves and dying back of the branches, is said by one investigator to be due to a toxic material secreted by the causal fungus.

Cause. Bleeding canker is caused by the fungus *Phytophthora cactorum*, which is closely related to the fungus that causes basal canker. *P. cactorum* is also responsible for the crown canker disease of dogwoods. The senior author has produced infection in dogwood with the fungus isolated from Rhode Island maples. The same organism kills the growing tips of rhododendrons during rainy seasons and is known to attack a large number of other trees, including apple, apricot, cherry, peach, and plum.

Control. As with basal canker, no effective control measures are known.

ANTHRACNOSE. *Kabatiella apocrypta.* In rainy seasons this disease may be serious on sugar and silver maples and on boxelder and to a lesser extent on other maples. The spots are light brown and irregular in shape. They may enlarge and run together, causing the death of the entire leaves. Leaves partially killed appear as if scorched. Another anthracnose caused by *Discula* attacks maples also.

Control. Spraying three times at 2-week intervals with bordeaux mixture or some other copper fungicide or with zineb, Benlate, or Zyban starting when the leaves begin to unfurl in spring, will provide control.

LEAF SPOT (PURPLE EYE). *Phyllosticta minima.* The spots are ¼ inch or more in diameter, more or less irregular, with brownish centers and purple-brown margins. The black pycnidia of the fungus develop in the center of the spots. This disease is most severe on red, sugar, and silver maples but also occurs on Japanese, Norway, and sycamore maples.

Control. Spray with zineb, fixed copper, Captan, or Benlate three times at 2-week intervals, starting when the leaves are unfolding from the buds.

BACTERIAL LEAF SPOT. *Pseudomonas aceris.* In California the leaves of Oregon maple (*A. macrophyllum*) may be spotted by a bacterium. Spots vary from pinpoint dots to areas ¼ inch in diameter. They first appear as though water-soaked and are surrounded by a yellow zone; later they turn brown and black.

Control. Preventive sprays containing a copper fungicide applied in early spring should control this organism.

LEAF BLISTER. *Taphrina sacchari.* The lesions produced by this fungus are circular or irregular in shape (Fig. III-70), pinkish or buff on the underside, and ochre or buff above. Sugar and black maple are most susceptible. Blistering, curling, and blighting of other maples are produced by different species: *Taphrina lethifer* affects mountain maple; *T. aceris*, Rocky Mountain hard maple; *T. dearnessii*, red maple; and *T. carveri*, silver maple. *Taphrina* spores are released from diseased leaves in summer and lodge in bud scales where they lie dormant until new leaves emerge the following spring, at which time infection occurs.

Control. Leaf blisters on valuable specimens can be prevented by applying dormant lime sulfur or Ferbam spray just before growth starts in spring.

BULL'S-EYE SPOT This name was coined by the senior author to describe a spot that occurs on red, silver, sugar, and sycamore maples in particularly shaded spots in the eastern United States. The spots show a distinct target pattern with layers of concentric rings (Fig. III-71). The causal organism is *Cristulariella depraedeus* or *C. moricola.*

Control. Control measures have not been developed.

TAR SPOT. *Rhytisma acerinum.* Street maples are seldom infected with this fungus, but red maples in forests may be prematurely defo-

Fig. III-70. Maple leaf blister.

liated. The spots are irregular, shining black tarlike discolorations up to ½ inch in diameter, developed on the upper sides of the leaves. The dark color is due to the masses of brown or black mycelium.

Control. Rake up and discard the leaves in fall. If necessary spray with a copper fungicide or Ferbam when buds are opening, and, when the infection is severe, repeat spraying several times at 2-week intervals.

OTHER FUNGUS LEAF SPOTS. *Rhytisma punctatum* causes a minute black spotting on many species of maples. It is rare in the East but prevalent on the Pacific Coast. A number of other leaf spots that occur occasionally are caused by fungi belonging to the genera *Cercospora,*

Fig. III-71. Bull's-eye spot of sycamore maple.

Venturia, Didymosporina, Septoria, Cylindrosporium, and *Monochaetia.*

Control. Spray as for tar spot.

POWDERY MILDEWS. *Uncinula circinata, Microsphaera aceris,* and *Phyllactinia guttata.* These mildews, which are rarely serious, can be controlled by spraying with Bayleton, Benlate or wettable sulfur.

NECTRIA CANKER. *Nectria cinnabarina.* Cankers appear on twigs and branches and occasionally develop on the trunks to such an extent as to cause the death of weak trees. Reddish fungus fruiting bodies develop in large numbers. Although the fungus is most common on maples and lindens, it attacks a wide variety of other hardwood trees. Another fungus, *Nectria galligena,* causes the development of target cankers, accompanied by the formation of thick calluses, which later become diseased and leave an open wound. Cankers should be cut out well beyond the diseased areas. The vigor of the trees should be increased by fertilization.

EUTYPELLA CANKER. *Eutypella parasitica.* Boxelder and Norway, red, and sugar maples occasionally show another type of canker, which differs strikingly from that produced by either of the Nectria fungi. The cankers are irregularly circular and contain broad, slightly raised concentric rings of callus tissue. The cankered tissue is firmly attached to the wood with heavy, white-to-buff, fan-shaped wefts of fungus tissue under the bark near the margins. Tiny black fungus bodies are present in the centers of the old cankers.

Other cankers may be caused by the following fungi: *Botryosphaeria dothidea, Diaporthe* spp., *Cytospora* sp., *Fusarium solani, Nectria* sp., *Botryosphaeria obtusa, Septobasidium fumigatum, Valsa sordida, V. ambiens* subsp. *leucostomoides, Steganosporium ovatum,* and *Cryptosporiopsis* sp.

Control. Cankers on trunks can rarely be eradicated by surgical methods once they have become very extensive. Some cankers weaken the tree enough that trunk breakage can occur during strong winds. Those on branches can be destroyed by removing and discarding the affected members. Removal of dead branches, avoidance of unnecessary injuries, and maintenance of trees in vigorous condition by fertil-

ization and watering are probably the best means known of preventing canker formation.

PHOMOPSIS BLIGHT. *Phomopsis acerina.* Dying of Norway maples along city streets has been attributed to infection by this fungus. Galls and burls of red and sugar maple stems and branches have been attributed to *Phomopsis* species.

Control. No controls have been developed.

GANODERMA ROT. *Ganoderma lucidum.* Rapid decline and death of many trees growing along city streets were found by the senior author to be caused by this fungus. It forms large, reddish fruit bodies with varnishlike coating at the base of the infected tree or on its surface roots (Figs. III-72, 73, and 74). Red and Norway maples appear to be most susceptible.

Control. No control is known for diseases of this type. Planting trees in deep, fertile soil, avoiding bark and root injuries, and fertilizing and watering properly will reduce chances of infection.

TRUNK DECAY. In New England sugar maples tapped for their sap are subject to a serious trunk decay caused by the fungus *Valsa leucostomoides.* In longitudinal section the affected areas appear as truncated cones with pale yellow centers bordered by deep olive or greenish black streaks. Wood decay (Chapter 12) is especially damaging to red maple. Fungi such as *Ganoderma applanatum, Climacodon septentrionale, Cerrena unicolor,* and *Oxyporus populinus* may be involved.

Control. Tap holes should be sprayed or painted with a fungicide at the time the spiles are pulled. The trees should be tapped on the diagonal or up and down the butts.

ARMILLARIA ROOT ROT. See Chapter 12.

XYLARIA ROOT ROT. *Xylaria mali.* The infected root appears black with a carbonaceous crust. The fruiting body of *Xylaria,* the dead man's finger fungus, is a dark, upright mass which protrudes through the soil. There is no control for Xylaria root rot.

BACTERIAL WETWOOD. See Chapter 12.

WILTS. Already discussed in detail in Chapter 12, wilt is perhaps the most important vascular disease of maples. Positive diagnosis of this disease is possible only by collecting adequate

Fig. III-72. The senior author is examining a dying Norway maple infected with the fungus *Ganoderma lucidum*. He is holding a live branch cut from the healthy side of this tree. The fungus bodies are at the base of the tree.

specimens of discolored sapwood beneath the bark of the trunk near the soil line and culturing the material in the laboratory. The *Verticilium* fungus, however, is not the only fungus capable of discoloring the sapwood of maples.

SAPSTREAK. *Ceratocystis coerulescens.* This killing disease of sugar maples was first described from North Carolina in 1944 and has since been reported from New England and the lake region of the Midwest. Typical symptoms include thinning crowns with undersized chlorotic foliage, followed by death of the tree.

Trunk xylem tissues are discolored. Sapstreak also affects tuliptree in woodlands.

Control. No effective control has been developed.

MOSAIC. Virus diseases of maple have been reported.

Abiotic Disease

MAPLE DECLINE. Weakening and death of sugar maples along heavily traveled highways in the northeastern United States has been attributed

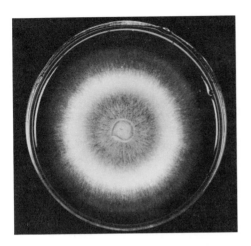

Fig. III-73. A culture of the *Ganoderma* fungus growing on a synthetic medium in the laboratory.

to excessive use of salts (sodium and calcium chlorides) during the winter months. See Chapter 10. Other stress-inducing factors such as drought, restricted root space, or injuries can initiate decline. Canker and decay fungi and insect borers may attack weakened trees.

LEAF SCORCH. See Chapter 10.

GIRDLING ROOTS. See Chapter 10.

Insects and Mites That Attack Leaves

Besides the leaf-eating insects described below, there are others that attack maple leaves. These include the elm sawfly, *Cimbex americana;* brown-tail moth, *Nygmia phaeorrhoea;* gypsy moth (see under Elm) in certain eastern states; white-marked tussock moth (see under Elm); American dagger moth, *Acronicta americana;* saddled prominent caterpillar, *Heterocampa guttivitta;* and the oriental moth, *Cnidocampa flavescens.* These are readily controlled by spraying with Sevin, *Bacillus thuringiensis,* or methoxychlor. The spring and fall cankerworms, *Paleacrita vernata* and *Alsophila pometaria* (also known as inchworms, and measuring-worms), and bagworms (see under Juniper), also are controlled in this way. Yellow-necked caterpillar, io and cecropia moth larvae, green mapleworm, and elm spanworm all feed on maple foliage as caterpillars (see Chapter 11).

BOXELDER BUG. *Leptocoris trivittatus.* The leaves of boxelder are injured by the sucking of young bright red bugs. The adult stage is a stout, ½-inch-long, grayish black bug with three red lines on the back (Fig. III-75). All stages of this pest are clustered on the bark and branches in early fall. The four-lined plant bug, *Poecilocapsus*

Fig. III-74. Close-up of the *Ganoderma* fungus at the base of a dying maple tree.

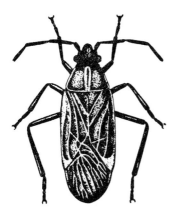

Fig. III-75. Boxelder bug.

lineatus, also feeds on maple, causing circular brown leaf spots. A few individuals of this yellow and black striped insect can cause important damage.

Control. Spray bark with Sevin when clusters of bugs appear in late May or early June. Because this bug feeds on seeds of pistillate (female) trees, one should plant only staminate (male) trees.

FOREST TENT CATERPILLAR. *Malacosoma disstria.* These caterpillars, bluish with a row of diamond-shaped white spots along the back (Fig. III-76), feed individually on leaves of maple, birch, oak, and poplar, but do not form a tent

in the forks of smaller branches as do the eastern tent caterpillars *(M. americana),* which have a white stripe down the back. The egg masses look the same, about ½ inch long, completely surrounding the twig, and having a brown, varnished appearance. Fall webworm (see under Ash) also attacks maple.

Control. They can be easily controlled by spraying with Dylox, Sevin, *Bacillus thuringiensis,* Diazinon, or methoxychlor.

GREENSTRIPED MAPLEWORM. *Dryocampa rubicunda.* These caterpillars are pale yellowish green, 1½ inches long, with large black heads; the stripes along the back are alternately pale yellowish green and dark green. The insects prefer red and silver maples. A close relative, the orange striped oakworm (see under Oak), also feeds on maple

Control. They can be easily controlled by spraying with Sevin in summer when larvae are active.

BUFFALO TREEHOPPER. *Stictocephala bubalus.* This insect damages maples by scarring their twigs during egg laying.

LEAFHOPPERS. Japanese leafhopper (see under Mountain-Ash) and rose leafhopper, *Edwardsiana roseae,* attack maple. The Norway maple may become seriously infested with yellowish leafhoppers *(Alebra albostriella);* they also cause swellings on the twigs by depositing eggs under the young bark. Leaf symptoms appear as a

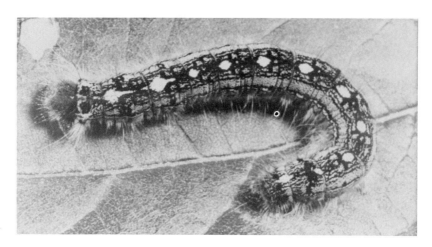

Fig. III-76. Forest tent caterpillar.

chlorotic stippling or flecking, resembling ozone injury.

Control. Spray in early spring with Sevin or Diazinon.

LEAF STALK BORER. *Nepticula sericopeza.* Though these small borers more commonly infest the fruit of Norway maples, they may attack the leaf stalks in June. The borers tunnel in the stalks, causing a black discoloration about ½ inch from the base. The lower end of the stalk is somewhat enlarged. These borers occasionally cause severe defoliation. The adult is a minute moth with spinelike hairs on the surface of its wings.

Control. Spray with a dormant miscible oil in early spring to destroy the cocoons. If this is not done, spray with methoxychlor in late May.

PETIOLE BORER. *Caulocampus acericaulis.* This smooth yellowish sawfly larva causes leaf fall of sugar maple. It tunnels in the upper end of the leaf stalk about ½ inch from the blade. The leaf blades fall off in May and June and sometimes the leaf stalk itself falls. Only the leaves of the lower branches are usually seriously infested.

Control. This pest is rarely serious enough to warrant control measures, but fallen leaves should be raked up and destroyed.

MAPLE LEAF CUTTER. *Paraclemensia acerifoliella.* Much defoliation of sugar maple and beech trees results from the work of this small caterpillar, which is about ¼ inch long. It cuts out small sections of the leaves and forms a case in which it hides while it feeds. It skeletonizes a ringlike portion, the center of which may fall out.

Control. Rake up and discard all leaf litter in fall to kill the hibernating pupae, or apply Sevin to the foliage.

MAPLE TRUMPET SKELETONIZER. *Epinotia aceriella.* Red and sugar maple leaves are folded loosely by small green larvae which develop inside a long trumpetlike tube. Fruit tree leaf roller (see under Crabapple) also feeds on maple.

Control. Methoxychlor, Sevin, or Diazinon sprays applied in mid-July will control this pest.

NORWAY MAPLE APHID. *Periphyllus lyropictus.* The leaves of Norway maple are subject to heavy infestation by these hairy aphids, which are greenish with brown markings. Like other aphids, they secrete large amounts of honeydew. The leaves attacked become badly wrinkled, discolored, and reduced in size. Defoliation may follow. Several other species, including *Drepanaphis acerifoli, Periphyllus aceris,* and *P. negundinis,* also occur on maples.

GREENHOUSE THRIPS. *Heliothrips haemorrhoidalis.* This insect feeds on maple leaves with its piercing–sucking mouthparts, causing stippling, glazing, and browning of leaves. It produces large quantities of small, dry excrement droplets.

Control. Spray in early summer with malathion, Orthene, Diazinon, or Sevin, being careful to cover the undersides of the leaves.

OCELLATE LEAF GALL. *Cecidomyia ocellaris.* These galls are less than ½ inch in diameter, with cherry red margins; they are easily confused with purple eye leaf spot. They occur on red maple but are usually not injurious. The little maggots, which may be seen at the center of the spot, soon drop off.

BLADDER-GALL MITE. *Vasates quadripes.* The upper surfaces of maple leaves are often covered with small, green, wartlike galls, which later turn blood red. These are caused by the feeding of this mite. If the galls are very numerous, the leaves become deformed.

Control. Spray with dormant oil plus ethion in late April. Kelthane spray in spring when the buds show green color and again 2 weeks later should also protect the foliage.

OTHER MITES. Other maple galls are caused by the mites *Vasates aceriscrumena* and *Eriophyes elongatus.* The galls are about ⅕ inch long, tapering at both ends (fusiform). They develop on the upper side of the leaf and, when numerous, render the foliage unsightly. Spray with lime sulfur at blossom time or with Kelthane in early June.

Several other species of *Eriophyes* produce large blotches along the veins or between them. Brilliant purple, red, or pink minute blisterlike or pilelike growths are caused by the mites. A dormant strength lime sulfur spray just before leaf buds open, or a Kelthane spray when the buds show green color, is recommended for control of these mites.

The mite *Paratetranychus aceris* occasionally

infests maple leaves, causing mottling and premature yellowing. A spray containing Acaraben or Kelthane when these mites are noticed in early June will give good control.

Scale Insects

MAPLE PHENACOCCUS. *Phenacoccus acericola.* The leaves of sugar maple may be covered on the underside by cottony masses that envelop the females in July (Fig. III-77). The males collect in the crevices of the bark and give it a white, chalky appearance. There are several generations a year. Predacious insects feed on the eggs of this scale.

Control. Spray with malathion, Diazinon, or Sevin in spring just as the buds burst. Repeat the spray in early August.

COTTONY MAPLE SCALE. *Pulvinaria innumerabilis.* Silver maples are especially susceptible to infestation by this scale. Smaller branches are often covered with large white cushionlike masses. The scale itself is brown, from ⅛ to ¼ inch in diameter; the whole mass is about ½ inch in diameter. The cottony mass may contain as many as 500 eggs in June. The young move out and infest the leaves; later they migrate to the branches, being found especially on their undersides.

Cottony maple leaf scale, *P. acericola,* attacks maples and other trees throughout the eastern United States. Cottony-cushion scale (see under Acacia), with its large cottony white and fluted egg sac, and cottony taxus scale, another soft scale, also attack maple.

Control. Spray with lime sulfur in early spring when trees are dormant. Malathion or Sevin may be applied during midsummer when the young are crawling.

OTHER SCALES. Several other scales attack maples. Among them are the terrapin, gloomy, Japanese, black, obscure, walnut, calico, ivy, and banded oystershell scales. Comstock mealybug also attacks maple.

Control. These may be controlled by spraying with dormant miscible oil or lime sulfur. The latter is safer. Malathion, Orthene, or Sevin sprays are most effective when the young scales are crawling.

Wood Borers

Limbs infested with borers are the first to break after a heavy snowstorm or after they have been coated with ice. Five or six species of woodboring insects infest maples

FLATHEADED BORER. *Chrysobothris femorata.* Trees in poor vigor are attacked by this light yellow larva, 1 inch long, which builds flattened galleries below the bark, often girdling the tree. The adult is a dark coppery brown beetle, ½ inch in length, which deposits eggs in bark crevices in June and July. The pest overwinters in the wood in the pupal stage.

This borer, sometimes referred to as flatheaded apple tree borer, may attack newly transplanted maples in the nursery. A similar pest, *C. mali,* the Pacific flatheaded borer, is also a threat to newly transplanted trees.

Control. Increase the tree's vigor by fertilizing and watering, or protect recently transplanted trees with crepe wrapping paper. Spray the trunk and branches with Dursban or Lindane in mid-May and repeat twice more at 2-week intervals. A precisely timed single insecticide spray is very effective if applied three weeks after full expansion of the first red maple leaves, which occurs shortly after adult beetle emergence. Borers active in the trees can be killed

Fig. III-77. Maple phenacoccus *(Phenacoccus acericola)* on sugar maple leaf.

by inserting a flexible wire in the borer holes.

LEOPARD MOTH BORER. *Zeuzera pyrina.* This brown-headed borer is white or slightly pinkish and marked by blackish spots on the body. Full-grown borers may be 2 to 3 inches long and nearly ½ inch in diameter. They winter in the tunnels they make. Besides their cylindrical burrows they make wide cavities which so weaken the limbs that they break off very readily. Soft maples are especially susceptible. Adult female moths are white with dark spots and are weak fliers.

Control. Preventive treatments include spraying the trunk and main branches with Lindane or methoxychlor in late May and repeating four times at 2-week intervals. On valuable young trees that are already infested, the boring caterpillars can be killed by inserting into their tunnels a flexible wire or pastes containing nicotine or a few drops of carbon disulfide, and then sealing the openings with putty or chewing gum. Adult females should be caught and destroyed.

METALLIC BORER. *Dicerca divaricata.* The larvae of these brass- or copper-colored beetles frequently invade the limbs of peach, cherry, beech, maple, and other deciduous trees. The beetles live 2 or 3 years as borers in the trees and lay their eggs in August and September. The adults have been known to cause much defoliation.

Control. Infested limbs should be cut and destroyed to get rid of the grubs before the adult beetles emerge to lay their eggs.

PIGEON TREMEX. *Tremex columba.* This insect, also called horntails, bores round exit holes in the bark about ¼ inch in diameter. The larvae usually attack trees which are dying because of the attacks of fungi or other species of borers.

Control. Insert Bortox into the burrows and seal the openings.

SUGAR MAPLE BORER. *Glycobius speciosus.* The appearance of dead limbs among leafy branches or dead areas on branches and trunks is evidence of this insect's presence.

Control. Methoxychlor or Lindane spray applied to the trunk and branches in mid-August will control this pest.

Other Pests

TWIG PRUNER. See oak twig pruner, under Hickory.

CARPENTERWORM. See under Ash.

JAPANESE BEETLE. See under Linden.

LINDEN LOOPER. See under Linden.

WHITEFLY. See under Dogwood.

WOOLLY ALDER APHID. See under Alder.

CAMBIUM MINER. See under Holly.

BIRCH LACE BUG. See under Birch.

LEAF MINER. See Chapter 11.

SQUIRRELS. The red squirrel, *Tamiasciurus hudsonicus,* bites the bark of young red and sugar maples to drink the sap that flows from the wounds. Cankers develop as a result of fungi which enter the wounds.

NEMAS. Dieback of sugar maples in woodland areas of Wisconsin has been associated with heavy infestations of several species of root nemas. In Massachusetts the dagger nema, *Xiphinema americanum,* is associated with a decline of sugar maples.

Control. Controls have not been developed.

MAYTEN TREE *(Maytenus)*

This native of Chile, which reaches medium size, displays weeping habit. Mayten tree withstands hot weather in regions with mild winters, and it tolerates salty soils. It has few pest problems. Nigra scale, *Saissatia nigra,* attacks mayten tree but is largely controlled by natural parasites.

MESQUITE *(Prosopis)*

A tree for hot, dry regions, mesquite tolerates drought and alkaline soil.

MIMOSA. See Silk-Tree.

MOUNTAIN-ASH *(Sorbus)*

Because mountain-ashes belong to the rose family, they are subject to most of the same diseases and insects common in this family. Cultivars of European mountain-ash are generally more available than American mountain-ash cultivars.

Bacterial and Fungus Diseases

CANKER. *Cytospora chrysosperma, C. massariana, C. microspora,* and *Fusicoccum* sp. These four fungi occur occasionally on mountain-ash, especially on weakened trees.

Control. Prune severely affected branches and fertilize. The latter practice may increase the tree's susceptibility to fire blight, however, if highly nitrogenous fertilizers are used.

LEAF SPOTS. The alternate stages of the rust fungi *Gymnosporangium aurantiacum, G. cornutum, G. globosum, G. tremelloides, G. nelsoni, G. nootkatense,* and *G. libocedri* occur on juniper or incense cedar species. On mountain-ash leaves, circular, light yellow, thickened spots first appear during summer. Later, orange cups develop on the lower surfaces of these spots. *Diplocarpon mespili, Phyllosticta sorbi,* and *Venturia inaequalis* also cause leaf spots.

Control. Where mountain-ashes are highly prized, and where practicable, remove the alternate hosts. If such removal is impractical, valuable trees can be protected by sprays recommended for rust control of flowering crabapples.

CROWN GALL. See Chapter 12.

FIRE BLIGHT. See Chapter 12. This disease is sometimes very damaging.

SCAB. See under Crabapple.

OTHER DISEASES. Mountain-ash is subject to many other diseases. These are discussed under rosaceous hosts in other parts of this book.

Abiotic Disease

AIR POLLUTION. European mountain-ash is sensitive to ozone. See Chapter 10.

Insects

APHIDS. The rosy apple and woolly apple aphids frequently infest this host. See Chapter 11.

Control. Spray with malathion.

JAPANESE LEAFHOPPER. *Orientus ishidae.* This insect causes a characteristic brown blotching bordered by a bright yellow margin, the yellow zone merging into the color of the leaf.

Control. Spray with dimethoate, Sevin, or malathion in late May or early June.

MOUNTAIN-ASH SAWFLY. *Pristiphora geniculata.* Green larvae with black dots feed on mountain-ash leaves from early June to mid-July, leaving only the larger veins and midribs. The adults, yellow with black spots, deposit eggs on the leaves in late May.

Control. Spray with Diazinon, methoxychlor, or Sevin when the leaves are fully expanded.

PEAR SLUG. See under Cherry.

PEAR LEAF BLISTER MITE. *Eriophyes pyri.* Tiny brownish blisters on the lower leaf surface and premature defoliation result from infestations of the pear leaf blister mite, a tiny, elongated, eight-legged pest, $1/125$ inch long. The pests overwinter beneath the outer bud scales. Eggs are deposited in spring in leaf galls, which develop as a result of feeding and irritation by the adult.

Control. Spray the trees with lime sulfur in late fall or early spring or with Kelthane early in the growing season.

MOUNTAIN-ASH SPIDER MITE. See Chapter 11.

ROUNDHEADED BORER. *Saperda candida.* Trees are weakened and may be killed by the roundheaded borer, a light yellow, black-headed, legless larva 1 inch long. Galleries in the trunk near the soil level, frass at the base of the tree, and round holes of the diameter of a lead pencil in the bark are typical signs of this pest. The adult is a beetle ¾ inch long, brown, with two white longitudinal stripes on the back. It emerges in April and deposits eggs on the bark.

Control. Spray trunk and branches thoroughly with Thiodan starting in June and repeating three times at 3-week intervals.

DOGWOOD BORER. See borers, under Dogwood.

AMERICAN PLUM BORER. See under Planetree.

FLATHEADED BORER. See under Maple.

SCURFY SCALE. *Chionaspis furfura.* A grayish scurfy covering on the bark indicates the presence of this pest. The scale covering the adult female is pear-shaped, gray, and about $1/10$ inch long. The pest overwinters as purple eggs under the female scale.

Control. Spray in late spring, before the leaf

buds open, with dormant oil plus ethion, or in mid-May and again in early June with malathion or Sevin.

SCALES. Several other species of scale insects infest mountain-ash: black, globose, frosted, walnut, cottony maple, oystershell, and San Jose.

Control. Dormant oil or lime sulfur sprays, followed by malathion or Sevin sprays during the growing season, will control these insects.

BIRCH LACE BUG. See under Birch.

MULBERRY *(Morus)*

Red mulberry, native to North America, is unusually resistant to diseases and pests.

Diseases

BACTERIAL BLIGHT. *Pseudomonas mori.* Water-soaked spots appear on leaves and shoots; they later become sunken and black. The leaves are distorted, and the shoots have black stripes. The leaves at the twig tips wilt and dry up.

Control. Some control is obtainable on young trees by pruning dead shoots in autumn and spraying with bordeaux mixture the following spring. Streptomycin sprays may also be effective.

BACTERIAL WETWOOD. This can be a serious problem on mulberry. See Chapter 12.

LEAF SCORCH. Caused by xylem-inhabiting bacteria, leaf scorch occurs on red mulberry in the East. See Chapter 12.

LEAF SPOTS. *Cerospora moricola, C. missouriensis,* and *Cercosporella mori.* The leaves of mulberry are spotted by these fungi in very rainy seasons.

Control. The *Cercosporella* fungus can cause defoliation of older trees. Valuable specimens should be sprayed with a copper fungicide if leaf spots are serious.

POPCORN DISEASE. *Ciboria carunculoides.* This disease, known only in the southern states, is largely confined to the carpels of the fruit. It causes them to swell and remain greenish, and it interferes with ripening.

Control. The disease is of little importance. It does not lessen the value of the tree as an ornamental.

FALSE MILDEW. *Mycosphaerella mori.* The foliage of mulberries grown in the southern states may suffer severely from attacks of this mildew. It appears in July as whitish, indefinite patches on the undersides of the leaves. Yellowish areas then develop on the upper sides. The fungus threads that will produce spores emerge from the stomata on the underside and spread out to form a white, cobweblike coating; the general appearance is that of a powdery mildew. The asexual spores are colorless, each composed of several cells. The infected leaves fall to the ground, and the overwintering or ascocarpic stage matures in spring on these leaves.

Control. Gather and destroy all fallen leaves in autumn. Spray with bordeaux mixture as soon as the mold appears in July.

CANKERS. *Cytospora* sp., *Dothiorella* sp., *D. mori, Gibberella baccata,* f. *moricola, Nectria* sp., and *Stemphyllium* sp. These six species of fungi may cause cankers on twigs and branches and dieback of twigs. They can be distinguished only by microscopic examination or laboratory tests.

Control. Prune and destroy dead branches. Keep trees in good vigor by watering and fertilizing.

POWDERY MILDEWS. *Phyllactinia guttata* and *Uncinula geniculata.* The lower leaf surface is covered by a white, powdery coating of these fungi.

Control. Valuable specimens can be protected by occasionally spraying with Benlate or wettable sulfur.

Abiotic Disease

AIR POLLUTION. Mulberry is sensitive to sulfur dioxide. See Chapter 10.

Insects and Related Pests

CITRUS FLATID PLANTHOPPER. *Metcalfa pruinosa.* This very active insect, also called mealyflata, is ¼ inch long with purple-brown wings and is covered with white woolly matter.

Control. Spray with pyrethrum or rotenone or a combination of the two.

CERAMBYCID BORER. *Dorcaschema wildii.* In the South this borer mines large areas of cambium and tunnels into the wood of mulberry trees. Branches or even entire trees are girdled and killed.

Control. Spray trunks and branches of valuable trees with methoxychlor twice at monthly intervals, starting in mid-May.

SCALES. The following scales infest mulberry: California red, cottony maple, hickory lecanium, ivy, glover, dictyospermum, Florida wax, greedy, olive parlatoria, peach, San Jose, and soft.

Control. Spray with malathion or Sevin to control the crawling stage of these pests.

OTHER PESTS. The Comstock mealybug and the mulberry whitefly also infest this host.

In dry seasons the two-spotted mite, *Tetranychus urticae,* may be abundant and injurious. Leaves are mottled and yellow.

Control. Malathion or Sevin sprays are effective in controlling the mealybug and whitefly. To control the mite, valuable specimens should be sprayed with chlorobenzilate, Kelthane, or Tedion after the fruits are gone.

AMERICAN PLUM BORER. See under Planetree.

PLANTHOPPER. See under Cherry.

NORFOLK ISLAND PINE *(Araucaria)*

This beautiful evergreen tree does well outdoors in the warmer parts of the United States. It is also grown indoors as a potted plant.

Diseases

BLIGHT. *Cryptospora longispora.* The lower branches are attacked first, and the disease gradually spreads upward. As the entire branch becomes infected, the tip end becomes bent. The limbs die and then break off at the tip ends. Plants 5 or 6 years old are soon killed by this disease.

Control. The infected branches should be pruned off and destroyed as soon as discovered. Seeds of *Araucaria excelsa* imported from Norfolk Island Territory, near New Zealand, frequently harbor the causal fungus. Hence they are dipped in sulfuric acid by plant quarantine inspectors before being released to nurserymen.

CROWN GALL. *Agrobacterium tumefaciens.* This host has been proved experimentally to be susceptible. A typical gall is smooth and up to 1 inch in diameter.

Control. Prune infected branches.

BLEEDING CANKER *Botryosphaeria rhodina* and *Dothiorella* sp. Two fungi are associated with this disease in Hawaii, where the Norfolk Island pine is grown for Christmas trees and is used as windbreaks and as an ornamental.

Control. Control measures have not been developed.

Insects

MEALYBUG. *Pseudococcus aurilanatus, Planococcus citri,* and *Pseudococcus ryani.* The first species listed is the golden mealybug; the second, the citrus mealybug; and the third, the cypress mealybug. All are serious pests of Norfolk Island pine in California.

Control. Spray with malathion or Sevin whenever mealybugs are seen.

SCALES. Five species of scales are known to attack this host: araucaria *(Eriococcus araucariae),* pure white, feltlike, oval sacs enclosing bodies and eggs of females; black araucaria *(Lindingaspis rossi),* an almost black species resembling the Florida red scale; chaff *(Parlatoria pergandii),* a circular to elongate, smooth, semitransparent, brownish gray kind; Florida red (see under Camellia), an armored, $1/12$ inch, circular, reddish brown to nearly black scale; and soft *(Coccus hesperidum),* a flat, soft, oval, yellowish brown, $1/8$-inch-long species.

Control. See Chapter 11.

OAK *(Quercus)*

There are many native oaks throughout North America. Many species are desirable trees for large spaces. Oaks are generally durable and longlasting. The open-branched habit of species like northern red oak, bur oak, and white oak facilitates passage of utility wires; these species suffer less topping than the pin oak. Some oaks are sensitive to alkaline soil conditions, becoming chlorotic.

Diseases

BASAL CANKER. *Phaeobulgaria inquinans.* This fungus frequently develops in crevices of the sunken bark that overlies basal cankers. It enters through open wounds but then invades the surrounding bark and sapwood and eventually girdles and kills the tree. The mature fruit bodies are cup- or saucer-shaped and grow from short stems that extend into the bark. When moist, the fruiting bodies look and feel like rubber.

Control. The same as for the canker disease described below.

STRUMELLA CANKER. *Urnula craterium (imperfect, Conoplea).* Although primarily a disease of forest oaks, canker occasionally affects red and scarlet oaks in ornamental plantings. American beech, chestnut, red maple, tupelo, and pignut and shagbark hickories are also susceptible. Several types of cankers, depending on the age of the tree and the rate of growth of the causal fungus, are produced on the trunks. Smooth-surfaced, diffuse, slightly sunken cankers are common on young trees with a diameter of 3 to 4 inches. Cankers on older trees have a rough surface ridged with callus tissue. Open wounds may be present in the center of large cankers as a result of secondary decay and shedding of the bark.

HYPOXYLON CANKER. Recognized by the gray or tan crusty fungus tissue (stroma) over the cankered area, canker caused by *Hypoxylon atropunctatum* appears on trees under stress from drought, construction damage, or other injuries. Trees weakened by *Hypoxylon* can be a safety hazard in urban areas.

OTHER CANKERS. *Nectria galligena,* cause of nectria canker (see canker, under Walnut), also attacks oak. Cankers on red oaks *(Quercus borealis maxima)* were found to be caused by the fungus *Fusarium solani;* those on stressed oaks growing at high temperatures, by *Lasiodiplodia theobromae;* those on scarlet oak and live oak *(Q. virginiana)* by *Endothia parasitica;* and those on pin oak *(Q. palustris)* by *E. gyrosa.*

Canker-rots of oak are caused by *Inonotus hispidus, Phellinus spiculosus,* and *Irpex mollis.* Canker-rots can result in affected trees being a hazard in urban areas. Smooth-patch canker disease is caused by *Aleurodiscus oakesii.*

Control. Prune dying and dead branches to eliminate an important possible source of in-

oculum. Remove small cankers on the trunk by surgical means, and fertilize and water the trees to improve their vigor.

ANTHRACNOSE. *Apiognomonia quercina.* This fungus, whose asexual stage is *Discula quercina,* is common in the northern states especially on white oak, and also red oaks, American elm, and black walnut. Rainy weather favors the disease, which may cause defoliation. Weak trees, if defoliated, may die. The spots on the leaves run together, causing the appearance of a leaf blotch or blight. The dead areas follow the veins or are bounded by larger veins. These blotches are light brown. Infection may occur in midsummer.

Control. Three applications of zineb, fixed copper, or Benlate will control this disease: the first when leaves unfurl, the second when leaves reach full size, and the third 2 weeks later.

LEAF BLISTER. During cool, wet springs almost all species of oak are subject to the leaf blister disease. Circular, raised areas ranging up to ½ inch in diameter are scattered over the upper leaf surface, causing a depression of the same size on the lower surface (Fig. III-78). The upper surface of the bulge is yellowish white, and the lower yellowish brown. The leaves remain attached to the tree, and there is rarely any noticeable impairment in their functions.

Cause. Leaf blister of oaks is caused by the fungus *Taphrina coerulescens.* New infections in spring are caused by spores from overwintered leaves and possibly by spores lodged in bud scales and on twigs.

Control. A single application at bud-

Fig. III-78. Leaf blister of oak.

swelling time in spring of any of the following fungicides, applied with a power sprayer so as to coat buds and twigs thoroughly, will give control: Captan, Ferbam, Fore, maneb, or zineb. Dilute as directed by the manufacturer.

LEAF SPOTS. Like all other trees, oaks are subject to a number of leaf spot diseases. These rarely cause much damage to the trees, inasmuch as they become numerous late in the growing season. The following fungi are known to cause leaf spots: *Actinopelte dryinum, Cylindrosporium microspilum, Dothiorella phomiformis, Elsinoë quercus-falcatae, Gloeosporium septorioides, G. umbrinellum, Leptothyrium californicum, Marssonina martini, M. quercus, Microstroma album, Monochaetia monochaeta, Phyllosticta tumericola, P. livida, Septogloeum querceum, Septoria quercus,* and *S. quercicola.*

Control. Gathering and destroying all fallen leaves is usually sufficient to keep down most outbreaks in seasons of normal rainfall. Valuable oaks may be protected by several applications of Benlate, copper, or zineb sprays at 2-week intervals, starting in early spring when the leaves unfold.

POWDERY MILDEW. In the southern and western states *Sphaerotheca lanestris* is the most troublesome mildew producer. It forms a white mealy growth on the undersides of the leaves; this later turns brown. The entire surface of the leaf, as well as the tip ends of the twigs, may be covered with the brown feltlike mycelium.

Other powdery mildew fungi affecting oaks include *Brasiliomyces trina, Microsphaera* spp. and *Phyllactinia guttata.*

Control. Leaf-infecting powdery mildew fungi can be controlled with Bayleton or Benlate.

RUST. *Cronartium quercuum.* The alternate hosts of this fungus are species of pine, such as *Pinus rigida* and *P. banksiana,* to which some damage is done by the development of gall-like growths. On oak leaves small yellowish spots first appear on the undersides; later brown, bristlelike horns of spores develop.

Control. Little damage is done to the oak; hence no control has been found necessary.

TWIG BLIGHTS. Chestnut oak and, at times, red and white oaks in ornamental plantings are affected by twig blights.

Symptoms. A sudden blighting of the leaves on twigs and branches scattered over the tree is the most striking symptom. The dead twigs and small branches, with their light brown, dead leaves still attached, are readily visible at some distance from the tree because of the sharp contrast with unaffected leaves near them.

Diseased bark becomes sunken and wrinkled, and the sapwood beneath is discolored.

Cause. Several species of fungi are capable of causing twig blight: *Botryosphaeria quercuum* (imperfect, *Dothiorella quercina*), *Coryneum kunzei, Diplodia longispora, Pseudovalsa longipes,* and *Botryosphaeria obtusa.* Most of the fungi overwinter in dead twigs and in cankers on larger branches. During rainy weather of the following spring, spores in large numbers ooze from the dead areas and are splashed onto young shoots and into bark injuries, where they germinate and cause new infections.

Control. During summer, prune infected twigs and branches to sound wood. As a rule, cutting to about 6 inches below the visibly infected area will ensure removal of all fungus-infected tissue in the wood. In severely weakened trees, considerably more tissue must be removed, inasmuch as the fungus may penetrate down the branch for a distance of 2 or more feet.

Valuable trees should be fertilized and watered and their foliage covered with a combination spray containing methoxychlor and a copper fungicide to destroy leaf-chewing insects and prevent infections by leaf-spotting fungi.

WILT. This most highly publicized disease of oaks in recent years is causing some concern to arborists, nurserymen, tree owners, and lumber interests in the Middle West. The disease has been found from Kansas and Nebraska eastward to western New York state, and from Minnesota southward to Texas. It is most damaging to live oaks in the South and to red oaks in the upper Midwest, but white oaks are also susceptible.

Symptoms. A dull, pale, water-soaked look at the edges of leaves progresses toward the midrib and base. Leaves may quickly drop from the tree and turn brown or bronze. Live oak leaves may show brown midribs and veins. Spring infections may result in less leaf drop, with brown leaves remaining attached. Fungal mats are found under the bark of infected trees in the spring.

Cause. The fungus *Ceratocystis fagacearum* is known to cause wilt. When first described in the literature in 1943, it was known as *Chalara quercina.* The fungus causes infections primarily through wounds made by people or other agents. It is readily transmitted by tools used by arborists and foresters. The most susceptible period appears to be during spring wood development in the tree.

The causal fungus is spread by root grafts and by several insects and related pests, including fruit flies, Nitidulid beetles, the flatheaded borer *Chrysobothris femorata,* and the mite *Garmania bulbicola.* This fungus has also been recovered from a species of bark beetle and from the two-lined chestnut borer. Two species of oak bark beetles, *Pseudopityophthorus minutissimus* and *P. pruinosus,* were found capable of carrying the oak wilt fungus. Tools and climbing spurs used by lumberjacks, foresters, or arborists are other ways by which the fungus is spread. Squirrels are also believed to be vectors. Tree species belonging to the red oak group appear to be the most susceptible, whereas the white oak is markedly resistant. In nature the fungus has been found on Chinese chestnuts and related genera, and it has been transmitted experimentally to a wide variety of trees, including apple.

Control. No effective control is known. For the present, eradication and destruction of infected specimens and infected firewood is being advocated. Where diseased trees are not removed and destroyed, infection centers develop. Mechanically or chemically severing root grafts can halt further spread. Because insect transmission occurs in spring and the oak wilt fungus appears to be most infectious early in the growing season when the new spring wood vessels are developing, it is suggested that oaks be pruned only in fall and winter.

ARMILLARIA ROOT ROT. *Armillaria mellea.* Oaks are a favored host of this disease, which is discussed in Chapter 12.

ROOT AND BUTT ROT. *Ganoderma lucidum (Polyporus lucidus).* This wound rot fungus causes internal decay of the base of the tree without necessarily killing it. It infects through dead and scarred roots and butts and sometimes produces large fruiting bodies at the tree base.

Hericum erinaceus, Inonotus dryadeus, Polyporus fissilis, Laetiporus sulphureus, and *Pleurotus sapidus* also cause butt rot. *Corticium galactinum* may cause root rot of urban trees; and *Phymatotrichum omnivorum,* cause of Texas root rot, attacks seedlings.

WOOD DECAY. Oaks are subject to attacks by a number of fungi that cause wood decays. Fungi such as *Globifomes graveolens, Phellinus everhartii, Stereum gausapatatum, S. subpileatum,* and *Inonotus andersonii* attack living trees. The butt rot fungi *Polyporus fissilis, Laetiporus sulphureus, Hericum erinaceus,* and *Pleurotus sapidus* also decay wood. The same tree may show infection by more than one of these fungi, although in most cases only one is involved.

Control. Once wood decays have become extensive, little can be done to check their advance. Trees may live for a long time, however, despite the presence of these decays. Except to eliminate a breeding place for vermin, cavity work is rarely done where extensive decay occurs.

RINGSPOT. Plant pathologists have found virus-like particles in black and blackjack oak leaves exhibiting ringspot symptoms. The symptoms were found in trees with dead crowns as well as in vigorous saplings.

Control. Control measures have not been developed.

DECLINE. Oak decline following periods of stress such as construction injury, drought, and insect defoliation has been observed on landscape trees.

Abiotic Disease

CHLOROSIS. Leaf yellowing between the veins of oaks is often the result of iron deficiency. In severe cases, tree growth may be stunted, dead areas may develop on leaves, and trees may die. The problem is often referred to as iron chlorosis or lime-induced chlorosis. Chlorosis occurs on trees growing in neutral or alkaline soils (pH 7.0 or above). Pin oaks grown in many midwestern locations are subject to this disorder.

Control. Iron can be added to the soil in the form of iron chelates, but high pH soils may still tie up the iron. Iron as iron sulfate, iron chelate, or iron citrate may be sprayed on the foliage, implanted into the trunk (Medicaps), or injected via feeder tubes (Mauget). Foliage will turn green in response to these treatments, and the effects may last several months to a few years.

Sulfur should be applied to the root zone to decrease the soil pH. Depending on soil tests, 60–100 pounds/1000 square feet can be applied, and once the soil pH has been changed, the effects should be long-lasting. Elemental sulfur used for soil pH change is more effective if it is worked into the soil; thus planting site soil pH should be tested before planting so that sulfur can be incorporated.

Avoid using alkaline hard water for irrigation. Remove masonry materials from soil in oak planting sites, or avoid planting sites located near masonry walls or other construction.

AIR POLLUTION. Pin, scarlet, and white oaks are sensitive to ozone. See Chapter 10.

Insects

GALLS. There are hundreds of kinds of galls on oak leaves, twigs, and flowers, and a special manual is needed to identify them. Galls are caused by many species of flies, nonstinging wasps, and mites, some of which have larvae that feed and grow completely protected inside the galls. The life cycles of gall insects and mites vary according to the species involved. The pests overwinter either on the trees or on the ground. The adults emerge in spring and travel to the leaves and twigs, in which they deposit eggs. The young that hatch from the eggs then mature in the galls which form around them.

One of the most beautiful galls is the wool sower (Fig. III-79), produced by the gall wasp *Callirytis seminator*. It is found on white, chestnut, and basket oaks. Most kinds of galls on oaks rarely affect the health of the trees. However, some such as the gouty gall (Fig. III-80), the horned gall, and the oak–potato gall (Figs. III-81 and 82), affect the health as well as the appearance of the tree by killing branches. Another common gall is the oak–apple gall (Fig. III-83).

Control. Normally, controls are not needed;

however, before growth starts in spring, spray the trees either with dormant lime sulfur or with a dormant miscible oil to destroy some of the pests overwintering on the branches. Valuable trees should also be given a Sevin or a methoxychlor–Kelthane spray in mid-May and again in mid-June. Heavily infested branches should be pruned and destroyed before the adults emerge.

OAK PHYLLOXERID. *Phylloxera* sp. Phylloxerids are very small aphidlike insects that attack oak buds and developing leaves, causing the leaves to curl and twist, thus marring their normal appearance. They are usually found in clusters.

OAK LEAF APHID. *Myzocallis* sp. See Chapter 11.

GIANT BARK APHID. See aphids, under Sycamore.

WOOLLY APHIDS. *Stegophylla* sp. They can be found on several species of oak.

MAPLE LEAFHOPPER. See leafhoppers, under Maple.

GOLDEN OAK SCALE. *Asterolecanium variolosum.* Shallow pits are formed in the bark by these circular, greenish gold scales. Sometimes referred to as pit scales, they attain a diameter of only 1/16 inch. Infested trees have a ragged, untidy appearance. Young trees may be killed outright, the lower branches dying first.

Control. A superior miscible oil or ethion plus oil spray applied in early spring will destroy most of the pests. Orthene, Diazinon, or Sevin sprays in May and June will control the crawler stage. A Cygon spray in August will control the adult stage.

LECANIUM SCALES. *Lecanium corni* and *L. quercifex.* Branches and twigs infested with these pests are covered with brown, down-enveloped, hemispherical scales. The adults are 1/6 inch in diameter. The winter is passed in the partly grown stage. *L. quercitronis* infests several species of oaks in the eastern states and live oak in California.

Control. Spray with superior miscible oil or ethion plus oil before the buds open in spring, or with malathion, Sevin, Orthene, Diazinon, or methoxychlor in early summer when the young scales are moving about.

OAK GALL SCALE. *Kermes pubescens.* Large, globular scale insects, about ⅛ inch in diame-

Fig. III-79. Wool sower gall caused by the gall wasp *Callirytis seminator*.

ter, infest the twigs, leaf stalks, and midribs and tend to gather on the young buds. Kermes scales are illustrated in Figure III-84. A red and brown mottling characterizes the pest. The leaves become distorted and many of the twigs are killed, but the tree as a whole is not usually seriously injured.

Control. A dormant oil or ethion plus oil spray applied in March or early April is very effective. Orthene, Diazinon, methoxychlor, malathion, or Sevin sprays in late spring are effective against the crawler stage.

OBSCURE SCALE. *Melanaspis obscurus.* Tiny, circular, dark gray scales, 1/10 inch in diameter, may occasionally cover the bark of twigs (Fig. III-85) and branches. The pest overwinters in the partly grown adult stage. Heavy infestations kill branches and weaken trees.

Control. Spray with dormant oil, oil plus ethion, or winter-strength lime sulfur before growth starts in spring, or with Diazinon, malathion, methoxychlor, Orthene, or Sevin in midsummer. Obscure scales are difficult to control because they settle close to one another and their collective waxy coating sheds pesticides. In addition, crawler emergence occurs over a long time period.

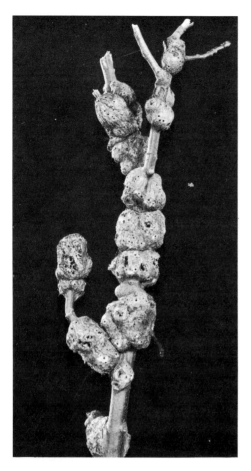

Fig. III-80. Gouty oak gall caused by the gall wasp *Andricus punctatus.*

PURPLE SCALE. *Lepidosaphes beckii.* The female of this species has an elongated, oyster-shaped, slightly curved brown or purplish body. This species infests citrus trees primarily, but it occurs also on many other species in California. Cottony-cushion scale (see under Acacia) attacks oak, as does cottony maple (see under Maple) and California red scale.

Control. Spray with Diazinon, methoxychlor, malathion, or Sevin when the young crawling stage is present.

OAK LACE BUG. *Corythucha arcuata.* The leaves of many oaks, particularly the white oak species, may turn whitish gray and become defoliated when heavily infested with this lace bug. The top surface of the leaf shows epidermal chlorotic flecks or stippling and the undersurface may be covered with tiny black spiny nymphs and adults with lacy, transparent wing covers. Tiny black tarlike splotches also present are insect fecal wastes.

Control. In May, when bugs appear, and again in July, if needed, spray with malathion, Orthene, Sevin, methoxychlor, or Guthion.

OAK MITE. *Oligonychus bicolor.* See Chapter 11.

STANFORD WHITEFLY. *Tetraleurodes stanfordi.* This pest infests coastal oaks in California.

PERIODICAL CICADA. See under Crabapple. Oaks can be attacked.

Borers

TWO-LINED CHESTNUT BORER. *Agrilus bilineatus.* Large branches are killed on some trees as a result of the formation of tortuous galleries underneath the bark by the two-lined chestnut borer, a white flat-headed larva ½ inch long. The adult, a slender greenish black beetle ⅜ inch long, appears in late June. The larvae pass the winter beneath the bark. Weakened European hornbeam, beech, and American chestnut are also attacked by this pest. A slightly larger relative, *A. acutipennis,* attacks members of the white oak group.

Control. Cut down and destroy all badly infested trees. Increase the vigor of the remaining trees by fertilizing and watering. Lindane or methoxychlor sprays applied to the bark of trunk and branches monthly from May to August will also help.

FLATHEADED BORER. See under Maple.

TWIG BORERS. *Agrilus angelicus.* The oak twig girdler is responsible for patches of dead foliage of California live oak. The oak twig pruner (see under Hickory) and the twig girdler (see under Hackberry) also infest oak.

Control. Prune, rake, and destroy girdled twigs.

CARPENTERWORM. See under Ash. This insect produces unsightly scars on oaks.

OAK CLEARWING BORER. *Paranthrene simulans.* This 1-inch-long, black-headed gray borer

Fig. III-81. Oak–potato gall. **Fig. III-82. Inset:** Adult of *Neuroterus batatus*, the cause of oak–potato gall.

Fig. III-83. Oak–apple gall caused by a species of gall wasps.

Fig. III-84. Scale *(Kermes pubescens)* on bur oak. The scale insects closely resemble the oak buds in the axils of the leaves and at the tips of twigs.

attacks the lower trunk of red and white oaks in the East. Sap spots and eventually clumped frass appear from galleries ⅓ inch in diameter. Adults are black and orange-banded beelike clear-winged moths.

RED OAK BORER. *Enaphalodes rufulus.* Adults are 1-inch-long gray beetles with long antennae, and larvae are shiny and white with tiny legs. The larva mines under the bark the first year and tunnels into the wood the second year, producing frass.

WHITE OAK BORER. *Goes tigrinus.* The 1- to 1½-inch legless whitish larva feeds on wood of young trees. The adult beetle is white and brown mottled, almost 1 inch long.

OAK TIMBERWORM. *Arrhenodes minutus.* The whitish larvae of this weevil attack wounds of mature trees.

COLUMBIAN TIMBER BEETLE. *Corthylus columbianus.* Black to reddish brown beetles are less than ¼ inch long and construct galleries 1/10

inch in diameter. Even vigorous trees may be attacked.

PIN-HOLE BORERS. *Platypus* and *Xyleborus* spp. These brown ambrosia beetles are ⅛ inch long and attack weakened, stressed, and dying trees or trees with injuries.

ROOT BORERS. *Prionus imbricornis* and *P. laticollis.* Several years of feeding by creamy white larvae on oak roots causes tree decline. Larvae grow to 3 inches, and dark brown adult beetles are 1½ inches long. Stressed trees are most likely to be attacked.

OTHER BORERS. Oaks may be invaded by other borers, including the carpenterworm (see under Ash); the leopard moth borer (see under Maple); dogwood borer (see borers, under Dogwood); red-headed ash borer (see under Ash); beech borer and oak branch borer, *Goes pulverulentis* and *G. debilis;* and oak stem borer, *Aneflormopha subpubescens.*

Leaf-Eating Insects

There are so many species of insect larvae that feed on oak leaves that it is impossible to list them all. The following are some of the more important.

OAK BLOTCH LEAF MINERS. *Lithocollertis hamadryadella* and *L. cincinnatiella.* The solitary leaf miner makes pale blotches on many kinds of oak leaves. One leaf miner larva is found in each blotch. The gregarious leaf miner makes similar blotches on leaves of white oak, each blotch containing 10 or more larvae.

Control. Rake up and destroy fallen leaves because the insect overwinters in the mines of such leaves. Spray with Sevin, Diazinon, malathion, or Orthene in spring when the young miners hatch out of the eggs and again in early summer.

YELLOW-NECKED CATERPILLAR. *Datana ministra.* The leaves are chewed by this caterpillar, a black and yellowish white-striped larva 2 inches long. The adult female moth is cinnamon brown with dark lines across the wings, which have a spread of 1½ inches. Eggs are deposited in batches of 25 to 100 on the lower leaf surface. The insect hibernates in the pupal stage in the soil. The walnut caterpillar (see under Walnut) also feeds on oak.

Fig. III-85. Obscure scale on red oak.

Control. Spray the leaves with Dursban, *Bacillus thuringiensis,* methoxychlor, or Sevin when the caterpillars are young.

PIN OAK SAWFLY. *Caliroa lineata.* The lower surfaces of pin oak leaves are chewed by ⅜- to ½-inch-long greenish larvae or slugs, which leave only the upper epidermal layer of cells and a fine network of veins. Injured leaves turn a golden brown, and when the feeding by the larvae is extensive, the injury can be readily distinguished at a considerable distance. The adult stage is a small, shining-black, four-winged insect ¼ inch long.

Control. Spray the leaves with Diazinon or Sevin when the larvae begin to feed in early summer. Direct the spray primarily to the lower surfaces of the leaves. This pest appears to prefer the upper parts of the tree.

OAK SAWFLY. *Periclista* sp. This insect occasionally feeds on oak.

SADDLEBACK CATERPILLAR. *Sibine stimulea.* This broad, spine-bearing, red caterpillar with a large green patch in the middle of its back attains an inch in length and occasionally chews the leaves of ornamental and streetside oaks. The adult is a small moth with a wingspread of 1½ inches. The upper wings are dark reddish brown; the lower, a light grayish brown.

Control. The same as for the yellow-necked caterpillar.

OAK SKELETONIZER. *Bucculatrix ainsliella.* The leaves of red, black, and white oaks may be skeletonized by a yellowish green larva, ¼ inch long when fully grown. The adult, a moth with a wing expanse of ⁵⁄₁₆ inch, is creamy white, more or less obscured by dark brown scales.

Control. Spray with Diazinon, methoxychlor, or Sevin when the larvae begin to feed in early June and when the larvae of the second brood begin to feed in August. Rake up and destroy leaves to get rid of cocoons.

ASIATIC OAK WEEVIL. *Cyrtepistomus castaneus.* Severe damage to the leaves of oaks and chestnuts can be caused by a deep red or blackish weevil, ¼ inch long, with scattered green scales which have a metallic luster. Its long antennae are characteristic of only one other species in this country, the longhorned weevil, *Calomycteris setarius.* Like the elm leaf beetle and the boxelder bug, they move into homes in fall to hibernate.

Control. Methoxychlor sprays on foliage at the time the weevils begin to feed provide good control.

OAK LEAF TIER. *Croesia purpurana.* Buds and leaves are chewed by the larval stage of this pest.

Control. Spray with Sevin, malathion, or Diazinon no later than April 15 in the latitude of New York or before the buds begin to break.

OAK WEBWORM. *Archips fervidanns.* The larva spins a dense webbing around the feeding site.

CALIFORNIA OAKWORM. *Phryganidia californica.* The larvae of this moth feed on and defoliate many kinds of oak trees in California.

CALIFORNIA TUSSOCK MOTH. *Hemerocampa vetusta.* This West Coast insect feeds on oak.

MAPLE TRUMPET SKELETONIZER. See under Maple.

WALKINGSTICK. *Diapheromera femorata.* The walkingstick defoliates oaks in the East. Nymphs and adults are slender, with long thin legs and antennae. They resemble twigs of their host when still, and are 2½ to 3 inches long. Chemical control is occasionally needed, since their presence can be a nuisance in high-use areas.

OAK LEAFROLLER. *Archips semiferanus.* The ⅘-inch black-headed, green larvae fold or roll individual leaves together, forming a protected enclosure. They feed on leaves and cause defoliation. Chemical controls may be needed to protect high-value trees.

GYPSY MOTH. See under Elm.

CALIFORNIA TENT CATERPILLAR. See under Strawberry-tree.

FOREST TENT CATERPILLAR. See under Maple.

EASTERN TENT CATERPILLAR. See under Willow.

ELM SPANWORM. This destructive pest attacks red and white oaks. See under Elm.

FALL CANKERWORM. See under Elm.

ORANGESTRIPED OAKWORM. *Anisota senatoria.* The black 2-inch larva with eight narrow yellow longitudinal stripes feeds in late summer, consuming leaf blades. The adult is a yellowish red moth.

VARIABLE OAKLEAF CATERPILLAR. *Heterocampa manteo.* The yellowish green 1½-inch larvae feed on leaves in early summer and again in early fall. The adult is a gray moth with dark forewing lines.

Control. Sprays of *Bacillus thuringiensis,* Sevin, or Diazinon as the caterpillars begin to feed generally are effective.

OTHER FOLIAGE FEEDERS. Among other insects that chew oak leaves are the caterpillars of the io moth, *Automeris io;* the cecropia moth, *Hyalophora cecropia;* the luna moth, *Actias luna;* the American daggar moth, *Acronicta americana;* and the satin moth, *Stilpnotia salicis.* The spring cankerworm, *Paleacrita vernata;* green fruitworm, *Lithophane antennata;* linden looper, *Erannus tiliaria;* Japanese weevil, *Pseudocneorhinus bifasciatus;* and pinkstriped and spiny oakworms, *Anisota virginiensis* and *A. stigma,* also feed on oaks.

Control. All these can be controlled by spraying with Sevin, *Bacillus thuringiensis,* or Diazinon when the larvae begin to feed.

Other Pests

OAK MITE. *Oligonychus* and *Eotetranychus* spp. Mottled yellow foilage results from the sucking of the leaf juices by the oak mite, a yellow, brown, or red eight-legged pest. The mite usually overwinters in the egg stage.

Control. Spray with dormant oil or lime sulfur during early spring, or with Kelthane or Tedion in mid-June and again 2 to 3 weeks later.

OLIVE *(Olea)*

This Mediterranean native is planted in mild regions such as California. It transplants easily and fruitless types such as the cultivar 'Swan Hill' are available.

Diseases

LEAF SPOTS. *Cycloconium oleaginum, Gloeosporium olivarum, Cercospora cladosporioides,* and *Asterine oleina.* These fungi cause leaf spots of olive.

OLIVE KNOT. *Pseudomonas syringae* pv. *savastanoi.* This is a bacterial knot.

ROOT ROTS. *Armillaria mellea* and *Phymatotrichum omnivorum.* These fungi cause root rots of olive.

WILT. *Verticillium albo-atrum.* This fungus causes a devastating wilt of olive.

Insects

SCALES. A California scale *(Parlatoria pittospori),* and black, greedy, ivy, latania, California red, dictyospermum, and euonymus scales (see Chapter 11) attack olive. Comstock mealybug also feeds on olive.

AMERICAN PLUM BORER. See under Planetree.

ORCHID-TREE *(Bauhinia)*

Several species of orchid-trees are grown as spectacular flowered lawn specimens and as framing for small houses in Florida and other warm parts of the country. Orchid-tree prefers warm, well-drained soil, and it does not tolerate heat and drought.

Diseases

LEAF SPOT. *Colletotrichum* and *Phyllosticta* spp. The former has been reported from Texas and the latter from Florida.

Control. Controls are usually unnecessary.

Insects

The Cuban may beetle, *Phyllophaga bruneri,* two species of mealybugs—citrus and long-tailed—and 19 species of scales attack orchid-trees.

Control. Methoxychlor sprays should be effective against the adult stage of the Cuban may beetle. Malathion or Sevin sprays will control mealybugs and the crawler stage of the scales. A summer oil emulsion is also effective against mealybug and scales.

OSAGE-ORANGE *(Maclura)*

Osage-orange is typically a small, thorny tree adaptable to a wide range of sites. Occasional specimens, sometimes multitrunked, grow to 60 feet with an irregular crown. Where it has space, its large spreading trunks and branches provide an interesting effect.

Diseases

LEAF SPOTS. Leaves are occasionally spotted by four specimens of fungi: *Cercospora maclurae, Ovularia maclurae, Phyllosticta maclurae,* and *Septoria angustissima.*

Control. Leaf spots are rarely serious enough to warrant control measures.

OTHER DISEASES. Osage-orange is susceptible to a leaf blight caused by *Sporodesmium maclurae,* rust by *Physopella fici,* Verticillium wilt, and *Phymatotrichum omnivorum* root rot.

The wood of osage-orange was found to contain about 1 percent of the chemical 2,3,4,5-tetrahydroxystilbene, which was toxic to a number of fungi. This may explain why this host is remarkably resistant to decay-producing fungi.

Insects

SCALES. Five species of scale insects infest osage-orange: cottony-cushion, cottony maple (see under Maple), European fruit lecanium, Putnam, and San Jose.

The citrus mealybug and the citrus whitefly also attack this tree.

Control. Malathion or Sevin sprays will control the scales.

PAGODA-TREE, JAPANESE *(Sophora japonica)*

This member of the legume family produces clusters of white, showy flowers in summer. It should be more widely planted as a specimen tree. Japanese pagoda-trees do well in cities. The single-stemmed form 'Regent,' should be used along city streets. The falling flowers in summer will stain automobiles, and hence this tree should not be planted in parking lots or similar areas.

Winter injury and branch breakage can be problems.

Diseases

CANKER. *Fusarium lateritium.* Oval, 1- to 2-inch cankers, with definite, slightly raised, dark red-brown margins and light tan centers occasionally appear on this host. When the cankers completely girdle the stem, the distal portion dies. The sexual stage of this fungus is *Gibberella baccata.* Another fungus, *Cytospora sophorae,* is also found on dead branches. Both fungi are usually associated with frost damage in late fall or early spring.

Control. Prune diseased or dead branch tips to sound wood and apply a copper fungicide several times during the growing season.

DAMPING-OFF. *Pellicularia filamentosa.* In Connecticut and the southern states this fungus causes a damping-off of seedlings.

Control. Use steam-pasteurized soil or soil treated with PCNB (Terraclor).

TWIG BLIGHT. *Nectria cinnabarina and Diplodia sophorae.* Two fungi cause cankers and dieback of twigs of this host.

Control. The same as for canker.

OTHER FUNGUS DISEASES. Japanese pagoda-tree is also subject to powdery mildew caused by *Microsphaera,* root rot by *Phymatotrichum omnivorum,* leaf spot by *Phyllosticta sophorae,* and rust by *Uromyces hyalinus.*

Control. Controls are unnecessary.

VERTICILLIUM WILT. See Chapter 12.

Insects

SCALES. Cottony-cushion *Icerya purchasi* and long soft *Coccus elongatus* have been reported on Japanese pagoda-tree.

Control. Malathion or Sevin sprays will control the crawler stage in late spring and early summer.

OTHER PESTS. A nema, *Meloidogyne* sp., and mistletoe, *Phoradendron flavescens,* also attack Japanese pagoda-tree.

Control. Control measures are not available.

PALMS (Palmaceae)

Diseases

BACTERIAL WILT. *Xanthomonas sp.* A wilt and trunk rot of coconut and Cuban royal palm is caused by this bacterium. At first the lower leaves turn gray and wilt. This is followed by gummosis of the trunk, discoloration of the vascular tissues, and finally collapse of the crown.

Control. No control measures are known.

LETHAL YELLOWING. This highly fatal disease, caused by a mycoplasmalike organism, was first observed in Jamaica nearly a century ago. In recent times it has been responsible for the death of hundreds of thousands of coconut palms (*Cocos nucifera*) in southern Florida. First observed in the Key West area in 1955, it has now spread throughout much of Florida and into south Texas. The vector for this disease appears to be a planthopper.

An early symptom is the appearance of dead tips of the inflorescences when they emerge from the spathes. Leaf yellowing of the lower fronds follows. Another symptom is "shelling," that is, premature dropping of coconuts, regardless of size. The disease progresses until the bud becomes necrotic and the tree dies.

In Florida lethal yellowing has also resulted in death of *Veitchia, Arikuryroba, Washingtonia,* and *Pritchardia* palms.

Control. No practical control is presently available, although cessation of symptom development and resumption of normal growth has been obtained by injecting affected coconut palm trees with oxytetracycline HCL (Terramycin) antibiotic.

The rate of lethal yellowing spread reportedly decreases in trees injected with oxytetracycline at 4-month intervals, as compared with untreated trees. The treatment may be economically viable for coconut palms grown primarily for ornamental purposes.

One promising method of coping with lethal

yellowing is to use resistant varieties. Seed nuts of resistant palms have been imported from Jamaica to be used as replacements. 'Malayan Dwarf', 'Fiji Dwarf', and 'Panama Tall' coconut palms are some of the more resistant kinds presently available. Other resistant or tolerant palms include areca, cabbage, MacArthur, paurotis, pindo, pygmy date, queen, royal, and solitaire.

BUTT ROT. *Ganoderma zonatum.* This is one of the most serious fungus diseases of palms in Florida. The spores of the fungus enter through wounds at the base of the tree made by lawn mowers or other instruments. The lower leaves die and new leaves are stunted. Typical fruiting bodies of the fungus appear at the base of the infected tree near the soil line.

Control. Control measures have not been developed. Affected trees should be removed and destroyed. Clean or steam-pasteurized soil should replace soil in the vicinity of the diseased tree.

FALSE SMUT, LEAF SCAB. *Graphiola phoenicis.* Palms belonging to the genera *Arecastrum, Arenga, Howea, Phoenix, Roystonea,* and *Washingtonia* suffer from a parasite that causes a yellow spotting of the leaves and the formation of numerous small black scabs or warts. The outer parts of these fruiting bodies are dark, hard, and horny. From the inner parts protrude many long, flexuous, sterile hyphae. Within the inner membrane of the structure, powdery yellow or light brown masses of spores arise. The exact place of this unique fungus in the system of classification is not known. Severely infected leaves soon die.

In an experimental planting of date palms in Texas, the variety Kustawy from Iraq developed only light infections from this fungus. In Texas this disease is most troublesome in areas with consistently high humidities.

Control. Cut out and destroy infected leaves or leaf parts at the first sign of the disease, and spray with a copper fungicide. On greenhouse palms, control insects with insecticides rather than by syringing, a practice that helps to spread false smut.

LEAF BLIGHTS. *Pestalotiopsis palmarum* and *P. palmicola.* The first of these fungi infects *Cocos romanzoffiana (plumosus)* and other palms

in the axils of the leaves and where leaflets are attached to the leaf stalk. The fungus penetrates into the deeper tissues, causing a brown discoloration. Gray-brown spots develop on the leaf blades and run together to form large blotches. Brown septate spores with characteristic appendages develop in black masses on the upper parts of the leaves.

Blight caused by *P. palmicola* is reported to be very destructive in Florida to *Phoenix, Cocos,* and *Washingtonia.* The blight begins at the tips and works toward the leaf stalk, killing the tissues, which turn brown.

Control. Copper fungicides will probably control these diseases.

Still another serious leaf blight of *Washingtonia* is caused by the fungus *Cylindrocladium macrosporium.* Numerous small dark brown spots with light-colored margins disfigure the leaves.

Along the Pacific Coast in southern California the fungus *Penicullium vermoeseni* causes three diseases which result in a great loss of ornamental palms: leaf base rot of *Phoenix,* bud rot of *Washingtonia,* and trunk canker of *Cocos romanzoffiana.* Among the effects are a successive decay of the leaf bases from the oldest to the youngest of the tightly folded bud leaves, and the weakening and breaking of the trunk.

Control. Control measures have not been developed.

LEAF SPOT. *Stigmina palmivora.* This leaf spot is especially serious in greenhouses where insufficient light is provided. The spots are small, round, yellowish, and transparent. They often run together to form large irregular gray-brown blotches, which may result in death of the leaf. The spores are long, club-shaped, many-celled, and brown. They are formed on tufts of short basal cells.

Control. The same as for false smut.

STEM AND ROOT ROTS. Species of several genera of fungi, among them *Pythium, Fusarium, Phymatotrichum,* and *Armillaria,* have been reported as causing root disease of palms in Florida and several western states.

The *Pythium* root rot is accompanied by yellowing and wilting of the leaves one after another, until the bud falls from the top of the plant. *Fusarium* has been associated with stem and root rot of the royal palm *(Roystonea).*

Armillaria mellea has been thought to cause root rot of the date palm *(Phoenix)* in California.

In certain palm houses numerous fruiting bodies of *Xylaria schweinitzii* have been found growing from roots of *Howea* and other palms. Since another species of *Xylaria* is known to cause root rot of apple, the poor condition of some greenhouse palms possibly may be due to this fungus. No control has been suggested. As a precautionary measure, however, all black, club-shaped fruiting bodies about 2 inches long should be removed.

In Florida a stem rot of *Chamaedorea seifrizii* is caused by *Gliocladium vermoeseni*. Controls for this rot have not been developed.

BUD ROT AND WILT. *Phytophthora* sp. Wilting and bud rot of the coconut palm have been at various times referred to different species of *Phytophthora*, such as *P. palmivora, P. arecae, P. faveri.*

Control. When this disease occurs in ornamental plantings, it is controlled by cutting out the crown of the infected plant.

BLACK SCORCH AND HEART ROT. *Ceratocystis paradoxa.* This disease may be destructive to date palms of plantations and ornamental gardens in California. It is found in all structures of the plant except the roots and stems. Lesions are dark brown or black, hard carbonaceous, having the appearance of a black scorch. The fungus gains entrance even in the absence of wounds. The disease is most serious as a terminal bud rot.

Control. All infected fronds, leaf bases, and flower parts should be pruned out and destroyed and the cuts disinfested. Copper sprays give promise of control.

Insects and Related Pests

PALM APHID. *Cerataphis variabilis.* In its wingless form this aphid is often mistaken for a whitefly because it is dark and disclike with a white fringe. The green peach aphid, *Myzus persicae,* also infests palms.

Control. Spray with malathion or Sevin when the aphids appear.

PALM LEAF SKELETONIZER. *Homaledra sabalella.* In Florida this is a major pest, feeding on the leaves of many species of palms under a protective web of silk. Leaves are blotched, then shrivel and die.

Control. Cut out and destroy infested fronds, or spray with Sevin.

MEALYBUGS. Four species of mealybugs infest palms: citrus, ground, long-tailed, and palm.

Control. All but the ground mealybug can be controlled with malathion or Sevin sprays. The ground mealybug can be destroyed by drenching the soil around the roots it infests with Diazinon solution.

SCALES. Palms are extremely susceptible to many species of scale insects. At least 23 species of scales have been recorded on this plant family, particularly those grown outdoors. See Chapter 11.

Control. Sevin or malathion sprays will control the crawler stage of scales.

THRIPS. *Heliothrips haemorrhoidalis, H. dracaenae,* and *Hercinothrips femoralis.* Although these species have been reported on palms, they are not considered serious pests. They are not troublesome where scale insects and mealybugs are controlled.

MITES. Three species of mites—banksgrass, privet, and tumid spider—infest palms.

Control. Chlorobenzilate, or Kelthane sprays will control these pests.

FULLER ROSE BEETLE. See under Camellia.

PAPER-MULBERRY *(Broussonetia)*

Diseases

CANKER. *Fusarium solani.* Branch dieback results from cankers produced by this fungus. The disease was first found in Ohio in 1965. The same fungus also infects cottonwood, red oak, and sweetgum.

Control. Prune infected branches. Keep trees in good vigor by fertilizing and by watering during dry spells.

ROOT ROT. *Phymatotrichum omnivorum.* This root rot, common on many plants in the southern states, has been reported from Texas on the paper-mulberry tree.

Control. Control measures have not been developed.

OTHER DISEASES. Among the other diseases reported from the southern states are a dieback and canker caused by the fungus *Nectria cinnabarina*, a leaf spot by *Cercosporella mori,* a mistletoe disease by *Phoradendron flavescens,* and root knot by the nema *Meloidogyne incognita.*

Control. These diseases are seldom serious enough to warrant control measures.

PAULOWNIA. See Empress-Tree.
PEACH, FLOWERING. See Cherry.

PEAR, ORNAMENTAL *(Pyrus)*

The flowering pear *(Pyrus calleryana* 'Aristocrat', 'Bradford', 'Capitol', 'Redspire', 'Whitehouse', and others) is said to be virtually free of diseases. Callery pears *(P. communis),* usually nicely shaped medium-sized trees with many landscape uses, have been touted as resistant to fire blight, a serious disease of pears grown for fruit.

Diseases

FIRE BLIGHT. In some seasons, all cultivars planted in a region have been observed with fire blight (Fig. III-86), some at seriously damaging levels. See Chapter 12.

CANKER. Trees with cankers on their trunks produced by a species of *Coniothyrium* were found in Maryland.

PECAN *(Carya illinoensis)*

A large nut tree of the South and some lower midwestern areas of the United States, pecan is seldom used in landscapes, perhaps because of its susceptibility to disease. Pecan disease and insect problems are similar to those of hickory. It is subject to zinc deficiency in alkaline soil.

Diseases

BROWN LEAF SPOT. *Cercospora fusca.* This leaf spot is common throughout the pecan-growing areas in the South. Downy spot, caused by the fungus *Mycosphaerella caryigena,* also occurs in the same localities.

Control. Spray with a low-lime bordeaux mixture as recommended by the plant pathologist in the state where trees are affected.

Fig. III-86. Callery pear fire blight.

CROWN GALL. *Agrobacterium tumefaciens.* The roots of young pecan trees may be infected by this bacterial disease. Occasionally older trees in orchards are also infected.

Control. Remove galls from roots of young infected trees. Treat with Gallex as recommended by the manufacturer. See Chapter 12.

SCAB. *Cladosporium effusum.* This is one of the most destructive diseases of pecans in the Southeast. The fungus attacks leaves, shoots, and nuts.

Control. Dodine sprays will control this disease. The local state plant pathologist at the state college of agriculture can provide details on dilution and timing.

OTHER DISEASES. Pecans are subject to a number of diseases that infect other species of *Carya.* They are not treated in detail here because the pecan is not an important ornamental tree. See Hickory.

Insects

BORERS. A large number of borers infest pecans. Following are the most important: dogwood, flatheaded apple tree, pecan, pecan carpenterworm, shot-hole, twig girdler, and twig pruner.

Control. As with borers on other trees, keep ornamental trees in vigorous condition by feeding and watering. Particularly valuable ornamental specimens can be protected from most borers by spraying the trunks and branches with Lindane or methoxychlor at the time the insects are most vulnerable. State entomologists will provide information on concentrations and application dates.

PEPPERIDGE TREE. See Tupelo.

PEPPER-TREE *(Schinus molle)*

This native of South America grows in mild regions, being tolerant to some frost, poor, dry soils, and wind. It is considered a messy tree because it sometimes sheds twigs and leaves.

Diseases

Heart rot caused by the fungi *Ganoderma applanatum, Polyporus dryophilus, P. farlowii, Laetiporus sulphureus,* and *Trametes versicolor* have been reported from California. These fungi are difficult to control.

Root rot caused by *Armillaria mellea* and wilt by *Verticillium albo-atrum* also affect peppertree. The controls for these fungus diseases are discussed in Chapter 12.

Insects

The omnivorous looper caterpillar, citrophilus mealybug, citrus thrips, and thirteen species of scales including black, hemispherical, cottonycushion, greedy, and ivy infest the pepper-tree.

Control. Caterpillars are controlled with Sevin sprays; mealybugs, scales, and thrips with malathion.

PERSIAN PARROTIA *(Parrotia)*

This native of Iran is relatively pest-free.

PERSIMMON *(Diospyros)*

American persimmon *(D. virginiana),* native to the eastern United States tolerates a wide range of soils and climates. Persimmons in the wild and those grown for their fruits are subject to a great number of diseases and pests. Ornamentals, however, are relatively free of problems.

Diseases and Insects

LEAF SPOT. *Cercospora fuliginosa.* This fungus causes small black spots.

WILT. Wilt, caused by the fungus *Acremonium diospyri,* has been a serious problem of persimmons in the Tennessee Valley.

TWIG BLIGHT. *Botryosphaeria diplodia.* Infection causes twig dieback.

SCALES. White peach, European fruit lecanium, and greedy scales, as well as Comstock and long-tailed mealybug, attack persimmon.

AMERICAN PLUM BORER. See under Planetree.

TWIG GIRDLER. See under Hackberry.

FULLER ROSE BEETLE. See under Camellia.

ASIATIC OAK WEEVIL. See under Oak.

PINE *(Pinus)*

Forty different pines are native to North America. Their size, shape, and adaptability to landscape use vary.

Although pines fare better than hemlock, fir, or spruce in the environment of larger cities, they grow best in small towns and in the country. Among the species most commonly used in ornamental plantings are Austrian pine, mugho pine, Scotch pine, and white pine. Each is subject to a number of fungus diseases and insect pests.

Many pines have a generally pyramidal shape which becomes flat-topped as they mature. Although some species will live 80, 100, or 200 years in good native sites, landscape pines often become aged, or flat-topped, in much less time owing to generally poor urban growing conditions. Scotch and Austrian pines are a dubious choice in landscapes where Diplodia tip blight and pine wilt diseases prevail.

Diseases

CANKERS. A large number of canker diseases occur on all pines. These are more common on trees in forests than in ornamental plantings.

The fungus *Ascocalyx abietina* (formerly called *Scleroderris* and *Gremeniella*) has caused serious losses of red and Scotch pines in the northeastern states and southeastern Canada in the past several years. The disease, called Scleroderris canker, causes needles of infected shoots to discolor at the base and drop.

Infection usually starts in the lower branches and then moves upward. Cankers develop along the main stem, eventually girdling the stem and causing death of the entire tree. Scolytid beetles are suspected of being vectors of the fungus. Control measures suggested for the dieback, or tip blight, of pines (see below) might be helpful in combating this disease.

The fungus *Tympanis pinastri* attacks red pine and, to a lesser extent, white pine. It produces elongated stem cankers with or without definite margins and with depressed centers that become roughened and open after 2 or 3 years. Each canker centers at a node, indicating that the fungus enters the branch at the base of the lateral branch. Trees in weakened condition are most susceptible.

Other cankers on pines are associated with the fungi *Dasyscypha pini, Atropellis pinicola, A. piniphila, A. tingens, Caliciopsis pinea, Fusarium lateritium* f. *pini, Fusarium moniliforme* var. *subglutinans,* the southern pitch canker fungus, *Sphaeropsis* sp., *Phomopsis* sp., *Valsa* spp. (imperfect *Cytospora*), and *Leucostoma kunzei.*

Control. Removal of dead and weak branches, avoidance of bark injuries, and fertilization to increase the vigor of the trees are suggested. Surgical treatment of cankers should be at-

tempted only on valuable specimens. Sprays of Daconil 2787 can be applied in spring.

CENANGIUM TWIG BLIGHT. *Cenangium ferruginosum.* The twigs of exotic species of pine and white pine may be killed by this fungus. Early symptoms are the dying of terminal buds and reddening of the needles. Infection rarely spreads beyond the current season's growth.

Control. Affected twigs should be pruned to sound wood and destroyed. Because twig blight is usually severe only on weakened trees, fertilization and watering are suggested for valuable pines.

DIEBACK, TIP BLIGHT. Since the late 1930s, dieback or tip blight has become a disease of major importance on several of the most desirable pines for ornamental plantings. Austrian pine is, without question, most susceptible to the disease. Scotch, red, mugho, Western yellow, and white pines are decreasingly susceptible in the order named. Douglas-fir and blue spruces are occasionally attacked.

Symptoms. The most prominent symptoms are stunting of the new growth and browning of the needles. The lower branches are most heavily infected, although in wet springs the branches over the entire tree may show browned tips.

Tip blight is often confused with troubles caused by low temperatures, drought, winter drying, and pineshoot moth injury. It can be differentiated from these by the presence of small, black, pin-point fungus bodies, which break through the epidermis at the base of the needle (Fig. III-87). These bodies also occur on affected twigs and cone scales but are less

Fig. III-87. Single needle cluster of Austrian pine greatly magnified to show fruiting bodies of the causal fungus, *Sphaeropsis sapinea.*

readily discernible than on needles, especially if the examination is made soon after the needles turn yellow and wilt. Placing such material in a moist atmosphere for a day or so will induce the black bodies to break through the epidermis, thus becoming visible.

Cause. The fungus *Sphaeropsis sapinea,* formerly called *Diplodia pinea,* causes tip blight. The generic name is used to distinguish this disease from twig blights caused by other microorganisms. The pest overwinters in infected needles, twigs, and cones. In spring, the small fruiting bodies release egg-shaped, light brown spores, which are splashed by rain and wind to the newly developing needles. The fungus grows down through the needles and into the twigs, where it destroys tissues as far as the first node.

Sphaeropsis sapinea has also been associated with a root and root-collar rot of red pine in nurseries, where such trees were completely destroyed. By inserting bits of the fungus into wounded bark near the root collar, the senior author has induced the same disease on Austrian, mugho, Scotch, and white pines, on Douglas-fir, and on Norway spruce. The fungus may follow an infestation of the spittle bug.

The fungus *Sirococcus conigenus* also causes a shoot blight of red pine *(Pinus resinosa)* in the Great Lake states.

Control. No cure is possible for plants that are infested at the base. The use of steam-pasteurized soil is suggested for avoiding root and stem base infections.

Control is possible for the tip blight phase of the disease on older trees. As soon as the blight is noticed, the infected needles, twigs, and cones should be pruned to sound tissues and destroyed. Pruning should be done when the branches are dry, because there is less danger of spreading the spores by contact with the operator and with tools. Where infection has been particularly severe, preventive fungicides are also recommended. Bayleton, Daconil 2787, Benlate, or fixed copper sprays should be applied very early in spring, starting when the buds swell and continuing twice at 10-day intervals. In rainy springs, a fourth application about 7 days after the third may be necessary. As a supplementary treatment, fertilization of the trees in fall or early spring is suggested.

LEAF CASTS. *Bifusella linearis, B. striiformis, Elytroderma deformans, Ploioderma (Hypoderma) desmazierii, H. hedgecockii, H. lethale, H. pedatum, H. pini, Hypodermella sp. Lophodermium seditiosum, Lophodermella spp., Cyclaneusma (Naemacyclus) minus, and Mycosphaerella dearnessii (Scirrhia acicola)* (Fig. III-88). Many species of fungi cause leaf yellowing and, at times, premature defoliation of pines primarily in nurseries and in forest and Christmas tree plantings.

Control. Leaf cast diseases rarely cause sufficient damage to large trees to warrant preventive practices. They can be controlled on small trees and on nursery stock by applying bordeaux mixture, 4-4-50, or any other copper spray, Daconil 2787, or maneb when the needles are half grown and again about 2 weeks later. Because such sprays do not readily adhere to pine foliage, casein soap or some other material should be added as a spreading agent. Zineb sprays also provide good control. Sprays for *Lophodermium* control are applied in late summer.

NEEDLE BLIGHT. *Dothistroma septospora.* This fungus causes slightly swollen, dark spots or bands on 1-year-old needles (Fig. III-89) of planted ponderosa and Austrian pines in late summer. The distal portion of the swollen needle turns light brown and dies. Severely affected trees show sparse foliage as a result of the premature dropping of needles. The disease is favored by cool, moist weather.

Control. Bordeaux mixture or any copper fungicide, Daconil 2787, maneb, or Benlate is applied in spring at budbreak to protect previous season's needles, and a second time in early summer to protect current-year needles.

COMANDRA BLISTER RUST. *Cronartium comandrae.* This rust is widespread on at least sixteen species of hard pines in the United States and has recently become severe on lodgepole pine. Spindle-shaped swellings are formed on branches and trunks of young trees, but on older trunks they may be constricted. Trunk cankers seldom exceed 3½ feet in length.

Control. Control measures have not been developed for this rust.

NEEDLE RUST. *Coleosporium asterum.* Needle rust may be more or less serious. This rust is

Fig. III-88. Brown spot needle blight of pine.

Fig. III-89. *Dothistroma* fruiting bodies on pine needles.

Fig. III-90. Pine needle rust.

most common on pitch pine *(P. rigida)* in the eastern states, but other species such as Austrian, loblolly, mugho, ponderosa, red, Scotch, and Virginia are also susceptible. The blisters break out as pustules (Fig. III-90), which open to discharge the bright orange-colored aeciospores of the rust. Infestation may be serious enough to cause defoliation. The pustules are from ¹⁄₁₆ to ⅛ inch high. The peridium or bounding layer remains visible after the spores have been discharged. The spores from these pustules infect either aster or goldenrod, on which plants the fungus forms golden or rust-colored pustules during summer. The rust is often able to winter on the crown leaves of goldenrod and asters, so that the rust can perpetuate itself on these hosts; but the rust on pine must first infect goldenrod or aster before it can be carried to the pine again. It is not perennial in the pine, as is blister rust.

Control. The destruction of wild asters and goldenrod near valuable pines and spraying or

dusting the pines with sulfur early in the season will provide control.

SCRUB PINE NEEDLE RUST. *Coleosporium pinicola. Pinus virginiana* and occasionally other species are seriously infected with this rust, which breaks out in spring with reddish pustules on the needles. The pustules, up to ½ inch long, fade and become inconspicuous later. Defoliation may occur. This is a short-cycle rust, having no alternate host stage; the spores that develop in the pustules germinate there and shed secondary spores (basidiospores), which in spring reinfect the pines in the vicinity.

It is not a serious rust on ornamental pines except on *P. virginiana.* No control measures have been suggested. Many other species of *Coleosporium* also cause pine needle rusts.

EASTERN GALL RUST. *Cronartium quercuum.* This is a gall-forming stem blister rust. *Pinus sylvestris, P. rigida, P. banksiana,* and *P. virginiana* are all very susceptible to it. A characteristic distortion or kink of the trunk

Fig. III-91. Eastern gall rust of pine.

(Fig. III-91) is formed 3 or 4 feet from the ground. The rust may cause the growth of a deep canker instead of galls. Since the rust may appear at the base of a tree near the ground, it may be inconspicuous, even though the infected area may be a foot across. More or less spherical galls 1 or 2 inches in diameter appear on the branches; eventually the parts above the galls are killed. These galls may be so numerous on jack pines as to kill the trees. They should not be mistaken for galls caused by insects. The infection of trunks of small trees of scrub pine results in very striking spherical galls which entirely encircle the trunks. Galls 5 or 6 inches in diameter are not unusual.

The rust has species of oak as alternate host. Numerous long horns of brownish or dark brown spores are formed on the infected oak leaves.

The basidiospores, developed on these horns, reinfect pines in spring.

WESTERN GALL RUST. *Endocronartium harknesii.* Stem galls caused by this fungus produce spores which reinfect other nearby pines.

Control. Other than pruning, no practical controls for these rusts have been suggested.

FUSIFORM RUST. *Cronartium quercuum* f. sp. *fusiforme.* A related oak/pine rust, fusiform rust, damages landscape and forest pines in the South. Pruning infected cankers, where practical, is advised, and fertilization should be avoided until trees are 10 years old, to reduce infection of young, succulent tissue.

SWEETFERN RUST. *Cronartium comptoniae.* This rust can attack stems of several pine species. Sweetfern is the alternate host.

WHITE PINE BLISTER RUST. This disease is probably the most important disease of white pine in the northeastern United States. Other pines having five needles in the leaf cluster, such as Western white pine and sugar pine, are also susceptible. Blister rust also attacks both wild and cultivated forms of currants and gooseberries but produces less damage on these than on pines.

Symptoms. On pines, reddish brown, drooping needles appear on one or more dead branches. Rough, swollen cankers are found at the bases of affected branches or on the trunk. In late spring, orange-yellow blisters filled with powdery spores break through the bark of the cankered areas (Fig. III-92). These blisters are most conspicuous from April to June and are the most positive sign of the disease. After that time they may be eaten by squirrels or covered with resin. The disease is fatal when trunk cankers or branch cankers completely girdle the affected parts.

The symptoms are less conspicuous on currants and gooseberries, where minute, orange-yellow pustules or brown, curved tendrils appear on the lower surface. Most of the leaves are stunted when the pustules become numerous.

Cause. White pine blister rust is caused by the fungus *Cronartium ribicola.* Winter is passed in cankered bark of living pines. The orange-yellow spores produced in these cankers in late spring are blown to susceptible currant and

Fig. III-92. Blister rust of white pine caused by the fungus *Cronartium ribicola*.

gooseberry leaves, where they produce infec- tion. A new type of spore is produced on these leaves in fall. This type is blown to pines, where they infect the needles to complete the life cycle. After the fungus enters the needle it penetrates slowly downward into the twig and eventually the trunk. The whole process takes place slowly: several years may elapse from the needle infection stage to the appearance of the orange-yellow blisters on the branch or trunk.

Control. Because currants and gooseberries are absolutely essential for the survival of the fungus, blister rust can be controlled by eradi- cation of these undesirable alternate hosts. However, even where both plants are grown near each other, the environment may not be

suitable for infection. Control measures are needed only in high-hazard areas, such as those having cool, moist summers. Local control of the disease can be accomplished by destroying all currants and gooseberries within 900 feet of the pines. Furthermore, the European or culti- vated black currant should not be grown within a mile of pines. This species, which is more susceptible to rust than are other currants and wild gooseberries, has been found to be the chief agent in the extensive spread and estab- lishment of the disease in previously uninfected areas.

Many valuable pines showing moderate in- fection can be saved by prompt surgical treat- ments. Branches with cankers at least 4 inches away from the trunk should be removed. Cankers closer than 4 inches or on the trunk may kill the tree, so excessive surgery may not be justified. Forest geneticists have been breeding pines for resistance to this disease. *Pinus peuce,* one of the five-needle pines, is said to be resistant to blister rust.

OTHER RUSTS. Many other rusts, too numerous to mention in this book, also occur on pines. These appear on all the aboveground parts of the tree, producing many and varied symptoms. They rarely become important parasites, however, on pines in ornamental plantings.

PINE WILT. A nematode, *Bursaphalenchus xylophilus,* causes wilt of pine in the Midwest. The nematode inhabits epithelial cells of resin ducts and infected pines turn pale green, then turn brown and die. Nematodes are carried from diseased to healthy trees by longhorn beetles, which utilize dead and dying trees for breeding purposes. Emerging beetles may carry large numbers of nematodes in their breathing pores to healthy tree branch sites where nematodes enter feeding wounds. In some circumstances, wilt-causing blue-stain fungi such as *Ceratocystis ips* and *C. pilifera* are associated with the nematode. In such cases, it is not certain that nematodes are causing wilt. Landscape Scotch and Austrian pines are especially susceptible to wilt.

WESTERN DWARF MISTLETOE. *Arceuthobium campylopodum.* Several species of pine in the western United States are subject to the debilitating effects of this pest.

LATE DAMPING-OFF AND ROOT ROT. *Pythium debaryanum, Cylindrocladium scoparium, Fusarium discolor, Pellicularia filamentosa, and Phomopsis juniperovora.* In nurseries, coniferous seedlings too old to be susceptible to ordinary damping-off of very young seedlings are subject to a root rot and a top damping-off by five fungi. Seedlings of *Pinus banksiana* and *P. resinosa* are especially susceptible.

Control. Disinfest seeds and then plant them in steam-pasteurized soil. Under some conditions Arasan or PCNB (Terrachlor) soil drenches give good control of some damping-off fungi and can be substituted for steaming.

LITTLE LEAF. *Phytophthora cinnamomi.* This highly destructive disease occurs on shortleaf

and loblolly pines in forest plantings of the southeastern United States. The disease is most easily recognized in its advanced stages when the crown is sparse and ragged in appearance, and the branches, lacking the mass of normal foliage, often assume an ascending habit. Leaves are only half their normal length. Trees die prematurely.

Control. Ornamental pines in home plantings appear to escape this disease even in areas where the causal fungus is known to be present in the soil.

WOOD AND ROOT DECAY. Because pine wood contains a high percentage of resins, wood decays are not as frequent as in hardwoods and in some other evergreens. Some of the fungi associated with wood and root decay are *Armillaria mellea, Haematostereum sanguinolenta, Heterobasidion annosum, Phellinus pini, Fomes officinalis, F. roseus, Inonotus circinans,* and *Phaeolus schweinitzii.*

Control. No control is feasible where extensive decay occurs. Wounds should be avoided and the trees kept in good vigor.

ROOT DECLINE. *Verticicladiella procera.* White pine root decline is a problem in the Ohio River valley and mid-Atlantic states. It causes a dark brown canker under the bark at the base of the tree. Symptoms also include resin flow at the tree base and reduced shoot growth. *V. wageneri* causes a serious black stain root disease of pine and other conifers in the West.

Control. Avoid planting white pines in wet sites.

Abiotic Diseases

WHITE PINE DECLINE. A needle blight of white pine *(P. strobus)* which appears as a browning of entire needles or tips of needles of the current season's growth has been prevalent throughout the northeastern United States. Needle browning is often accompanied by wrinkling of bark (Fig. III-93) on branches and thin, weak foliage. The primary cause of the trouble is unknown, although in some areas it has been associated with high soil pH, high soil clay content, and soil compaction.

STUNT. Stunting and death of 5- to 40-year-old red pines *(P. resinosa)* may be due to poor

Fig. III-93. Branch wrinkling associated with white pine decline.

soil drainage. To avoid injury, select a well-drained planting site at the start.

AIR POLLUTION. Most pines are sensitive to ozone, and when damage occurs it is sometimes called emergence tipburn (see Chapter 10). White and ponderosa pines are sensitive to sulfur dioxide.

SALT. White pines planted within 50 feet of heavily used highways may be severely injured or killed by the salt used to melt snow and ice. Trees growing close to salt water may also suffer. Spraying trees in such locations with an antidesiccant on a mild day in December and again in February will reduce injury.

GIRDLING ROOTS. Pines also show this problem. See Chapter 10.

NEEDLE YELLOWING and DROP. In some areas pine needles are shed at the end of their second growing season. These second-year needles may all yellow at once (Fig. III-94), giving tree owners cause for alarm. This yellowing is normal and merely precedes shedding. After the discolored needles drop, the tree is green again.

Insects and Other Animal Pests

PINE BARK ADELGID. *Pineus strobi.* Though commonly called the white pine bark louse or aphid, this insect is actually an adelgid. The insects usually work on the undersides of the limbs and on the trunk from the ground up. They may be recognized by the white cottony material (Fig. III-95) that collects in patches wherever they are present. In the eastern states white pine may be seriously injured. The winter is passed in the egg stage, the eggs being protected by the cottony covering. The eggs hatch in spring, and soon the young insects may be seen crawling about on the trunk and branches. Several generations are developed each summer.

Control. Apply a superior dormant oil or ethion plus oil spray in early spring, and spray toward the end of April with malathion, Thiodan, or Diazinon. Respray with any of these in mid-May and again in early July if some adelgids reappear.

WHITE PINE APHID. *Cinara strobi.* These aphids

Fig. III-94. Normal yellowing of previous season's white pine needles.

feed on the smooth bark of the twigs and smaller branches of young trees and cause a winter injury which results from drying out of the twigs. Sometimes several hundred aphids are clustered together. Sooty mold often develops on the honeydew.

Powdery pine needle and spotted pine aphids (*Eulachnus* spp.) also feed on pine.

Control. The same as for pine bark adelgid.

PINE LEAF CHERMID. *Pineus pinifoliae.* This insect infests species of pines and spruces. It winters over on pine and then moves in spring to spruces, on which it causes terminal galls. In summer the aphids move to white pine, on which they give birth to nymphs. These then suck out the sap from the new shoots, causing either the development of undersized leaves or the death of the new shoots.

Control. Spray the white pines with malathion just after the first galls open on the spruces in June.

PINE WEBWORM. *Tetralopha robustella.* Mass-

es of brown frass at the ends of terminal twigs result from infestations of the webworm, a yellowish brown larva with a black stripe on each side of the body, which is ⅘ inch long.

Control. Spray with Dylox, methoxychlor, or Sevin in mid-June when the needles are young and before they are webbed. A second application in early August may be necessary.

BARK BEETLES. *Scolytidae* (*Ips* spp. and *Dendroctonus* spp.). Trees in weakened or dying condition are subject to infestation by bark beetle larvae, small worms that mine the bark and engrave the sapwood. On the bark surface where beetles emerge, tiny "shot-holes" varying from ¹⁄₂₀ to ⅓ inch in size (depending on the species) are present.

Control. No effective control measure is known. Trees should be kept in vigorous condition by fertilizing and watering. Severely infested trees should be cut down and destroyed or, if left standing, should be debarked.

EUROPEAN PINE SHOOT MOTH. *Rhyacionia buo-*

Fig. III-95. Pine bark adelgid secretions on trunk and branches of white pine.

liana. This is a very serious pest of mugho pines and red pines in ornamental plantings. Austrian, Scotch, and Japanese black pines also may be badly damaged. The caterpillars, by attacking the tip ends of young shoots, cause them to turn over and become deformed and killed or the lateral buds to be blasted. The caterpillars can be found working in the tip ends in May or the early part of June. Their presence is usually indicated by quantities of resin. The moths that are the adult stage of this caterpillar emerge about June 15 and lay their eggs in August on the new buds.

Control. For small plantings and low trees, hand-picking of infested shoots and buds is probably the most satisfactory method. An early spring (mid-April) spraying with Cygon, Guthion, or Sevin, followed by a late June spraying, gives good control.

NANTUCKET PINE MOTH. *Rhyacionia frustrana.* This species attacks two- and three-needled pines in the eastern United States. A

Fig. III-96. Larvae of the sawfly *Neodiprion sertifer*.

relative, the subtropical pine tip moth, spreads pine pitch canker disease in the south.

Control. Spray in May and July with Cygon, Dylox, Orthene, or Sevin.

SAWFLIES. *Diprion similis.* The caterpillars of the introduced pine sawflies are about 1 inch long and have black heads and greenish yellow bodies with double brown stripes down the middle of the back. They feed on the leaves of various species of pine during May and June and later again during September. Four impor-

tant species are southern, *Neodiprion excitans;* the red-headed, *N. lecontei;* the European pine, *N. sertifer* (Fig. III-96); and the jack-pine sawfly, *N. pratti banksianae*. In severe infestations pines may be entirely defoliated. There are over a dozen species of sawflies whose larvae feed on pines.

Control. As soon as an infestation is observed, spray the trees with methoxychlor or Sevin. The red-headed sawfly has broods through the season and hence may require spray applications in June, July, and August. The European pine sawfly in forest plantings has been controlled by airplane spraying of a virus suspension capable of infecting the larvae.

PINE FALSE WEBWORM. *Acantholyda erythrocephala.* The larvae of this pest are about ¾ inch long, greenish to yellowish brown. They feed on the leaves and tie the masses of excreta and leaf pieces together into loose balls (Fig. III-97).

Control. The same as for pine webworm.

PINE ROOT COLLAR WEEVIL. *Hylobius radicis.* Austrian, mugho, red, and Scotch pines on Long Island and in southeastern New York may be severely damaged by this borer. The symp-

Fig. III-97. Work of the pine false webworm *(Acantholyda erythrocephala)* in a branch of pitch pine *(Pinus rigida);* infested branch is at the left.

toms on infested trees are sickly and dead foliage and masses of pitch around the base of the trunk 3 or 4 inches below the soil surface.

Control. No effective control is available.

PITCH MOTH. *Vespamima sp.* Weakened pines may be attacked by the larvae of these clearwinged moths. Pitch and frass accumulate on the surface of the tree.

PINE NEEDLE MINER. *Exotelia pinifoliella.* Needles of ornamental pines turn yellow and dry up as a result of mining by a ⅕-inch-long larva.

Control. Spray with Sevin or methoxychlor in June. A second application in summer when the yellow-brown moths are flying about is suggested where this pest is particularly prevalent.

PINE NEEDLE SCALE. *Chionaspis pinifoliae.* Pine needles may appear nearly white (Fig. III-98) when heavily infested with this scale, an elongated insect ⅒ in long, white with a yellow spot at one end. The pest overwinters in the egg state under female scales. The eggs hatch in May. The black pine needle scale, *Aspidiotus californicus,* is also widespread on pine and is

Fig. III-98. Pine needle scale on white pine.

often found in association with pine needle scale.

Control. Spray with dormant oil or ethion plus oil in April, or with Cygon or malathion in mid-May and again in late July.

PINE TORTOISE SCALE. *Toumeyella numismaticum.* This cherry red or reddish brown scale, about ⅛ inch long, attacks jack pine in reforested areas, at times destroying 50 percent of the trees. It is closely related to *T. pini,* which occurs on Scotch and mugho pines in some eastern states. *T. pinicola,* the irregular pine scale, attacks many pine species in California.

Control. A dormant lime sulfur spray will kill this scale. The crawling stage may be killed in early summer with Diazinon, malathion, or Sevin.

RED PINE SCALE. *Matsucoccus resinosae.* This species now infests red pines in the area of southeastern New York and southwestern Connecticut. The current season's growth on infested trees consists of yellowed needles, which turn brick red; later the tree dies. The young and adult stages, very difficult to detect, are yellow to brown in color and are hidden in the bark or inside the needle clusters.

Control. Severely infested trees should be cut down and destroyed. Spray less severely infested ones with Cygon early in June and repeat in early September.

WOOLLY PINE SCALE. *Pseudophillipia quaintancii.* This scale, resembling a mealybug, feeds on several pine species.

PINE SPITTLEBUGS. *Aphrophora parallela.* These insects are perhaps most common on Scotch pine, but white pine also is seriously attacked at times. They cause injury to smaller twigs by drawing the sap from them. They form a foamy matter about the young insects, which gives the branches a whitish appearance (Fig. III-99). The adult insects are sometimes ½ inch long, grayish brown in color, and resemble small frogs. Another species, the Saratoga spittlebug, *A. saratogensis,* seriously damages jack and red pines.

Control. Spray with malathion, Dursban, or Sevin in mid-May and again in mid-July, directing the spray forcefully to hit the little masses of spittle that cover the insect.

Fig. III-99. Spittlebug on Austrian pine.

WHITE PINE SHOOT BORER. *Eucosma gloriola.* The whitish caterpillars are about ½ inch long. They burrow down the centers of the lateral shoots, causing them to wilt and to die back several inches. The insect spends the winter in the soil and appears as a moth in spring.

Control. Cut off and destroy infested branches as soon as they are discovered. This, of course, is practical only for young trees

WHITE PINE TUBE MOTH. *Argyrotaenia pinatubana.* Greenish yellow larvae make tubes by tying the needles together side by side and squarely eating off the free end. White pines in the East and lodgepole and whitebark pines in the Rocky Mountain region are susceptible.

Control. Spray with Diazinon or Imidan in early May and again in mid-July, if needed. This insect is not normally a serious pest.

PALES WEEVIL. *Hylobius pales.* The bark of young white, red, and Scotch pines may be chewed by a night-feeding reddish brown to black weevil ⅓ inch long. Young trees may be girdled completely.

Control. Spray the bark of seedlings and

other conifers and the twigs of larger conifers with methoxychlor in April

WHITE PINE WEEVIL. *Pissodes strobi.* This weevil is a very common pest of white pine in estates and lawns where trees are planted as ornamentals. It is also injurious in forests of white pine. The larvae feed on the inner bark and the sapwood of the leading branches and terminal shoots of the main trunks. The leader is girdled and killed, and the branches that grow out to replace the leader are more or less distorted. The beetles begin to emerge in July, leaving characteristic holes in the bark. They are about ¼ inch long, reddish brown and somewhat white-mottled. The larvae are pale yellowish grubs about ⅓ inch long. This insect also attacks spruce, as shown in Figure III-116.

Deodar weevil, *P. nemorensis,* vectors the pitch canker fungus in the South.

Control. Cut out and destroy all infested branches to kill the insects before they emerge as beetles. Spray valuable trees with Lindane in April.

Where the central leader is destroyed, a new leader may be encouraged to develop by tying the next lower lateral shoot in an erect position with a small stick and a soft rope. Pruning the dead leader branch at a 45 degree angle, rather than straight across, will encourage the development of a new leader.

ZIMMERMAN PINE MOTH. *Dioryctria zimmermani.* The bark of twigs and branches on many species of pines is invaded by white to reddish yellow or green ¾-inch larvae. The adult moth is reddish gray in color and has a wingspread of 1 to 1½ inches. Branch tips turn brown, and the entire tops of trees may break off as a result of boring by the larvae.

Control. Dylox or Thiodan spray applied just before the larvae emerge from hibernation in spring and another application in early August will provide control. Control of adult moths can be accomplished using Thiodan, Cygon, Dylox or Dursban in July.

BLACK TURPENTINE BEETLE. *Dendroctonus terebrans.* This pest (Fig. III-100) has caused the death of Austrian pine *(P. nigra austriaca),* Japanese black pine *(P. thunbergii),* and pitch pine *(P. rigida)* in recent years on Long Island, New York, and in Massachusetts. Once con-

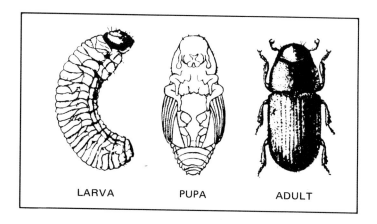

LARVA   PUPA   ADULT

Fig. III-100. Top: Life stages of the beetle. Actual size is approximately ¼ inch. **Bottom:** Pitch tubes of the black turpentine beetle at the base of a pine tree.

sidered a secondary invader which attacked pine weakened by some other agent, it is capable of infesting apparently healthy pines and killing them.

The beetles are ⅕ to ⅜ inch long, robust, reddish brown to black. They bore into the lower 4 feet of the trunk, making galleries ½ to 1 inch broad which extend downward for several inches to several feet. Resin flows out of the injured area forming so-called pitch tubes, which are readily visible on the bark surface (Fig. III-100). Destruction of the inner bark and wood by the larval stage eventually results in the death of the infested tree.

Control. Maintaining susceptible pine species in good vigor by fertilizing and watering is one way of warding off attacks by this beetle.

PINE LOOPER. *Lambdina athaseria pellucidaria.* This 1-inch-long moth larva is periodically a serious defoliator of red pine and pitch pine in the eastern United States.

PINE TUSSOCK MOTH. *Dasychira plagiata.* The colorful larvae are defoliators of jack, white, and red pine in the Midwest.

SPRUCE SPIDER MITE. See under Spruce.

PITTOSPORUM *(Pittosporum)*

Some pittosporum species may develop into small trees. They are relatively pest-free.

PLANETREE, LONDON *(Platanus acerifolia)* *

The London planetree is one of the most commonly planted trees in cities in the northeastern United States because it is more tolerant to air pollutants and other unfavorable growing conditions than most trees. However, it does not do well where extremely low winter temperatures occur. Many trees 3 to 5 inches in trunk diameter were killed or severely injured in New York City as a result of the severe 1977 winter.

Fungus Diseases

CANKERSTAIN. Thousands of London planetrees have died from this disease in the eastern

---

* Diseases and pests of the American planetree or buttonwood *(Platanus occidentalis)* are discussed under Sycamore.

United States since the early 1930s. The disease has also occurred in some of the southern and midwestern states (Fig. III-101).

Symptoms. Conspicuous reduction in both amount and size of foliage is the most obvious symptom. This is followed within a year or two by complete death of the tree.

Sunken cankers appear on trunks, large limbs, and occasionally on small limbs (Fig. III-102). The cankers frequently have longitudinal cracks and roughened bark. Bluish black or brown discolorations appear on freshly exposed bark over the cankers. The callus tissue formed at the margin of a canker usually dies early. Dark-colored streaks extend from the cankers inward through the wood to the central pith, and frequently radiating web-shaped brown streaks extend from the pith outward through the sound wood in other regions (Fig. III-103). The rays are mostly dark-colored. As the disease progresses, there is a gradual thinning of the leaves, which are smaller than usual. The disease is spread largely by pruning and is confined mainly to landscape trees. The sycamore, *P. occidentalis,* is less susceptible to this disease.

Cause. Cankerstain is caused by the fungus *Ceratocystis fimbriata* f. sp. *platani.* Several closely related species cause blue stain in lumber. There is some evidence that as many as five species of nitidulid beetles are capable of disseminating the causal fungus.

The fungus enters the trunk or branches through bark injuries and grows toward the center. From this point it grows outward radially along various lines, forming cankers wherever it reaches the bark. Large numbers of spores are formed beneath recently killed bark or on freshly exposed wood surfaces.

Many cankers center around injuries made by saw cuts, pole pruner cuts, rope burns, accidental saw scratches, and injuries by climbers' boots. Observations indicate that cankers may also start from pruning cuts that have been painted with a wound dressing containing fungus-infested sawdust. The fungus can apparently survive in some of the commonly used asphalt wound dressings. The disease has been experimentally transmitted from infected to healthy trees by means of contaminated pruning saws.

Fig. III-101. London planetree affected by the cankerstain disease.

It is highly probable that the use of contaminated tools and tree paints constitutes the chief means of spread on streetside and shade trees. Injuries to the roots and trunk during curb and sidewalk installation and those made by vandalism are also important foci for fungus invasion. Natural bark fissures and frost cracks may also constitute entrance points.

Control. Diseased trees should be removed and destroyed as soon as the diagnosis is confirmed. All injuries to sound trees should be avoided. Saws and other implements used in

Fig. III-102. Trunk of London planetree showing infection by the cankerstain fungus *Ceratocystis fimbriata* f. *platani*.

pruning planetrees should be thoroughly disinfested by washing in denatured alcohol or some other strong disinfestant after use on each tree in the more heavily infested zones. Because of the possibility of spreading the fungus by means of infested wound dressings, it is best not to apply a dressing on fresh wounds.

Plant pathologists once believed that the disease could not be spread by way of contaminated saws if pruning was done in winter, from December 1 to February 15. It is now known that this is not the case. Hence all pruning tools should be sterilized with 70 percent denatured alcohol after use on each tree regardless of the season.

BOTRYOSPHAERIA CANKER. *Botryosphaeria dothidea.* This fungus, like *Verticillium alboatrum* and *Armillaria mellea*, is far more widespread than most professional arborists and nurserymen realize. It has long been known to cause cankers and dieback of redbud, but the senior author showed that its asexual stage, *Dothiorella*, caused a highly destructive disease of London planetrees in New York City. By cross-inoculation tests, he proved that the same fungus can infect other important shade trees, including sweetgum. The fungus also has been isolated from naturally infected sweetgums showing branch dieback and discolored wood. Among other trees known to be susceptible to this fungus are apple, avocado, beech, flowering cherry, flowering dogwood, Japanese persimmon, hickory, horsechestnut, pecan, poplar, quince, sourgum, sycamore maple, and willow.

Many years ago, the senior author, working with Mr. E. C. Rundlett, then arboriculturist for the New York City Department of Parks, recorded that leaf fires beneath streetside trees made the trees more susceptible to infection by the conidial stage *Dothiorella gregaria*. He also pointed out that this is true of infections by the cankerstain fungus discussed earlier.

Control. No control is possible once the fungus has invaded the main trunk. When infections are limited to the branches, pruning well below the cankered area may remove all the infected material. Wounds and damage to the bark by fires should be avoided. Burning of leaves and other materials is now prohibited in most municipalities. The trees should be kept in good vigor by fertilizing, watering during dry spells, and spraying to control leaf-chewing and leaf-sucking insects.

POWDERY MILDEW. *Microsphaera platani.* London planetrees are especially susceptible to attacks of powdery mildew. Leaves and young twigs are covered with a whitish mold to such an extent that much of the foliage is destroyed. The weather seems to determine the seriousness of the attack; during some seasons no mildew appears, while in others the disease is serious, especially on young trees of a size for transplanting.

Control. In nursery plantings it is practical to spray trees with Benlate or wettable sulfur for this disease. Individual specimens are rarely sprayed to control mildew.

BLIGHT, ANTHRACNOSE. See blight, under Sycamore.

Fig. III-103. Cross section of planetree trunk infected by *Ceratocystis fimbriata* f. *platani*. The discoloration shows the spread of the fungus through the wood.

Abiotic Disease

DOG CANKER. Injury to trees planted along streets where it is customary to walk dogs is confined to the lower 2 feet of the trunk. Many trees up to 6 inches in trunk diameter may be killed in this way.

Control. Placing a metal collar around trees visited by male dogs will help to eliminate cankers, but the dogs' urine will still seep into the soil and root area to cause severe damage to the roots and premature death of the tree.

Insects

AMERICAN PLUM BORER. *Euzophera semifuneralis.* London planetrees, particularly streetside specimens whose bark has been damaged, are particularly susceptible to this pest. Damage to the inner bark and cambial regions by the larval stage, a dusky white, pinkish, or dull brownish green caterpillar, may be so extensive that the tree dies prematurely. Wild cherry, mountain-

ash, and all kinds of fruit trees are also susceptible.

Control. Avoid damaging trees and keep them in good vigor by fertilizing and watering. Spray the main trunk with Thiodan three times at 3-week intervals, starting in mid-May.

LACE BUG. *Corythucha ciliata.* Parts of the leaves turn pale yellow and many drop prematurely as a result of the sucking by the lace bug. The adult, 1/8 inch long, white and with lacelike wings (Fig. III-104), deposits eggs on the lower leaf surface. There are several generations a year. The adult passes the winter in bark crevices and other sheltered places in the vicinity of the trees. London planetree is much more susceptible to this pest than is sycamore.

Control. As soon as the insects appear on the leaves in May and July, spray with Sevin, malathion, Guthion, or Orthene. Some control is possible with a dormant spray applied just before growth begins in spring.

SYCAMORE TUSSOCK MOTH. See under Sycamore.

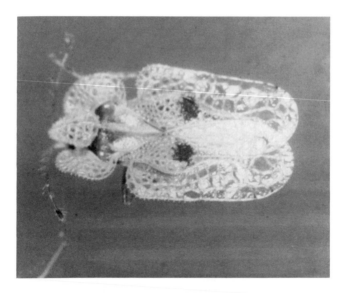

Fig. III-104. Lace bug adult.

OTHER INSECTS. London planetrees may be attacked by other insects; these are discussed under Sycamore.

PLUM, FLOWERING See Cherry.

POINCIANA *(Delonix)*

Flame-tree or flamboyant-tree is a striking red-flowered tree grown in frost-free areas of the United States.

Diseases

CANKER. This tropical tree is subject to relatively few diseases. Perhaps the most destructive is canker, caused by the fungus *Botryosphaeria dothidea.*

Control. Branches that show cankers should be pruned and destroyed.

OTHER DISEASES. Other diseases reported on this host are crown gall caused by the bacterium *Agrobacterium tumefaciens,* anthracnose by a fungus belonging to the genus *Gloeosporium,* and root rots by two fungi, *Clitocybe tabescens* and *Phymatotrichum omnivorum.*

Control. Control measures are rarely necessary.

Insects

LESSER SNOW SCALE. *Pinnaspis strachani.* The lesser snow scale frequently infests this host. The female is pear-shaped, white, semitransparent, sometimes speckled with brown. The male is narrow and white. Silk-tree, hackberry, palm, citrus, avocado, and other trees in the Deep South are infested.

Control. Spray with malathion or Sevin to control the crawler stage.

POPLAR *(Populus)*

Poplars are fast-growing trees hardy in temperate regions. Some, such as cottonwood, are adapted to the relatively harsh growing conditions of the Great Plains. Some, such as Lombardy poplar, are short-lived because of their high susceptibility to disease. Poplars are generally not suitable for urban areas because their weak branches break easily and they require much maintenance to retain an attractive appearance. Some poplars are commonly known as aspens.

Diseases

Several destructive diseases affect the trunks and limbs of poplars. The generic name of the

Fig. III-105. The Lombardy poplars on the left died as a result of infection by the fungus *Discosporium populea.*

causal fungus is used to differentiate the various types of cankers produced. A dieback, which is not clearly understood, and a number of leaf diseases, which rarely cause much damage, also occur on poplars.

DOTHICHIZA CANKER. The dying and dead Lombardy poplars *(P. nigra italica)* in the eastern United States are ample evidence of this destructive disease. Black and eastern cottonwoods and balsam, black, and Norway poplars also may be affected, although these species are much less susceptible than Lombardy poplar (Fig. III-105).

Symptoms. Elongated, dark, sunken can-

Fig. III-106. Trunk of Lombardy poplar showing elongated sunken cankers caused by the fungus *Discosporium populea.*

kers occur on the trunk, limbs, and twigs (Fig. III-106). The bark and cambium in the cankers are destroyed, and the sapwood is invaded and discolored. When the cankers completely girdle the trunks or branches, the distal portions die. Trees less than 4 years old are usually killed outright in a relatively short time. Older trees do not succumb as rapidly but are readily disfigured and become virtually worthless as ornamentals or windbreaks.

Cause. The fungus *Discosporium (Dothichiza) populea* causes this disease. The sexual stage of this fungus is *Cryptodiaporthe populea.* It forms small raised pustules on diseased bark, from which cream-colored masses of spores ooze out in April and May. The spores are splashed onto leaves and into bark injuries, where they germinate and infect these parts. Invasion often advances down the leaf petiole into the twig, causing death of the latter.

Control. No effective control measures are known. Wounds of all sorts should be avoided. Pruning of diseased parts, as suggested for most canker diseases, does not appear to help control this disease; in fact, it often spreads it.

Inasmuch as the leaves are attacked and the fungus enters the twigs through infected leaves, some investigators have recommended repeated applications of copper sprays to reduce leaf infections. As a rule, individual trees are not sufficiently valuable to justify four or more applications of a fungicide each year. Such a practice may be warranted, however, in nurseries where the disease threatens young stock.

The only hope lies in the development of resistant varieties. The Japan poplar *(P. maximowiczii)* appears to show some resistance. Fremont and black cottonwoods, Carolina and gray poplars, and trembling aspen are considered resistant. Conflicting reports exist concerning the susceptibility of the Simon poplar *(P. simonii);* some people assert that it is very resistant; others, that it is very susceptible.

CYTOSPORA CANKER. Another destructive canker disease of poplar, Cytospora canker, primarily affects trees poor in vigor. Among the factors that most commonly contribute to the weakened condition are drastic pruning and the unfavorable environment of city soils. Carolina and silver-leaf appear to be the most susceptible of the poplars. Maples, mountain-ash, and willows are also susceptible.

Symptoms. Brown, sunken areas covered with numerous red pustules first appear on young twigs. The fungus moves down the stem and invades larger branches or even the trunk. Here, circular cankers are formed, which continue to expand until the entire member is girdled and the parts above die.

Cause. The fungus *Cytospora chrysosperma*, a stage of the fungus *Valsa sordida,* is the causal agent. It lives on dead branches but can attack live branches and trunks when the tree is weakened. Spores are developed in the black pinpoint fruiting structures on dead wood. They ooze from these bodies and are splashed by rains or carried by birds and insects to bark wounds or weakened and dead branches, where they germinate and infect.

Control. Because Cytospora canker is primarily a disease of weak trees, the most effective preventive is the maintenance of the trees in high vigor by fertilization, watering, and the control of insect and fungus parasites of the leaves. In addition, dead and dying branches should be removed and all unnecessary injuries avoided. 'Noreaster' hybrid poplar and 'Platte,' 'Mighty Mo,' and 'Ohio Red' cottonwoods are resistant to this disease.

CRYPTOSPHAERIA CANKER. This disease is destructive to aspens and poplars throughout the Great Plains.

Symptoms. A stem canker causes branch and tree mortality. Sapwood discoloration and decay are also associated with this disease.

Cause. *Cryptosphaeria populina* (imperfect, *Libertella* sp.) causes canker and decay.

Control. Prune infected branches.

HYPOXYLON CANKER. A highly destructive disease, Hypoxylon canker occurs less commonly than Dothichiza and Cytospora cankers. In the northeastern states it affects quaking and large-toothed aspens and balsam poplar. It is most harmful to young trees and is more common on forest trees than on ornamentals.

Symptoms. Gray cankers of varying sizes appear along the trunk but never on the branches. The color changes to black as the outer bark falls away from the surface of the canker. When the bark is peeled off, blackened sapwood is visible around the edge of the canker. Wefts of fungus tissue, resembling the chestnut blight fungus but different in color, are also visible beneath the peeled bark. Many cankers attain a length of several feet and cause considerable distortion of the trunk before complete girdling of the trunk and death occur.

Cause. This disease is caused by the fungus *Hypoxylon mammatum*. Spores are believed to initiate infections through bark injuries.

Control. This disease is so highly contagious and destructive that infected trees should be cut down and destroyed as soon as the diagnosis is confirmed. Injuries should be avoided as much as possible.

SEPTORIA CANKER AND LEAF SPOT. A serious stem canker disease of hybrid poplars, Septoria canker may also infect leaves of all species of native poplars.

Symptoms. Leaves on young shoots or lowermost branches of native poplars are spotted early in the season. Cankers appear later on the stems of hybrid poplars with black, balsam, and cottonwood parentage. When cankers girdle the stem, the distal portion dies.

Cause. *Mycosphaerella populorum* causes leaf spot and stem canker. The asexual stage of this fungus is *Septoria musiva,* hence the common name for the disease. Fruit bodies of the Septoria stage are frequently found in the cankered areas.

Control. The most effective way of combating the canker stage of this disease is to use hybrid poplar clones that have proved to be naturally resistant. The leaf spot stage on ornamental poplars can be prevented by periodic applications of a copper fungicide or benomyl.

FUSARIUM CANKER. *Fusarium solani.* This disease occurs on *P. deltoides* in the South, Middle West, and Canada.

Control. Control measures have not been developed.

PHOMOPSIS CANKER. *Phomopsis macrospora.* Cottonwood suffers dieback and canker caused by this fungus.

SOOTY BARK CANKER. *Encoelia pruinosa.* This disease is especially damaging to aspen in the Rocky Mountain States.

BRANCH GALL. Small globose galls up to 1½ inches in diameter occasionally occur at the bases of poplar twigs. Primarily, the bark is hypertrophied, although some swelling of the woody tissues also occurs. Twigs and some branches may be killed, but the disease rarely becomes serious.

Cause. Small black pinpoint fruiting bodies of the fungus *Macrophoma tumefaciens* are embedded in the bark of the gall, especially along the fissures. Fungus penetration is initiated in the developing buds in early spring, and the swelling of the tissue probably results from some stimulant secreted by the fungus.

Control. Pruning the galled branches and dead twigs to sound wood and destroying the removed wood are usually sufficient to hold the fungus in check.

LEAF RUSTS. Several species of rust fungi attack poplars in the eastern United States. These produce yellowish orange pustules, usually on the lower leaf surface. The two species most common in the East are *Melampsora medusae* and *M. abietis-canadensis*, which require alternate hosts to complete their life cycles. The alternate host of the former is larch, and that of the latter, hemlock. Two other leaf rusts are caused by *M. albertensis*. and *M. occidentalis*.

Control. Except for the rust fungus *M. medusae*, which causes heavy losses of young cottonwoods *(P. deltoides)* in nurseries in the Midwest and South, leaf rusts rarely cause enough damage to necessitate special control measures. Several rust-resistant cottonwood clones are available.

LEAF BLISTER. Brilliant yellow to brown blisters of varying sizes occasionally appear on poplar leaves following extended periods of cool, wet weather. The blisters, which may be more than an inch in diameter, result from stimulation of the leaf cells by the fungus *Taphrina populina.* Another species, *T. johansoni,* causes a deformity of the catkins.

Control. Spraying the trees in early spring with Ferbam will control leaf blister. Such a practice is suggested only where particularly valuable specimens are involved.

LEAF SPOTS. *Cercospora populina, C. populicola, Linospora tetraspora, Marssonina populi, M castagnei, M. brunnea, Mycosphaerella populicola, Apioplagiostoma populi, Phyllosticta alcides, Venturia populina,* and *V. tremulae* (imperfect, *Pollaccia*). The latter two also cause a shoot blight. Many species of fungi, including *Mycosphaerella populorum* mentioned under the Septoria canker disease, cause leaf spots of poplars. Of these, *Marssonina,* which produces brown spots with a darker brown margin and premature defoliation, is by far the most com-

mon. It also invades the leaf petioles and twigs, causing tan spots with dark margins. A leaf blotch caused by *Septonina podophyllina* also affects poplars.

Control. The first requisite in the control of leaf spot diseases is the gathering and destruction of all fallen leaves to remove an important source of inoculum. Poplars are rarely sprayed with fungicides, but bordeaux or any ready-made copper fungicide may be used early in spring on valuable trees.

DIEBACK. A dieback of the top and complete death of Lombardy poplar, cottonwood *(P. deltoides)*, and the goat willow *(Salix caprea)* have been reported on trees in the vicinity of Washington, D.C., and in western Tennessee. On Lombardy poplar, the wood appears first water-soaked, then red, and finally brown. The tree dies when the entire cross section of the trunk is stained brown.

Cause. The cause of this trouble in not known. Bacteria have been isolated from the margins of the discolored area. There is a strong possibility that the water-soaked appearance and the subsequent staining actually result from extremely low winter temperatures and that the bacteria enter after the injury occurs.

Control. No control measures are known.

POPLAR INKSPOT. *Ciborinia whetzelii.* Thick round or elongated black spots form on leaves and then drop out, leaving holes. Aspen, Lombardy, and Carolina poplars are most subject to inkspot, whereas the large-toothed aspen *(P. grandidentata)* appears to be immune.

Control. Where the fungus attacks valuable specimen poplars or nursery stock, the use of bordeaux mixture or some fixed copper is suggested.

HEART ROT. *Phellinus igniarius* and *Phellinus tremulae* are the most important of many poplar heart rotters. See Chapter 12.

POWDERY MILDEW. *Uncinula adunca.* This is a common superficial disease, appearing as a white mildew on both sides of the leaves; usually the damage is not serious.

Control. Spray with Benlate, Bayleton, or wettable sulfur if the specimens are valuable and the disease is serious.

WETWOOD. Poplars are especially susceptible to this condition. See Chapter 12.

CROWN GALL. *Agrobacterium tumefaciens.* See Chapter 12.

MOSAIC VIRUS. Poplars have been reported as susceptible.

Abiotic Disease

AIR POLLUTION. Many poplars are sensitive to sulfur dioxide and very sensitive to ozone. See Chapter 10.

Insects

APHIDS. *Pemphigus populitransverus* and *Mordwilkoja vagabunda.* The former produces galls on the leaf petioles of certain poplars; the latter is responsible for convoluted galls at the tips of twigs. Several other species of leaf- and bark-infesting aphids such as *Pterocomma* and giant bark occur on this host.

Control. Valuable ornamental specimens can be sprayed with lime sulfur when the trees are dormant or with malathion just as the leaves unfurl in early spring.

BRONZE BIRCH BORER. See under Birch.

POPLAR BORER. *Saperda inornata* and *S. populea.* Swollen areas on small branches usually result from attack by the poplar borer, a white larva 1 inch long when fully grown. Hollowed out stem galls weaken and deform the young tree and damaged branches may break off. The adult female is a bluish gray beetle, with black spots and yellow patches, and is 1/3 to 1/2 inch in length. Egg-laying scars made in late spring on young stems are brown and horseshoe shaped. This pest also attacks willow.

Control. Remove and destroy badly infested trees and branches. Dig out and destroy young larvae in spring.

MOTTLED WILLOW BORER. See Mottled Willow Borer under Willow.

HORNET MOTH. *Aegeria apiformis.* This clear-winged moth larva is an important wood borer pest of poplar in some locations in the northern United States.

CARPENTERWORM. See under Ash.

AMERICAN PLUM BORER. See under Planetree.

FLATHEADED and LEOPARD MOTH BORER. See under Maple.

COTTONWOOD TWIG BORER. *Gypsonoma haimbachiana.* This caterpillar feeds in growing cottonwood shoots.

TWIG GIRDLER. See under Hackberry.

COTTONWOOD LEAF BEETLE. *Chrysomela scripta.* This light green and black-spotted larva skeletonizes cottonwood leaves. Willows are also attacked by this pest. *C. interrupta* also feeds on poplar leaves.

IMPORTED WILLOW LEAF BEETLE. See under Willow.

RED-HUMPED CATERPILLAR. *Schizura concinna.* The leaves are chewed by clusters of red-humped caterpillars, yellow and black striped forms with red heads and red humps. The adult, a grayish brown moth with a wingspread of 1 1/4 inches, deposits masses of eggs on the lower leaf surface in July. Winter is passed in a cocoon on the ground.

Control. Spray with *Bacillus thuringiensis* or Sevin in midsummer when the caterpillars are small.

SATIN MOTH. *Stilpnotia salicis.* The larval stage, black with conspicuous irregular white blotches, feeds on the leaves of poplar, willow, and sometimes on oaks in late April or May. The adult, satin-white and with a wing expanse up to 2 inches, emerges in July.

Control. Spray with Sevin in early June and again in early August.

TENTMAKER. *Ichthyura inclusa.* The leaves are chewed and silken nests appear on twigs as a result of infestations of the tentmaker, a black larva with pale yellow stripes, which attains a length of 1 1/4 inches at maturity. The adult female is a moth with white-striped gray wings. The pupae overwinter in fallen leaves.

Control. Remove and destroy nests or spray with Sevin when the larvae are small.

FOREST TENT CATERPILLAR. See under Maple.

FRUIT TREE LEAF ROLLER. See under Crabapple.

GYPSY MOTH, MOURNING-CLOAK BUTTERFLY. See under Elm.

WHITE-MARKED TUSSOCK MOTH. See under Elm.

IO MOTH. See under Sycamore.

FALL WEBWORM. See under Ash.

EASTERN TENT CATERPILLAR. See under Willow.

SCALES. Many species of scale insects infest poplars. Among these are black, cottony maple, European fruit lecanium, greedy, lecanium, oystershell, San Jose, soft, terrapin, walnut, and willow.

Control. Valuable specimens infested with scales should be sprayed with lime sulfur or miscible oil in early spring just before growth starts, or with malathion or Sevin in mid-May and again in mid-June to control the crawling stages.

COMSTOCK MEALYBUG. See under Catalpa.

HONEY LOCUST PLANT BUG. See under Locust, Honey.

PORT ORFORD CEDAR. See False Cypress.

QUINCE, FLOWERING *(Chaenomeles)*

Bacterial and Fungus Diseases

CROWN GALL. *Agrobacterium tumefaciens.* This disease occasionally affects this host.

Control. See Chapter 12.

FIRE BLIGHT. *Erwinia amylovora.* Flowering quince, like other rosaceous hosts, is subject to this disease.

Control. Destroy nearby neglected and unwanted pear and apple trees. Spray with an antibiotic at mid-bloom stage.

BROWN ROT. *Monilinia fructicola* and *M. laxa.* These fungi are usually more destructive on fruit trees than on the ornamental varieties. They cause a leaf blight and a blossom and twig blight of flowering quince.

Control Periodic applications of Captan or wettable sulfur sprays during the early growing season are effective.

RUST. *Gymnosporangium clavipes.* This bright orange-colored rust is more commonly found on fruiting trees than on the flowering varieties. It attacks the fruit as well as the leaves and young twigs. It also attacks apples and hawthorns. The alternate host is the common red cedar, *Juniperus virginiana.* Though very destructive to the common quince, it does little damage to the cedar. (See Juniper and also under Hawthorn.)

Another rust, *G. libocedri*, also attacks flowering quince leaves.

Control. Spray periodically with Ferbam or wettable sulfur during the growing season.

LEAF SPOTS. *Diplocarpon mespili* and *Cercospora cydoniae.* These leaf spots can become troublesome and cause premature defoliation in rainy seasons.

Control. Spray with Benlate or zineb early in the growing season.

OTHER DISEASES. Among other diseases occasionally found on flowering quince are twig blight caused by *Botryosphaeria dothidea;* cankers by *Nectria cinnabarina*, *Phoma* sp., and *Botryosphaeria obtusa;* and root-knot nema, *Meloidogyne* sp.

Control. These diseases are rarely serious enough on flowering quince to warrant control measures.

Insects

COTTON APHID. *Aphis gossypii.* Young leaves and upper ends of tender branches may be heavily infested with these aphids. Apple aphid also feeds on quince.

Control. Spray with malathion when the insects appear.

YELLOW-NECKED CATERPILLAR. See under Oak.

FOREST TENT CATERPILLAR. See under Maple.

JAPANESE BEETLE. See under Linden.

SCALES. Cottony-cushion, pitmaking, pittosporum, Japanese wax, black, greedy, and California red scales feed on flowering quince.

REDBUD *(Cercis)*

Redbud is recognized by early spring flowers which appear on bare branches, just before flowering dogwood blooms. Redbud is tolerant of shade; sun; and alkaline, acid, and moist soils.

Diseases

CANKER. *Botryosphaeria dothidea.* This, the most destructive disease of redbud, also affects many other trees and shrubs. On redbud the cankers begin as small sunken areas and increase slowly in size. The bark in the center of the canker blackens and cracks along the edges. The wood beneath the cankered area becomes discolored. When the canker girdles the stem, the leaves above wilt and die. The causal fungus is easily recovered from discolored wood by standard tissue culture techniques in the labo-

Fig. III-107. Verticillium wilt of redbud showing vascular discoloration.

ratory. The causal fungus was formerly known as *Botryosphaeria ribis*.

Control. Prune and destroy branches showing cankers. Surgial excision of cankered tissue on the main stem is occasionally successful if all infected bark and wood are removed. Periodic applications of a copper fungicide during the growing season may help to prevent new infections.

LEAF SPOTS. A conspicuous leaf spot disease occurs on redbud throughout the range of this tree in the eastern half of the United States. Scattered circular to angular brown spots, about ¼ inch in diameter, appear on the leaves early in summer. The spots increase in size until they attain a diameter of nearly ½ inch by fall.

The fungus *Mycosphaerella cercidicola* produces circular to angular spots with dark brown borders on the leaves. Spores from overwintered fallen leaves initiate spring infections. The fungi *Cercosporella chionea*, *Plasmopara cercidis*, *Kabatiella*, and *Phyllosticta cercidicola* also cause leaf spots on redbud.

Control. Valuable specimens can be protected by periodic applications of copper or dithiocarbamate fungicides in late spring and early summer.

WILT. *Verticillium albo-atrum.* This fungus causes a serious wilt disease (Fig. III-107) in landscape plantings. See Chapter 12.

OTHER DISEASES. Among other fungus diseases of redbud are root rots caused by *Clitocybe*

tabescens, Ganoderma lucidum, and *Phymatotrichum omnivorum.*

Abiotic Disease

AIR POLLUTION. Redbuds are sensitive to ozone. See Chapter 10.

In Mississippi, branch dieback attributed to automobile exhaust fumes has been noted.

Insects

CATERPILLARS. The larval stage of the California tent caterpillar, *Malacosoma californicum*, and the grape leaf folder, *Desmia funeralis*, occasionally infest redbuds.

Control. Spray with *Bacillus thuringiensis* or Sevin when the young caterpillars begin to chew the leaves.

COTTON APHID. *Aphis gossypii.* The young leaves and upper ends of tender branches may be heavily infested with these aphids.

Control. Spray with malathion or Sevin as soon as the aphids appear on the leaves, and repeat in 2 weeks if necessary.

LEAFHOPPER. An *Erythronema* species causes stippling of redbud leaves (Fig. III-108).

TWO-MARKED TREEHOPPER. See under Hop-Tree.

SCALES. Eleven species of scales infest the twigs and branches of redbud, including European fruit lecanium, greedy, ivy, and white peach.

Fig. III-108. Redbud leafhopper injury.

Control. Dormant lime sulfur or oil sprays usually are sufficient to control scales. Severely infested trees should also be sprayed in May and June with malathion or Sevin.

OTHER INSECTS. The Rhabdopterus beetle, *Rhabdopterus deceptor;* the green fruitworm, *Lithophane antennata;* the redbud leafroller, *Fascista cercerisella;* and the greenhouse whitefly (see whiteflies, under Citrus) all attack redbud.

Control. Sevin sprays will control these pests.

ASIATIC OAK WEEVIL. See under Oak.

REDWOOD *(Sequoiadendron and Sequoia)*

These two West Coast natives, the legendary biggest and tallest trees, are grown in landscapes where they do not attain the immense size of those in native groves. Coast redwood requires more fog and moisture than giant redwood.

Diseases

CANKER. *Dermatea livida.* Cankers develop on the bark, in which brown to black fruiting bodies develop. These contain one-celled color-less spores. Another canker is caused by the fungus *Botryosphaeria dothidea.*

Control. No satisfactory control measures have been developed.

NEEDLE BLIGHT. *Chloroscypha chloramela* and *Mycosphaerella sequoiae.* Redwood leaves may be blighted by these fungi. The giant sequoia *(Sequoiandendron giganteum)* is susceptible to two other leaf-blighting fungi, *Cercospora sequoiae* and *Pestalotiopsis funerea,* and to twig blight by *Phomopsis juniperovora.*

Control. Small, importantly placed trees infected with these fungi should be sprayed with a copper or a dithiocarbamate fungicide.

ARMILLARIA ROOT ROT. See Chapter 12.

Insects

CEDAR TREE BORER. *Semanotus ligneus.* The larvae of this beetle make winding burrows in the inner bark and sapwood of redwood, cedars, arborvitae, Douglas-fir, and Monterey pine, occasionally girdling and killing the trees. The adult beetle is ½ inch long, black with orange and red markings on its wing covers.

Control. Young trees should be kept in good vigor by fertilizing and watering when necessary. Sprays containing methoxychlor applied to the trunk and branches when the adult beetles emerge should provide control.

SEQUOIA PITCH MOTH. *Vespamima sequoiae.* The cambial region of redwoods and many other conifers in the West is mined by an opaque, dirty-white larva. The adult moth resembles a yellowjacket wasp because it is black and has a bright yellow segment at its lower abdomen.

Control. This pest is rarely abundant enough to warrant control measures.

MEALYBUGS. Three species of mealybugs may infest redwoods: citrus, *Planococcus citri;* cypress, *Pseudococcus ryani;* and yucca, *Puto yuccae.*

Control. Spray young, valuable trees with malathion. Repeat in a few weeks if the first spray does not provide good control.

SCALES. Two scale insects have been reported on this host: greedy, *Aspidiotus camelliae;* and oleander, *A. hederae.*

Control. Spray with malathion or Sevin when the young scales are crawling about.

REDWOOD, DAWN *(Metasequoia)*

This tree is a native of China, and although present in this hemisphere 15 million years ago, it was reintroduced as recently as 1945 by the late Dr. E. D. Merrill of the Arnold Arboretum. It resembles bald cypress in appearance.

Diseases

CANKER. *Botryosphaeria dothidea.* This is the only disease known to occur on dawn redwood (see Botryosphaeria canker, under Planetree). Early symptoms include the wilting of leaves on individual lateral branches. Cankers develop on the trunk at the base of the infected branches. Resin exudes from the infected area.

The canker disease also occurs on *Sequoiadendron giganteum* and on *Sequoia sempervirens* planted outside the native range.

Control. Control measures have not been developed.

RUSSIAN-OLIVE *(Elaeagnus)*

Russian-olive, an excellent tree for very difficult situations, is relatively free of fungus parasites and insect pests. This low-growing tree tolerates cold temperatures.

CANKERS. *Lasiodiplodia theobromae, Nectria cinnabarina (Tubercularia), Fusicoccum elaeagni, Fusarium sp., Phomopsis arnoldiae,* and *Phytophthora cactorum.* Cankers on the branchers and trunk of Russian-olive may be caused by any of these fungi. The first-named fungus has killed large numbers of Russian-olives in shelterbelts in Nebraska and other Great Plains states. Bark, cambium, and phloem tissues are killed in strips along the trunk and major branches. Infection is rapid up and down stems but slow around them. Hence complete girdling and death may take several years.

Phomopsis appears to be more prevalent in the central and northeastern states.

Control. No control measures have been developed; however, trees should be managed for optimum vigor.

LEAF SPOTS. Several fungi including *Cercospora carii, C. elaeagni, Phyllosticta argyrea, Septoria argyrea,* and *S. elaeagni* cause spots on leaves of this host.

Control. These fungi are rarely severe enough to warrant the use of protective fungicides.

RUST. *Puccinia caricis-shepherdiae* with the alternate host *Carex* and *P. coronata* f. *elaeagni* with the alternate host *Calamagrostis* occur on Russian-olive in the midwestern United States.

Control. These diseases are not important enough to warrant control measures.

WILT. *Verticillium albo-atrum.* This disease was first found on Russian-olive in the West.

Control. See Chapter 12.

OTHER DISEASES. Russian-olive is occasionally affected by crown gall caused by *Agrobacterium tumefaciens* and hairy root by *A. rhizogenes.*

Control Control measures are rarely necessary.

Insects

OLEASTER-THISTLE APHID. *Capitophorus braggii.* This pale yellow and green aphid, which lives on thistles during the summer, overwinters on Russian-olive.

Control. Control measures are unnecessary.

SCALES. *Parlatoria oleae* and *Lepidosaphes beckii.* The olive (see under Kentucky Coffee Tree) and the purple (see under Oak) scales occasionally infest this host. Many other species of scales, including European peach, ivy, and dictyosperm, have also been recorded on Russian-olive.

Control. Spray with malathion or Sevin in late May and again in late June to control the crawler stage of scales.

COMSTOCK MEALYBUG. See under Catalpa.

SASSAFRAS *(Sassafras)*

Sassafras, an excellent tree for landscape plantings, is extremely susceptible to Japanese beetle infestations. Like American holly, it is dioecious, that is, the pistillate and staminate flowers grow on different trees. The slate-blue fruits are ornamental and provide food for birds. It is native to the eastern United States, where it grows in acid soils.

Diseases

CANKER. Two fungi, *Nectria galligena* and *Botryosphaeria obtusa,* are frequently associated with branch cankers (Fig. III-109).

Control. Pruning affected branches to sound wood is suggested.

LEAF SPOTS. *Actinopelte dryina, Diplopeltis sassafrasicola, Glomerella cingulata, Metasphaeria sassafrasicola, Phyllosticta illinois-* ensis, and *Stigmatophragmia sassafrasicola.* These six species of fungi cause leaf spots on sassafras. They are rarely serious enough to warrant control measures.

OTHER DISEASES. Powdery mildew caused by the fungus *Phyllactinia guttata,* wilt by a species of *Verticillium,* shoestring root rot by *Armillaria mellea,* and a disease of the yellows type have also been recorded on sassafras. The last causes bunching and fasciation of the branch tips, leaf rolling, and leaf dwarfing.

This host is also subject to curly top caused by the beet curly top virus and to aster yellows caused by a mycoplasmalike organism.

Insects

JAPANESE BEETLE. (Fig. III-110.) See under Linden.

PROMETHEA MOTH. *Callosamia promethea.* The leaves are chewed by the promethea moth larva, a bluish green caterpillar that grows to 2 inches in length. The adult female has reddish brown wings, near the tips of which eyelike spots are visible; the wingspread is nearly 3 inches. The cocoons of this pest are suspended in the trees.

The caterpillars of the hickory horned devil, *Citheronia regalis,* the io (see under Sycamore), gypsy moth (see under Elm) and the polyphemus moth, *Antheraea polyphemus,* also chew sassafras leaves.

Control. Spray with *Bacillus thuringiensis,* Imidan, or Sevin when the caterpillars are small.

SASSAFRAS WEEVIL. *Odontopus calceatus.* This small snout weevil begins to feed as

Fig. III-109. Cankers on sassafras stem caused by the fungus *Nectria galligena.*

Fig. III-110. Japanese beetles feeding on a sassafras leaf.

soon as the buds break and before the leaves have expanded, making numerous holes in the leaves. The female adults then deposit eggs on the midribs of the leaves, and the larvae which hatch from these mine into the leaves to cause blotches. Tuliptree and magnolia are also susceptible.

Control. Sevin sprays applied when the adults begin to feed in June, or after the eggs hatch in late May, should provide control.

SCALES. The oystershell and San Jose scales occasionally infest sassafras. Valuable specimens can be protected from these pests with two applications of malathion or Sevin sprays, one in May and the other in June. Dormant oil or ethion plus oil sprays applied in April kill the eggs.

OTHER SCALES. In addition to the two scales mentioned above, sassafras is occasionally infested by the European fruit lecanium, *Lecan-*

ium corni; terrapin, *Lecanium nigrofasciatum;* Florida wax, *Ceroplastis floridensis;* and pyriform *Protopulvinaria pyriformis.*

Control. The crawler stage of scales can be controlled by spraying with Sevin or malathion.

SCREW-PINE *(Pandanus)*

Diseases

LEAF BLOTCH. *Melanconium pandani.* Large leaf spots up to 2 inches wide and 3 or 4 inches long may develop from the leaf margin inward. Black fruiting pustules develop in a light gray zone along the inner margin of the spot. Similar spots may develop along the leaf blade at any point.

Control. Prune off infected leaves and spray with a copper fungicide. Badly infected plants should be destroyed.

LEAF SPOTS. *Heterosporium iridis, Melanconium pandani, Phomopsis sp., and Volutella mellea.* These fungi have been reported on screw-pine.

Control. Leaf spots are rarely serious enough to warrant control measures.

Insects

The long-tailed mealybug, *Pseudococcus adonidum,* and thirteen species of scales attack screw-pine.

Control. Spray with Sevin or malathion whenever these pests appear.

SEQUOIA. See Redwood.

SERVICEBERRY *(Amelanchier)*

Serviceberry, also known as Juneberry, shadblow, and shadbush, blooms early, has good fall color, and is an excellent low-growing tree of the rose family. As such, it is subject to many of the diseases and insects common to members of that family.

Diseases

FIRE BLIGHT. *Erwinia amylovora.* This bacterial disease, which is serious on certain hawthorns and other pomaceous trees, occasionally occurs on serviceberry.

Control. See Chapter 12.

FRUIT ROTS. *Monilinia amelanchieris* and *M. fructicola.* The fruits of serviceberry are rotted in rainy seasons by these fungi.

Control. Infections on particularly valuable specimens can be prevented by periodic applications of Captan or wettable sulfur sprays.

LEAF BLIGHT. *Diplocarpon mespili.* Only occasionally is serviceberry attacked by this fungus. It is more serious on hawthorn and apple.

Control. See under Hawthorn.

MILDEWS. *Erysiphe polygoni, Phyllactinia guttata,* and *Podosphaera clandestina.*

Control. These three species of powdery mildew fungi can be controlled with Benlate, sulfur, or Bayleton sprays.

RUST. More than 10 species of rust occur on serviceberry, with alternate hosts on juniper, white cedar, and incense cedar.

Control. Same as for hawthorn rusts.

WITCHES' BROOM. Several species of serviceberry are subject to the witches' broom disease, which rarely causes much damage.

Symptoms. The production of many lateral branches from a common center results in a bunching or broomlike effect. The lower surfaces of leaves attached to such twigs are covered with sooty fungus growth.

Cause. The fungus *Apiosporina collinsii* causes witches' broom. It penetrates the twigs in spring and stimulates the production of the many lateral branches. The black growth of the fungus on the leaves contains small, round fruiting bodies in late summer. These produce spores for the following season's infections.

Control. Prune and destroy affected leaves and twigs.

Insects and Related Pests

BORERS. A number of borers, including the lesser peach tree, *Synanthedon pictipes;* apple bark, *Thamnosphecia pyri;* roundheaded apple tree, *Saperda candida;* and the shot-hole, *Scolytus regulosus,* occasionally attack serviceberry.

Control. Keep trees in good vigor by fertilizing and watering. Spraying the trunk and branches with methoxychlor at 2- to 3-week intervals during late spring and early summer will control these borers.

LEAF MINER. *Nepticula amelanchierella.* Broad irregular mines, especially in the lower half of the leaf, result from infestation by the larvae.

Control. See birch leaf miner, under Birch.

PEAR LEAF BLISTER MITE. See under Mountain-Ash.

PEAR SLUG SAWFLY. See pear slug, under Cherry.

WILLOW SCURFY SCALE. See under Willow.

CAMBIUM MINER. See under Holly.

WOOLLY APPLE AND WOOLLY ELM APHIDS. Serviceberry roots are injured by these pests. See aphids, under Elm.

MITES. Spider and eriophyid mites attack serviceberry. See Chapter 11.

SHADBLOW, SHADBUSH. See Serviceberry.

SILK-OAK (*Grevillea*)

Silk-oak, a native of Australia, has become naturalized in southern Florida. It is fast growing, but its wood is brittle. This tree prefers well-drained soil.

Diseases

Few diseases have been recorded for this tree. These include dieback, said to be caused by a species of *Diplodia,* leaf spot by the alga *Cephaleuros virescens,* root-knot by the nema *Meloidogyne* sp., and root rot by *Phymatotrichum omnivorum.*

Control. Measures are rarely taken to control the diseases on this host.

Insects

Leaf-eating insects on silk-oak include the tobacco budworm, *Heliothis virescens;* and the omnivorous looper, *Sabulodes caberata.* Sucking insects include the citrophilus and grape mealybugs, at least nineteen species of scales, and two species of mites—avocado red and Grevillea.

Control. Leaf-eating insects can be controlled with Sevin sprays; mealybugs and scales with malathion; mites with Kelthane.

SILK-TREE *(Albizzia)*

Silk-tree *(Albizzia julibrissin),* also known as mimosa, is used as a street and lawn tree in the South and as an ornamental in protected places along the Atlantic Coast as far north as Boston. It is a fast-growing, relatively short-lived tree. Its most destructive disease is wilt, and its most destructive insect the webworm.

Diseases

WILT. This highly destructive disease is extremely prevalent from Maryland to Florida and along the Gulf Coast into Louisiana. The senior author found the first cases in New Jersey many years ago and more recently found diseased trees at the New York Botanical Garden in New York City. This is the northernmost known location of the disease at this time.

Symptoms. Wilting of the leaves, which soon die, is the most striking symptom. A brown ring of discolored sapwood, usually in the current annual ring of the roots, stem, and branches, is another positive symptom. Bleeding along the stems occurs occasionally. Large numbers of spores are formed in the lenticels on the trunk and branches of affected trees, at times even before actual wilting of the leaves occurs. Such spores may account for widespread outbreaks of the disease. Trees usually die within a year or so of infection.

Cause. The fungus *Fusarium oxysporum* f. *perniciosum* is known to cause wilt. The fungus lives in soil and spreads in this medium. It also can be carried over in seed collected from diseased trees. There are two races of this fungus. One will infect silk-tree and not *Albizzia procera,* a silk-tree relative. The other, from Puerto Rico, will infect *A. procera* and not the silk-tree. *A. lophantha* and *A. kalkora* are susceptible to wilt; *A. lebbek,* to just one race; and *A. thorelii* and *A. pudica* are immune to wilt.

Control. Dead and dying trees should be cut down and destroyed to avoid the spread of the disease. Until recently the only control known was to use wilt-resistant trees. The cultivar 'Union' is resistant.

The cultivars 'Tryon' and 'Charlotte', at first resistant to Fusarium wilt, are now considered susceptible, probably because new fungal races have appeared.

OTHER FUNGI. Several other fungi are reported on silk-tree. *Nectria cinnabarina* (imperfect, *Tubercularia vulgaris*) causes dieback and canker; *Thyronectria austro-americana,* destructive to honey locust, is weakly pathogenic to silk-tree twigs. *Ganoderma lucidum, Armillaria mellea,* and *Clitocybe tabescens* are associated with root rots. The rust fungus, *Sphaerophragmium* sp., has been reported on *A. lebbeck* in Florida.

Insects

WEBWORM. *Homadaula anisocentra.* The larvae of this pest, also called mimosa webworm, appear on the foliage in early spring. They are also highly destructive to honey locust. The larvae are ⅗ inch long when fully grown and are gray-brown, sometimes pinkish, with five narrow, white, lengthwise stripes. At first they feed together in a web, but later they spread throughout the tree, tying the leaves in conspicuous masses and skeletonizing them. When mature, they drop to the ground on silken threads and spin cocoons in various cracks and crevices. The moths, gray with a wingspread of about ½ inch, appear in June to lay eggs on the silk-tree flowers and leaves.

Control. When the young caterpillars are just beginning to feed in early July and in August, spray with Orthene, Sevin, or Diazinon. The spraying must be repeated several times, because there are as many as four generations of this pest each year in some areas.

SCALES. Lesser snow (see under Poinciana), Asiatic red, and several other scales infest silk-tree.

JAPANESE WEEVIL. See under Holly.

BAGWORM. See under Juniper.

ALBIZZIA PSYLLID. See under Acacia.

OTHER INSECTS. Blister beetle, thornbug, and citrus mealybug also infest this host.

NEMAS. *Meloidogyne arenaria* and *Trichodorus primitivus.* The former, known as the peanut root-knot nema, causes small knots or galls on silk-trees in the South. The latter, known as the stubby root nema, infests *A.*

julibrissin roots in Maryland and azaleas in California.

Control. See under Boxwood.

SILVERBELL *(Halesia)*

Also called Carolina silverbell, snowdrop tree, and wild olive, this tree is an attractive native of the southeastern United States. Silverbell is subject to few diseases, and no insects of importance are reported.

Diseases

LEAF SPOT. *Cercospora halesiae.* In rainy seasons, brown circular spots are caused by this fungus.

Control. This leaf spot rarely becomes serious enough to warrant periodic applications of a copper fungicide.

WOOD DECAY. *Polyporus halesiae.* A decay of the wood has been reported from Georgia.

Control. Wood decay fungi are difficult to control. Avoid bark wounds and keep the tree in good vigor.

SMOKE-TREE *(Cotinus)*

The smoke-tree, a close relative of ordinary sumac, is occasionally used in ornamental plantings. When it is grown in soils where nitrates tend to become deficient, its leaves show red spots or blotches. The application of a commercial fertilizer high in nitrogen usually helps overcome this trouble. It is drought-tolerant and easy to grow.

Diseases

WILT. *Verticillium albo-atrum.* This is the only serious disease of smoke-tree.

Control. See Chapter 12.

RUST. Two rusts have been reported on this tree: *Puccinia andropogonis* f. *onobrychidis* and *Pileolaria cotini-coggyriae.* The latter, found in Georgia several years ago, produces conspicuous spots with hypertrophied centers surrounded by dead tissue.

Control. These diseases are not serious enough to warrant control measures.

LEAF SPOTS. Four species of fungi are known to cause leaf spots on smoke-tree: *Cercospora rhoina, Pezizella oenotherae, Septoria rhoina,* and *Gloeosporium* sp.

Control. Valuable specimens can be protected with copper fungicides.

POWDERY MILDEW. *Erisyphe cichoracearum.* This fungus also attacks smoke-tree.

Insects

OBLIQUE-BANDED LEAF ROLLER. *Choristoneura rosaceana.* The leaves may be mined and rolled in June by pale yellow larvae. The adult moth is reddish brown and has a wingspread of 1 inch, with the front wings crossed by three distinct bands of dark brown.

Other trees subject to this pest are apple, apricot, ash, birch, cherry, dogwood, hawthorn, horsechestnut, linden, maple, oak, peach, pear, plum, and poplar.

Control. Spray early in June with Sevin.

SAN JOSE SCALE. *Aspidiotus perniciosus.* This scale occasionally infests smoke-tree.

Control. Malathion or Sevin sprays will control the crawler stage.

SNOW-IN-SUMMER *(Melaleuca)*

This Australian native is attractive when in bloom. Adapted to mild climates, it faces few serious pests.

SORREL-TREE *(Oxydendrum)*

Also called sourwood, this tree is a native of the eastern United States and grows to medium height, preferring acid soils. Its flowers are attractive to bees.

Diseases

TWIG BLIGHT. *Sphaerulina polyspora.* This fungus occasionally causes blighting of leaves at the tips of the branches. Trees injured by fire or in poor vigor appear to be most subject to this disease.

Control. Pruning infected twigs is usually sufficient to keep this disease under control. Fertilize and water when necessary to keep trees in good vigor.

LEAF SPOTS. *Cercospora oxydendri* and *My-cosphaerella caroliniana*. These two fungi occasionally spot the leaves of sorrel-tree. The latter produces reddish or purple blotches with dry brown centers in midsummer.

Control. Spray with a copper fungicide, starting when the leaves are fully expanded and repeating in 2 weeks.

Insects

No insects of any consequence attack this tree.

SOURGUM. See Tupelo.

SOURWOOD. See Sorrel-Tree.

SPRUCE *(Picea)*

Native spruces grow in the northern United States and in Canada. Besides being valuable forest trees, the spruces are highly prized as ornamentals. They thrive in moist, sandy loam soil and can tolerate more shade than most other conifers. The varieties most commonly used in ornamental plantings are Norway spruce and selected cultivars of Colorado blue spruce. They are subject to a number of destructive fungus diseases and insect pests.

Diseases

CYTOSPORA CANKER. The most prevalent disease on Norway spruce and Colorado blue spruce is known as Cytospora canker. Koster's blue spruce and Douglas-fir are also subject to it, but to a lesser extent. Although the disease occurs on young trees, those over 15 years old appear to be most susceptible.

Symptoms. The most striking symptom is the browning and death of the branches, usually starting with those nearest the ground and slowly progressing upward. Occasionally branches high in the tree are attacked, even though the lower ones are healthy.

The needles may drop immediately from infected branches or may persist for nearly a year, eventually leaving dry brittle twigs that contrast sharply with unaffected branches (Fig. III-111). White patches of pitch or resin may appear along the bark of dead or dying branches (Fig.

III-112). Cankers occur in the vicinity of these exudations. They are not readily discernible but can be found by cutting back the bark in the area that separates diseased from healthy tissues. In the cankered area, tiny, black, pinpoint fruiting bodies of the causal fungus are a positive sign of the disease. Trunk cankers may cause complete or partial girdling of some spruces.

Cause. The fungus *Leucocytospora (Leucostoma) kunzei* causes this disease. The fruiting structures protrude slightly during wet weather but are concealed beneath the bark scales during dry spells (Fig. III-113). Under the former conditions they release long, curled, yellow tendrils, which contain millions of spores. Wind and rain splash the spores to bark wounds on branches, where infection takes place. When the infection spreads completely around the branch, the distal portions die as a result of the girdling. The fungi *Valsa* (imperfect, *Cytospora*) *abietis*, *V. friesii*, and *V. pini* also attack spruce, but are not as important as *Leucostoma kunzei*.

Control. Infected branches cannot be saved. They should be cut off a few inches below the dead or infected parts, or at the point of attachment to the main stem. Because of the danger of spreading spores to uninfected branches, pruning should never be undertaken while the branches are wet. Inasmuch as the available evidence indicates that the fungi enter the branches only through wounds, bark injuries by lawn mowers and other tools should be avoided. The disease appears to be most prevalent on trees in poor vigor. Consequently, trees should be fertilized at least every few years and watered during dry spells, and the soil should be improved to increase or maintain their vigor.

SIROCOCCUS SHOOT BLIGHT. *Sirococcus conigenus*. This fungus causes cankers and tip dieback of the current season's growth. Foliage of affected twigs turns yellow, then reddish brown, and then drops. Small black fruiting bodies form on the leafless dead shoots.

Control. Prune and destroy affected branches.

NEEDLE CASTS. *Lophodermium piceae, Isthmiella crepidiformis, Lirula macrospora,* and *Rhizosphaera kalkhoffi.* The lower branches of Colorado blue, Engelmann, red, black, and Sitka

Fig. III-111. Blue spruce affected by *Cytospora* canker disease. Note the dead leafless branches, which contrast with unaffected branches.

spruces may be defoliated by one of these fungi. The leaves are spotted and turn yellow before they drop.

Control. Spray valuable ornamental specimens with a copper fungicide, Benlate, or Daconil 2787 in June and July.

RUSTS. *Chrysoomyxa ledi* f. *cassandrae, C. empetri, C. ledicola, C. chiogenis, C. arctostaphyli, C. piperiana, C. roanenis, and C. weirii.* Several species of rust fungi occur on spruce needles. Each requires an alternate host to complete its life cycle. On the spruce, the fungi appear as whitish blisters on the lower leaf surface. Affected needles turn yellow and may drop prematurely. The names of the alternate hosts for each of the fungi listed can be obtained from state or federal plant pathologists.

Control. Removal of the alternate hosts in the vicinity of valuable spruces is suggested. Protective sprays containing sulfur or Ferbam are rarely justified.

WITCHES' BROOM. Eastern dwarf mistletoe, *Arceuthobium pusillium*, causes compact masses of branches to form on black spruce.

WOOD DECAY. Spruces are subject to a number of wood decays caused by various fungi. Among those most commonly associated with various types of decay are *Haematostereum sanguinicola, Phaeolus schweinitzii, Laetiporus sulphureus, Inonotus tomentosus, Fomitopsis pinicola,* and *Phellinus pini.* Root decay is caused by *Armillariella mellea.*

Control. Little can be done once these decays become extensive. Avoidance of wounds and fertilization to increase vigor are suggested preventive measures.

Fig. III-112. Close-up of *Leucostoma* infected branch showing white, resinous exudation.

Insects and Other Animal Pests

SPRUCE GALL ADELGID. *Adelges abietis.* Elongated, many-celled, cone-shaped galls, less than 1 inch long (Fig. III-114), result from feeding and irritation by this pest. Trees are weakened and distorted when large numbers of these galls are formed. Norway spruce is most seriously infested; white, black, and red spruces are less so. In spring the adult wingless females deposit eggs near where the galls later develop. Immature females hibernate in the bud scales. Spruce gall adelgid is sometimes called spruce gall aphid.

Control. Spray the tips of twigs and bases of buds with ethion plus oil in April when the trees are still dormant, or with Sevin in May. Extreme caution must be used when applying oil sprays to spruces. Oil will remove blue spruce needle color and too much oil kills spruces.

COOLEY SPRUCE GALL ADELGID. *Adelges cooleyi.* This insect is sometimes referred to as Cooley spruce gall aphid. Galls ranging from ½ to 2½ inches in length on the terminal shoots of blue, Englemann, and Sitka spruce are produced by feeding and irritation by an adelgid closely related to that which infests the Norway spruce. The adult female overwinters on the bark near the twig terminals and deposits eggs

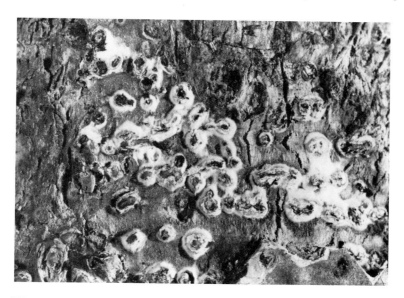

Fig. III-113. Fruiting bodies of the canker fungus *Leucocytospora (Leucostoma) kunzei.*

Fig. III-114. Galls of spruce gall adelgid, *Adelges abietis,* on Norway spruce.

in that vicinity in spring. On Douglas-fir the adelgid insect resembles a woolly aphid.

Control. Because Douglas-fir is an alternate host for this pest and is also injured by it, avoid planting that species near spruces. Spray with malathion or Sevin in spring just before new growth emerges.

APHIDS. The green spruce aphid, *Elatobium abietinum,* and the pine leaf adelgid, *Pineus pinifoliae,* also infest spruce leaves. The balsam twig aphid, discussed under Fir, at times attacks this host.

Control. Sevin and malathion sprays are very effective against these species.

SPRUCE BUD SCALE. *Physokermes piceae.* Globular red scales, about ⅛ inch in diameter, may occasionally infest the twigs of Norway spruce. The young crawl about in late July, and the winter is passed in the partly grown adult stage. Hemlock scales (see hemlock Fiorinia scale and hemlock scale, under Hemlock) and

pine needle scale (see under Pine) also attack spruce, especially trees in poor vigor.

Control. Before growth starts in spring, spray most spruces with a superior type dormant oil. For blue spruce, which is sensitive to this type of spray, use malathion or Cygon.

SPRUCE BUDWORM. *Choristoneura fumiferana.* One of the most destructive pests of forest and ornamental evergreens, this pest attacks spruce and balsam firs in forest plantations and also infests ornamental spruce, balsam fir, Douglas-fir, pine, larch, and hemlock. The opening buds and needles are chewed by the caterpillar, which is dark reddish brown with a yellow stripe along the side. The adult female moths, dull gray, marked with brown bands and spots, emerge in late June and early July.

WESTERN SPRUCE BUDWORM. *C. occidentalis.* This pest defoliates various conifers in western North America.

JACK PINE BUDWORM. *C. pinus.* This causes considerable damage to jack pine and to a lesser extent to red and white pine in the Great Lake states.

Control. Forest areas can be protected from these budworms with Sevimol sprayed from airplanes. Ornamental spruces, pines, firs, hemlocks, and larches can be sprayed with Sevin just as the buds burst in spring.

SPRUCE EPIZEUXIS. *Endothemia aemula.* Ornamental spruces in the Northeast may be attacked by small brown larvae with black tubercles which web needles together and fill them with excrement.

Control. Spray with Sevin or Diazinon when the larvae begin to feed in late summer.

SPRUCE NEEDLE MINER. *Taniva albolineana.* The light greenish larvae web the leaves together and mine the inner tissues. They enter through small holes at the bases of the leaves, and as they feed inside, the leaves turn brown. The adult is a small grayish moth.

Other species of spruce needle miners, *Recurvaria piceaella* and *Epinotia nanana,* also mine the leaves of spruce, killing groups of needles along the stem.

Control. Spray with Sevin or Diazinon in mid-May and again in mid-June.

SAWFLIES. *Pikonema alaskensis* and *Diprion hercyniae.* The former, known as the yellow-

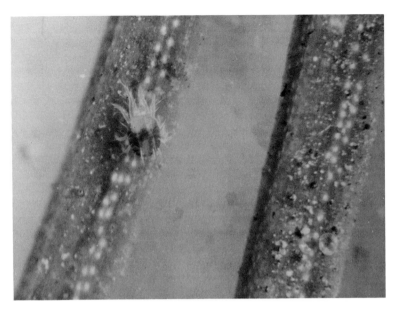

Fig. III-115. Spruce spider mite, *Oligonychus ununguis.*

headed sawfly, may sometimes defoliate trees completely. Ordinarily, however, infestation results merely in a ragged appearance with damaged needles, thin, brown, and shriveled by the larval feeding. The latter, the European spruce sawfly, feeds on the old foliage, and consequently the trees are not killed although the growth may be stunted. The larvae overwinter in cocoons on the ground, where they are subject to attack by mice.

Control. Spray the trees with Sevin as soon as the caterpillars are noticed.

SPRUCE SPIDER MITE. *Oligonychus ununguis.* Yellow, sickly needles, many of which are covered with a fine silken webbing, indicate a severe infestation of the spider mite, a tiny pest only 1/64 inch long (Fig. III-115). The young are pale green; the adult female is greenish black. Winter is passed in the egg stage on the twigs and the needles. Spider mites cause greatest damage in hot, dry seasons. In addition to spruces, arborvitae, junipers, and hemlock are also attacked. Hemlock eriophyid mite (see under Hemlock) also attacks spruce.

Control. Spray with lime sulfur or with superior dormant oil in early spring just before new growth emerges. Or spray in mid-May with Acaraben, Vendex, Orthene, Kelthane or Tedion to kill the young mites of the first generation. Repeat in September if necessary. Blue spruces should not be sprayed with dormant oils.

WHITE PINE WEEVIL. See under Pine. The work of the weevil is easily distinguished from that of the pine shoot moth by the greater length of the shoots killed and by the holes in the bark from which the beetles have emerged (Fig. III-116).

Control. See under Pines.

BAGWORM. See under Juniper.

PITCH MOTH. See under Pine.

STEWARTIA *(Stewartia)*

These natives of Asia, which grow in the South, have attractive bark, flowers, and foliage. They are relatively problem-free.

STRAWBERRY-TREE *(Arbutus)*

This low-growing evergreen requires frequent pruning. Another *Arbutus* species, madrone, is

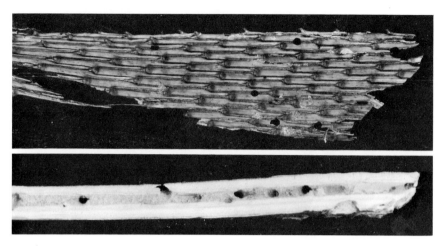

Fig. III-116. Work of the white pine weevil, *Pissodes strobi,* in a spruce branch. On the top may be seen the small holes through which the adults emerge; on the bottom is a twig spit to show their work inside.

a slow-growing, medium-sized tree of the Pacific coast.

Diseases

LEAF SPOTS. *Septoria unedonis* and *Elsinoë mattirolianum.* The former fungus produces small brown spots on leaves of *Arbutus unedo* in the Pacific Northwest, and the latter a spot anthracnose in California. In addition, madrone is infected by *Ascochyta hanseni, Cryptostictis arbuti, Didymosporium arbuticola, Exobasidium vacinii, Mycosphaerella arbuticola, Phyllosticta fimbriata, Pucciniastrum sparsum,* and *Rhytisma arbuti.*

Control. If only a few plants are involved, pick off the infected leaves; if necessary, spray with wettable sulfur.

CROWN GALL. *Agrobacterium tumefaciens.* This bacterial disease occurs occasionally on strawberry-tree in California and Connecticut.

Control. Remove and destroy infected parts.

BASSAL CANKER. Crown thinning is caused by *Phytophthora cactorum* trunk infections.

Insects

CALIFORNIA TENT CATERPILLAR. *Malacosoma californicum.* This species occasionally feeds on the leaves of strawberry-tree in California. It makes large tents in the trees like those of the eastern tent caterpillar.

Control. Spray with *Bacillus thuringiensis,* Dylox, methoxychlor, or Sevin when the caterpillars are small.

SCALES. *Saissetia oleae, Aspidiotus camelliae,* and *Coccus hesperidum.* These three species of scales—black, greedy, and brown soft—are known to attack strawberry-tree in the West.

Control. Spray with malathion or Sevin from time to time to control the young crawling stages of these pests.

SUMAC *(Rhus)*

African sumac, which has a weeping growth form, is adapted to hot, dry climates with relatively mild winter temperatures. It is susceptible to few pests.

SWEETGUM *(Liquidambar)*

Sweetgum is native to the eastern and southern United States and to Mexico. It grows best in rich clay or loam soils, but it will tolerate a wide variety of sites. Winter hardiness has been a problem for some cultivars growing in the

Ohio River valley region. Because sweetgums have become widely used, their diseases and insect pests have received more than the usual amount of attention.

Diseases

BLEEDING NECROSIS. The senior author studied the bleeding necrosis disease of sweetgum in New Jersey and on Staten Island, New York, as far back as 1941.

Symptoms. The most striking symptom, readily visible at some distance, is the profuse bleeding of the bark, usually at the soil line or a few feet above, but occasionally observed as high as 20 feet up on the main trunk and lower branches. The exudation looks like heavy motor oil poured over the bark.

The condition of the inner bark and sapwood beneath the oozing area, however, presents a more positive symptom of the disease. Such bark is dark reddish brown with an occasional pocket containing a white crystalline solid. The outer layer of sapwood may be brown or olive green in color.

Infected trees may exhibit undernourished foliage and more or less extensive dying-back of terminal branches by the time profuse bleeding is visible on the trunk.

Cause. The fungus *Botryosphaeria dothidea* causes bleeding necrosis, but when first reported the fungus was listed under the asexual stage *Dothiorella*. There is some evidence that raising the grade around the tree makes it more subject to this fungus.

Drought may also predispose trees to bleeding necrosis.

Control. No control measures are known. Removal and destruction of diseased trees is suggested.

LEAF SPOTS. *Actinopelte dryina, Cercospora liquidambaris, C. tuberculans, Gloeosporium nervisequam, Exosporium liquidambaris, Leptothyriella liquidambaris, and Septoria liquidambaris.* These seven fungi may occasionally spot the leaves of this host.

Control. Leaf spots are rarely serious enough to warrant control measures.

OTHER DISEASES. Several other important diseases of sweetgum have been reported in recent years. For most, the causal agent is still unknown. One, leader dieback, was associated with drought and secondary fungi.

Similarly, the disease known as sweetgum blight has caused death of many trees in Maryland and adjacent states. At this writing no causal organism has been associated consistently with the disease.

Control. Control measures are unavailable.

Abiotic Diseases

COLD INJURY. Record-setting cold winter temperatures can kill phloem and cambium tissues of sweetgum, resulting in dieback and decline. See Chapter 10.

AIR POLLUTION. Sweetgum is sensitive to ozone. See Chapter 10.

Insects

SWEETGUM WEBWORM. *Salebria afflictella.* The leaves are tied and matted together by small larvae.

Control. Spray with Sevin before the leaves are matted together.

CATERPILLARS. The leaves of sweetgum are occasionally chewed by the caterpillars listed below, as well as by those of several species of moths including the luna, *Actias luna;* polyphemus, *Antheraea polyphemus;* and promethea, *Callosamia promethea.*

Control. Spray with *Bacillus thuringiensis,* Imidan, or Sevin when caterpillars begin to feed.

FALL WEBWORM. See under Ash.

FOREST TENT CATERPILLAR. See under Maple.

RED-HUMPED CATERPILLAR. See under Poplar.

BAGWORM. See under Juniper.

ASIATIC OAK WEEVIL. See under Oak.

AMERICAN PLUM BORER. See under Planetree.

SWEETGUM SCALE. *Aspidiotus liquidambaris.* This scale is occasionally found on sweetgums in the eastern United States. Calico scale also attacks sweetgum (Fig. III-117).

Control. Spray with a superior miscible oil in spring before growth begins and with malathion or Sevin in early June.

COTTONY-CUSHION SCALE. See under Acacia.

WALNUT SCALE. See under Walnut.

Fig. III-117. Calico scale on sweetgum.

SYCAMORE *(Platanus occidentalis)*

This tree, also known as buttonwood and American planetree, is found from Maine to Minnesota, Florida to Texas. It is occasionally planted as a street tree but is not as desirable as the London planetree *(Platanus acerifolia).*

Diseases

BLIGHT, also known as anthracnose and scorch, is the most serious disease of sycamore. Two western species, *Platanus racemosa* and *P. wrighti,* are also susceptible. The disease is locally severe in some sections of the United States every year (Fig. III-118). The London planetree *(P. acerifolia),* though much more resistant, shows some injury during epidemic years. The destructiveness of this disease is often underestimated. While a single attack seldom results in serious harm, repeated annual outbreaks will eventually so weaken the tree that it becomes susceptible to borer attack and winter injury.

Symptoms. The first symptom, often confused with late frost injury, is the sudden browning and death of single leaves or of clusters of leaves as they are expanding in spring. Later, brown, dead areas along and between the veins appear in other leaves. The dead tissues assume a triangular form as a result of the death of the veinal tissues. The leaf falls prematurely when several lesions develop or when the leaf stem is infected. Many trees are completely defoliated and remain bare until late summer, when a new crop of leaves is formed. Small twigs are also attacked and killed. Cankers appear on the leaf spurs and on twigs as sunken areas with slightly raised margins (Fig. III-119). When these completely girdle the infected parts, the distal portions, including the leaves, are

Fig. III-118. Left: A sycamore tree lost nearly all its leaves in June as a result of anthracnose. **Right:** The same tree with a new set of leaves by early October.

Fig. III-119. Twig canker of sycamore anthracnose. Note fruiting bodies protruding from the bark.

killed. This is known as the shoot blight stage. Moderately infected trees show clusters of dead leaves scattered over the entire tree; these contrast sharply with the unaffected foliage.

Cause. Blight is caused by the fungus *Apiognomonia veneta.* The fungus has several distinct types of spores (Fig. III-120). It may overwinter in the vegetative stage (mycelium)

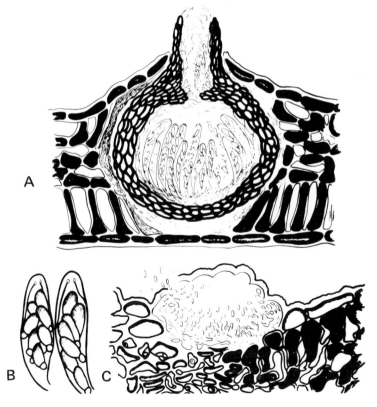

Fig. III-120. A, Cross section of sycamore leaf showing fruiting body of the fungus *Apiognomonia veneta.* B, Close-up of so-called ascospores. C, The summer spore *Discula platani* stage of the fungus.

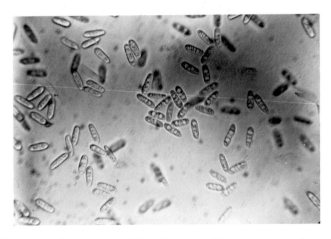

Fig. III-121. The spores of the summer *(Discula)* stage of the anthracnose fungus as they appear through a microscope.

in fallen infected leaves and in twig cankers. In the South it passes the winter in the spore stage on dormant buds. Initial infections of the young leaves in spring may originate from spores formed in overwintered leaves on the ground, from spores developed on infected twigs, and from spores on the dormant buds. Additional crops of spores for late infections are formed on the lower surfaces of the newly infected leaves in early spring (Fig. III-121).

The prevalence and severity of attack are governed by weather conditions; frequent rains and cool temperatures favor rapid spread.

The severity of the shoot blight stage is largely governed by the average temperature during the 2-week period following the emergence of the first leaves. If the average temperature during this period is below 55°F (13°C), the injury will be severe. If it is between 55 and 60°F (16°C), it will be less severe, and if it is over 60°F, little to no injury will occur.

Control. Control measures are rarely attempted on sycamores growing in open fields or in woodlands. They are justified, however, on particularly valuable specimens and on those used for shade and ornament.

All fallen leaves and twigs should be gathered and discarded in autumn to destroy the overwintering mycelium, which produces spores for the following spring's infections. Infected spurs and dead twigs should be pruned, whenever

feasible, and destroyed. To protect valuable specimens, spray the leaves with maneb, Zyban, Cyprex, Benlate, fixed copper, or zineb when they unfurl, when they reach full size, and a third time 2 weeks later.

Trees suffering from repeated attacks should be heavily fertilized in the fall or the following spring to increase their vigor.

OTHER DISEASES. Several diseases of minor importance, but which can be locally serious, include leaf spots caused by the fungi *Mycosphaerella platanifolia, M. stigmina-platani, Phloeospora multimaculans, Phyllosticta platani,* and *Septoria platanifolia.*

Control. These can be controlled by adopting the spray schedule suggested for the control of blight.

In the South, bark and twig cankers associated with the fungi *Botryosphaeria rhodina, Diaporthe scabra,* and *Hypoxylon tinctor* have been reported.

Control. Control measures are rarely adopted.

POWDERY MILDEW. *Microsphaera platani* and *Phyllactinia guttata.* Powdery mildew normally occurs late in the growing season and does little damage. See Chapter 12.

Abiotic Disease

AIR POLLUTION. Sycamore is very sensitive to ozone. See Chapter 10.

Insects

APHIDS. *Longistigma caryae.* The giant bark aphid, up to ¼ inch in length, frequently attacks the twigs of sycamore and may gather in clusters on the undersides of the limbs. These insects are also called planetree aphids. They are the largest species of aphids known and exude great quantities of honeydew.

Another species, *Drepanosiphum platanoides,* infests maples in addition to sycamores throughout the country.

Control. Spray with malathion, or Sevin.

SYCAMORE PLANT BUG. *Plagiognathus albatus.* This bug in its adult stage is ⅛ inch long, tan or brown in color, with dark eyes and brown spots on the wings. The young bugs are yellow-green, with conspicuous reddish brown eyes. These pests suck out the plant juices on the upper sides of sycamore and London planetree leaves, presumably leaving a poisonous material which results in yellowish or reddish spots. As the leaves grow, the injured areas drop out, leaving holes, and by midseason the leaves may appear tattered and yellowed.

Control. Malathion sprays applied in early May and again 2 weeks later will control this insect.

LACE BUG. *Corythucha ciliata.* See under Planetree.

SYCAMORE TUSSOCK MOTH. *Halisodota harrisii.* The caterpillar stage of this moth is yellow and has white to yellow hairs on its body. It occasionally becomes abundant on sycamores in the northeastern United States. White-marked tussock moth (see under Elm) also attacks sycamore.

Control. Sevin sprays will provide control.

IO MOTH. *Automeris io.* The larvae of this moth feeds on sycamore. The larva is covered with spines, which sting when touched.

ASIATIC OAK WEEVIL. See under Oak.

JAPANESE BEETLE. See under Linden.

FLATHEADED BORER. See under Maple.

SCALES. Several species of scale insects, including black, cottony maple, grape, oystershell, sycamore, and terrapin, occasionally infest the sycamore and other species of *Platanus.*

Control. Malathion or Sevin sprays in late spring when the young scales are crawling about are effective. A dormant spray containing a superior miscible oil should be applied in early spring where the infestations are unusually heavy.

OTHER PESTS. *Platanus* species are infested by other pests, including bagworms, borers, mites, whiteflies, root-knot nema (*Meloidogyne* sp.), and several other species of parasitic nemas.

Control. Control measures are rarely needed.

TANOAK (Lithocarpus)

Tanoak grows in the western United States, where it is used occasionally as a landscape tree. It is especially susceptible to *Armillaria mellea,* which causes basal breakage of the tree through root and butt rot.

TREE-OF-HEAVEN (Ailanthus)

Tree-of-heaven, also called ailanthus, is never used as a streetside tree but thrives in the larger cities where no other tree will grow. It frequently sprouts in parks and yards of large cities. Only the seed-bearing kinds should be planted, as the male or pollen-bearing form has an offensive odor when it blooms. It prefers natural to alkaline soil and moderate winters. It is tolerant to most diseases.

Diseases

WILT. *Verticillium albo-atrum.* This is the most destructive fungus parasite of this host. Many trees have been killed by it in the Philadelphia–New York environs since the late 1940s.

Control. See Chapter 12.

ARMILLARIA ROOT ROT. *Armillaria mellea.* This fungus also has been destroying this tree in the northeastern United States in recent years.

Control. See Chapter 12.

LEAF SPOT. *Cercospora glandulosa, Phyllosticta ailanthi,* and *Gloeosporium ailanthi.* Three leaf-spotting fungi occasionally attack this tree.

Control. Control measures are never applied for leaf spots.

TWIG BLIGHT. *Fusarium lateritium.* This fungus also causes branch and twig cankers on Japanese pagoda-tree. The sexual stage of this fungus is *Gibberella baccata.*

Fig. III-122. Cynthia moth, *Samia cynthia*. The larval stage of this insect feeds voraciously on ailantus (tree-of-heaven) foliage.

Control. Control measures are rarely necessary.

OTHER DISEASES. Other fungus diseases causing dieback or cankers of ailanthus include *Botryosphaeria dothidea, Cytospora ailanthi, Coniothyrium insitivum, Nectria coccinea, Botryosphaeria obtusa,* and *B. rhodina.*

Control. Control measures are rarely needed.

Insects

CYNTHIA MOTH. *Samia cynthia.* The larvae of this moth can completely defoliate a tree-of-heaven in a few days. Mature larvae are 3½ inches long, light green, and covered with a glaucous bloom. The adult moth is beautifully colored and has a wingspread of 6 to 8 inches (Fig. III-122). A crescent-shaped white marking is present in the center of each of the grayish brown wings.

Control. Spray with Sevin in June when the caterpillars begin to chew the leaves.

AILANTHUS WEBWORM. *Atteva punctella.* Olive-brown caterpillars with five white lines feed in webs on the leaves in August and September. Adult moths have bright orange forewings, with four crossbands of yellow spots on a dark blue ground.

Control. Spray with Sevin in August.

OTHER INSECTS. Among other caterpillars which attack this host, particularly in cities, are the fall webworm, *Hyphantria cunea,* and the white-marked tussock moth, *Hemerocampa leucostigma.* Another species, the pale tussock moth, *Halisidota tessellaris,* attacks a wide variety of deciduous trees.

Control. Methoxychlor or Sevin sprays will control these pests.

Oystershell scale and citrus whitefly occasionally infest ailanthus.

Control. Sevin or malathion sprays will control these pests.

TULIP POPLAR. See Tuliptree.

TULIPTREE *(Liriodendron)*

Native to the eastern United States, this tree is also called yellow poplar or tulip poplar. It is a large tree that needs space. Tuliptree does not tolerate drought or construction grade changes.

Fungus Diseases

CANKERS. *Botryosphaeria dothidea, Cephalosporium* sp., *Fusarium solani, Myxosporium* sp., *Nectria magnoliae,* and *Nectria* sp. These six species of fungi cause cankers in tuliptree. The first mentioned is perhaps the most destructive.

Control. Prune and cart away infected branches. Valuable trees should be sprayed with

Fig. III-123. Anthracnose of tuliptree leaves caused by the fungus *Gloeosporium liriodendri.*

a copper fungicide and fertilized and watered to increase their vigor.

LEAF SPOTS. *Cylindrosporium cercosporioides, Gloeosporium liriodendri* (Fig. III-123), *Mycosphaerella liriodendri, M. tulipiferae, Phyllosticta liriodendrica,* and *Ramularia liriodendri.* These six species of fungi cause leaf spots of tuliptree.

Control. Leaf spots rarely become sufficiently destructive to warrant control measures other than the gathering and destroying of infected leaves. The use of copper sprays may be justified on valuable trees.

POWDERY MILDEWS. *Phyllactinia guttata* and *Erysiphe polygoni.* Two species of fungi produce a white coating over the leaves of this host (Fig. III-124).

Control. Where particularly valuable small trees are affected, spray with Benlate or wettable sulfur.

ROOT AND STEM ROT. *Cylindrocladium scoparium.* This disease has been associated with decline of large tuliptrees in Georgia and North Carolina. This may be the same cause of a root and lower trunk rot that has led to the decline of landscape tulip poplar in other areas.

Control. Control measures have not been developed.

SAPSTREAK. *Ceratocystis coerulescens.* Sapstreak occasionally occurs on tuliptree.

Control. No effective control has been developed.

WILT. *Verticillium albo-atrum.* This disease

can be very damaging to landscape tulip poplars. See Chapter 12.

Abiotic Diseases

LEAF YELLOWING. Starting in midsummer, during dry, hot periods, many tuliptree leaves turn yellow and drop prematurely. The yellowing is due to climatic conditions and not to any destructive organism. The leaves of recently transplanted trees exhibit yellowing more frequently than those of well-established trees. Very often, small, angular, brownish specks appear on the leaves between the leaf veins as a preliminary stage of yellowing and defoliation.

AIR POLLUTION. Tulip poplars are very sensitive to ozone. See Chapter 10.

Insects

TULIPTREE APHID. *Macrosiphum liriodendri.* This small green aphid secretes copious quantities of honeydew. Hence leaves of other plants growing beneath the tree are coated with honeydew, which is then overrun by the sooty mold fungus (Fig. III-124).

Control. Late spring and early summer applications of Orthene, malathion, or Thiodan will control this aphid.

TULIPTREE SCALE. *Toumeyella liriodendri.* Trees may be killed by heavy infestations of oval turtle-shaped, often wrinkled brown scales, ⅓ inch in diameter (Figs. III-125 and 126). The

Fig. III-124. Tuliptree sooty mold **(left)**; powdery mildew **(right)**.

Fig. III-125. Left: Adult stage of tuliptree scale. **Fig. III-126. Right:** Crawling stage of tuliptree scale, which appears in late summer.

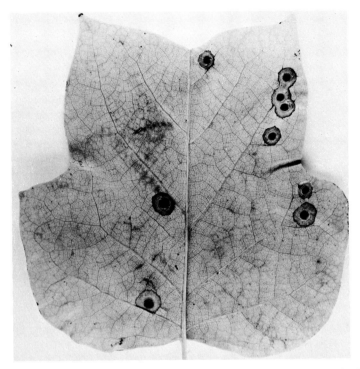

Fig. III-127. Tuliptree spot gall caused by the gallfly *Thecodiplosis liriodendri.*

lower branches, which are usually the first to die, may be completely covered with scales. Like the tuliptree aphid, this scale insect secretes much honeydew. The secretion drops on leaves and is soon covered by sooty black mold. Tuliptree scales overwinter as partly grown young. They grow rapidly until they mature in August. Young scales are produced in late August, at which time the old females dry up and fall from the tree.

Control. Spray the trees with a dormant miscible oil or ethion plus oil on a relatively warm day in late March or early April before the buds open. Or spray with malathion, Orthene, methoxychlor, or Sevin in late August or early September when the young are crawling about, and repeat the spray 2 weeks later.

OTHER SCALE INSECTS. In addition to the very common and destructive tuliptree scale, this host is also subject to oystershell, walnut, and willow scale.

Control. Spray with miscible oil in early spring followed by malathion in late May.

TULIPTREE SPOT GALL. *Thecodiplosis liriodendri.* A gallfly produces purplish spots about 1/8 inch in diameter on the leaves (Fig. III-127). These are frequently mistaken for fungus leaf spots.

Control. The damage is more unsightly than detrimental to the tree. Hence sprays are rarely used to control this pest.

SASSAFRAS WEEVIL. See under Sassafras.

ASIATIC OAK WEEVIL. See under Oak.

FLATHEADED BORER. See under Maple.

TUNG *(Aleurites)*

Native to China, tung is used as a source of oil and occasionally as a street tree. Tung normally does not tolerate freezing, and it requires deep, moist soils.

Foliage Diseases

Angular leaf spot *(Mycosphaerella aleuritidis)* and thread blight *(Ceratobasidium stevensii)* are the two most important. The latter kills foliage and the leaves are left hanging by fungus threads.

Root Rot

Clitocybe tabescens causes root rot similar to *Armillaria* but without rhizomorphs.

Mineral Deficiencies

Zinc, copper, and manganese deficiencies cause leaf chlorosis and malformation.

TUPELO (Nyssa)

Tupelo, also known as black gum, sourgum, and pepperidge tree, makes a fine lawn or park tree but because it is difficult to transplant, it is not commonly used. Its fall coloration is outstanding. It is native to the eastern United States.

Diseases

CANKER. *Botryosphaeria dothidea, Fusarium solani, Nectria galligena, Strumella coryneoidea,* and *Septobasidium curtisii.* These five species of fungi produce stem cankers on tupelo.
Control. Prune and destroy infected branches. Keep trees in good vigor by watering and fertilizing when necessary.
LEAF SPOTS. *Mycosphaerella nyssaecola.* Irregular purplish blotches, which later enlarge to an inch or more in width, are commonly scattered over the upper surfaces of the leaves of young tupelo trees in the southeastern states. Minute black fruiting bodies of the fungus are visible in the infected areas.
Three other fungi, *Cercospora nyssae, Actinopelte dryina,* and *Ceratobasidium stevensii,* also cause leaf spots and blights.
Control. Copper fungicides are suggested as preventive sprays. Raking and destroying infected leaves in fall should also help reduce infections.
RUST. *Aplospora nyssae.* This rust fungus occasionally infects tupelos from Maine to Texas.
Control. Control practices are not warranted.

WILT. Tupelo is susceptible to Verticillium wilt. See Chapter 12.

Insects

TUPELO LEAF MINER. *Antispila nyssaefoliella.* Leaves are first mined by the larval stage. When mature, the larvae cut oval sections out of the leaves and fall to the ground with the severed pieces.
Control. Spray with malathion in May when the adult moths emerge. Repeat within 10 days.
TUPELO SCALE. *Phenacaspis nyssae.* This scale is nearly triangular in shape, flat, and snow white.
Control. A dormant oil spray in early spring will provide control.
OTHER PESTS. Two other pests have been reported on tupelo: the azalea sphinx moth and San Jose scale.
Control. Malathion or Sevin sprays will control the larval stage of the former and the crawling stage of the latter.

TURKISH FILBERT (Corylus)

A drought-tolerant tree, it grows well in almost any type of soil. It has few pest problems.

WALNUT (Juglans)

Black walnut and butternut are rarely used as street trees. They require exacting soil conditions, are subject to wind and ice damage, are untidy, and their nuts tempt children to cause damage to the trees.

In ornamental plantings, the black walnut may produce detrimental effects on other types of plants growing nearby. The authors have observed extensive damage to native and hybrid rhododendrons, mountain laurel, and to tomatoes when these plants were grown in close proximity to black walnut roots. Other observers have reported similar damage to apple trees, alfalfa, potato, and a number of perennial ornamental plants growing near black walnut and butternut as well. In most instances, the removal of the nut trees resulted in the disappearance of the harmful effects once the roots had decomposed completely.

A number of important diseases and insect pests occur on the different species of walnut.

Diseases

CANKER. This disease is also common on apple; aspen; beech; birch; mountain, red, and sugar maples; dogwood; butternut; pignut hickory; and red, white, and black oaks.

Symptoms. Scattered, often numerous, rough, sunken, or flattened cankers with a number of prominient ridges of callus wood arranged more or less concentrically on the trunks or branches are characteristic symptoms. The size of the cankers varies according to age; on larger branches and on the trunk some attain a length of 4 feet and a width of 2½ feet and have as many as twenty-four concentric ridges. Complete girdling and death of the distal portions may occur within a few years on smaller branches, and in 20 or more years on large trunks. The concentric pattern of the canker is less evident on twigs and smaller branches.

On the dead bark and wood in late fall and winter, small, globose, red masses—the fruiting structures of the causal fungus—are barely visible to the unaided eye.

Cause. Canker is caused by the fungus *Nectria galligena*. In late winter and in spring, the red fruiting bodies produce spores, which are forcibly ejected into the air. These are then carried by wind and rain to bark injuries made by insects, fungi, ice, low temperatures, wind, rubbing plant parts, or dead branch stubs. The spores germinate and produce mycelium, which penetrates the bark and wood. Penetration progresses slowly, usually about ½ inch of tissue being killed annually around the developing canker.

Control. All badly diseased trees should be felled and the cankered tissues cut out and destroyed.

Valuable trees only mildly infected can be saved by removing the cankers by surgical methods. In addition, the trees should be fertilized, watered, and sprayed to control leaf pests and ensure rapid callusing.

Sites exposed to full sunshine should be used for new walnut plantings, inasmuch as the fungus does not appear to thrive under such conditions.

BRANCH WILT. *Hendersonula toruloidea (Exosporina fawceti).* This fungus kills English walnut branch bark and wood, resulting in withering and death of leaves on infected branches. It spreads down the branch, ultimately invading larger limbs. Infection begins at a wound such as sunscald. Avoid unnecessary wounds.

DIEBACK. Though black walnut occasionally dies back, this disease is most prevalent and destructive on butternut and the Japanese walnut (*J. ailantifolia*).

Symptoms. On butternut the smaller branches first die back, the dying back progressing slowly until the main branches are involved. The bark on the affected branches changes from the normal greenish brown to reddish brown and finally to gray. Small black pustules soon cover the dead bark. These disappear within a year or two, leaving small, irregular holes in the loose outer bark.

Cause. The fungus *Sirrococcus clavigignenti-juglandacearum* causes elliptical cankers with dark discoloration under the bark and eventual death of the tree. Spores are exuded in a tan, sticky mass from the black pycnidia on the bark to initiate new infections. The fungus *Melanconis juglandis* is often found as a secondary invader.

Control. Severely affected trees should be removed and destroyed. Where the disease is still confined to the upper ends of the branches, pruning to sound wood will help check further spread. In addition, fertilization, watering, sanitation, and insect and leaf-disease control measures should be adopted to help the tree regain its vigor.

BACTERIAL BLIGHT. English and Persian walnuts grown for ornament and for nut production in the eastern United States are susceptible to this destructive blight.

Symptoms. Small, water-soaked spots, which turn reddish brown, appear on the young, unfolding leaves in spring. The spots are usually isolated, but they may coalesce, causing considerable distortion of the leaves. Black, sunken lesions also appear on the twigs, which may be completely girdled, causing death of the parts above. Infection of the husks usually results in premature dropping of the fruit or in browning and decay of any fruit that remains attached.

Cause. The bacterium *Xanthomonas campestris* pv. *juglandis* causes blight. It is believed to overwinter mainly in diseased buds and is splashed by rain to various parts of the tree in spring. Pollen from blighted catkins may also spread disease.

Control. Cut out and destroy badly infected shoots. Spray with streptomycin, diluted at 50 parts per million parts of water, plus a spreader–sticker, as flower buds open, at full bloom, and at petal fall.

BARK CANKER. *Erwinia nigrifluens.* This bacterial disease, first found in California in 1955 on Persian walnut, appears as irregular, large, shallow, dark brown necrotic areas in the trunk and scaffold branches. The cankers enlarge in summer and are inactive in winter.

A phloem canker of *Juglans regia* in California is attributed to the bacterium *Erwinia rubrifaciens.*

Control. Control measures have not been developed for these diseases.

BROWN LEAF SPOT. *Gnomonia leptostyla.* Both butternut and walnut may be badly spotted by this walnut anthracnose fungus. The leaflets are attacked early in summer and are marked by irregular dark brown or blackish spots, somewhat similar to those caused on elm leaves by a closely related fungus. Much premature defoliation may result. *Marssonina juglandis* is the conidial stage of brown leaf spot. *Ascochyta juglandis, Cercospora juglandis, Phleospora multimaculans, Marssonina californica,* and *Cylindrosporium juglandis* also occasionally spot leaves. *Cristulariella pyramidalis* forms brown spots with white concentric rings (sometimes referred to as bull's-eye or zonate leaf spot). Complete defoliation of black walnut from this fungus has been reported from the Midwest.

Control. Because the leaves harbor the causal fungi, all fallen leaves should be gathered and destroyed to eliminate this important source of inoculum. Black walnuts and butternuts of ornamental value should be sprayed periodically with Benlate, Cyprex, Zyban, maneb, mancozeb, or zineb. The first application should be made when the buds start to open, the second about 10 days later, and the third when the leaves are fully grown. Nitrogen fertilization reduces the damage caused by this disease.

YELLOW LEAF BLOTCH. *Microstroma juglandis.* This disease is also called white mold, or downy leaf spot. This fungus causes a yellow blotching on the upper sides of the leaves. The snow white coating of fungus growth on the underside is composed of enormous numbers of spores, which spread the disease. It also causes the witches' broom disease of shagbark hickory.

Control. Same as for brown leaf spot.

TRUNK DECAY. The trunks of black walnut and butternut often show a white or a brown decay of the heartwood. The former is caused by the so-called false tinder fungus, *Phellinus igniarius,* which forms hard, gray, hoof-shaped fruiting bodies, up to 8 inches in width, along the trunk in the vicinity of the decay. The latter is caused by the fungus *Laetiporus sulphureus,* which forms soft, fleshy, shelflike fruiting structures that are orange-red above and brilliant yellow below. The fruiting bodies become hard, brittle, and dirty-white as they age. These fungi usually enter the trunk through bark injuries and dead branch stubs.

Control. Once these decays become extensive, little can be done to eradicate them. Cleaning out the badly decayed portions is suggested. Initial infections can be kept at a minimum by maintaining the vigor of the trees and by protecting the wounds as soon as they are formed.

WALNUT BUNCH DISEASE. This disease, formerly called witches' broom, was once thought to be caused by a virus, but is now known to be a mycoplasmalike organism. It is characterized primarily by the appearance of brooms or sucker growths on main stems and branches, tufting of terminals, profuse development of branchlets from axillary buds, leaf dwarfing, and, at times, death of the entire tree (Fig. III-128). These symptoms vary from mild to severe and are particularly pronounced on Japanese walnut. Butternut, Persian walnut, and eastern black walnut are also susceptible but to a lesser degree. The causal organism has been experimentally transmitted by grafting.

Control. Walnut bunch can be transmitted by grafting. An insect vector has not yet been

Fig. III-128. Walnut bunch in Japanese walnut tree, caused by a mycoplasmalike organism.

associated with the disease, and hence no control measures have been developed.

OTHER DISEASES. A leaf spot and blasting of the nutlets of *J. mandchurica* in California is attributed to the bacterium *Pseudomonas syringae,* which also attacks lilacs. Blackline, a disorder characterized by formation at the graft union of a narrow, dark brown, corky layer of nonconducting tissue that results in girdling, is suspected of being caused by the cherry leafroll virus.

Control. Control measures are rarely needed.

Insects and Related Pests

WALNUT APHID. *Chromaphis juglandicola.* This pale yellow species is common on the undersides of English walnut leaves on the West Coast. It secretes large quantities of honeydew or aphid honey.

The giant bark aphid (see aphids, under Sycamore) occasionally infests butternut and walnut.

Control. Spray with malathion.

SYCAMORE PLANT BUG. See under Sycamore.

WALNUT CATERPILLAR. *Datana integerrima.*

Trees are defoliated by this caterpillar, which is covered with long white hairs and which grows to 2 inches in length. The adult female moth has a wingspread of 1½ inches. Its dark buff wings are crossed by four brown lines. Eggs are deposited in masses on the lower sides of the leaves in July. Winter is passed in the pupal stage in the soil. Walnut caterpillar also feeds on hickory, oak, and honey locust.

OTHER CATERPILLARS. The following caterpillars also attack walnut: hickory horned devil, orange-tortrix, omnivorous looper, red-humped, yellow-necked, hickory tussock moth, fall webworm, and fruit tree leaf roller.

Control. Spray with *Bacillus thuringiensis,* Diazinon, methoxychlor, or Sevin when the larvae are small.

BUTTERNUT WOOLLY WORM. *Eriocampa juglandis.* This larva feeds on hickory, walnut, and butternut. The white, cottony tufts covering the worm give it a distinctive appearance.

AMERICAN PLUM BORER. See under Planetree.

LEOPARD MOTH BORER. See under Maple.

WALNUT LACE BUG. *Corythucha juglandis.* This pest is occasionally abundant on park trees, causing a bad spotting of the foliage.

Control. Spray early infestations with malathion or Sevin.

WALNUT SCALE. *Quadraspidiotus juglansregiae.* Part or all of a tree may be severely weakened by masses of round brown scales, ⅛ inch in diameter with raised centers. The adult female is frequently encircled by young scales. The pest overwinters on the bark.

OTHER SCALES. Nut trees are susceptible to many other scales, including black, California red, calico, cottony-cushion, greedy, oystershell, Putnam, scurfy, tuliptree, hickory lecanium, obscure, and white peach.

Control. To control walnut and other scales, spray with lime sulfur when the trees are dormant, or with malathion or Sevin in late spring and again in early summer.

MITES. The following species of mites infest *Juglans:* European red, black walnut pouch gall, platani, southern red, and walnut blister. The last-mentioned, known scientifically as *Eriophyes erinea,* causes yellow or brown feltlike galls on the undersides of leaves.

Control. Spray twigs and bark with Diazinon, Acaraben, or Kelthane before new growth begins in spring. A second application should be made to the newly formed leaves in late spring.

WILLOW *(Salix)*

Willow are rarely used as street trees or in locations where the breaking off of their weak-wooded branches is likely to cause injury. Some species, however, are used extensively in landscape plantings, especially around ponds and streams.

Diseases

LEAF BLIGHT. Blight, or scab, is the most destructive disease of willows in the United States. It has already killed hundreds of trees in the northeastern states and appears to be spreading westward and southward. It is almost always associated with black canker disease, and the two fungi together are responsible for the damage.

Symptoms. Soon after growth starts in spring, a few small leaves turn black and die. Later, all the remaining leaves on the tree suddenly wilt and blacken, as if they had been burned by fire. Cankers on twigs may result from the leaf infections. Following rainy periods, dense olive-brown fruiting structures of the causal fungus appear on the undersides of blighted leaves, principally along the veins and midribs.

Cause. Leaf blight is caused by the fungus *Venturia saliciperda.* It passes the winter in infected twigs, on which it produces spores during early spring. The spores are washed by rain onto the newly developing leaves and initiate early infections. Such leaves produce large numbers of spores of the asexual stage, *Pollaccia saliciperda,* which are responsible for the severe blighting of the remaining leaves later in the season.

Control. Pruning of dead twigs and branches will eliminate the most important source of inoculum for early-season infections. Spraying with mancozeb, maneb, or zineb three or four times at 10-day intervals, starting when the leaves begin to emerge in spring, will usually give control. These practices are recommended only for valuable specimens. A number of species, including weeping, bay-leaved, osier, purple, and pussy willows, appear resistant to blight. These should be planted in place of the more susceptible species, such as the crack and the heart-leaved willow. Golden willow has been reported by some investigators to be susceptible, by others to be resistant.

BLACK CANKER. Closely associated with leaf blight is the disease known as black canker.

Symptoms. The symptoms of black canker resemble those of leaf blight but usually appear later in summer. Dark brown spots, many of which show concentric markings, appear on the upper leaf surface. Whitish gray or gray elliptical lesions, with black borders, appear subsequently on the twigs and stems. Clusters of minute black fruiting bodies develop in the stem lesions. Successive attacks over a 2- or 3-year period usually result in death of the entire tree.

Cause. Black canker is caused by the fungus *Glomerella miyabeana.* Two types of spores are produced on the stem lesions: one exuding from fruiting bodies as small pink masses in early summer; and the other released from tiny black spherical bodies in early spring. The fungus lives over the winter on diseased twigs in the latter type of fruiting bodies.

Control. Pruning and spraying as suggested for leaf blight are recommended. The use of resistant varieties also offers a means of avoiding this destructive disease. The bay-leaved, the osier, and the weeping willow appear to be resistant, whereas the crack, the heart-leaved, the white, the purple, and the almond willow appear to be susceptible. As in the case of leaf blight, golden willow has been reported by some observers to be extremely susceptible, and by others to be resistant.

CYTOSPORA CANKER. Willows are subject to this canker disease, caused by the fungus *Cytospora chrysosperma,* the perfect stage of which is *Valsa sordida,* already discussed under Poplar. The control measures listed under that host hold true also for willows. In addition, the use of resistant varieties offers a means of avoiding serious trouble. One investigator has observed that the disease occurs rarely on black willow or on peach-leaf willow, whereas crack and golden willows appear to be extremely susceptible.

OTHER CANKERS. Willows are susceptible to other canker diseases caused by *Botryosphaeria dothidea, Cryptodiaporthe salicella, Cryptomyces maximus, Discella carbonacea, Diplodina* sp., *Leucostoma niveum,* and *Macrophoma* sp.

Control. Control measures are rarely needed when trees are provided with good growing conditions.

GRAY SCAB. *Sphaceloma murrayae.* This disease affects many species of willow, including *Salix fragilis, S. lasiandra,* and *S. lasiolepis.* Round, irregular, somewhat raised, grayish white spots with narrow, dark brown margins appear on the leaves. Affected portions of the leaves frequently drop away.

Control. The same as for leaf spots, described below.

LEAF SPOTS. *Ascochyta salicis, Asteroma capreae, Cercospora salicina, Cylindrosporium salicinum, Marssonina* sp., *Myriconium comitatum, Phyllosticta apicalis, Ramularia rosea, Septogloeum salicinum,* and *Septoria didyma.* At least ten species of fungi cause leaf

spots on willow. Some may also cause premature defoliation.

Control. Because most leaf spot fungi overwinter on diseased fallen leaves, gathering and destroying the leaves is a recommended practice. Valuable specimens should be sprayed with Ferbam, zineb, or copper fungicide when leaf spots cause considerable defoliation annually.

POWDERY MILDEW. *Uncinula adunca, Phyllactinia guttata.* Leaves infected with mildew are covered with a whitish feltlike mold. This develops chains of white spores which are shed in clouds. The little black fruiting bodies formed later in the season are characterized by microscopic appendages curled at the end like a shepherd's crook. This is not a serious disease of willows, but it may cause some loss of leaves.

Control. Valuable willows infected with this fungus can be protected by an occasional application of Benlate or Bayleton spray. Sprays containing wettable sulfur also can be used to control this mildew. Bordeaux mixture, recommended for several other diseases of willow, is also helpful.

RUST. *Melampsora* sp. Three or four species of rust attack willow leaves, causing lemon yellow spots on the lower surfaces. Later in the season the spore-bearing pustules are dark colored. The disease may be severe enough to cause dropping of the leaves. Of the rust-causing fungi, one species. *M. paradoxa,* has the larch as the alternate host; *M. abieti-capraearum* has the balsam fir; and *M. arctica* lives part of the year on a saxifrage.

Control. Although rust infections are not considered serious, they may result in heavy defoliation of young trees. Gathering and destroying fallen leaves will help prevent serious outbreaks. The use of copper, Bayleton, or sulfur fungicide as a preventive is suggested only in rare instances.

TAR SPOT. *Rhytisma salicinum.* These spots are usually jet black, very definitely bounded, about ¼ inch in diameter, somewhat raised about the surface of the leaf. They are more common on maple.

Control. Since this fungus winters on the old leaves, care should be taken, where the disease is serious, to rake up and destroy dead leaves. Spray the shrubs or trees early in May with bordeaux mixture or wettable sulfur.

WITCHES' BROOM. Mycoplasma-like organism. Early breaking of axillary buds and subsequent growth of numerous spindly, erect branches with stunted leaves on *Salix rigida* are characteristic symptoms of this disease. The witches' brooms die the winter after they are formed.

Control. Ways of preventing this disease have yet to be developed.

WOOD DECAY. Species of *Daedalea, Phellinus,* and *Trametes* cause decay of living willow trees. Their characteristic fruiting bodies protrude through the bark. Unnecessary wounding should be avoided.

BACTERIAL TWIG BLIGHT. *Pseudomonas saliciperda.* The leaves turn brown and wilt, and blighted branches die back for some distance. Brown streaks can be seen in sections of the wood. The parasite winters in the cankers, so that the young leaves are infected as soon as they unfold. The disease has so far been serious only in New England. There it has caused the death of a large number of trees by serious defoliation. The damage has been confused with frost injury.

Control. Pruning as many infested twigs as possible and spraying as suggested for leaf blight are the recommended control measures.

CROWN GALL. *Agrobacterium tumefaciens.* See Chapter 12.

Insects

APHIDS. Several species of aphids infest willows, the most common being the giant bark, *Longistigma caryae.*

Control. Spray valuable specimens with malathion, Diazinon, or Orthene.

EASTERN TENT CATERPILLAR. *Malacosoma americanum.* The leaves are chewed and small silken nests are formed in branch crotches by the tent caterpillar, a yellow-haired, black form with white lines down the back. It attains a length of nearly 2 inches. It prefers to feed on wild cherry but will feed on apple, peach, and plum in addition to willow. When these trees are scarce, the insects will also defoliate ash,

beech, birch, elm, maple, oak, poplar, and many shrubs. The adult female, a fawn-colored moth, has a wing expanse of 2 inches. Eggs are laid in cylindrical clusters on twigs in July.

Forest tent caterpillar (see under Maple) also attacks willow.

Control. Remove and destroy egg clusters and nests when they appear, and spray with *Bacillus thuringiensis,* Orthene, methoxychlor, Dylox, or Sevin when new twig growth is 2 to 4 inches long in spring.

FOLIAR-FEEDING CATERPILLARS. Dagger moth *(Acronicta americana).* This hairy white caterpillar is common but causes little damage. other hosts include maple and oak. Gypsy moth, mourning-cloak butterfly, and white-marked tussock moth (see under Elm), red-humped caterpillar (see under Poplar), io moth (see under Sycamore), and fall webworm (see under Ash) all feed on willow.

MOTTLED WILLOW BORER. *Cryptorhynchus lapathi.* Swollen and knotty limbs may be produced by mottled willow borers, white legless larvae ½ inch long, which eat through the cambium, sapwood, and heartwood. The adult beetle, ⅓ inch long, has a long snout and grayish black, mottled wing covers. Eggs are laid in fall, and the larvae overwinter in tunnels beneath the bark. This pest is also known as the poplar and willow borer.

Control. Spray the trunk thoroughly during the latter part of August with methoxychlor.

BORERS. The following attack willow: dogwood borer (see under Dogwood) and the leopard moth and flatheaded borer (see under Maple).

POPULAR BORER. See under Poplar.

BASKET WILLOW GALL. *Rhabdophaga salicis.* Swollen, distorted twigs may be produced by yellowish, jumping maggots of the basket willow gall midge. The adults appear in early spring.

Control. Prune and destroy infested twigs.

PINE CONE GALL. *Rhabdophaga strobiliodes.* Cone-shaped galls at the branch tips that hinder bud development are produced by small maggots. The adult, a small fly, deposits eggs in the opening buds. The larvae hibernate in cocoons inside the galls.

Control. Remove and destroy galls in the

fall, or spray thoroughly with malathion when the buds are swelling in spring.

IMPORTED WILLOW LEAF BEETLE. *Plagiodera versicolora.* These beautiful metallic blue beetles, about ⅛ inch long, live through the winter under the bark scales and in the dead leaves about the tree. They emerge and lay their lemon yellow eggs in early June. The ugly larvae or grubs feed on the undersides of the foliage, leaving only a network of veins. The adult beetle develops during July and produces a second brood in August. This beetle is so much smaller than the Japanese beetle (*Popillia japonica*) that it can scarcely be mistaken for it. In the vicinity of New York both beetles cause serious damage.

Control. Spraying with Orthene, Sevin or methoxychlor in late May will control this pest. A second spraying in late June may be necessary for the second brood.

WILLOW FLEA WEEVIL. *Rhynchaenus rufipes.* The overwintering adult beetle emerges in the middle of April. During the latter part of May it excavates a circular mine on the underside of the leaf and it deposits its eggs. The adults feed on the foliage, causing it to become brown and dry. The larvae begin to mine the leaves about the middle of June. By the end of July, where infestation is heavy, the trees appear as if scorched by fires.

California casebearer, *Coleophora sacramanta* is also a leaf miner of willow.

Control. To control the adult weevil, spray in late May with malathion, and in late June, spray with Sevin to control the larvae.

LEAF BEETLES. *Chysomela scripta* and *C. interrupta* skeletonize willow leaves.

ASIATIC OAK WEEVIL. See under Oak.

WILLOW LACE BUG. *Corythucha mollicula.* Willow leaves may be severely mottled and yellowed by this sucking insect.

Control. Malathion or Sevin sprays in late spring provide excellent control.

WILLOW SHOOT SAWFLY. *Janus abbreviatus.* The female lays its eggs in the shoots in early spring. It then girdles the stem, preventing further growth of the twig, which wilts and dies. The young borers feed in the pith of the shoots, which die eventually.

Leaf gall sawflies (*Pontania* sp.) make spher-

ical, reddish galls, ⅓ inch in diameter, on willow.

Control. Prune and destroy infested twigs. Spray with Sevin or Diazinon when the larvae are small.

WILLOW SCURFY SCALE. *Chionaspis salicisnigrae.* Branches and even small trees may be killed by heavy infestation of this pest, a pear-shaped white scale ⅛ inch long. The insect passes the winter in the egg stage underneath the scale of the female.

Other scale insects that attack willows are black, California red, cottony-cushion, cottony maple, European fruit, greedy, lecanium, hickory lecanium, azalea, latania, Glover, dictyospermum, obscure, oystershell, Putnam, soft, and terrapin.

Control. Spray with lime sulfur or a dormant oil in spring when trees are still dormant, then with Orthene, Diazinon, malathion, or Sevin in May and again in June to control the crawling stage.

BAGWORM. See under Arborvitae.

CALIFORNIA TENT CATERPILLAR. See under Strawberry-Tree.

CARPENTER WORM. See under Ash.

HEMLOCK LOOPER. See under Hemlock.

GIANT HORNET. See giant hornet wasp, under Franklin-Tree.

OMNIVOROUS LOOPER. See caterpillars, under Acacia.

WALNUT CATERPILLAR. See under Walnut.

WINGNUT, CAUCASIAN *(Pterocarya)*

This Chinese native is related to walnut and is relatively free of pest problems.

YELLOW POPLAR. See Tuliptree.

YELLOWWOOD *(Cladrastis)*

This medium-sized tree is native to Kentucky and North Carolina. Narrow crotch angles make yellowwood subject to breakage.

Few diseases and only one insect pest have been recorded on this host. A powdery mildew caused by *Phyllactinia guttata*, canker by *Botryosphaeria dothidea*, wilt by *Verticillium albo-atrum*, and a decay of the butt and roots of living trees by *Polyporus spraguei* and *Xylaria mali* occur occasionally. An unidentified species of scale may occasionally infest yellowwood. Control measures are rarely necessary.

YEW *(Taxus)*

Yews are among the most useful evergreens for ornamental purposes. They withstand city conditions better than most other evergreens. The English yew, the Japanese yew, and a large number of other species and varities are now planted extensively.

Diseases

NEEDLE BLIGHT. *Herpotrichia nigra* and *Sphaerulinia taxi.* Blighting of needles of the host in the western United States may be caused by these fungi.

Control. This disease is rarely destructive enough to warrant control measures.

TWIG BLIGHT. Several fungi are associated with a twig blight of yew during rainy seasons. One of the most common is *Phyllostictina (Phoma) hysterella*, whose perfect stage is *Physalospora gregaria*. Others are *Pestalotiopsis funerea* and a species of *Sphaeropsis*.

Control. Prune diseased twigs and destroy them. Spray with a copper fungicide several times at 2-week intervals during rainy springs.

ARMILLARIA ROOT ROT. *Armillaria mellea.* This disease appears in soil formerly occupied by apple or oak trees. Affected plants wilt and die; white wefts of fungus mycelium are present beneath the bark at the base of the plant. See Chapter 12.

Phytophthora cinnamomi also causes root rot as well as dieback of yews, especially in wet locations.

OTHER FUNGUS DISEASES. Among other fungus diseases of yews are a root rot caused by *Pythium* sp. and premature leaf drop in which a species of *Alternaria* is implicated.

Control. Control measures have not been developed.

Abiotic Diseases

The most prevalent trouble on yews is dieback. The plants first turn yellow at the growing tips;

this is followed by general yellowing, wilting, and death. Several months may elapse from the time the first symptoms appear to the complete wilting and death of the plant. The below-ground symptoms are a decay of the bark on the deeper roots; the affected bark sloughs off readily.

This trouble is not caused by parasitic organisms. Studies by the senior author showed that it is associated with unfavorable soil conditions. In nearly every case investigated, the soil was very acid, pH 4.7 to 5.4, in addition to being heavy and poorly drained. It has been observed that yews could be killed by immersing their roots in water for 32 to 64 hours and then drying out the soil.

Control. Improve drainage by embedding tile in the soil, or move plants to a more favorable area. Add ground limestone, according to the recommendation of a soil specialist, to increase the pH to about 6.5.

TWIG BROWNING. This condition is caused not by fungus parasites but by snow and winter damage. The combined effect of heavy snow cover and low temperatures causes many small twigs at the ends of branches to turn brown in late winter and early spring. Ice falling from rooftops can also injure yew twigs, causing browning of individual twigs in spring.

Control. Shake off heavy snow covers with a bamboo rake as soon as the snowfall stops. Prune browned branches in late spring.

Insects and Other Animal Pests

BLACK VINE WEEVIL. *Otiorhynchus sulcatus.* This pest, also known as the taxus weevil, is by far the most serious one on yews. It is so named because of its color and because it attacks grapes in Europe. In the United States, it also attacks false cyrpess, hemlocks, and such broad-leaved plants as rhododendrons and azaleas (see Figs. III-129, 130, and 131). The leaves of yew turn yellow, and whole branches or even the entire plant may die when the roots are chewed by the larval stage, a white-bodied, brown-headed pest, ⅜ inch long. As few as eight larvae are capable of killing a large-sized yew. The adult, a snout beetle, ⅜ inch long, feeds on the foliage of yews and other hosts at

Fig. III-129. Black vine weevil adults.

night. The edges of such leaves are scalloped. Eggs deposited in the soil during July and August hatch into larvae, which feed on the roots of susceptible hosts.

Control. See under strawberry root weevil, below.

STRAWBERRY ROOT WEEVIL. *Otiorhynchus ovatus.* This insect is related to the black vine weevil and has similar habits but is smaller, measuring ⅕ to ¼ inch in length. It feeds on hemlock, spruce, and arborvitae in addition to yews. It often wanders into homes in search of

Fig. III-130. The larval stage of the black vine weevil causes severe injury to and even death of yews.

Fig. III-131. Rhododendron leaves chewed by adult black vine weevils.

Fig. III-132. Mealybug *Dysmicoccus wistariae* on yew.

hibernating quarters and thus may become a nuisance.

Control. Spraying the lower parts of susceptible plants and the soil surface beneath them in mid-June and again in mid-July with Orthene, Vydate L., Turcam, or Guthion will provide control. In container-grown plants Furadan, as a soil drench, will control black vine and strawberry weevils.

TAXUS MEALYBUG. *Dysmicoccus wistariae.* First reported from a New Jersey nursery in 1915, this pest has become increasingly prevalent in the northern United States. The ⅜-inch-long bug, covered with white wax, may completely cover the trunk and branches of yews (Fig. III-132). Young mealybugs overwinter in bark crevices and mature in June. Although all species of *Taxus* are susceptible, those with

Fig. III-133. The scale *Lecanium fletcheri* on yew. Some of the scales have been opened to show eggs.

dense foliage such as *T. cuspidata nana* and *T. wardii* are preferred hosts. This pest has also been seen on apple, linden, and rhododendron, but it probably does not breed on these plants.

Control. Good control is obtainable by spraying with malathion, Sevin, or Cygon when the young stage is crawling about. In the area of Long Island and Connecticut, the latter part of May is the proper time; in cooler regions the application should be made a week or two later.

Ethion plus oil or oil sprays made in early April are also beneficial.

GRAPE MEALYBUG. *Pseudococcus maritimus.* This pest occasionally infests the Japanese yew. *P. longispinus,* the long-tailed mealybug, and *P. comstocki,* Comstock mealybug, also attack yews.

Control. Where these mealybugs infest aboveground parts of yews, spray with malathion.

SCALES. Seven species of scale insects infest yews: cottony taxus, California red, Asiatic red, dictyospermum, Fletcher (Fig. III-133), oleander, and purple. The first-mentioned, *Pulvinaria floccifera,* is ⅛ inch long, light brown, and hemispherical. In spring it produces long, narrow, fluted, cottony egg masses.

Control. Apply a superior dormant oil spray while the plants are dormant in early spring, when the temperature is well above freezing. In July and again in September, spray with Cygon, Diazinon, or Sevin.

TAXUS BUD MITE. *Cecidophyopsis psilaspis.* The growing tips of yews are enlarged and may

Fig. III-134. Termite injury of yew.

be killed. New growth is distorted. As many as a thousand mites can be found in a single infested bud.

Control. Spray with Kelthane or Thiodan in early May and repeat within 2 weeks if necessary.

ANTS AND TERMITES. The black carpenter ant, *Camponotus pennsylvanicus,* and the eastern subterranean termite, *Reticulitermes flavipes,* occasionally infest the trunks of older yews. Both are capable of excavating the trunk and making their nests therein. We have observed earthen termite tunnels extending several feet up the stem; *Taxus* limbs associated with the termite activity turn brown (Fig. III-134).

Control. Apply Diazinon in liquid form around the trunk base and soil surface.

NEMAS. *Criconemoides and Rotylenchus* spp. Several species of nemas attack the roots of *Taxus.*

Control. See under Boxwood.

ZELKOVA, JAPANESE *(Zelkova)*

This tree is touted as an elm substitute, having a similar leaf. Although it is susceptible to Dutch elm disease, it nevertheless survives infections.

NECTRIA CANKER. *Nectria cinnabarina.* This fungus has been reported on this host. See under Maple.

ELM LEAF BEETLE (see under Elm) and CALICO SCALE feed on Zelkova.

Selected Bibliography

Anonymous. 1960. Index of plant diseases in the United States. U.S.D.A. Agriculture Handbook 165, Washington, D.C. 531 pp.

Anonymous. 1979. A guide to common insects and diseases of forest trees in the northeastern United States. U.S.D.A. Forest Service, Northeast Area State and Private Forests Publication NA-FR-4, Washington, D.C. 127 pp.

Anonymous. 1985. Insects and diseases of trees in the south. U.S.D.A. Forest Services Southern Region General Report R8-GR5, Washington D.C. 98 pp.

Drooz, A. T. 1985. Insects of eastern forests. U.S.D.A.

Forest Service Misc. Publ. 1426, Washington, D.C. 608 pp.

Hepting, G. H. 1971. Diseases of forest and shade trees of the United States. U.S.D.A. Forest Service Agriculture Handbook 386, Washington, D.C. 658 pp.

Horst, R. K. 1982. Westcott's plant disease handbook, 4th ed. Van Nostrand Reinhold, New York. 803 pp.

Hudler, G. W. 1984. Diseases of maples in eastern North America. Cornell Tree Pest Leaflet A-13, A Cornell Cooperative Extension Publication, Ithaca, N.Y. 13 pp.

Johnson, W. T., and H. H. Lyon. 1976. Insects that feed on trees and shrubs. Cornell University Press, Ithaca, New York. 464 pp.

Jones, R. K., and R. C. Lambe (eds.). 1982. Diseases of woody ornamental plants and their control in nurseries. North Carolina Agricultural Extension Service AG-286, Raleigh. 130 pp.

Partyka, R. E., J. W. Rimelspach, B. G. Joyner, and S. A. Carver. 1980. Woody ornamentals: Plants and problems. Chemlawn Corporation, Columbus, Oh. 427 pp.

Peace, T. R. 1962. Pathology of trees and shrubs. Clarenden Press, Oxford, England. 722 pp.

Peterson, G. W. 1981. Pine and juniper diseases in the Great Plains. U.S.D.A. Forest Service General Technical Report RM-86, Rocky Mountain Forest and Range Experiment Station, Fort Collins, Co. 47 pp.

Peterson, J. L. 1982. Diseases of holly in the United States. Holly Society of America Bull. 19, Baltimore, Md. 44 pp.

Pirone, P. P. 1978. Diseases and pests of ornamental plants, 5th ed. Wiley, New York. 566 pp.

Riffle, J. W., and G. W. Peterson (eds.). 1986. Diseases of trees in the Great Plains. U.S.D.A. Forest Service General Technical Report RM-129, Washington, D.C. 149 pp.

Rose, A. H., and O. H. Lindquist. 1982. Insects of eastern hardwood trees. Can. For. Serv. For. Tech. Rep. 29. 304 pp.

Sinclair, W. A., H. H. Lyon, W. T. Johnson. 1987. Diseases of trees and shrubs. Cornell University Press, Ithaca and London. 574 pp.

Solomon, J. D., F. I. McCracken, R. L. Anderson, R. Lewis, Jr., F. L. Oliveria, T. H. Filer, and P. J. Barry. 1980. Oak pests: A guide to major insects, diseases, air pollution and chemical injury. U.S.D.A. Forest Service General Report SA-GR11, Washington, D.C. 69 pp.

Stipes, J. R., and R. J. Campana (eds.). 1981. A compendium of elm diseases. American Phytopathological Society, St. Paul, Minn. 96 pp.

Index

Abamectin, 278
Abgrallaspis: ithacae, 229, 365; *pini,* 365
Abies, 352–54
Abiotic disease. *See* Damage, nonparasitic
Abiotic leaf scorch, 192
Abscission, 25, 28
Acacia, 248, 291; diseases, 291; insects, 292
Acacia, 291–92
Acalitus fagerinea, 231
Acantholyda erythrocephala, 434
Acaraben, 278
Acecap, 278, 281; implant, pesticide, 270
Acephate, 278, 281
Acer, 394–404; *macrophyllum,* 396; *negundo,* 57; *platanoides,* 35, 251; *rubrum,* 34, 251; *saccharinum,* 57, 251; *saccharum,* 35, 251
Acid deposition, 206
Acidity, soil, 41
Acid rain, 205–6
Acme Bordeaux Mixture, 284–85
Acremonium diospyri, 423
Acronicta americana, 400, 417, 478
Actias luna, 417, 463
Actinopelte dryina, 409, 452, 463, 472
Adelges: abietis, 459; *cooleyi,* 459–60; *lariciatus,* 383; *laricis,* 383; *piceae,* 228, 354; *tsugae,* 365
Adelgids, 228
Adventitious buds, 24
Aegeria apiformis, 447
Aeration, soil, 34, 36, 48, 177, 184, 190
Aesculus, 374–75; *octandra,* 57
Aesthetic injury levels, 265
Aglaospora anomala, 386
Agrilus: acutipennis, 413; *angelicus,* 413; *anxius,* 219, 306; *bilineatus,* 219, 413; *difficilis,* 389
Agrimycin, 287

Agri-Strep, 287
Agrobacterium: radiobacter, 244, 268; *rhizogenes,* 451; *tumefaciens,* 243, 268, 318, 329, 332, 358, 366, 407, 422, 442, 447, 448, 451, 462, 478
Ailanthus, 467–68; *altissima,* 57
Ailanthus webworm, 468
Air, soil, 33–36
Air pollution, damage from, 204–7
Alaska cedar. *See* False cypress
Albizzia, 455–56
Albizzia psyllid, 292
Alcoholic flux, 244
Alder, 211, 248, 292; diseases, 292–93; insects, 293–94
Alder borer, 294
Alder flea beetle, 293
Alder lace bug, 226, 293, 330
Alder leaf miner, 225
Alder psyllid, 294
Aldicarb, 281, 283
Alebra albostriella, 401
Aleurites, 471–72
Aleurodiscus, 372; *amorphus,* 353, 382; *griseocana,* 383; *oakesii,* 408
Aleurothrixus floccosus, 328
Algal diseases, 312–13
Algal spot, 393
Aliette, 283
Alkalinity, soil, 41
Almond, 249, 251; flowering, 318–24
Alnus, 292–94
Alpine fir. *See* Fir
Alpine larch. *See* Larch
Alsophila pometaria, 223, 347, 400
Alternaria, 480; *catalpae,* 316; *tenuis,* 392
Altica ambiens, 293

Aluminum, 41
Aluminum sulfate, 44
Amathes c-nigrum, 314–15
Ambrosia beetle, 339
Amelanchier, 453–54
American arborvitae, 35. *See also* Arborvitae
American Association of Nurserymen, 6, 13
American basswood. *See* Linden
American beech. *See* Beech
American boxwood. *See* Boxwood
American chestnut. *See* Chestnut
American dagger moth, 400, 417
American elm. *See* Elm
American Forestry Association, 13
American Forests, 14
American holly. *See* Holly
American hornbeam. *See* Hornbeam
American larch. *See* Larch
American linden. *See* Linden
American National Standards Institute, 129
American Nurseryman, 14, 270
American planetree. *See* Sycamore
American plum borer, 321, 355, 386, 441
American Society of Consulting Arborists (ASCA),
    6, 12
Ammonia, 42, 206
Ammonium, 104
Anaerobic soil conditions, damage from, 184
Anastrepha ludens, 357
Anchor, soil as, 47–48
Aneflormopha subpubescens, 415
Angiosperm, 21, 29–30
Anion, 39
Anisota: senatoria, 223, 417; *stigma,* 417; *virgi-
    niensis,* 417
Annual rings, 23
ANSI Z-133, 129
Antheraea polyphemus, 452, 463
Anthophila pariana, 363
Anthracnose, 249–51
Antispila nyssaefoliella, 472
Antitranspirant, 68, 70, 191, 194
Antrodia juniperina, 379
Ants, 484; damage from, 202–3
Aonidiella aurantii, 229
Aphid, 226–28
Aphis: craccivora, 228, 356; *fabae,* 356; *gossypii,*
    448, 449; *pomi,* 324, 330
Aphrophora: parallela, 435; *saratogensis,* 435
Apical meristem, 16
Apiognomonia, 249; *quercina,* 408; *veneta,* 465–66
Apioplagiostoma populi, 446
Apiosporina: collinsii, 454; *morbosa,* 320
Aplospora nyssae, 472
Appearance, and pruning, 115–16

Apple, 35, 187, 249, 251, 258. *See also* Crabapple,
    flowering
Apple and thorn skeletonizer, 363
Apple aphid, 324, 330, 345, 362, 448
Apple bark borer, 454
Apple grain aphid, 362
Apple leaf blotch miner, 362–63
Apple mosaic virus, 304
Apple scab, 329
Apple tree borer, 363
Apricot, 249, 251
Arasan, 287
Araucaria, 407
Araucaria scale, 407
Arbor Age, 14, 270
Arboricultural Journal, 14
Arborists: organizations, 12–14; publications, 14
Arborvitae, 35, 294; abiotic diseases, 294; diseases,
    294; insects and other pests, 295
Arborvitae aphid, 295
Arborvitae leaf miner, 381
Arborvitae weevil, 295, 351, 381
Arbotect 20–S, 283
Arbutus, 461–62
Arceuthobium, 237, 339; *campylopodum,* 354, 430;
    pusillium, 458
Archips, 308; *argyrospilus,* 330, 355; *cerasivor-
    anus,* 322; *fervidanns,* 417; *negundanus,* 294;
    semiferanus, 417
Argyresthia, 381; *cupressella,* 332; *thuiella,* 295
Argyrotaenia: citrana, 292; *pinatubana,* 436
"Aristocrat" flowering pear, 242
Arizona cypress. *See* Cypress
Armillaria, 258, 381, 420, 472; *mellea,* 260, 291,
    299, 316, 321, 328, 332, 334, 353, 358, 372,
    410, 418, 421, 423, 430, 440, 452, 455, 458,
    467, 480
Armillaria root rot, 258–61
Armored scale, 229–30
Aronia, 351
Arrhenodes minutus, 415
Arthropoda, 217
Artificial support, types, 141–45
Ascocalyx abietina, 424
Ascochyta: cornicola, 334; *hansena,* 462; *juglandis,*
    474; *paulowniae,* 350; *salicis,* 477
Ash, 185, 187, 196, 211, 248, 249, 254, 295;
    abiotic diseases, 297; diseases, 295–97; insects
    and other pests, 297–98. *See also* Mountain-
    ash
Ash borer, 267, 297
Ash flower gall, 231, 297–98
Ash midrib gall midge, 298
Ash plant bug, 298
Ash ringspot virus, 236

Ash sawfly, 297
Ash yellows, 295
Asiatic chestnut, 326. *See also* Chestnut
Asiatic garden beetle, 324
Asiatic oak weevil, 327, 387, 417
Asiatic red scale, 455, 483
Aspen, 211, 248. *See also* Poplar
Asphondylia ilicicola, 371
Aspidiotus: californicus, 435; *camelliae,* 451, 462; *hederae,* 229, 451; *juglansregiae,* 354; *liquidambaris,* 463; *perniciosus,* 323–24, 456; *uvae,* 366
Associated Landscape Contractors of America, 6, 13–14
Asterine oleina, 418
Asterolecanium: arabidis, 298; *pustulans,* 328; *puteanum,* 372; *variolosum,* 411
Asteroma: capreae, 477; *tiliae,* 384
Asteromella fraxini, 296
Asterosporium hoffmanni, 302
Aster yellows, 452
Athel tamarisk, 298
Atlas cedar. *See* Cedar
Atropellis: pinicola, 424; *piniphila,* 424; *tingens,* 424
Atteva punctella, 468
Auger: in fertilizer application, 102; soil, 166
Aulacaspis rosea, 355
Australian-pine, 298; diseases, 299; insects, 299. *See also* Pine
Australian-pine borer, 299
Austrian pine. *See* Pine
Automeris io, 223, 347, 417, 467
Autumn frosts, 185
Avermectin B1, 278
Avid, 278
Avocado, 251, 299; diseases, 299; insects, 299
Avocado red mite, 315, 358, 454
Azalea. *See* Rhododendron
Azalea bark scale, 363
Azalea scale, 479
Azalea sphinx moth, 472
Azinphos-methyl, 280
Azodrin, 278

Bacillus: popillae, 266; *thuringiensis,* 266, 278
Backfilling, 80–82
Bacteria, 235; overwintering, 240
Bacterial diseases: cherry, 318–21; flowering crabapple, 328–30; maple, 395–99; mountain-ash, 405
Bacterial leaf spot, 320
Bacterial scorch, 192, 206, 246–47
Bacterial wetwood, 244–46, 406

Bactericides, 283–88
Bactospeine, 278
Bactur, 278
Bagworm, 328, 366, 387
Bakthane, 278
Bald cypress, 299; diseases, 299–300; insects, 300. *See also* Cypress; False cypress
Baliosus ruber, 384
Balkan pine. *See* Pine
Ball and burlapped trees, 79, 82; preparation of hole for, 78; transplanting, 68, 71–75
Balsam fir. *See* Fir
Balsam fir sawfly, 354
Balsam gall midge, 353–54
Balsam twig aphid, 228, 353, 460
Balsam woolly adelgid, 228
Balsam woolly aphid, 354
Banana, 300
Band damage, 203–4
Banded ash borer, 219
Banded oystershell scale, 403
Banrot, 284
Banyan, 300
Bare-root trees, 79, 81–82; preparation of hole for, 78; transplanting, 68, 70–71
Bark, 23; torn from trunk, 146
Bark scale, 332
Barnacle scale, 299, 357, 363
Basic Copper Sulfate, 284
Basi-Cop, 284
Basket willow gall, 478
Basswood. *See* Linden
Basswood leaf miner, 384
Basswood looper, 384
Bauhinia, 418
Bayleton, 284
Baytcx, 278
Bean aphid, 356
Beech, 141, 177, 187, 211, 248, 253, 300; diseases, 300–302; insects, 302–3
Beech bark disease, 301–2
Beech blight aphid, 302
Beech borer, 413, 415
Beech leaf miner, 303
Beech scale, 228, 302
Beet curly top, 452
Beetle, 225–26
Bendiocarb, 281
Beneficial insects, 267–68
Benjamin fig, 352
Benlate, 284
Benomyl, 284
Berry midge, 371
Betula, 303; *papyrifera,* 35; *populifolia,* 35
Bidrin, 278, 280

Bifenthrin, 281
Bifusella: abietis, 353; *faullii,* 353; *linearis,* 425; *striiformis,* 425
Big-cone spruce. *See* Spruce
Biological control of insects and diseases, 265–68
Biotrol, 278
Birch, 35, 177, 211, 248, 249, 253, 258, 293, 303; abiotic disease, 304; diseases, 303–4; insects and other pests, 304–8
Birch aphid, 294
Birch borer, 306
Birch lace bug, 306, 372
Birch leaf miner, 221, 225, 305–6
Birch sawfly, 308
Birch skeletonizer, 306
Black araucaria scale, 407
Black blister beetle, 371
Black carpenter ant, 484
Black cherry, 35, 58, 211. *See also* Cherry
Black citrus aphid, 315, 328
Black gum. *See* Tupelo
Blackheaded budworm, 366
Black knot, 320
Black Leaf Bordeaux Powder, 284–85
Blackline, 475
Black locust. *See* Locust, black; Locust, honey
Black maple, 251. *See also* Maple
Black oak. *See* Oak
Black peach aphid, 324
Black pecan aphid, 367
Black pineleaf scale, 229, 339
Black pine needle scale, 435
Black root rot, 370
Black scale, 229, 300, 303, 311, 314, 324, 328, 330, 357, 358, 372, 380, 387, 392, 393, 403, 406, 418, 423, 447, 448, 462, 467, 476, 479
Black spot, 331
Black spruce. *See* Spruce
Black turpentine beetle, 436–38
Black vine weevil, 315, 480–81
Black walnut, 34, 58. *See also* Walnut
Bladder-gall mite, 402
Bleeding canker, 301, 395–96, 407
Bleeding necrosis, 463
Blister beetle, 455
Blister mite, 230
Blitex 80 MN, 286
Blotch, 303
Blue ash. *See* Ash; Mountain-ash
Blue cypress. *See* Cypress
Blue holly. *See* Holly
Blue spruce, 35. *See also* Spruce
Blumeriella jaapii, 319
Bonide Benomyl, 284
Bonsai, 116

Bordeaux mixture, 284–85
Bordo, 284–85
Bor-dox, 284–85
Borer, 219–21; increment, 167; oak, 413–15
Boron, 40, 41, 96, 108, 109
Bortox, 297
Botryodiplodia hypodermia, 343
Botryosphaeria: dothidea, 296, 299, 316, 329, 332, 334, 343, 355, 358, 368, 384, 398, 440, 442, 448, 450, 451, 463, 468, 472, 477, 479; *obtusa,* 292, 304, 329, 398, 409, 448, 452, 468; *quercuum,* 409; *rhodina,* 407, 466, 468; *ribis,* 449; *tsugae,* 364
Botryosphaeria canker, 440
Botrytis, 313; *cinerea,* 312, 334, 339, 370
Boxelder, 57, 248. *See also* Maple
Boxelder bug, 400–401
Boxwood, 248, 308; diseases, 308–9; insects and other pests, 310–11; nonparasitic diseases, 309–10
Boxwood leaf miner, 221, 310
Boxwood mite, 311
Boxwood psyllid, 310
Boxwood webworm, 310–11
Brachycaudus persicae, 324
Brachys aeruginosus, 303
Bracing: flexible, 143–45; rigid, 141–43
"Bradford" pear, 141
Branch-bark ridge, 125–26
Branch collar, 125–26
Branch shedding, 25
Brasiliomyces trina, 248, 409
Bravo, 285
Breakage, susceptibility to, 141
Bridge grafting, 186
Broadcast fertilizer application, 100–101
Broken limbs, 146
Bromacil, 210
Bromo-O-Gas, 282
Bronze birch borer, 219, 303, 304, 306
Broussonetia, 421–22
Brown elm scale, 349
Brown-headed ash sawfly, 297
Brown leaf spot, 422
Brown rot, 255
Brown soft scale, 299, 357, 462
Brown-tail moth, 400
Brown wood borer, 297, 302, 327, 386
Brozone, 282
Bucculatrix: ainsliella, 417; *canadensisella,* 306
Buckeye, 57, 249. *See also* Horsechestnut
Buckthorn, 248, 311; insects, 311
Bud, 24–25; and pruning, 116
Bud mite, 230
Budworm, 339

Buffalo treehopper, 298, 322, 401
Bull's-eye spot, 396, 474
Bunch disease, 367
Bundle sheath, 27
Burrowing nematode, 300
Bursaphalenchus xylophilus, 430
Bursera, 357
Butterfly, 218, 222, 348
Butternut. *See* Walnut
Butternut woolly worm, 224–25, 475
Buttonwood. *See* Sycamore
Butt rot, 255

Cabling, 141, 143–45
Caeoma faulliana, 353
Cajeput, 311; insects, 312
Calamagrostis, 451
Calaphis betulaecolens, 304
Calcium, 39, 40, 41, 95, 107
Calcium chloride, 210–11
Calcium polysulfide, 286
Caliciopsis pinea, 424
Calico scale, 229, 330, 336, 349, 403, 463, 476, 484
California boxelder, 251. *See also* Maple
California casebearer, 322–23
California laurel, 312; insect pests, 312; leaf diseases, 312. *See also* Laurel
California laurel aphid, 228, 312
California oakworm caterpillar, 358, 417
California prionus borer, 297, 358
California red scale, 229, 292, 300, 311, 314, 328, 330, 358, 372, 394, 407, 413, 418, 448, 476, 479, 483
California tent caterpillar, 224, 298, 322, 449, 462
California tussock moth, 417
Caliroa: cerasi, 225, 322; *lineata*, 416
Callery pear, 422
Calligrapha scalaris, 384
Callirhopalus bifasciatus, 372
Callirytis seminator, 411
Callosamia promethea, 452, 463
Calomycteris setarius, 417
Caloptila syringella, 297
Cambium, 22–23
Cambium miner, 371
Camellia, 312: abiotic diseases, 313–14; fungus and algal diseases, 312–13; insects and other pests, 314–15; virus diseases, 313
Camellia: japonica, 312–15; *sasanqua*, 312–15
Camellia parlatoria scale, 314
Camellia scale, 314
Cameraria cincinnatiella, 222
Camperdown elm. *See* Elm

Camphor scale, 315, 349, 360
Camphor thrips, 315
Camphor-tree, 251, 315; diseases, 315; insects, 315
Camponotus pennsylvanicus, 484
Canadian hemlock, 35. *See also* Hemlock
Canker. *See names of specific trees*
Canker rot, 255
Cankerstain, 438–40
Cankerworm, 203, 223, 327, 347
Capillary water, 36
Capitophorus braggii, 451–52
Capnodium, 331, 394
Captan, 285
Carbamate, 285
Carbaryl, 281
Carbofuran, 280, 282
Carbon, 40
Carbon dioxide, 29; in soil air, 33–34
Carex, 451
Carob, 251, 315
Carolina hemlock. *See* Hemlock
Carolina silverbell. *See* Silverbell
Carpels, 29–30
Carpenterworm, 297, 350
Carpinus, 373–74
Carulaspis: carueli, 380–81; *juniperi*, 229
Carya, 366–68; *illinoensis*, 422
Case bearer, 305, 345
Casparian strip, 18
Castanea, 324–27; *japonica*, 326; *mollissima*, 326, 327; *pumila*, 325
Casuarina, 298–99
Catalpa, 196, 248, 249, 251, 253, 315–16; abiotic disease, 316; diseases, 316; insects, 316–17
Catalpa, 315–17; *bignonioides*, 57, 316; *speciosa*, 57
Catalpa midge, 316, 317
Catalpa sphinx, 317
Caterpillar, 223–24
Cation, 39, 41, 42
Cation exchange capacity, 41
Caucasian wingnut, 479
Caulocampus acericaulis, 225, 402
Cavity filling, 141, 147–48
Cecidomyia: catalpae, 317; *ocellaris*, 402
Cecidophyopsis psilaspis, 483–84
Cecropia moth, 308, 363, 384, 400, 417
Cedar, 211, 317; diseases, 317; insects, 317. *See also* Cedar, incense; False cypress
Cedar, incense, 317–18; diseases, 318; insects and other pests, 318. *See also* Cedar
Cedar, Port Orford, 350
Cedar-apple rust, 376–77
Cedar-apple rust gall, 329
Cedar-hawthorn rust, 360, 377–78

Cedar-quince rust, 360, 377
Cedar tree borer, 450–51
Cedrus, 317–18
Cellulose, 29, 40, 257
Celtis, 358–60
Cenangium: abietis, 353; *ferruginosum*, 424
Cenangium twig blight, 424
Cephaleuros virescens, 291, 313, 393, 454
Cephalosporium, 468
Cerambycid borer, 406–7
Cerataphis variabilis, 421
Ceratobasidium stevensii, 315, 329, 331, 357, 392, 471, 472
Ceratocystis, 198; *coerulescens*, 399, 469; *fagacearum*, 410; *fimbriata platani*, 438; *ips*, 430; *paradoxa*, 421; *pilifera*, 430; *ulmi*, 341
Ceratomia catalpae, 317
Ceratonia, 315
Cercidiphyllum, 381
Cercis, 448–50; *canadensis*, 35
Cercospora, 291, 331, 357, 397; *afflata*, 373; *apiifoliae*, 361; *carii*, 451; *catalpae*, 316; *chionanthi*, 355; *citri-grisea*, 328; *cladosporioides*, 418; *condensata*, 389; *confluens*, 361; *cornicola*, 334; *cydoniae*, 448; *elaeagni*, 451; *fraxinites*, 296; *fuliginosa*, 423; *fusca*, 422; *gigantea*, 328; *glandulosa*, 467; *gymnocladi*, 381; *halesiae*, 456; *ilicicola*, 369; *ilicis*, 369; *juglandis*, 474; *laburni*, 355; *liquidambaris*, 463; *lumbricoides*, 296; *lythracearum*, 331; *maclurae*, 418; *meliae*, 327; *microsora*, 383; *missouriensis*, 406; *moricola*, 406; *nyssae*, 472; *olivacea*, 389; *oxydendri*, 457; *populicola*, 446; *populina*, 446; *psidii*, 357; *pteleae*, 373; *pulvinula*, 369; *rhoina*, 456; *salicina*, 477; *sequoiae*, 332, 379, 450; *spegazzini*, 358; *sphaeriaeformis*, 344; *subsessilis*, 327; *Texensis*, 296; *theae*, 313; *thujina*, 294, 332; *tuberculans*, 463; *viticis*, 318
Cercosporella: celtidis, 358; *chionea*, 449; *mirabilis*, 361; *mori*, 406, 422
Cercospora blight, 379
Ceroplastes floridensis, 331, 453
Cerrena unicolor, 398
Chaenomeles, 448
Chaff scale, 314, 357, 393, 407
Chalara: elegans, 370; *quercina*, 410
Chamaecyparis, 350–52
Chamaedorea seifrizii, 421
Chaste-tree, 318; diseases, 318; insects, 318
Chelated micronutrients, 108–9
Chemical growth retardants, 136
Chemical injuries, 207–13
Chemicals. *See* Fertilizers; Pesticides
Chem-Neb, 286

Chem-Zineb, 287
"Cherokee Chief" dogwood, 35
Cherry, 58, 187, 211, 249, 251, 258, 318; abiotic disease, 321; bacterial and fungus diseases, 318–21; insects and other pests, 321–24; mycoplasma diseases, 321; virus diseases, 321
Cherry leaf miner, 323
Cherry leafroll virus, 345, 475
Chestnut, 248, 251, 253, 258, 324; diseases, 324–27; insects and other pests, 327
Chestnut blight, 324–26; control by hypovirulent virus, 268
Chestnut borer, 218, 301
Chestnut oak. *See* Oak
Chilopsis, 332–33
Chinaberry, 248, 327; diseases, 327; insects, 327
China tree, 251
Chinese elm. *See* Elm
Chinese holly. *See* Holly
Chinese jujube, 327
Chinese juniper. *See* Juniper
Chinese obscure scale, 300
Chinese pistachio, 327
Chinese tallow tree, 327–28
Chinquapin. *See* Chestnut
Chionanthus, 355
Chionaspis: furfura, 405–6; *pinifoliae*, 435
Chipco 26019, 285
Chisel, 167
Chloride ions, 39
Chlorinated hydrocarbons, 282
Chlorine, 40, 206
Chlorobenzilate, 278
Chlorophyll, 27, 40
Chlor-O-Pic, 282
Chloropicrin, 282
Chloroplast, 27
Chloroscypha chloramela, 450
Chlorosis, 178; damage from, 213–14
Chlorothalonil, 285
Chlorpyrifos, 279
Chokeberry, 351
Choke cherry. *See* Cherry
Choristoneura: fumiferana, 460; *occidentalis*, 460; *pinus*, 460; *rosaceana*, 456
Chromaphis juglandicola, 475
Chrysomyxa: arctostaphyli, 458; *chiogenis*, 458; *empetri*, 458; *ledi cassandrae*, 458; *ledicola*, 458; *piperiana*, 458; *roanenis*, 458; *weirii*, 458
Chrysobothris: femorata, 219, 403, 410; *mali*, 403; *tranquebarica*, 299
Chrysoclista linneela, 385
Chrysomela: interrupta, 447, 479; *scripta*, 447, 479
Chrysomphalus: aonidum, 314; *dictyospermi*, 229; *ficus*, 229; *perseae*, 354

Ciboria carunculoides, 406
Ciborinia: camelliae, 312; *whetzelii*, 446
Cicada, periodical, 330, 413
Cimbex americana, 384, 400
Cinara: pseudotsugae, 339; *sabinae*, 379; *strobi*, 431–32; *thujafilina*, 295
Cinders, damage from, 211
Cinnamomum, 315
Citheronia regalis, 452
Citocybe tabescens, 332
Citricola scale, 349
Citrophilus mealybug, 423, 454
Citrus, 328; diseases, 328; insects, 328
Citrus aphid, 315, 372
Citrus blast, 328
Citrus canker, 328
Citrus flatid planthopper, 406
Citrus mealybug, 299, 312, 407, 418, 421, 451, 455
Citrus root weevil, 328
Citrus scale, 328
Citrus thrips, 423
Citrus whitefly, 393, 418, 468
Cladosporium, 382; *effusum*, 422; *epiphyllum*, 386; *fasciculatum*, 392
Cladrastis, 35, 43, 479–80; *lutea*, 35
Clasterosporium cornigerum, 373
Clastoptera: arizonana, 292; *juniperina*, 381
Clay, 38–39
Climacodon septentrionalis, 367, 398
Climate zone, 55
Clitocybe root rot, 315
Clitocybe tabescens, 260, 291, 299, 328, 331, 334, 357, 358, 372, 442, 449, 455, 472
Coal cinders, damage from, 211
Coccomyces hiemalis, 319
Coccus elongatus, 419; *hesperidum*, 357, 407, 462
Cockscomb gall, 345
Cockspur hawthorn. *See* Hawthorn
Coconut palm, 419
Cocoon, 219
C-O-C-S, 285
CODIT, 257
Codling moth, 267
Cold, damage from, 184–91
Cold-hardiness, 53–55, 186–87
Cold lightning, 194
Coleophora: caryaefoliella, 367; *fuscedinella*, 305; *laricella*, 383; *sacramenta*, 322–23, 479; *ulmifoliella*, 345
Coleosporium, 427; *asterum*, 425–27; *pinicola*, 427
Colletotrichum, 308, 374, 392, 418; *gloeosporioides*, 299, 303, 312, 334
Colloidal clay, 39
Collybia, 258; *velutipes*, 375

Colopha: ulmicola, 345; *ulmisaccula*, 345
Colorado blue spruce, 35. *See also* Spruce
Columbian timber beetle, 415
Columnar shape, 116
Comandra blister rust, 425
Commercial fertilizers, 97–98
Common birch aphid, 304
Compaction, soil, 48, 51, 154, 178–79
Companion cells, 23
Components, soil, 33–40
Comptonia, 350
Comstock mealybug, 228, 311, 316–17, 324, 328, 349, 372, 393, 403, 407, 418, 423, 483
Cones, 30
Coniophora, 353; *puteana*, 365
Coniothyrium, 343, 422; *fuckelii*, 392; *insitivum*, 468; *ulmi*, 344
Conk, 255, 258
Construction damage, 155, 175–81
Construction sites, preservation at, 148–55
Container-grown trees, 75–80, 183, 185
Contarinia: candensis, 298; *juniperina*, 380
Controlled-release fertilizers, 98–99
Cooley spruce gall adelgid, 459–60
Copper, 40, 108, 109
Copper Bordo, 284–85
Copper 53 Fungicide, 284
Copper Hydro Bordo, 284–85
Copper hydroxide, 286
Copper oxychloride, 285
Copper oxychloride sulfate, 285
Copper spray, 200
Copper sulfate, 284; damage from, 213
Coprantol, 285
Coral spot canker, 356–57
Cork, 23
Cork cambium, 23
Cork tree, 251, 328
Cornelian cherry, 34. *See also* Cherry
Cornus, 333–39; *florida*, 35, 55, 333; *mas*, 34
Coromate, 285
Cortex, 18, 23
Corthylus columbianus, 415
Corticium galactinum, 410
Corylus, 472
Corynebacterium ilicis, 368
Coryneum: cardinale, 318; *kunzei*, 409; *thujinum*, 294; *tumoricola*, 344
Corythucha: arcuata, 413; *bellula*, 363; *ciliata*, 226, 441, 467; *cydoniae*, 363; *juglandis*, 226, 475–76; *mollicula*, 479; *pallipes*, 306; *pergandei*, 226, 293; *ulmi*, 346
Cosetacus camelliae, 315
Cotinus, 456
Cotoneaster, 241

Cotton aphid, 448, 449

Cottonwood, 35, 211. *See also* Poplar

Cottonwood leaf beetle, 447

Cottonwood twig borer, 447

Cottony-cushion scale, 229, 292, 299, 303, 311, 324, 328, 330, 332, 354, 360, 372, 387, 393, 403, 413, 418, 419, 423, 448, 476, 479

Cottony maple leaf scale, 228, 372, 403

Cottony maple scale, 229, 303, 311, 324, 330, 336, 349, 360, 363, 386, 387, 392, 403, 406, 407, 413, 418, 447, 467, 479

Cottony taxus scale, 314, 372, 403, 483

Council of Tree and Landscape Appraisers, 6, 14

Cowpea aphid, 228, 328, 356, 358, 392

Crabapple, flowering, 35, 187, 248, 253, 328; abiotic disease, 330; bacterial and fungus diseases, 328–30; insects and other pests, 330. *See also* Apple

Crack, frost, 187–88

Crapemyrtle, 248, 330–31; diseases, 331; insects, 331

Crapemyrtle aphid, 331

Crataegus: crusgalli, 361, 362–63; *lavallei*, 35; *phaenopyrum*, 35, 361

Criconema, 311

Criconemoides, 311, 484

Crimean linden, 251. *See also* Linden

Cristulariella: depraedeus, 396; *moricola*, 396; *pyramidalis*, 474

Croesia purpurana, 417

Croesus latitarsus, 308

Cronartium: comandrae, 425; *comptoniae*, 428; *quercuum*, 409, 427–28; *quercuum fusiforme*, 428; *ribicola*, 428

Cross-pollination, 30–31

Crotch: and artificial support, 140; bracing, 142

Crown gall, 243–44, 268, 318, 320, 329, 332, 352, 358, 366, 407, 422, 447, 448, 451, 462, 478

Cryphonectria: cubensis, 358; *parasitica*, 326

Cryptococcus, 303; *fagisuga*, 228, 301, 302

Cryptodiaporthe: castanea, 327; *populea*, 444; *salicella*, 477

Cryptomeria, 331; diseases, 331

Cryptomeria, 331

Cryptomyces maximus, 477

Cryptorhynchus lapathi, 478

Cryptosphaeria canker, 445

Cryptosphaeria populina, 445

Cryptospora longispora, 407

Cryptosporiopsis, 398

Cryptosporium macrospermum, 353

Cryptostictis, 334; *arbuti*, 462

Cuban laurel thrips, 328

Cuban may beetle, 418

Cucumbertree, 331; diseases, 331

Cultivated quince. *See* Quince, flowering

Cupressus, 331–32

Cup shakes, 188–89

Curculio auriger, 327

Curculio proboscideus, 327

Cuscuta, 389

Cut, pruning, 125–29

Cuticle, 26

Cutworm, 314–15

Cyclaneusma minus, 425

Cycloconium oleaginum, 418

Cygon, 278–79

Cylindrocarpon, 353

Cylindrocladium, 369; *macrosporium*, 420; *scoparium*, 430, 469

Cylindrosporium, 398; *betulae*, 303; *brevispina*, 361; *cercosporiodes*, 469; *crataegi*, 361; *dearnessi*, 372; *defoliatum*, 358; *fraxini*, 296; *juglandis*, 474; *microspilum*, 409; *salicinum*, 477; *solitarium*, 386; *tenuisporium*, 344

Cynthia moth, 384, 468

Cypress, 34, 331–32; diseases, 332; insects, 332. *See also* Bald cypress; False cypress

Cypress aphid, 332

Cypress bark beetle, 318

Cypress mealybug, 318, 332, 407, 451

Cypress moth, 300, 332, 366

Cypress scale, 318

Cypress tip moth, 318, 332, 381

Cypress webber, 295

Cyprex, 285

Cyrtepistomus castaneus, 417

Cythion, 280

Cytophoma pruinosa, 295

Cytospora, 302, 364, 398, 406, 424; *abietis*, 353; *ailanthi*, 468; *ambiens*, 343; *annulata*, 296; *cenisia littoralis*, 332; *chrysosperma*, 405, 444, 477; *cincta*, 320; *gleditschiae*, 389; *leucostoma*, 320; *massariana*, 405; *microspora*, 405; *pini*, 353; *sophorae*, 419

Cytospora canker, 134, 239, 332, 444–45, 457, 477

Daconil 2787, 285

Daedalea, 478; *ambigua*, 389; *elegans*, 389

Dagger moth, 478

Dagger nematode, 236

Damage, nonparasitic: from air pollution, 204–7; from chemicals, 207–13; chlorosis, 213–14; from construction, 148–55, 175–81; from gas, 183–84; from girdling roots, 181–83; from hail, 196; from leaf scorch, 192–94; from lichens, 198–200; from lightning, 194–96; from low temperature, 184–91; mechanical, 196–98; from rodents, 201; from sapsuckers, 201–2; to sidewalks, 62; from sunscald, 191–92; from termites and ants, 202–3; from tree bands, 203–4; from vines, 200–201

Dasanit, 282

Dasineura gleditschiae, 390

Dasychira plagiata, 438

Dasyscypha: ellisiana, 339, 382; *pini,* 424; *pseudotsugae,* 339

Datana: integerrima, 475; *ministra,* 223, 415–16

Date palm. *See* Palm

Dawn redwood, 451. *See also* Redwood

D-D Soil Fumigant, 282

Dead man's finger fungus, 321, 389

Decay, 255–58

Decurrent shape, 116

Defoliation, 206

Degenerate scale, 314

Dehorning, 120

Delonix, 442

Dendroctonus, 432; *terebrans,* 436–38

Deodar weevil, 317, 436

Dermatea: balsamea, 364; *livida,* 450

Dermea pseudotsugae, 339

Desert willow, 332–33. *See also* Willow

Desiccation, 189

Desmia funeralis, 449

Determination of tree value, 7–12

Deuterophoma ulmi, 343

Diagnosis: girdling roots, 183; importance, 161–63; procedures, 164–66; questionnaire, 170–73; tools, 166–69; and tree diagnostician, 159–61

Diapheromera femorata, 417

Diaphnocoris chlorionis, 226, 391

Diaporthe, 398; *alleghaniensis,* 304; *ambigua,* 375; *ancostoma,* 386; *cinerescens,* 352; *citri,* 328; *crustosa,* 368; *eres,* 368; *scabra,* 466

Diaspis carueli, 352

Diatrypella oregonensis, 292

Diazinon, 279

Dibrom, 279

Dicamba, 207, 208

Dicerca divaricata, 404

Dichomeris: ligulella, 322: *marginella,* 381

Dicofol, 280

Dicrotophos, 278

Dictyospermum scale, 229, 292, 299, 314, 328, 358, 372, 407, 418, 452, 479, 483

Didymascella chamaecyparissi, 350

Didymosphaeria oregonensis, 292

Didymosporina, 398

Didymosporium arbuticola, 462

Dieback, 178. *See also diseases of specific trees*

Diflubenzuron, 279

Digging, 70–76

Dimethoate, 278

Dimilin, 279

Dinocap, 286

Dioryctria: abietella, 295; *zimmermani,* 436

Diospyros, 423; *virginiana,* 57, 423

Dipel, 278

Diplocarpon mespili, 360, 405, 448, 454

Diplodia, 368, 454; *infuscans,* 296; *longispora,* 409; *pinea,* 425; *sophorae,* 419

Diplodia tip blight, 134

Diplodina, 477

Diplopeltis sassafrasicola, 452

Diprion: hercyniae, 460–61; *similis,* 434

Dipterex, 279

Discella carbonacea, 477

Discosporium populea, 444

Discula, 249, 295, 334, 396; *betulina,* 304; *quercina,* 408

Disease: agents, 234–40; biological control, 268; development factors, 240; symptoms, 233–34; pruning for control, 134–35. *See also* Parasitic disease

Dismicoccus wistariae, 228

Disparlure, 267, 348

Disulfoton, 279

Di-Syston, 279

Dithane M-22, 286

Dithane M-45, 286

Dithane Z-78, 287

Dodder, 389

Dodine, 285

Dog canker, 441

Dogwood, 35, 177, 248, 333; diseases, 333–35; insects, 335–39

Dogwood borer, 198, 219, 267, 321–22, 330, 335, 367–68, 415, 422–23

Dogwood club gall, 335–36

Dogwood sawfly, 339

Dogwood scale, 336

Dogwood twig borer, 335, 347

Dolgo crabapple, 35. *See also* Crabapple, flowering

Dorcaschema wildii, 406–7

Dormant oil sprays, 279

Dormant twig spray, 108–9

Dothichiza canker, 443–44

Dothiorella, 381, 389, 406, 407, 440, 463; *fraxinicola,* 296; *gregaria,* 440; *mori,* 406; *phomiformis,* 409; *quercina,* 409

Dothistroma septospora, 425

Douglas-fir, 339; diseases, 339; insects, 339–40. *See also* Fir

Douglas-fir beetle, 339

Douglas-fir tussock moth, 383

Dowfume MC-2, 282

Dowfume MC-33, 282

Downy spot, 422

Drainage, soil, 36, 48, 78

Drench, fertilizer application, 101

Drepanaphis, 375; *acerifoli*, 402

Drepanosiphum platanoides, 467

Dressings, pruning wounds, 129–30

Drill holes, for fertilizer application, 101–3

Drought, 38, 206, 365

Drummond maple, 251. *See also* Maple

Dryocampa rubicunda, 401

Dursban, 279

Dusky birch sawfly, 308

Dutch elm disease, 239, 340–43; resistant hybrids, 343

Dwarf mistletoe, 237, 430

Dycarb, 281

Dylox, 279

Dynaspidiotus britannicus, 372

Dysmicoccus wistariae, 482–83

Eacles imperialis, 332, 352

Eastern gall rust, 427–28

Eastern hemlock. *See* Hemlock

Eastern red cedar. *See* Cedar; Juniper

Eastern subterranean termite, 357, 484

Eastern tent caterpillar, 224, 298, 308, 322, 330, 401, 478

Ecdytolopha insiticiana, 387

Echinodontium, 258, 299; *tinctorum*, 365

Ectotrophic mycorrhizae, 19

Edema, 358

Edwardsiana roseae, 363, 401

Ehrhornia cupressi, 332

Elaeagnus, 451–52

Elaphinoides villosus, 368

Elatobium abietinum, 228, 460

Elder, 57

Elm, 141, 177, 185, 196, 211, 248, 249, 251, 253, 340; abiotic disease, 345; diseases, 340–45; insects and other pests, 345–50

Elm bark beetle, 341, 346–47

Elm borer, 347

Elm calligrapha, 294, 350, 384

Elm case bearer, 345

Elm cockscomb gall, 345

Elm lace bug, 346

Elm leaf aphid, 345

Elm leaf beetle, 225, 346

Elm leaf miner, 225, 346

Elm Research Institute, 270

Elm sawfly, 350, 384, 400

Elm scurfy scale, 349

Elm spanworm, 223, 302, 308, 347, 367, 384, 400

Elm spider mite, 350

Elm yellows, 344

Elsinoe: corni, 334; *fawcetti*, 328; *floridae*, 334; *ilicis*, 369; *magnoliae*, 393; *mangiferae*, 394; *mattirolianum*, 462; *quercus-falcatae*, 409

Elytroderma deformans, 425

Empoasca, 375; *fabae*, 308

Empress-tree, 58, 196, 258, 350; diseases, 350

Enaphalodes rufulus, 415

Enchenopa binotata, 373

Encoelia pruinosa, 445

Endocronartium harknesii, 428

Endodermis, 18

Endosulfan, 281

Endothemia aemula, 460

Endothia gyrosa, 302, 408; *parasitica*, 268, 326, 408

Endotrophic mycorrhizae, 19

Englerulaster orbicularis, 369

English elm, 251. *See also* Elm

English hawthorn, 241. *See also* Hawthorn

English holly. *See* Holly

English walnut, 243. *See also* Walnut

English yew, 243. *See also* Yew

Ennomos subsignarius, 223, 347

Enterobacter cloacae, 246

Entomosporium mespili, 360

Environment: effect of changing, 154–55; importance of diagnosis, 161

Eotetranychus, 417; *hicoriae*, 368; *matthyssei*, 350

Epargyreus clarus, 224

Epicoccum: nigrum, 392; *purpurascens*, 355

Epicormic growth, 175, 178

Epidermis, 18, 26

Epinotia: aceriella, 402; *nanana*, 460; *subviridis*, 295, 332

Equipment: for applying pesticides, 270–74; for pruning, 123–25

Erannis tiliaria, 223, 384, 417

Erineum, 231

Eriocampa juglandis, 224–25, 475

Eriococcus araucariae, 407

Eriophyes: betulae, 306–8; *calaceris*, 231; *celtis*, 231, 359; *elongatus*, 231, 402; *erinea*, 476; *fraxinivorus*, 231; *fraxinoflora*, 297; *pyri*, 405; *tiliae*, 384–85

Eriophyid mite, 230–31, 294, 303, 315

Eriosoma: americana, 345; *crategi*, 345; *langninosa*, 345; *lanigerum*, 228, 345; *pyricola*, 345; *rileyi*, 330, 345

Erwinia: amylovora, 242, 328–29, 448, 453–54; *nigrifluens*, 474; *nimmipressuralis*, 246; *rubrifaciens*, 474

Erysiphe: aggregata, 248, 292; *cichoracearum*, 248; *lagerstroemiae*, 248, 331; *liriodendri*, 248; *polygoni*, 248, 291, 386, 454, 469

Erythronema, 449

Essential elements. *See* Nutrient elements
Ethion, 279
Ethoprop, 282
Ethylene dibromide, 327
Etridiazole, 284, 287
Eucalyptus, 248, 258
Eucalyptus, 357–58
Eucalyptus longhorn borer, 358
·*Euceraphis betulae,* 294, 304
Eucommia, 360
Eucosma gloriola, 436
Eulachnus, 432
Euonymus scale, 314, 324, 372, 418
European alder leaf miner, 225, 294
European ash, 251. *See also* Ash; Mountain-ash
European beech. *See* Beech
European birch aphid, 304
European elm bark beetle, 346–47
European elm scale, 349
European fruit lecanium scale, 295, 303, 311, 330,
    349, 363, 386, 393, 418, 423, 447, 449, 453
European hackberry. *See* Hackberry
European hornbeam borers, 413
European juniper. *See* Juniper
European larch. *See* Larch
European larch canker, 381
European linden. *See* Linden
European linden bark borer, 385
European littleleaf linden, 35. *See also* Linden
European mountain-ash, 35. *See also* Mountain-ash
European peach scale, 324, 349, 452
European pine shoot moth, 432–33
European red mite, 230, 330
European white birch. *See* Birch
Eurytetranychus buxi, 311
Eutetrapha tridentata, 347
Euthoracaphis umbellulariae, 228, 312
Eutypella canker, 398
Eutypella parasitica, 398
Euzophera semifuneralis, 441
Evaluation of trees, 6
Evergreen, winter drying of, 189–91
Excavation: damage from, 178; and tree preserva-
    tion, 148–50
Excurrent shape, 116
Exobasidium: camelliae, 313; *vacinii,* 462
Exophoma magnoliae, 392
Exosporina fawceti, 473
Exosporium liquidambaris, 463
Exotelia pinifoliella, 435
Eyebolt installation, bracing, 145

Fabrella thujina, 294; *tsugae,* 364
Fagus, 300

Fall cankerworm, 223, 302, 347, 367, 400
Fall webworm, 224, 294, 297, 322, 350, 367, 468,
    475
False caterpillar, 218
False cypress, 350; diseases, 350–51; insects, 351–
    52. *See also* Bald cypress; Cedar; Cypress
Fascista cercerisella, 450
Fenarimol, 287
Fensulfothion, 282
Fenthion, 278
Fenusa: dohrnii, 225, 294; *pusilla,* 221, 225, 305;
    ulmi, 225, 346
Ferbam, 285
Fermate, 285
Fermate Ferbam Fungicide, 285
Fertility, soil, 41
Fertilizer: application methods, 100–104; applica-
    tion rates, 104–8; benefits, 91–92; determining
    need for, 92–94; foliar application, 108–9; for-
    mulations, 96–99; for newly transplanted trees,
    113; nutrient elements, 94–96; postplanting,
    88–89; timing, 113–14; treatment area, 99–
    100; trunk implants and injections, 109–12
Ficam, 281
Ficus, 352; *benghalensis,* 300
Fig, 352
Filbert, Turkish, 472
Fills, soil, damage from, 175–77
Filmfast, 288
Fiorinia: externa, 229, 365; *theae,* 314
Fiorinia hemlock scale, 229
Fir, 189, 211, 352; diseases, 352–53; insects and
    other pests, 353–54. *See also* Douglas-fir
Fire blight, 239, 241–43, 328–29, 360, 422, 453
Firewheel tree, 354
Fir flatheaded borer, 366
Fitness, 53
Flamboyant-tree. *See* Poinciana
Flame-tree. *See* Poinciana
Flatheaded apple tree borer, 297, 363, 423
Flatheaded borer, 219, 299, 330, 335, 386, 403–4,
    410
Flea beetle, 293, 314
Fletcher scale, 483
Flexible bracing, 143–45
Flooding, soil, 34–35
Florida red scale, 229, 314, 328, 357, 407
Florida wax scale, 314, 331, 357, 363, 393, 407,
    453
Flowering almond. *See* Cherry
Flowering cherry. *See* Cherry
Flowering crabapple. *See* Crabapple, flowering
Flowering dogwood. *See* Dogwood
Flowering peach, 35. *See also* Cherry
Flowering pear. *See* Pear

Flowering plum. *See* Cherry

Flowering prunus. *See* Cherry

Flowering quince. *See* Quince, flowering

Flowers, 30; and pruning, 118

Fluorides, 206

Foliage-feeding beetle, 225–26

Foliage-feeding caterpillar, 222–24

Foliar analysis, 93–94

Foliar application, fertilizer, 108–9

Folpet, 287

Fomes, 258, 297, 299, 389; *extensus,* 300; *fasciatus,* 393; *fomentarius,* 302, 304; *geotropus,* 300, 393; *meliae,* 355; *officinalis,* 430; *roseus,* 430

Fomitopsis, 258; *pinicola,* 458

Fore, 286

Forest tent caterpillar, 224, 298, 322, 350, 401

Formaldehyde, 254

Formulations, fertilizer, 96–99

Fosetyl-Al, 283

Four-lined plant bug, 226, 400–401

Four-spotted hawthorn aphid, 362

Four-spotted mite, 350

Franklinia, 43; *alatamaha,* 354

Franklin-tree, 43, 354; diseases, 354; insects, 354–55

Fraser fir. *See* Fir

Fraxinus, 295–98; *americana,* 34, 57

Fringe-tree, 67, 355; diseases, 355; insects, 355

Frost cankers, 189

Frost cracks, 187–88

Frosted scale, 305, 324, 330, 349, 363, 387, 406

Frost injury, 184–85

Frozen soil ball, 75

Fruit, 30; parthenocarpic, 30; and pruning, 118

Fruiting body, 255

Fruiting mulberry. *See* Mulberry

Fruitless mulberry. *See* Mulberry

Fruit tree leaf roller, 315, 330, 355, 367, 475

Fuller rose beetle, 314, 324, 328

Fumago vagans, 357

Function: of leaves, 29; of root system, 19; and tree selection, 55; of stems, 25–26

Fungicides, 283–88; damage from, 213

Funginex, 286

Fungus, 235–36, 312–13; and leaf injuries, 164; overwintering, 240

Fungus diseases: camellia, 312–13; cherry, 318–21; flowering crabapple, 328–30; London plane-tree, 437–40; maple, 395–99; mountain-ash, 405; tuliptree, 468–69

Furadan, 280, 282

Fusarium, 420, 451; *buxicola,* 308; *discolor,* 430; *lateritium,* 291, 355–56, 373, 419, 467; *moniliforme subglutinans,* 424; *oxysporum*

perniciosum, 455; *solani,* 398, 408, 421, 445, 468, 472

Fusarium canker, 445

Fusicladium dendriticum, 329

Fusicoccum, 405; *elaeagni,* 451

Fusiform rust, 428

Galasa nigrinodis, 310–11

Gall, 231, 243–44

Gall midge, 298, 353–54

Ganoderma, 258; *applanatum,* 291, 300, 304, 367, 398, 423; *lucidum,* 359, 365, 381, 389, 398, 410, 449, 455; *zonatum,* 420

Ganoderma rot, 359, 398

Garden plum, 251. *See also* Plum

Garmania bulbicola, 410

Gas injury, 183–84

Giant bark aphid, 228, 302, 304, 327, 345, 367, 447, 467, 475, 478

Giant hornet, 308, 339, 354–55

Gibberella: baccata, 291, 356, 373, 406, 419, 467; *baccata moricola,* 406

Ginkgo, 141, 253, 355; diseases, 355; insects and other pests, 355

Ginkgo, 355

Girdling roots, 67, 80, 181–83

Girdling wires, 196–97

Gleditsia, 389–92; *triacanthos,* 57; *triacanthos inermis,* 34

Gliocladium vermoeseni, 421

Globifomes graveolens, 410

Globose scale, 324, 406

Gloeosporium, 249, 315, 442, 456; *ailanthi,* 467; *aquifolii,* 369; *aridum,* 295; *catalpae,* 316; *crataegi,* 361; *fagi,* 302; *ilicis,* 368; *inconspicuum,* 344; *liriodendri,* 469; *musarum,* 300; *nervisequam,* 463; *revolutum,* 386; *septorioides,* 409; *ulmicolum,* 344; *umbrinellum,* 409

Glomerella: cingulata, 303, 312, 315, 326, 328, 355, 357, 374, 392, 394, 452; *miyabeana,* 476–77

Gloomy scale, 229, 311, 349, 360, 372, 403

Glover scale, 295, 311, 314, 393, 407, 479

Glycobius speciosus, 404

Glyphosate, 208–9

Gnomonia, 249; *caryae,* 366; *leptostyla,* 474

Gnomoniella fimbriella, 373

Goes: debilis, 415; *pulverulentis,* 415; *tigrinus,* 415

Golden arborvitae. *See* Arborvitae

Golden-chain, 355; diseases, 355–56; insects and other pests, 356

Golden-leaved beech. *See* Beech

Golden mealybug, 299, 407

Golden oak scale, 411

Goldenrain-tree, 187, 251, 258, 356; diseases, 356–57; insects, 357

Gossyparia spuria, 349

Gouge, 167

Gouty gall, 411

Grade change damage, 175–81; preventing, 150–54

Graft: bridge, 186; roots, 21

Grape colaspis, 314

Grape leaf folder, 449

Grape mealybug, 355, 356, 454, 483

Grape scale, 366, 368, 467

Graphiola phoenicis, 420

Graphiolitha molesta, 322

Grasshopper, 219

Gray birch, 35, 211. *See also* Birch

Great basin tent caterpillar, 266

Greedy scale, 229, 292, 314, 324, 327, 330, 357, 358, 372, 380, 387, 393, 407, 418, 423, 447, 448, 449, 451, 462, 476, 479

Green ash, 251. *See also* Ash

Green fruitworm, 417, 450

Greenhouse effect, 3

Greenhouse leaf tier, 315

Greenhouse thrips, 328, 358

Greenhouse whitefly, 315, 328, 450

Green mapleworm, 302, 339, 367, 400

Green peach aphid, 228, 315, 324, 421

Green ring mottle virus, 321

Green shield scale, 300, 357

Green spruce aphid, 228, 460

Greenstriped mapleworm, 401

Gregarious oak leaf miner, 222

Gremeniella, 424

Grevillea, 454

Grevillea mite, 454

Ground mealybug, 311, 421

Grounds Maintenance, 14, 270

Growth: chemical retardants of, 136; epicormic, 175, 178; essential elements of, 40–41; leaf, 27–29; and pruning, 118; root, 18–19; stem, 23–25; subnormal, and fertilizer need, 92–93

Grub, 218

Guadalupe cypress. *See* Cypress

Guard cells, 26

Guava, 357; diseases, 357; insects and other pests, 357

Guide for Establishing Values of Trees and Other Plants, 5–6

Gumbo-limbo, 357; diseases, 357; insects, 357

Gum-tree, 357; diseases, 358; insects and other pests, 358

Guthion, 280

Guying, 145; after transplanting, 83–85

Gymnocladus, 381

Gymnosperm, 21, 27, 30–31

Gymnosporangium, 376; *aurantiacum,* 405; *clavipes,* 360, 377, 448; *cornutum,* 405; *ellissii,* 350–51; *fraternum,* 351; *globosum,* 360–61, 378, 405; *juniperi-virginianae,* 329, 376–77; *libocedri,* 318, 405, 448; *nelsoni,* 405; *nidusavis,* 378; *nootkatense,* 405; *tremelloides,* 405

Gypchek, 280

Gypsonoma haimbachiana, 447

Gypsy moth, 222–23, 267, 294, 347–48, 363, 366, 383, 384, 400, 452

Gyrostroma austro-americana, 389

Hackberry, 141, 248, 249, 253, 358; diseases, 358–59; insects and other pests, 359–60

Hackberry nipple-gall, 359–60

Haematostereum, 353; *sanguinicola,* 458; *sanguinolenta,* 430

Hail injury, 196

Halesia, 43, 456

Halisidota: harrisii, 467; *tessellaris,* 468

Hamamelistes spinosus, 304–5

Hanson Manchurian elm. *See* Elm

Hardiness zone, 55

Hardpan, 51

Hardy rubber tree, 360

Hawthorn, 248, 249, 253, 360; diseases, 360–62; insects and other pests, 362–63

Hawthorn aphid, 345, 362

Heading back, 120–22

Health, pruning for, 115

Heart rot, 255

Heartwood, 21

Hedge maple, 251. *See also* Maple

Helicotylenchus, 311

Heliothis virescens, 454

Heliothrips: dracaenae, 421; *haemorrhoidalis,* 299, 402, 421

Hemerocampa: leucostigma, 468; *vetusta,* 349, 417

Hemiberlesia: latania, 229; *rapax,* 229

Hemicriconemoides gaddi, 315

Hemispherical scale, 314, 328, 357, 423

Hemlock, 177, 211, 363–64; abiotic disease, 365; diseases, 364–65; insects and other pests, 365–66

Hemlock borer, 365

Hemlock eriophyid mite, 366, 461

Hemlock fiorinia scale, 229, 365

Hemlock looper, 302, 354, 365

Hemlock sawfly, 366

Hemlock scale, 229, 365, 460

Hemlock woolly aphid, 365

Hendersonia, 358; *crataegicola,* 361; *magnoliae,* 392

Hendersonula toruloidea, 473

Herbicides, 207–10
Hercinothrips femoralis, 421
Hericium, 258; *erinaceus,* 302, 410
Herpotrichia nigra, 318, 480
Heterobasidion, 258; *annosum,* 379, 382, 430
Heterocampa: guttivitta, 400; *manteo,* 417
Heterosporium iridis, 453
Hexakis, 281
Hickory, 177, 196, 211, 248, 249, 253, 366; diseases, 366–67; insects, 367–68
Hickory bark beetle, 367
Hickory horned devil, 367, 452, 475
Hickory leaf stem gall aphid, 367
Hickory lecanium scale, 229, 303, 305, 324, 349, 360, 392, 407, 476, 479
Hickory tussock moth, 367, 384, 475
Hicks' yew, 35
Holly, 189, 248, 249, 253, 368; diseases, 368–71; insects and other pests, 371–72; nonparasitic diseases, 371
Holly leaf miner, 221, 371
Holly scale, 372
Homadaula anisocentra, 455
Homaledra sabalella, 421
Honey locust. *See* Locust, black; Locust, honey
Honey locust borer, 389–90
Honey locust plant bug, 226, 387, 391
Honey mushrooms, 259
Hop-hornbeam, 211, 248, 254, 372; diseases, 372; insects, 372
Hop-tree: diseases, 373; insects, 373
Hormones, 80, 130
Hornbeam, 248, 249, 254, 373; diseases, 373–74; insects, 374
Horned gall, 411
Hornet moth, 447
Hornet wasp, 354–55
Horntails, 404
Horsechestnut, 141, 187, 196, 249, 251, 374; diseases, 374–75; insects, 375
Hot bolts, lightning, 194
Howardia biclavis, 357
Humus, 41
Hurricane, 212
Hyalophora cecropia, 308, 417
Hyalospora aspidiotus, 353
Hydrated lime, 284
Hydraulic sprayers, 270–73
Hydrogen, 39, 40
Hydrogen chloride, 206
Hydrogen fluoride, 206
Hydrogen sulfide, 184
Hydroxyl ions, 39
Hygroscopic water, 36
Hylobius: pales, 436; *radicis,* 434–35

Hylurgopinus rufipes, 341, 346
Hymenochaete agglutinans, 292, 364
Hyphae, 235, 257
Hyphantria cunea, 224, 297, 468
Hypoderma: desmazierii, 425; *hedgecockii,* 425; *lethale,* 425; *pedatum,* 425; *pini,* 425
Hypodermella, 425; *laricis,* 382; *mirabilis,* 353; *nervata,* 353
Hyponectria buxi, 308
Hypoxylon, 375; *atropunctatum,* 408; *mammatum,* 445; *tinctor,* 466
Hypoxylon canker, 408, 445
Hyssop crabapple. *See* Crabapple, flowering

Icerya purchasi, 229, 292, 419
Ice storm, 196
Ichthyura inclusa, 447
Ilex, 368–72
Imbibition, 36
Imidan, 224, 280
Imperial moth caterpillar, 302, 332, 352
Implants, trunk: fertilizer, 109–12; pesticide, 269–70
Imported willow leaf beetle, 479
Incense cedar. *See* Cedar, incense
Inchworm, 347, 400
Incomplete metamorphosis, 219
Increment borer, 167
Industrial air pollutants, 206
In-field container-grown trees, 75–76
Inject-A-Cide, 280
Inject-A-Cide B, 280
Injection: fertilizer application by, 103–4; of pesticides, 269–70; trunk, 109–12
Injury. *See* Damage, nonparasitic
Inoculum, 242
Inonotus, 258; *andersonii,* 367, 410; *circinans,* 430; *dryadeus,* 410; *hispidus,* 408; *obliquus,* 304; *tomentosus,* 458
Insect(s): beneficial, 267–68; biological control, 265–68; and leaf injuries, 164; life cycle, 218–19; structure, 217–18. *See also* Insecticides; Insect types; Pest control; Pesticides
Insecticides, 278–82; damage from, 213
Insect types: aphids, 226–28; borers, 219–21; foliage-feeding beetles, 225–26; foliage-feeding caterpillars, 222–24; lace bugs, 226; leaf miners, 221–22; plant bugs, 226; sawflies, 224–25; scales, 228–30. *See also* Insect(s); Insecticides; Pest control; Pesticides
Integrated pest management (IPM), 264–65
International Society of Arboriculture (ISA), 6, 8, 13
Inversion layers, atmosphere, 204

Io moth, 223, 298, 302, 347, 400, 417, 452, 467
Iprodione, 285
Ips, 432
Irene araliae, 392
Iron, 40, 41, 96, 107–8, 109, 214
Iron chlorosis, 410
Ironwood, 372
Irpex mollis, 408
Irregular pine scale, 435
Irrigation, 38
Isobutylidene diuren (IBDU), 105
Isthmiella crepidiformis, 457
Ivy scale, 229, 299, 311, 328, 336, 358, 403, 407, 418, 423, 449, 452

Jacaranda, 375
Jacaranda, 375
Jack pine, 212. *See also* Pine
Jack pine budworm, 460
Jack-pine sawfly, 434
Janus abbreviatus, 479
Japanese apricot, 251. *See also* Cherry
Japanese beetle, 225–26, 266–67, 331, 350, 371, 383, 384
Japanese black pine. *See* Pine
Japanese cedar. *See* Cryptomeria
Japanese flowering cherry, 68. *See also* Cherry
Japanese holly. *See* Holly
Japanese keaki. *See* Zelkova, Japanese
Japanese leafhopper, 401, 405
Japanese lilac tree, 375
Japanese maple, 251. *See also* Maple
Japanese pagoda-tree. *See* Pagoda-tree, Japanese
Japanese persimmon, 251. *See also* Persimmon
Japanese red maple, 251. *See also* Maple
Japanese red pine. *See* Pine
Japanese scale, 403
Japanese snowbell, 376
Japanese wax scale, 311, 314, 366, 372, 448
Japanese weevil, 315, 372, 417
Japanese Zelkova. *See* Zelkova, Japanese
Journal of Arboriculture, 14
Journal of Forestry, 14
Juglans, 472–76; *nigra*, 34, 58
Jujube, Chinese, 327
Juneberry. *See* Serviceberry
June bug, 367
Juniper, 376; abiotic diseases, 379; diseases, 376–79; insects and other pests, 379–81
Juniper blight, 294
Juniper broom rust, 378
Juniper mealybug, 381
Juniper midge, 380
Juniper scale, 229, 318, 332, 352, 380–81

Juniper twig girdler, 381
Juniperus, 376–81; *chinensis* "Pfitzeriana," 35; *virginiana*, 35
Juniper webworm, 381

Kabatiella, 449; *apocrypta*, 396; *phoradendri umbellulariae*, 312
Kabatina juniperi, 379
Kabatina twig blight, 379
Karathane, 286
Karbam Black, 285
Katsura-tree, 254, 258, 381
Keaki, 343. *See also* Zelkova, Japanese
Kelthane, 280
Kentucky coffee tree, 251, 381; diseases, 381; insects, 381
Kermes pubescens, 411–12
Kermes scale, 228
Kocide 101, 286
Koelreuteria, 356–57
Kunzeana kunzii, 292
Kwanzan flowering cherry. *See* Cherry

Laburnum, 355–56
Laburnum vein mosaic, 356
Lace bug, 204, 219, 226, 293, 306, 346, 441, 467
Lacewings, 267
Lachnellula willkommii, 382
Ladybird beetle, 266, 267
Laetiporus sulphureus, 389, 410, 423, 458, 474
Lagerstroemia, 330–31
Lag-hook installation, 145
Lambdina: athasaria, 365; *athaseria pellucidaria*, 438; *fiscellaria*, 365
Lannate, 280
Larch, 258, 381; abiotic disease, 383; diseases, 381–83; insects, 383
Larch casebearer, 383
Larch sawfly, 383
Largetooth aspen, 211. *See also* Poplar
Larix, 381–83
Larvabrom, 282
Larvae, 218–19
Lasiodiplodia theobromae, 408, 451
Latania scale, 229, 292, 299, 314, 372, 418, 479
Lateral buds, 24–25
Laterals, pruning, 125
Laurel, 189, 312. *See also* California laurel
Lavalle hawthorn, 35. *See also* Hawthorn
Lawn mowers, damage from, 196–98
Lawson cypress. *See* False cypress
Leaching, 51
Leaf-eating insects: maple, 400–403; oak, 415–17

Leaf fall, 28–29
Leaffooted bug, 331
Leaf gall aphid, 304–5
Leaf gall sawfly, 479
Leaf miner, 221–22
Leaf primordia, 24
Leaf scorch, 371, 375, 406; abiotic, 38, 154, 192–94, 211; bacterial, 246–47
Leaf tissue analysis, 93–94
Leafy mistletoe, 237
Leaves, 26–29
Lecanium, 354; *cerasorum*, 229; *corni*, 411, 453; *fletcheri*, 295; *nigrofasciatum*, 453; *quercifex*, 411; *quercitronis*, 411
Lecanium scale, 228, 327, 363, 372, 411, 447, 479
Lens, hand, 169
Lenzites, 299, 353
Leopard moth borer, 302, 321, 327, 330, 404
Lepidosaphes: beckii, 413, 452; *ulmi*, 229, 298
Leptothyrium californicum, 409
Leptocoris trivittatus, 400
Leptoglossus, 331
Leptosphaeria, 332
Leptothyriella liquidambaris, 463
Lesion nematode, 236
Lesser peach tree borer, 219, 267, 321, 454
Lesser snow scale, 292, 300, 311, 328, 357, 442, 455
Lethal yellowing of palm, 236, 419–20
Leucocytospora kunzei, 364, 457
Leucostoma: kunzei, 339, 382, 424, 457; *niveum*, 477
Libocedrus, 317–18
Lichen, 198–200
Life cycle, insects, 218–19
Lightning: injury from, 146, 194–96; protection from, 146–47
Lignasan BLP, 286
Lignin, 29, 257
Lilac borer, 219, 267, 297
Lilac leaf miner, 297
Limbs: broken, 146; rubbing, 142–43; split, 142
Lime, 42, 43, 44–45, 51, 214; hydrated, 284
Lime-induced chlorosis, 410
Lime sulfur, 213, 286
Lindane, 280
Linden, 141, 187, 211, 248, 249, 251, 253, 383; diseases, 383–84; insects and other pests, 384–86
Linden aphid, 384
Linden borer, 385–86
Linden looper, 223, 350, 384, 417
Linden mite, 384–85
Linden moth, 347
Lindingaspis rossi, 407

Line clearance, pruning for, 135–36
Line pattern mosaic, 304
Liquidambar, 462–63
Liquid fertilizers, 98
Liriodendron, 468–71
Liriothrips floridensis, 315
Lirula macrospora, 457
Lithocarpus, 467
Lithocolletis: cincinnatiella, 415; *hamadryadella*, 222, 415
Lithophane antennata, 417, 450
Littleford elm. *See* Elm
Little leaf, 430
Live oak. *See* Oak
Location: street trees, 62–64; and tree value, 9–10
Locust, black, 35, 58, 248, 386; diseases, 386–87; insects and other pests, 387–89
Locust, honey, 34, 57, 141, 248, 253, 389; abiotic disease, 389; diseases, 389; insects and other pests, 389–92
Locust borer, 219, 387
Locust leaf miner, 221, 387
Locust twig borer, 387
Lombardy poplar, 116
London planetree. *See* Planetree, London
Longhorn borer, 358
Longistigma caryae, 228, 467, 478
Long soft scale, 299, 419
Long-tailed mealybug, 228, 299, 300, 328, 357, 358, 418, 421, 423, 453, 483
Looper, 223
Lophodermella, 425
Lophodermium: autumnale, 353; *juniperinum*, 318; *lacerum*, 353; *laricinum*, 382; *laricis*, 382; *piceae*, 457; *seditiosum*, 425
Lowering grade, 150–54
Low-growing vs. tall-growing trees, 60–62
Low temperature injuries, 184–91
Luna moth, 302, 417, 463
Lygus bug, 358
Lygus lineolaris, 226
Lymantria dispar, 222–23, 347–48

Maclura, 418; *pomifera*, 58
Macrophoma, 477; *candolei*, 308; *phacidiella*, 369; *tumefaciens*, 445–46
Macropsis fumipennis, 391
Macrosiphum: euphorbiae, 228; *liriodendri*, 469
Madrone, 461–62
Maggot, 218
Magicicada septendecim, 330
Magnesium, 39, 40, 41, 95–96, 107, 109
Magnolia, 248, 249, 251, 392; diseases, 392–93; insects, 393

Magnolia, 331, 392–93; *soulangiana*, 35

Magnolia scale, 393

Mahaleb cherry, 251. *See also* Cherry

Maidenhair-tree. *See* Gingko

Malacosoma: americana, 224, 401, 478; *californicum*, 224, 449, 462; *disstria*, 224, 401; *fragilis*, 266; *pluviale*, 363

Maladera castanea, 324

Malathion, 280

Malus, 35, 328–30; *dolgo*, 35

Mancozeb, 286

Maneb, 286

Manex, 286

Manganese, 40, 41, 92, 107–8, 109

Mangifera indica, 393–94

Mango, 393; diseases, 394; insects, 394

Manual for Plant Appraisers, 6

Manzate 200, 286

Maple, 141, 177, 185, 187, 189, 196, 211, 248, 249, 251, 253, 258, 394–95; abiotic disease, 399–400; fungus and bacterial diseases, 395–99; leaf-eating insects and mites, 400–403; other pests, 400–404; scale insects, 403; wood borers, 403–4

Maple bladder gall, 231

Maple decline, 399–400

Maple leaf cutter, 303, 402

Maple petiole borer, 225

Maple phenacoccus, 403

Maple trumpet skeletonizer, 303, 402

Maple worm, 339

Marginal fold-forming eriophyid mite, 323

Marlate, 280

Marmara, 371

Marssoniella juglandis, 366

Marssonina, 381, 477; *betulae*, 303; *brunnea*, 446; *californica*, 474; *castagnei*, 446; *fraxini*, 296; *juglandis*, 474; *martini*, 409; *populi*, 446; *quercus*, 409

Matsucoccus resinosae, 435

Mature trees, pruning, 131

Mauget system: fertilizer injection, 111–12; pesticide injection, 269–70

Mayten tree, 404

Maytenus, 404

MBC phosphate, 286

Mealybug, 228

Mealy flata, 322, 406

Measuring-worm, 347, 400

Mechanical injuries, 196–98

Mechanical tree digging, 75

Medullary rays, 23

Megacyllene: caryae, 367; *robiniae*, 219, 387

Melaleuca, 311–12, 456

Melampsora, 477; *abieti-capraearum*, 353, 477; *abietis-canadensis*, 364, 446; *albertensis*, 446; *arctica*, 477; *cerastii*, 353; *farlowii*, 364; *medusae*, 339, 382, 446; *occidentalis*, 446; *paradoxa*, 382, 477

Melampsorella caryophyllacearum, 353

Melampsoridium: aini, 293; *betulinum*, 303, 382; *carpinea*, 372

Melanaspis: obscura, 229; *obscurus*, 412; *tenebricosa*, 229

Melanconis juglandis, 473

Melanconium: betulinum, 304; *pandani*, 453

Melanophila fulvoguttata, 365

Melia, 327

Meliola, 394; *amphitrichia*, 392; *camelliae*, 312; *cryptocarpa*, 354; *magnoliae*, 392

Meloidogyne, 419, 448, 454, 467; *arenaria*, 455; *hapla*, 356; *incognita*, 300, 311, 315, 355, 357, 422

Melon aphid, 315, 339, 372

Meria laricis, 382

Meristem, 23–24, 27

Mesophyll, 27

Mesquite, 404

Metabolism, 40

Meta-cresol, 244

Metalaxyl, 287

Metallic borer, 404

Metamorphosis, 218–19

Metaphalaria ilicis, 372

Metasequoia, 451

Metasphaeria sassafrasicola, 452

Metcalfa pruinosa, 322, 406

Methane, 183, 184, 246

Methidathion, 281

Methomyl, 280

Methoxychlor, 280

Methyl bromide, 282

Metropolitan Tree Improvement Alliance, 14

Mexican fruit fly, 357

Mexican wax scale, 314

Microclimate, 4

Microcop, 284

Micronutrients, 40, 41, 96; chelated, 108–9

Microorganisms in soil, 40

Micropeltis alabamensis, 393

Microscope, 169

Microsphaera, 248, 409, 419; *aceris*, 398; *alni*, 315, 384; *americana*, 327; *caryae*, 367; *diffusa*, 248, 386; *elevata*, 316; *ellisii*, 372; *erineophila*, 302; *fraxina*, 296; *neglecta*, 344; *nemopanthis*, 369; *ornata*, 304; *penicillata*, 292; *platani*, 440, 466; *pulchra*, 334; *ravenelii*, 389

Microstroma: album, 409; *juglandis*, 367, 474

Microthyriella cuticulosa, 369

Mildew, powdery, 247–49

Milesia fructuosa, 353

Miller 658, 284

Miller's Systemic Fungicide, 284

Mimosa, 141. *See also* Silk-tree

Mimosa webworm, 455

Mindarus abietinus, 228, 353

Miner. *See* Leaf miner

Minerals, soil, 38–39

Mining scale, 299, 300, 357

Mist blowers, for pesticides, 273

Mistletoe, 237

Mite, 230–32

Miticides, 278–82

Mocap, 282

Moline elm. *See* Elm

Molt, 219

Molybdenum, 40, 96, 108, 109

Monarthropalpus buxi, 221, 310

Monilinia: amelanchieris, 454; *fructicola*, 320, 448, 454; *johnsonii*, 361; *laxa*, 320, 448

Monitoring for pests and diseases, 264–65

Monochaetia, 398; *desmazierii*, 344; *kansensis*, 327; *monochaeta*, 358, 366, 409; *unicornis*, 332

Monochaetia canker, 332

Monostichella robergei, 372, 373

Monterey cypress. *See* Cypress

Monterey pine. *See* Pine

Moralwilkoja vagabunda, 447

Morello cherry, 251. *See also* Cherry

Morestan, 280

Moreton Bay fig. *See* Fig

Morus, 58, 406–7; *alba*, 35

Mosaic virus, 236

Moth, 218, 219, 223. *See also* Gypsy moth

Mottled willow borer, 478

Mountain-ash, 35, 249, 404; abiotic disease, 405; bacterial and fungus diseases, 405; insects, 405–6. *See also* Ash

Mountain-ash sawfly, 224–25, 405

Mountain-ash spider mite, 405

Mountain maple. *See* Maple

Mourning-cloak butterfly, 224, 308, 348, 384

Mugho pine. *See* Pine

Mulberry, 58, 249, 254, 406; abiotic disease, 406; diseases, 406; insects and other pests, 406–7

Mulberry whitefly, 407

Mulching, 47, 51–52, 154, 190; postplanting, 86–87

Musa, 300

Mycodiplosis clavula, 335–36

Mycoplasma, 236

Mycorrhizae, 19–21, 40

Mycosphaerella: aleuritidis, 471; *arbuticola*, 312, 462; *caroliniana*, 457; *caryigena*, 422; *cercidi-*

cola, 449; *dearnessii*, 425; *effigurata*, 296; *fraxinicola*, 296; *liriodendri*, 469; *microsora*, 383; *milleri*, 392, 393; *moelleriana*, 358; *mori*, 406; *nyssaecola*, 472; *platanifolia*, 466; *populicola*, 446; *populorum*, 445; *sequoiae*, 450; *stigmina-platani*, 466; *tulipiferae*, 469; *ulmi*, 344

Myrica, 350

Myriconium comitatum, 477

Myrtle. *See* California laurel

Myxosporium, 468; *everhartii*, 334

Myzocallis kahawaluokalani, 331

Myzocallus tiliae, 384

Myzus persicae, 228, 324, 421

Naemacyclus minus, 425

Naled, 279

Nalepella tsugifoliae, 231, 366

Nantucket pine moth, 433–34

Narrow crotches, and artificial support, 140

National Arbor Day Foundation, 14

National Arborists Association (NAA), 6, 13

Natural gas, 183–84

Natural target pruning, 125

Nautical borer, 358

Nectria, 372, 398, 406, 468; *cinnabarina*, 296, 302, 343, 356–57, 374, 383, 389, 398, 419, 422, 448, 451, 455, 484; *coccinea*, 368, 468; *coccinea faginata*, 301; *desmazierii*, 308; *ditissima*, 291; *galligena*, 292, 302, 304, 312, 331, 335, 343, 366, 373, 386, 393, 398, 408, 452, 472, 473; *magnoliae*, 393, 468

Nectria canker, 393, 398, 408, 484

Nectricoccinea, 296

Needle blight, 352–53

Nemacur, 282–83

Nemaster, 282

Nematicides, 282–83

Nematode, 166, 236–37

Nemex, 282

Neoclytus acuminatus, 297

Neodiprion: excitans, 434; *lecontei*, 224–25, 434; *pratti banksianae*, 434

Neolecanium cornuparvum, 393

Neptula: amelanchierella, 454; *sericopeza*, 402

Nialate, 279

Nigra scale, 372

Nipaecoccus nipae, 299

Nipple-gall, 359–60

Nitrate ions, 39

Nitrogen, 40, 41, 51, 92, 95, 104–5, 243, 254; in soil air, 33

Nitrogen dioxide, 205

Noble fir. *See* Fir

Nodules, root, 21
Nonparasitic diseases: boxwood, 309–10; holly, 371
Norbac 84, 243–44
Norfolk Island pine, 407; diseases, 407; insects, 407
Northern catalpa. *See* Catalpa
Norway maple, 35, 211. *See also* Maple
Norway maple aphid, 402
Norway spruce, 35. *See also* Spruce
Nucleopolyhedrosis virus, 266
Nuculaspis californica, 229
Nudrin, 280
Nu-Film, 288
Nursery-grown trees, transplanting, 68
Nutrient elements, 39; availability in soil, 41–47;
    fertilizer, 94–96
Nygmia phaeorrhoea, 400
Nymph, 219
Nymphalis antiopa, 224, 348
Nyssa, 472

Oak, 141, 177, 187, 211, 248, 249, 251, 254, 258,
    407; abiotic disease, 410–11; borers, 413–14;
    diseases, 407–11; insects, 411–13; leaf-eating
    insects, 415–17; other pests, 417. *See also*
    Silk-oak; Tanoak
Oak-apple gall, 411
Oak bark beetle, 410
Oak blotch leaf miner, 415
Oak borer, 413–14
Oak branch borer, 415
Oak clearwing borer, 413–15
Oak gall scale, 411–12
Oak lace bug, 413
Oak leaf miner, 222
Oak leafroller, 417
Oak leaf tier, 417
Oak mite, 303
Oak phylloxerid, 411
Oak-potato gall, 411
Oak sawfly, 416
Oak skeletonizer, 417
Oak stem borer, 415
Oak timberworm, 415
Oak twig girdler, 322
Oak twig pruner, 322, 368
Oak webworm, 417
Oak wilt, 409–10
Oakworm, 223
Oberea tripunctata, 335, 347
Oberea ulmicola, 335
Oblique-banded leafroller, 267, 456
Obscure scale, 229, 327, 336, 349, 360, 368, 403,
    412, 476, 479
Ocellate leaf gall, 402

Odontia, 353
Odontopus calceatus, 452–53
Odontota dorsalis, 221, 387
Oidium mangiferae, 394
Oiketus: abbotii, 328; *townsendi*, 298
Oil spray, 279
Olea, 418
Oleander scale, 229, 292, 314, 315, 318, 324, 372,
    393, 451, 483
Oleaster-thistle aphid, 451–52
Oligonychus, 417; *ilicis*, 230, 372; *ununguis*, 230,
    461
Olive, 251, 418; diseases, 418; insects, 418
Olive knot, 418
Olive parlatoria scale, 314, 407
Olive scale, 381, 452
Omite, 280
Omnivorous looper, 292, 315, 355, 358, 393, 423,
    454, 475
Oncideres cingulata, 360
Ophiostoma ulmi, 239, 341
Orange-striped oakworm, 223, 417
Orange-tortrix caterpillar, 292, 315, 358, 475
Orchid-tree, 418; diseases, 418; insects, 418
Oregon maple. *See* Maple
Oregon myrtle. *See* California laurel
Organic amendments, 97
Organic concentrates, fertilizer, 96–97
Organic matter, soil, 39–40
Organisms, soil, 40
Organizations, arboricultural, 12–14
Orgyia: leucostigma, 223, 355; *pseudotsugata*, 339,
    383
Oriental arborvitae. *See* Arborvitae
Oriental fruit moth, 267, 322
Oriental moth, 400
Orientus ishidae, 405
Ormensis septentrionalis, 322
Ornalin, 286
Ornamental pear. *See* Pear, ornamental
Ornate aphid, 315
Orthene, 280–81
Orthocide, 285
Ortho Copper Fungicide, 284–85
Ortho Dry Spreader, 288
Osage-orange, 58, 251, 418; diseases, 418; insects,
    418
Osmosis, 36
Ostrya, 372
Otiorhynchus: ovatus, 315, 481–82; *sulcatus*, 315,
    480–81
Overwintering of bacteria and fungi, 240
Ovularia maclurae, 418
Oxamyl, 281–82, 283
Oxydemeton-methyl, 280

Oxydendrum, 43, 456–57
Oxygen, 29, 40, 184; in soil air, 33–34
Oxyporus populinus, 367, 398
Oxytetracycline, 287
Oxythioquinox, 280
Oystershell scale, 229, 298, 303, 311, 314, 324,
  330, 336, 349, 354, 360, 363, 372, 386, 387,
  406, 448, 453, 467, 468, 471, 476, 479
Ozone, 204

Pachypsylla: celtidis-mamma, 359–60; *celtidisve-*
  sicula, 360
Pacific dogwood. *See* Dogwood
Pacific flatheaded borer, 297, 358, 403
Paecilomyces buxi, 309
Pagoda-tree, Japanese, 251, 418–19; diseases, 419;
  insects, 419
Painted hickory borer, 367
Paleacrita vernata, 223, 347, 400, 417
Pales weevil, 436
Pale tussock, moth, 468
Palisade cells, 27
Palm, 419; diseases, 419–21; insects and other
  pests, 421
Palmaceae, 419–21
Palm aphid, 421
Palm leaf skeletonizer, 421
Palm mealybug, 421
PAN, 204–5
Pandanus, 453
Panonychus ulmi, 230, 330
Pantomorus cervinos, 314
Paper birch, 202, 211. *See also* Birch
Paper-mulberry, 421; diseases, 421–22
Paraclemensia acerifoliella, 402
Paradichlorobenzene, 321
Paradiplosis fumifex, 353–54
Parandra brunnea, 302
Paranthrene simulans, 413–15
Paraquat, 209
Parasites, 234–37; defined, 235; dissemination and
  survival, 237–40
Parasitic disease: agents, 234–40; anthracnose,
  249–51; Armillaria root rot, 258–61; bacterial
  scorch, 246–47; bacterial wetwood, 244–46;
  crown gall, 243–44; development factors, 240;
  fire blight, 241–43; overwintering of bacteria
  and fungi, 240; powdery mildews, 247–49;
  slime flux, 244–46; symptoms, 233–34; verti-
  cillium wilt, 251–54; wood decay, 255–58.
  See also Damage, nonparasitic; Disease
Parasitic plants, 237
Paratetranychus aceris, 402–3
Parenchyma, 21, 23

Parlatoria: oleae, 381, 452; *pergandii,* 407; *pittos-*
  pori, 418
Parrotia, 423
Parthenocarpic fruit, 30
Parzate, 287
Pathogen, defined, 235
Paulownia, 350; *tomentosa,* 58
Paul's scarlet thorn. *See* Hawthorn
Paving, 154; damage from, 179–81
Pawpaw, 254
PCNB, 287
Peach. *See* Cherry
Peach aphid, 315, 324
Peach bark beetle borer, 321
Peach leaf curl, 320
Peach scale, 324, 327, 336, 372, 407
Peach tree borer, 219, 267, 321
Peach yellows, 321
Peanut root-knot nema, 455
Pear, 141, 249, 251, 254
Pear, ornamental, 422; diseases, 422
Pear aphid, 345
Pear borer, 363
Pear leaf blister mite, 405
Pear psyllid, 228
Pear slug, 225, 322
Peat, 51
Pecan, 248, 254, 422; diseases, 422; insects, 422–
  23
Pecan borer, 423
Pecan carpenterworm, 423
Pecan cigar case bearer, 367
Pellicularia filamentosa, 368–69, 419, 430
Pemphigus populitransverus, 447
Penicillium vermoeseni, 420
Peony scale, 314
Pepperidge tree. *See* Tupelo
Pepper-tree, 251, 423; diseases, 423; insects, 423
Pepperwood, 312
Perenniporia, 258, 297
Periclista, 416
Pericycle, 18
Periderm, 23, 25
Peridermium ornamentale, 353
Periodical cicada, 330, 413
Periphyllus, 375; *aceris,* 402; *lyropictus,* 402; *ne-*
  gundinis, 402
Periploca nigra, 381
Peronospora parasitica, 327
Peroxyacetyl nitrate (PAN), 204–5
Persea, 299
Persian parrotia, 423
Persimmon, 57, 249, 251, 423; diseases, 423; in-
  sects, 423
Pesotum ulmi, 341

Pestalotiopsis: cryptomeriae, 331; *funerea,* 294, 299, 331, 450, 480; *maculans,* 313, 331; *mangiferae,* 394; *palmarum,* 420; *palmicola,* 420

Pest control, 263–64; biological, 265–68; chemical, 268–88; fungicides and bactericides, 283–88; insecticides and miticides, 278–82; integrated pest management (IPM), 264–65; nematicides, 282–83; pesticides, 268–78; pruning for, 134–35; spreading and sticking agents, 283–88. *See also* Insect(s); Insect types

Pesticides, 268–69; application methods and equipment, 269–75; labels, 275–77; preparing small quantities, 275; specific, 277–78; toxicity, 275–77

Pest threshold levels, 265

Petals, 29

Petiole, 26

Petroleum oils, 279

Pezicula carpinea, 373

Pezizella oenotherae, 456

Pfitzer juniper, 35. *See also* Juniper

pH, 41–47; and chlorosis, 214; meter, 169; and nutrients, soil, 41–45

Phacidiopycnis pseudotsugae, 339, 364

Phacidium: curtisii, 370; *infestans,* 294

Phaecryptopus gaeumannii, 339

Phaeobulgaria inquinans, 407–8

Phaeolus, 258; *schweinitzii,* 382, 430, 458

Phaltan, 287

Phellinus, 258, 297, 353, 478; *everhartii,* 410; *igniarius,* 302, 367, 446, 474; *laevigatus,* 304; *pini,* 382, 430, 458; *robiniae,* 387; *spiculosus,* 408; *tremulae,* 446

Phellodendron, 328

Phelloderm, 23

Phellogen, 23

Phenacaspis: nyssae, 374, 472; *pinifoliae,* 229

Phenacoccus acericola, 403

Phenamifos, 283

Pheromone traps, 266–67

Phleospora: celtidis, 358; *multimaculans,* 474; *pteleae,* 373; *robiniae,* 386

Phloem, 18, 22

Phloem necrosis, 236

Phloeosinus: canadensis, 295; *dentatus,* 379; *taxadii,* 300

Phloeospora multimaculans, 466

Phloeotribus liminaris, 321

Phlyctaena tiliae, 384

Phoma, 343, 448; *conidiogena,* 308; *hysterella,* 480

Phomopsis, 312, 331, 343, 381, 382, 393, 394, 424, 453; *acerina,* 398; *arnoldiae,* 451; *crustosa,* 368; *diatrypea,* 355; *imperialis,* 350; *juniperovora,* 294, 350, 378, 430, 450; *macrospora,* 445

Phomopsis canker, 445

Phomopsis lokoyae, 339

Phomopsis twig blight, 352, 378

Phoracantha semipunctata, 358

Phoradendron, 237; *flavescens,* 419, 422; *juniperinum libocedri,* 318

Phosmet, 280

Phosphate ions, 39

Phosphorus, 19, 40, 41, 89, 95, 105–7, 211

Photochemical smog, 204

Photosynthesis, 29

Phryganidia californica, 417

Phyllactinia: corylea, 331; *guttata,* 292, 296, 302, 304, 316, 327, 334, 344, 350, 355, 361, 367, 369, 372, 384, 386, 398, 406, 409, 452, 454, 466, 469, 477, 479

Phyllaphis fagi, 302

Phyllobius intrusus, 295

Phyllophaga, 367; *bruneri,* 418

Phyllosticta, 373, 418; *ailanthi,* 467; *alcides,* 446; *apicalis,* 477; *argyrea,* 451; *auerswaldii,* 308; *camelliae,* 313; *camelliaecola,* 313; *catalpae,* 316; *celtidis,* 358; *cercidicola,* 449; *chionanthi,* 355; *concomitans,* 369; *confertissima,* 344; *cookei,* 331, 393; *cornicola,* 334; *cytisii,* 355; *erysiphoides,* 333; *extensa,* 358; *faginea,* 302; *fimbriata,* 462; *ginkgo,* 355; *glauca,* 393; *gordoniae,* 354; *gymnocladi,* 381; *illinoisensis,* 452; *lagerstroemia,* 331; *liriodendrica,* 469; *livida,* 409; *maclurae,* 418; *magnoliae,* 393; *melaleuca,* 344; *minima,* 396; *mortoni,* 394; *paulowniae,* 350; *platani,* 466; *pteleicola,* 373; *robiniae,* 386; *sophorae,* 419; *sorbi,* 405; *terminalis,* 369; *tiliae,* 384; *tumericola,* 409; *virdis,* 296

Phyllostictina hysterella, 480

Phylloxera, 228

Phylloxera, 411; *caryaecaulis,* 367

Phymatotrichum omnivorum, 291, 316, 318, 334, 354, 357, 358, 373, 381, 410, 418, 419, 421, 442, 449, 454

Physalospora: fusca, 291, 357; *gregaria,* 480; *ilicis,* 368

Physokermes piceae, 460

Physopella fici, 418

Phytomycin, 287

Phytomyza: ilicicola, 221, 371; *ilicis,* 221, 371

Phytophthora, 328, 354, 395, 421; *arecae,* 421; *cactorum,* 301, 329, 333, 335, 375, 396, 451, 462; *cinnamomi,* 36, 299, 309, 313, 351, 353, 379, 430, 480; *citricola,* 395; *faveri,* 421; *ilicis,* 370; *inflata,* 343, 344; *lateralis,* 351; *palmivora,* 421; *parasitica,* 309, 386

Picea, 457–61; *abies,* 35; *pungens,* 35

Picfume, 282

Pigeon tremex, 404
Pignut hickory. *See* Hickory
Pikonema alaskensis, 460–61
Pileolaria cotini-coggyriae, 456
Pinching, 120
Pine, 177, 185, 189, 211, 258, 423; abiotic diseases, 430–31; diseases, 423–31; insects and other pests, 431–38. *See also* Australian-pine; Norfolk Island pine; Screw-pine
Pine bark adelgid, 228, 431
Pine butterfly, 354
Pine cone gall, 478–79
Pine false webworm, 434
Pine leaf adelgid, 460
Pine leaf chermid, 432
Pine looper, 438
Pine moth, 354
Pine needle miner, 435
Pine needle scale, 229, 318, 339, 354, 435, 460
Pine root collar weevil, 434–35
Pine sawfly, 224–25, 266
Pine shoot moth, 461
Pine spittlebug, 435
Pine tortoise scale, 435
Pine tussock moth, 438
Pineus: pinifoliae, 432, 460; *strobi*, 228, 431
Pine webworm, 432
Pine weevil, 339
Pine wilt, 423, 430
Pine wood nematode, 236–37
Pinkstriped oakworm, 417
Pinnaspis strachini, 328
Pin oak. *See* Oak
Pin oak sawfly, 416
Pinus, 423–38
Piperalin, 287
Pipron, 287
Piptoporus betulinus, 304
Pissodes: nemorensis, 317, 436; *strobi*, 436
Pistachio, 251. *See also* Chinese pistachio
Pistacia, 327
Pistillate flowers, 30
Pitch moth, 339, 435
Pith, 22
Pit-making pittosporum scale, 318
Pitmaking scale, 372, 448
Pitted ambrosia beetle, 339, 372
Pittosporum, 438
Pittosporum, 438
Pittosporum scale, 298, 448
Pityokteines sparsus, 354
Placosphaeria cornicola, 335
Plagiodera versicolora, 479
Plagiognathus albatus, 226, 467

Plagiosphaeria gleditschiae, 389
Planetree, London, 141, 177, 187, 189, 248, 249, 254, 258, 438; fungus diseases, 438–40; non-parasitic diseases, 441; insects, 441–42. *See also* Sycamore
Planetree aphid, 467
Planetree canker-stain, 239
Planococcus citri, 314, 407, 451
Plant bug, 219, 226, 243
Planting, 78–82
Plasmopara cercidis, 449
Platanus: acerifolia, 438–42; *occidentalis*, 35, 464–67
Platanus mite, 315, 358
Platypus, 415
Platytetranychus multidigituli, 391–92
Pleohaeta polychaeta, 249, 358–59
Pleurotus, 258, 297; *sapidus*, 410
Plioioderma desmazierii, 425
Plum, 249, 251. *See also* Cherry
Pod gall midge, 390
Podosesia: aureocincta, 219; *syringae fraxini*, 297; *syringae syringae*, 219, 297
Podosphaera: clandestina, 249, 319–20, 361, 454; *leucotricha*, 249, 329
Poecilocapsus lineatus, 226, 400–401
Poinciana, 442
Pollaccia, 446; *saliciperda*, 476
Pollarding, 120, 132–33
Pollen cones, 30
Polyphemus moth, 452, 463
Polyporus, 258, 297, 299, 353, 375; *catalpae*, 316; *dryophilus*, 423; *farlowii*, 423; *fissilis*, 410; *halesiae*, 456; *lucidus*, 410; *spraguei*, 350, 480
Ponderosa pine. *See* Pine
Pontania, 479
Popcorn disease, 406
Popillia japonica, 225, 384
Poplar, 58, 141, 177, 211, 249, 251, 258, 442; abiotic disease, 447; diseases, 442–47; insects, 447–48. *See also* Tuliptree
Poplar borer, 447
Poplar inkspot, 446
Populus, 58, 442–48; *deltoides*, 35
Poria, 299; *spiculosa*, 366
Porthetria dispar, 222–23
Port Orford cedar. *See* False cypress
Post oak. *See* Oak
Postplanting operations, 82–89
Potassium, 39, 40, 41, 89, 95, 105–7, 193, 211
Potato aphid, 228, 328, 339
Potato flea beetle, 371
Potato leafhopper, 308
Powdery mildew, 247–49
Powdery pine needle aphid, 432

Pratt Bordeaux Mix, 284–85

Pratylenchus: penetrans, 324; *pratensis,* 311

Praying mantis, 267

Preservation of trees: artificial support, 140; cavity filling, 147; and compaction, 154; at construction sites, 148–55; and construction injury, 155; environmental change, 154–55; preventing and treating structural damage, 139–40; sidewalk damage, 155; treating fresh wounds, 145–47

Prionoxystus robiniae, 297

Prionus: imbricornis, 415; *laticollis,* 415

Pristiphora: erichsonii, 383; *geniculata,* 224–25, 405

Prociphilus: californicus, 298; *imbricator,* 302; *tessellatus,* 293

Procriconema, 311

Profenusa collaris, 362–63

Prolate, 280

Promethea moth, 463

Prometon, 210

Propargite, 280

Prosopsis, 404

Protopulvinaria pyriformis, 453

Prune, 251. *See also* Cherry

Pruning: and chemical growth retardants, 136; conifers, 132; cut, 125–29; for disease control, 134–35; equipment, 123–25; for line clearance, 135–36; main stems, 127–29; mature trees, 131; natural, 25; need for, 115–16; newly transplanted trees, 130; postplanting, 85–86; safety, 129; scheduling, 136; schemes, 119–22; special shapes, 132; timing, 122–23; tree response to, 116–19; wound dressings, 129–30; young trees, 130–31

Pruning shears, 166–67

Prunus, 249. *See also* Cherry

Prunus, 258, 318–24; *persica,* 35; *serotina,* 35, 58; *subhirtella,* 35

Pseudaonidia duplex, 315

Pseudaulacaspis pentagona, 229, 373

Pseudocneorhinus bifasciatus, 315, 417

Pseudococcus, 299; *adonidum,* 314, 357, 453; *aurilanatus,* 407; *comstocki,* 228, 316–17, 483; *juniperi,* 381; *longispinus,* 228, 483; *maritimus,* 355, 356, 483, *ryani,* 295, 332, 407, 451

Pseudomonas: aceris, 396; *lauracearum,* 312; *mori,* 406; *saliciperda,* 478; *solanacearum,* 300; *syringae,* 328, 418, 475; *syringae syringae,* 320

Pseudonectria rousselliana, 308

Pseudoperonospora celtidis, 358

Pseudophillipia quaintancii, 435

Pseudopityophthorus: minutissimus, 410; *pruinosus,* 410

Pseudotsuga menziesii, 339

Pseudovalsa longipes, 409

Psidium guajava, 357

Psylla: buxi, 310; *floccosa,* 294; *pyricola,* 228; *uncatoides,* 292

Ptelea, 373

Pterocarya, 479

Pterocomma, 447

Publications about trees, 14

Puccinia: andropogonis onobrychidis, 456; *caricisshepherdiae,* 451; *coronata elaeagni,* 451; *sparganioides,* 296; *windsoriae,* 373

Pucciniastrum: goeppertianum, 353; *hydrangeae,* 364; *pustulatum,* 353; *sparsum,* 462; *vaccinii,* 364

Pulvinaria: floccifera, 483; *innumerabilis,* 229, 403

Punch bar, in fertilizer application, 102

Pupa, 219

Purple beech. *See* Beech

Purple scale, 372, 393, 413, 452, 483

Pussy willow, 35. *See also* Willow

Pustule scale, 328

Putnam scale, 303, 318, 349, 360, 363, 368, 386, 387, 418, 476, 479

Puto yuccae, 451

Pycnidia, 343

Pyriform scale, 453

Pyrrhalta luteola, 225, 346

Pyrus, 422; *calleryana,* 422; *communis,* 422

Pythium, 420, 480; *debaryanum,* 430

Quadraspidiotus: juglansregiae, 229, 476; *perniciosus,* 229

Quercus, 407–17; *borealis,* 35; *borealis maxima,* 408; *palustris,* 408; *virginiana,* 408

Questionnaire, diagnostic, 170–73

Quince, flowering, 251, 448; bacterial and fungus diseases, 448, insects, 448

Radopholus similis, 300

Ramularia: gracilipes, 334; *liriodendri,* 469; *rosea,* 477

Ravenelia opaca, 389

Recurvaria: apicitripunctella, 300; *piceaella,* 460; *stanfordi,* 332

Red ash. *See* Ash

Red-banded leafroller, 267

Red bay scale, 354

Red birch. *See* Birch

Redbud, 35, 258, 448; abiotic disease, 449; diseases, 448–49; insects, 449–50

Redbud leafroller, 450

Red cedar, 35, 211. *See also* Juniper

Red cedar bark beetle, 379

Red fern-leaved beech. *See* Beech
Red fir. *See* Fir
Red flowering dogwood, 35. *See also* Dogwood
Red-headed ash borer, 297, 347, 367–68, 386, 415
Red-headed pine sawfly, 224–25, 317, 434
Red-humped caterpillar, 223, 308, 322, 339, 349, 447, 475
Red-humped moth, 367
Red maple. *See* Maple
Red oak. *See* Oak
Red oak borer, 415
Red pine. *See* Pine
Red pine scale, 435
Red scale, 314
Red squirrel, 404
Redwood, 295, 450; diseases, 450; insects, 450–51
Redwood, dawn, 451; diseases, 451
Rehmiellopsis balsameae, 353
Replacement value, trees, 6
Reproductive organs, 29–31
Resins, 25
Resistant cultivars. *See names of individual trees*
Respiration, 29
Reticulitermes flavipes, 357, 484
Retinospora, diseases, 378
Rhabdocline pseudotsugae, 339
Rhabdogloeum hypophyllum, 339
Rhabdophaga: salicis, 478; *strobliodes,* 478–79
Rhabdopterus beetle, 314, 450
Rhabdopterus deceptor, 450
Rhamnus, 311
Rhizobium, 21
Rhizoctonia, 368
Rhizoecus falcifer, 311
Rhizomorph, 260
Rhizosphaera kalkhoffi, 457
Rhododendron, 189, 258
Rhopalosiphum nymphaeae, 324
Rhopobota naevana ilicifoliana, 371
Rhus, 462; *typhina,* 325
Rhyacionia: buoliana, 432–33; *frustrana,* 433–34
Rhynchaenus rufipes, 479
Rhytisma: acerinum, 396–97; *arbuti,* 462; *ilicinicola,* 369; *punctatum,* 397; *salicinum,* 477–78; *velatum,* 369
Rigid bracing, 141–43
River birch. *See* Birch
Road cuts, 150–51
Robinia, 386–89; *pseudoacadia,* 35, 58
Rocky Mountain hard maple. *See* Maple
Rocky Mountain juniper aphid, 379
Rodent damage, 201
Rogor, 278
Ronilan, 286
Root(s), 15–21; examination of, 165–66; girdling,

67, 80, 181–83; pruning, 80; severed, 148–49; and sidewalk lifting, 155
Root cap, 16
Root decline of pine, 430
Root hairs, 16, 38
Root-knot nema, 236, 355, 357, 370, 422, 467
Root rot, 255, 258–61
Root zone, 38
Rose beetle, 314, 324
Rose leafhopper, 322, 363, 401
Rosellinia: caryae, 366; *herpotrichioides,* 330
Rose scale, 355
Rosy apple aphid, 362, 405
Rotylenchus, 484
Roundheaded borer, 405, 454
Rovral, 285
Royal palm. *See* Palm
Rubber tree, hardy, 360
Rubbing limbs, 142–43
Rubigan, 287
Russian-olive, 251, 451; diseases, 451; insects, 451–52
Rust mite, 230
Rusty tussock moth, 302

Sabulodes caberata, 292, 355, 454
Saddleback caterpillar, 416
Saddled prominent caterpillar, 302, 400
Safety, pruning, 116, 129
Saissetia oleae, 229, 462
Salebria afflictella, 463
Salix, 476–79; alba, 35; *discolor,* 35
Salt damage, 210–11, 314
Salt spray damage, 211–13
Samia cynthia, 468
Sampling tube, soil, 166
Sand, 38
San Jose scale, 229, 292, 295, 303, 323–24, 330, 336, 349, 360, 363, 386, 387, 392, 406, 407, 418, 448, 453, 456, 472
Sanninoidea exitiosa, 219, 321
Saperda: candida, 405, 454; *inornata,* 447; *obliqua,* 294; *populea,* 447; *vestita,* 385–86
Sapium, 327–28
Saprophyte, 235, 239
Sapstreak, 399, 469
Sapsucker damage, 201–2
Sapwood, 21–22
Sapwood rot, 255
Saratoga spittlebug, 435
Sassafras, 141, 249, 251, 452; diseases, 452; insects, 452–53
Sassafras, 452–53
Sassafras weevil, 452–53

Satin moth, 417, 447
Saucer magnolia, 35. *See also* Magnolia
Savin juniper. *See* Juniper
Saw, for pruning, 167
Sawfly, 224–25
Scale, 228–30
Scaphoideus luteolus, 345
Scarlet oak. *See* Oak
Schinus molle, 423
Schizophyllum, 258
Schizoxylon microsporum, 343
Schizura: concinna, 223, 447; *ipomaeae,* 349
Schoene spider mite, 350
Schwedler's maple, 251. *See also* Maple
Science Systemic Fungicide, 287
Scirrhia acicola, 425
Scleroderris, 424
Sclerophoma, 369, 370
Sclerotinia sclerotiorum, 312, 313
Scoleconectria: balsamea, 353; *scolecospora,* 353
Scolytidae, 432
Scolytus: multistriatus, 341, 346; *quadrispinosus,*
    367; *rugulosus,* 321, 454
Scorch. *See* Leaf scorch
Scotch elm, 251. *See also* Elm
Scotch pine. *See* Pine
Screw-pine, 453; diseases, 453; insects, 453. *See*
    also Pine
Scrub pine needle rust, 427
Scurfy scale, 349, 363, 405–6, 476
Seed cones, 30
Seed mite gall, 306–8
Seiridium cardinale, 332
Selection, trees, 53–64
Semanotus igneus, 450
Senescence, leaf, 29
Sepals, 29
Septobasidium, 300; *curtisii,* 374, 472; *fumigatum,*
    398
Septogloeum: celtidis, 358; *profusum,* 344; *quer-*
    ceum, 409; *salicinum,* 477
Septonina podophyllina, 446
Septoria, 394, 398; *angustissima,* 418; *argyrea,*
    451; *besseyi,* 296; *betulicola,* 304; *carpinea,*
    373; *caryae,* 366; *chionanthi,* 355; *citri,* 328;
    cornicola, 334; *crataegi,* 361; *didyma,* 477;
    elaeagni, 451; *eleospora,* 355; *floridae,* 334;
    hicoriae, 366; *hippocastani,* 375; *ilicifolia,*
    369; *leucostoma,* 296; *Limonum,* 328; *liqui-*
    dambaris, 463; *magnoliae,* 393; *musiva,* 445;
    niphostoma, 393; *ostryae,* 372; *platanifolia,*
    466; *pteleae,* 373; *quercicola,* 409; *quercus,*
    409; *rhoina,* 456; *submaculata,* 296; *unedonis,*
    462
Sequoia, 450–51

Sequoia, 450–51
Sequoiadendron, 450–51
Sequoia pitch moth, 451
Serviceberry, 248, 249, 254, 453; diseases, 453–54;
    insects and other pests, 454
Setting, 78–80
Sevimol, 281
Sevin, 281
Sevin 4 Oil, 281
Sex attractants, 266–67
Shadblow. *See* Serviceberry
Shadbush. *See* Serviceberry
Shagbark hickory. *See* Hickory
Shaping trees, 116
Shearing, 120
Shears, pruning, 166–67
Shell D-D, 282
Shigo, Alex, 118, 257
Shigometer, 167–69
Shoestring root rot, 258
Shot-hole borer, 321, 363, 423, 454
Shoulder ring, 125
Siberian elm. *See* Elm
Sibine stimulea, 416
Sidewalk damage, 62–64, 155
Sieve cells, 23
Signs of disease, 233–34
Silicates, 39
Silicon, 41
Silk-oak, 454; diseases, 454; insects, 454. *See also*
    Oak
Silk-tree, 141, 196, 455; diseases, 455; insects,
    455–56
Silt, 38
Silverbell, 43, 456; diseases, 456
Silver maple, 57, 251. *See also* Maple
Silverspotted skipper, 224, 387
Siphonatrophia cupressi, 332
Sirococcus blight, 364, 457
Sirococcus: clavigignenti-juglandacearum, 473; *con-*
    igenus, 364, 425, 457
Site, and tree selection, 55
Size: and spacing, street trees, 62; and tree value,
    7–8
Slime flux, 244–46
Slippery elm, 251. *See also* Elm
Smaller European elm bark beetle, 346–47
Smog, 204
Smoke-tree, 248, 251, 456; disease, 456; insects,
    456
Smooth cypress. *See* Cypress
Snow, damage from, 196
Snowbell, Japanese, 376
Snowdrop tree. *See* Silverbell
Snow-in-summer, 456

Snow scale, 328
Snow-white linden moth, 347
Society of American Foresters, 14
Society of Municipal Arborists, 14
Sodium, 39, 41
Sodium chloride, 211
Soft scale, 228–29
Soil: as anchor, 47–48; compaction damage, 48, 51, 154, 178–79; components, 33–40; essential elements, 40–41; improvement for established trees, 51–52; improvement prior to planting, 50–51; and nutrient availability, 41–47; pH, 42–47; type, 41; urban, 48–50
Soil air, 33–36, 177
Soil fill damage, 175–77
Soil test, 93
Solenia anomala, 292
Solitary oak leaf miner, 221–22
Soluble salts meter, 169
Sooty bark canker, 445
Sooty mold, 228–29, 312, 328, 331, 357
Sophora japonica, 418–19
Sorbus, 404–6; aucuparia, 35
Sorrel-tree, 43, 456; diseases, 456–57
Sour cherry, 251. See also Cherry
Sourgum, 251, 472. See also Tupelo
Sourgum scale, 374
Sourwood, 456–57
Southern catalpa, 57. See also Catalpa
Southern cypress bark beetle, 300
Southern magnolia, 251. See also Magnolia
Southern pine sawfly, 434
Southern red mite, 230, 315, 358, 372
Southern root-rot nematode, 300, 311, 357
Spacing and size, street trees, 62
Spanish chestnut, 251. See also Chestnut
Specialized roots, 19–20
Species, and tree value, 8
Speckled alder, 211. See also Alder
Spectracide, 279
Sphaceloma, 313; magnoliae, 393; murrayae, 477; perseae, 299
Sphaerophragmium, 455
Sphaeropsis, 296, 334, 424, 480; malorum, 329; sapinea, 425; ulmicola, 343
Sphaerotheca: lanestris, 249, 409; pannosa, 249, 319–20; phytophila, 249; phytoptophila, 359
Sphaerulina: polyspora, 456; taxi, 480
Sphagnum peat, 51
Spider mite. See Mite
Spindle gall, 231, 323
Spiny oakworm, 417
Spiral nema, 311
Spirea aphid, 328

Split crotches, and artificial support, 140
Split limbs or trunks, bracing, 142
Spongy mesophyll, 27
Spores, fungus, 235–36
Sporodesmium maclurae, 418
Sporonema camelliae, 313
Sporothrix ulmi, 341
Spotted cutworm, 314–15
Spotted pine aphid, 432
Spotted tussock moth, 354
Spray compatibility, 278
Spraying, pesticides, 274–75; equipment, 270–74
Spray Stay, 288
Spreader-sticker, 288
Spreading agents, 288
Spreading yew, 35. See also Yew
Spring cankerworm, 223, 302, 327, 347, 400, 417
Spring wood, 23
Spruce, 177, 189, 211, 258, 457; diseases, 457–58; insects and other pests, 459–61
Spruce bud scale, 460
Spruce budworm, 339, 366, 460
Spruce epizeuxis, 460
Spruce gall adelgid, 459
Spruce leaf miner, 366
Spruce needle miner, 460
Spruce spider mite, 230, 339, 366, 461
Squirrels, 404
Staghorn sumac. See Sumac
Staking, 82–83
Stamens, 29–30
Stanford whitefly, 413
StanGuard, 278
Star magnolia, 251. See also Magnolia
Steganosporium ovatum, 398
Stegophora ulmea, 344
Stegophylla, 411
Stem, 21–26
Stemphyllium, 406
Stenocarpus, 354
Stereum, 258, 375; gausapatatum, 410; subpileatum, 410
Stewartia, 461
Stewartia, 461
Sticking agent, 288
Stictocephala bubalus, 298, 401
Stigmatophragmia sassafrasicola, 452
Stigmina palmivora, 420
Stilpnotia salicis, 417, 447
Stoma(ta), 26, 29
Strawberry root weevil, 315, 481–82
Strawberry-tree, 461–62; diseases, 462; insects, 462
Streetside trees, selection and problems, 58–64
Streptomycin, 242, 287

String trimmers, 196
Structural damage, preventing and treating, 139–40
Structure: of insects, 217–18; of leaves, 26–29; of roots, 15–21; of stems, 21–26; of tree reproductive organs, 29–31
Strumella canker, 408
Strumella coryneoidea, 302, 366, 372, 383, 472
Stubbing, 120
Stubby root nematode, 236, 455–56
Styrax, 376
Subdue, 287
Suberin, 23
Subnormal growth, and fertilizer need, 92–93
Subsoil, 51
Subterranean termite, 357
Suckers, 121, 175
Sugar, 29; translocation, 26
Sugarberry. *See* Hackberry
Sugar hackberry. *See* Hackberry
Sugar maple, 35, 211, 251. *See also* Maple
Sugar maple borer, 404
Sulfate ions, 39
Sulfur, 40–41, 43–45, 96, 107, 287
Sulfur dioxide, 42, 200, 205–6
Sulfuric acid, 206
Sumac, 462
Summer wood, 23
Sunscald, 154, 191–92
Superior oil, 279, 297–98
Support, tree: artificial, 141–45; postplanting, 82–85
Supracide, 281
Swamp oak. *See* Oak
Sweet birch. *See* Birch
Sweet cherry, 251. *See also* Cherry
Sweetfern rust, 428
Sweetgum, 141, 254, 462–63; abiotic diseases, 463; diseases, 463; insects, 463
Sweetgum scale, 463
Sweetgum webworm, 463
Swellings, 165
Swiss stone pine. *See* Pine
Sycamore, 248, 249, 251, 254, 464; abiotic disease, 466; diseases, 464–66; insects, 467. *See also* Planetree, London
Sycamore lace bug, 226
Sycamore maple. *See* Maple
Sycamore plant bug, 226, 467
Sycamore scale, 467
Sycamore tussock moth, 467
Symptoms of parasitic disease, 233–34
Synanthedon: pictipes, 219, 321, 454; *scitula,* 219, 335
Syringa, 375

Tall-growing vs. low-growing trees, 60–62
Tallow tree. *See* Chinese tallow tree
Talstar, 281
Tamarisk, Athel, 298
Tamarix, 298
Tamiasciurus hudsonicus, 404
Taniva albolineana, 366, 460
Tanoak, 467. *See also* Oak
Taphrina: aceris, 396; *amentorium,* 292; *carnea,* 303; *carveri,* 396; *coerulescens,* 408; *communis,* 320; *dearnessii,* 396; *deformans,* 320; *flava,* 303; *johansona,* 446; *lethifer,* 396; *macrophylla,* 292; *occidentalis,* 292; *populina,* 446; *robinsoniana,* 292; *sacchari,* 396; *ulmi,* 344; *virginica,* 372; *wiesneri,* 320
Tarnished plant bug, 226
Tar spot, 369–70, 389, 396–97, 477–78
Tatarian maple, 251. *See also* Maple
Taxodium, 299–300, 331n
Taxus, 480–84; *cuspidata,* 35; *media,* 35
Taxus bud mite, 483–84
Taxus mealybug, 228, 482–83
Taxus weevil, 480
T-B-C-S 53, 284
Tea olive, 251
Tea scale, 314, 328, 336, 372, 394
Tebuthiuron, 209
Tedion, 281
Telone, 282
Temik, 281, 283
Temperature: air, 184–91; soil, 47
Termites, 484; damage from, 202–3
Terracing, 150
Terraclor, 287
Terramycin, 287
Terrapin scale, 305, 324, 386, 403, 448, 453, 467, 479
Terrazole, 287
Terrocide 15D, 282
Testing of soil, 45
Tetradifon, 281
Tetraleurodes, 317; *mori,* 336–39; *stanfordi,* 413
Tetralopha robustella, 432
Tetranychus: canadensis, 350; *schoene,* 350; *urticae,* 230, 363, 366, 407
Texas persimmon, 251. *See also* Persimmon
Thamnosphecia pyri, 454
Thecodiplosis liriodendri, 471
Thiabendazole, 283
Thielaviopsis basicola, 370
Thinning out, 119–20
Thiodan, 220, 228, 281
Thiophanate-methyl, 284, 287
Thiram, 287

Thorn. *See* Hawthorn
Thornbug, 455
Thread blight, 329, 331, 471
Thuja, 294–95; *occidentalis*, 35
Thuricide, 224, 266, 278
Thylate, 287
Thyridopteryx ephemeraeformis, 379–80
Thyronectria austro-americana, 389, 455; *deni-grata*, 389
Tiles, drainage, 151–52
Tilia, 383–86; *cordata*, 35
Tinocallis ulmifolii, 345
Tip moth caterpillar, 332
Tip pruning, 120
Tobacco budworm, 454
Tobacco ringspot virus, 356
Tomato spotted wilt virus, 387
Tomostethus multicinctus, 297
Tools, diagnostic, 166–70
Topiary, 132–33
Topping, 120–22
Topsin-M, 287
Torula ligniperda, 304
Toumeyella: caryae, 229; *liriodendri*, 229, 393, 469–71; *numismaticum*, 435; *pini*, 435; *pinicola*, 435
Toxicity, pesticides, 275–77
Toxoptera aurantii, 372
Trace elements, 40–41, 94–96
Tracheids, 21
Trametes, 258, 375, 478; *versicolor*, 316, 350, 355, 423
Translocation, 26
Transpiration, 29, 68
Transplantability, 65–67
Transplanting, 65; digging trees, 70–76; and fertilizer, 113; planting trees, 78–82; postplanting operations, 82–89; preparing trees for, 68–70; and pruning, 130; season, 67–68; transporting trees, 76–77
Transplant shock, 66–67, 70, 75, 88
Transporting trees, 76–77
Treatment area, fertilizer, 99–100
Tree: leaves, 26–29; reproductive organs, 29–31; roots, 15–21; stems, 21–26; undesirable, 57–58
Tree band damage, 203–4
Tree-of-heaven, 57, 251, 253, 467; diseases, 467–68; insects, 468
Tree spade, 75, 78
Tree spikes, in fertilizer application, 103
Tree Tanglefoot, 202
Tremex columba, 404
Trenches, tree damage from, 149, 178
Trenching, for fertilizer application, 104

Triadimefon, 284
Trialeurodes vaporariorum, 328
Tribasic Copper Sulfate, 284
Trichlorfon, 279
Trichodorus primitivus, 455
Trichodothis comata, 392
Tricolor beech. *See* Beech
Trident maple, 251. *See also* Maple
Triforine, 286
Triton B1956, 288
Tropidosteptes, 298
Truban, 287
Trunk: bark torn from, 146; deep cavities, 141; examination, 164–65; implants and injections, 109–12; split, 142
Tsuga, 363–66; *canadensis*, 35
Tsugaspidiotus tsugae, 365
Tubercularia, 451; *ulmea*, 343; *vulgaris*, 389, 455
Tulip poplar. *See* Tuliptree
Tuliptree, 141, 177, 187, 196, 248, 249, 251, 468; abiotic diseases, 469; fungus diseases, 468–69; insects, 469–71. *See also* Poplar
Tuliptree aphid, 469
Tuliptree scale, 229, 299, 386, 393, 469–71, 476
Tuliptree spot gall, 471
Tung, 471; foliage diseases, 471–72; mineral deficiencies, 472; root rot, 472
Tunneling, for tree preservation, 149
Tupelo, 251, 472; diseases, 472; insects, 472
Tupelo leaf miner, 472
Tupelo scale, 472
Turcam, 281
Turkish filbert, 472
Tussock moth, 223, 308, 327, 330, 348–49, 354, 355
2, 4-D, 207-8
2-4-xylenol, 244
Two-lined chestnut borer, 218, 219, 301, 327, 372, 374, 410, 413
Two-marked treehopper, 373
Two-spotted mite, 230, 363, 366, 407
Tylose, 25
Tympanis pinastri, 424

Ugly nest caterpillar, 322
Ulmus, 340–50
Umbellularia, 312
Umbonia crassicarnis, 292
Uncinula, 249; *adunca*, 446, 477; *australiana*, 331; *circinata*, 398; *clintonii*, 350, 384; *flexuosa*, 375; *geniculata*, 406; *macrospora*, 344, 372
Undesirable trees, 57–58
Upright yew, 35. *See also* Yew
Urban soil, 48–50

Urea, 104–5
Urea formaldehyde (UF), 105
Uredinopsis: mirabilis, 353; *osmundae*, 353; *phegopteridis*, 353
Urnula craterium, 408
Uromyces hyalinus, 419

Valsa, 424; *abietis*, 339, 457; *ambiens leucostomoides*, 398; *chionanthi*, 355; *friesii*, 457; *leucostomoides*, 398; *pini*, 457; *sordida*, 398, 444, 477
Value of trees, 3–12; and preservation, 139
Vancide FE, 285
Vapam, 254, 283
Variable oak leaf caterpillar, 384, 417
Vasates aceriscrumena, 231, 402
Vasates quadripedes, 231, 402
Vein, 27
Vendex, 281
Venturia, 398; *inaequalis*, 329, 361, 362, 405; *populina*, 446; *saliciperda*, 476; *tremulae*, 446
Vertical mulching, 51–52, 154
Verticicladiella procera, 430
Verticillium, 308, 399, 452; *albo-atrum*, 253, 299, 343, 357, 375, 384, 393, 418, 423, 440, 449, 451, 456, 467, 469, 479; *dahliae*, 253
Verticillium wilt, 92, 251–54, 315, 316, 320, 343–44, 418, 419, 472; trees susceptible to, 251; trees tolerant to, 253–54
Vespamima, 435; *sequoiae*, 451
Vessel elements, 21
Vicam, 281
Vidden-D, 282
Vinclozolin, 286
Vine, 200–201
Viruses, 236
Viscum, 237
Vitex, 318
Volutella: buxi, 308; *mellea*, 453
Vorlan, 286
Vorlex, 283
Vydate, 283
Vydate L, 281–82

Walnut, 187, 248, 249, 472–73; diseases, 473–75; insects and other pests, 475–76
Walnut aphid, 475
Walnut bunch disease, 474–75
Walnut caterpillar, 302, 475
Walnut lace bug, 226, 475
Walnut scale, 229, 254, 324, 349, 354, 360, 363, 368, 372, 381, 386, 387, 392, 403, 406, 448, 471, 476

Washington hawthorn, 35. *See also* Hawthorn
Water, soil, 26, 36–38
Watering, 38; and leaf scorch, 194; postplanting, 86
Waterlily aphid, 324
Waterlogged soil, 34
Water sprouts, 25, 121
Water table, 36
Webber moth caterpillar, 332
Weedkillers, 207–10
Weeds, Trees, and Turf, 14, 270
Weeping beech. *See* Beech
Weeping cherry, 35. *See also* Cherry
Weeping fig. *See* Fig
Weeping purple beech. *See* Beech
Western catalpa, 57, 251. *See also* Catalpa
Western gall rust, 428
Western larch. *See* Larch
Western parsley caterpillar, 315
Western red cedar. *See* Juniper
Western spruce budworm, 460
Western tent caterpillar, 363
Western tussock moth, 349, 363
Western woolly aphid, 298
Western yellow pine. *See* Pine
West Indian peach scale, 373
Wetwood, 244–46, 446
White ash, 34, 57, 211. *See also* Ash
White-banded leafhopper, 345
White birch, 35. *See alo* Birch
White cedar. *See* Arborvitae; False cypress
White fir. *See* Fir
White flowering dogwood, 35. *See also* Dogwood
White-marked tussock moth, 223, 308, 327, 332, 348–49, 355, 367, 383, 384, 400, 467, 468
White oak. *See* Oak
White oak borer, 415
White peach scale, 229, 324, 327, 336, 355, 357, 423, 449, 476
White pine. *See* Pine
White pine aphid, 431–32
White pine blister rust, 428–30
White pine decline, 430
White pine shoot borer, 436
White pine tube moth, 436
White pine weevil, 339, 436, 461
White rot, 255
White spruce. *See* Spruce
White willow, 35. *See also* Willow
Wild cherry. *See* Cherry
Wilding trees, transplanting, 68–70
Wild olive. *See* Silverbell
Willow, 141, 177, 187, 249, 254, 476; diseases, 476–78; insects, 478–79. *See also* Desert willow
Willow flea weevil, 479

Willow lace bug, 479
Willow oak. *See* Oak
Willow scale, 386, 448, 471
Willow scurfy scale, 479
Willow shoot sawfly, 479
Wilt. *See* Verticillium wilt
Wilting, 38
Windstorm, 196
Wingnut. *See* Caucasian wingnut
Winter drying of evergreens, 189–91, 371
Winter injury, 185–87, 190–91
Witches' broom, 231
Witch-hazel leaf gall aphid, 304–5
Wood decay, 118–19, 255–58
Wood fiber, 21
Woolly alder aphid, 293
Woolly aphid, 298, 354, 365, 411
Woolly apple aphid, 228, 345, 362, 405, 454
Woolly beech aphid, 302
Woolly beech scale, 301–2
Woolly elm aphid, 345, 454
Woolly elm bark aphid, 345
Woolly hawthorn aphid, 345, 362
Woolly larch aphid, 383
Woolly pear aphid, 345
Woolly pine scale, 435
Woolly whitefly, 328
Woollyworm, 224–25
Wool sower gall, 411
Wounds: dressings, 129–30; reactions, 118–19; treating, 145–48
Wrapping, postplanting, 87–88, 192

Xanthomonas, 419; *campestris juglandis,* 474; *citri,* 328; *pruni,* 318–19

X-disease, 321
Xenochalepus dorsalis, 336
Xiphinema americanum, 404
Xylaria: mali, 321, 389, 398, 480; *schweinitzii,* 421
Xylaria root rot, 398
Xyleborus, 415
Xylem, 21
Xylemella fastidiosum, 246

Yaupon holly. *See* Holly
Yaupon psyllid, 372
Yellow-bellied sapsucker, 201–2
Yellow birch, 211. *See also* Birch
Yellow buckeye, 57. *See alo* Horsechestnut
Yellowing. *See* Chlorosis
Yellow-necked caterpillar, 223, 322, 327, 330, 367, 384, 387, 400, 415–16, 475
Yellow poplar. *See* Tuliptree
Yellowwood, 43, 141, 196, 249, 479–80
Yew, 480; abiotic diseases, 480; diseases, 480; insects and other pests, 480–84
Yucca mealybug, 451

Zelkova, Japanese, 254, 484
Zelkova serrata, 343
Zeuzera pyrina, 404
Zimmerman pine moth, 354, 436
Zinc, 40, 92, 107, 109, 112
Zineb, 287
Zinophos, 283
Ziziphus, 327
Zyban, 287–88